Principles
of Modern
Microbiology

Principles of Modern Microbiology

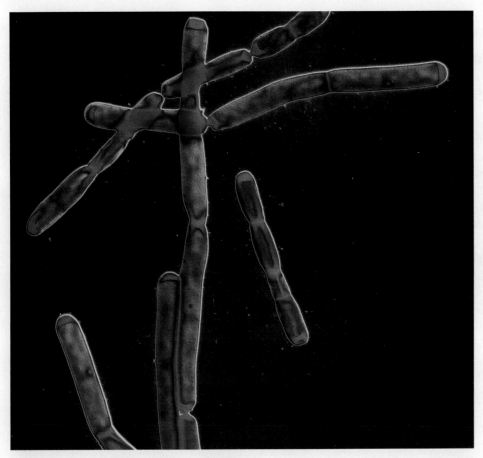

Mark L. Wheelis, PhD

University of California, Davis

JONES AND BARTLETT PUBLISHERS

Sudbury, Massachusetts

BOSTON   TORONTO   LONDON   SINGAPORE

World Headquarters

Jones and Bartlett Publishers
40 Tall Pine Drive
Sudbury, MA 01776
978-443-5000
info@jbpub.com
www.jbpub.com

Jones and Bartlett Publishers Canada
6339 Ormindale Way
Mississauga, Ontario L5V 1J2
CANADA

Jones and Bartlett Publishers
International
Barb House, Barb Mews
London W6 7PA
UK

Jones and Bartlett's books and products are available through most bookstores and online booksellers. To contact Jones and Bartlett Publishers directly, call 800-832-0034, fax 978-443-8000, or visit our website, www.jbpub.com.

Substantial discounts on bulk quantities of Jones and Bartlett's publications are available to corporations, professional associations, and other qualified organizations. For details and specific discount information, contact the special sales department at Jones and Bartlett via the above contact information or send an email to specialsales@jbpub.com.

Production Credits

Chief Executive Officer: Clayton Jones
Chief Operating Officer: Don W. Jones, Jr.
President, Higher Education and Professional
    Publishing: Robert W. Holland, Jr.
V.P., Design and Production: Anne Spencer
V.P., Manufacturing and Inventory Control: Therese
    Connell
V.P., Sales and Marketing: William J. Kane
Executive Editor, Science: Cathleen Sether
Acquisitions Editor, Science: Shoshanna Grossman
Managing Editor, Science: Dean W. DeChambeau
Associate Editor, Science: Molly Steinbach
Editorial Assistant: Briana Gardell
Senior Production Editor: Louis C. Bruno, Jr.
Production Assistant: Leah E. Corrigan
Senior Marketing Manager: Andrea DeFronzo
Text Design: Anne Spencer
Cover Design: Kate Ternullo
Illustrations: Elizabeth Morales
Photo Research Manager and Photographer:
    Kimberly L. Potvin

Associate Photo Researcher: Christine McKeen
Photo Researchers: Jennifer M. Ryan, Alison Meier,
    Christina Micek
Composition: Circle Graphics
Printing and Binding: Courier Kendallville
Cover Printing: Courier Kendallville
Cover Image: © Oliver Meckes/Nicole Ottawa/Photo
    Researchers, Inc.

About the cover: Color enhanced scanning electron micrograph of a *Bacillus anthracis,* a gram-positive species of bacteria that is highly pathogenic and causes the disease anthrax in humans. *B. anthracis* is transmitted to humans by contact with contaminated animal hair, hides, or excrement. The disease attacks either the lungs causing pneumonia (woolsorter's disease) or the skin producing severe ulceration (malignant pustule). Treatment with antibiotics such as penicillin may be effective if used promptly; untreated cases often progress to fatal septicemia.

Library of Congress Cataloging-in-Publication Data
Wheelis, Mark.
    Principles of modern microbiology / Mark Wheelis.
        p. cm.
    ISBN-13: 978-0-7637-1075-0 (alk. paper)
    1. Microbiology-Textbooks. I. Title.
    QR41.2.W52 2007
    616.9'041-dc22   2006013044

6048

Printed in the United States of America
11 10 09 08 07   10 9 8 7 6 5 4 3 2 1

This book is dedicated to Katy,
without whose love and support I could never have undertaken,
much less completed, a project so demanding.

Brief Contents

Contents

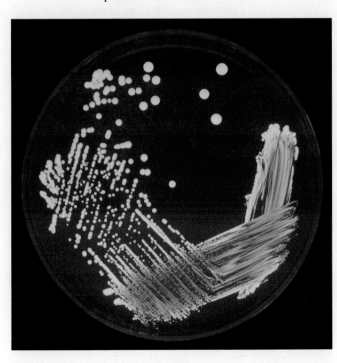

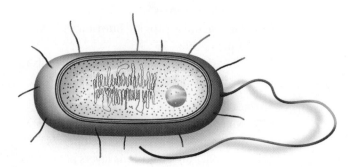

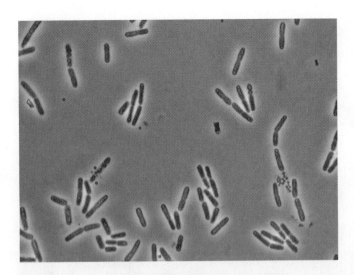

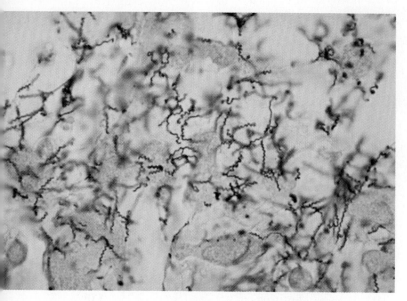

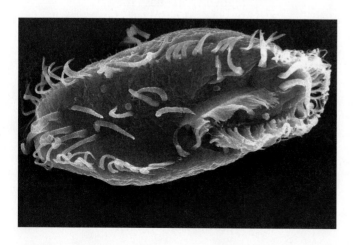

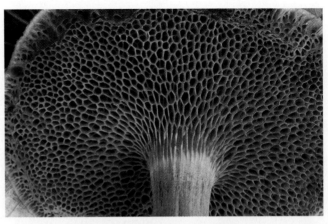

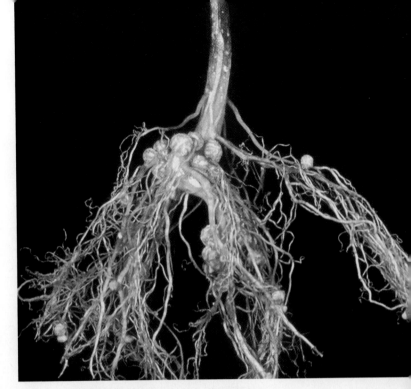

Preface

I have taught introductory microbiology repeatedly over the last 37 years. During that time I used a variety of textbooks: All are fine books, comprehensive and lavishly illustrated. They present a comprehensive overview of the field in considerable detail. Their completeness and detail, however, make them all very difficult to use in a single-term introductory course, even at the most selective universities.

My goal, therefore, was to write a text under 500 pages long that retains the intellectual rigor of the longer texts and the coherence of a text conceived and written for its intended audience. The price of this brevity is some selectivity, and instructors may find topics are omitted, or treated briefly, that they would like covered more fully. Nevertheless, I believe I have provided an overview of the field in sufficient detail to make sense. Where possible, I present biochemical and molecular mechanisms, as it is at this level that we most effectively explore the functioning of microbial cells.

One way that space can be saved without sacrificing substance is to pare down inessential parts that cover material that the student is expected to know on entry. I thus assume here that the student has had a year of biology (including cell and molecular biology) and general chemistry, and I do not repeat material from these prerequisite classes, nor do I define the technical vocabulary from them.

I have also elected to omit a glossary, instead defining each term in the text on first use. Therefore, for most technical terms, the first index entry will send the student to the definition.

In addition to brevity, I wanted a text that helped the student understand the basic concepts in microbiology without getting too distracted by the illustrative details. At the introductory level, students need the basics of the discipline, with minimal distraction from unnecessary details, exceptions, complications, etc. The richly detailed style of the current texts, which appeals to advanced students and to faculty, is a considerable impediment to learning by the beginner, particularly because it infuses the texts, and therefore cannot be avoided by judicious choice of assigned pages. Brevity helps avoid the distractions that come from too much detail but does not by itself help the student focus on principles. To help in this regard, each chapter is divided into many sections, and each section has a single major point to it. Each section head summarizes the concept of the section or alludes to it in some explicit way. As a result, the table of contents serves as a review of the important concepts. End-of-chapter questions are designed to guide synthesis and summary of the important information in the chapter and to develop quantitative abilities. Answers to the even-numbered questions are available to students on our website http://microbiology.jbpub.com/Wheelis. Answers to odd-numbered questions are available to faculty only.

My intent has been to produce a text that is conceptually sophisticated, coherent, authoritative, and demanding, but brief, clear, and conceptual. At the end, I hope that students will have a greatly enhanced appreciation of the importance and role of microbes in the global biosphere. This is, as the book repeatedly emphasizes, a planet on which microbes dominate—in number, biomass, diversity, and importance. If I have helped students to understand this, I am satisfied.

Ancillaries Accompanying the Text

To assist you in teaching this course and supplying your students with the best in teaching aids, Jones and Bartlett Publishers has prepared a complete ancillary package available to all adopters of *Principles of Modern Microbiology*. Additional information and review copies of any of the following items are available through your Jones and Bartlett sales representative.

For the Instructor

Instructor's ToolKit CD-ROM
Compatible with Windows and Macintosh platforms, the Instructor's ToolKit CD-ROM provides adopters with the following traditional ancillaries:

- Answers to the text's odd-numbered Study Questions.
- The *Test Bank* prepared by Teri Shors, University of Wisconsin, Oshkosh, is available as text files. The test bank contains over 2,500 questions. An additional set of test bank files is formatted for your own online courses using WebCT and Blackboard.
- The *PowerPoint® Lecture Outline Slides* presentation package provides lecture notes, graphs, and images for each chapter of *Principles of Modern Microbiology.* Instructors with the Microsoft PowerPoint software can customize the outlines, art, and order of presentation. The PowerPoint files have also been prepared in HTML format for use in online course management systems.
- The *PowerPoint Image Bank* provides the illustrations, photographs, and tables (to which Jones and Bartlett Publishers holds the copyright or has permission to reprint digitally) inserted into PowerPoint slides. With the Microsoft PowerPoint program, you can quickly and easily copy individual image slides into your existing lecture slides. If you do not own a copy of Microsoft PowerPoint or a compatible software program, a Microsoft PowerPoint Viewer is included on the CD-ROM.

For the Student

- The Web site we developed exclusively for *Principles of Modern Microbiology,* can be found at http://microbiology.jbpub.com/Wheelis. The site contains eLearning, a free on-line study guide with chapter outlines, research and reference links, answers to the text's even-numbered Study Questions, and links to microbiology news sources.
- *Alcamo's Laboratory Fundamentals of Microbiology, Eighth Edition,* is a series of over 30 multipart laboratory exercises providing basic training in the handling of microorganisms and reinforcing ideas and concepts described in the textbook.
- *Encounters in Microbiology* brings together "Vital Signs" articles from *Discover Magazine* in which health professionals use their knowledge of microbiology in their medical cases.

- *Guide to Infectious Diseases by Body System,* by Jeffrey C. Pommerville, Glendale Community College, is an excellent tool for learning about microbial diseases. Each of the 15 body-system units presents a brief introduction to the anatomical system and the bacterial, viral, fungal, or parasitic organism capable of infecting the system.
- *20th Century Microbe Hunters,* by Robert Krasner, Providence College, offers a dramatic portrayal of the achievements and lives of microbiologists such as Charles J. Nicolle (epidemic typhus), Barry Marshall and J. Robin Warren (*Helicobacter pylori*), Luc Montagnier and Robert Gallo (HIV), and Donald R. Hopkins (Guinea worm).
- *How Pathogenic Viruses Work,* by Lauren Sompayrac, is a concise summary of the basics of virology written in an understandable and entertaining manner. The book comprises nine lectures covering the essential elements of virus-host interactions with descriptive graphics, helpful mnemonic tactics for retaining the concepts, and brief lecture reviews. This is an ideal text for medical, science, and nursing students who want a review, or a simple explanation, of virology.
- *Microbiology Pearls of Wisdom,* by S. James Booth, is a review manual designed for those preparing for MCAT, VCAT, DCAT, USMLE Parts I, II and III and for students preparing for course exams. Through its use of a rapid-fire, question-and-answer format, this review of microbiology principles provides help for improving performance on microbiology written and practical examinations by offering students immediate gratification with the correct answer.

Acknowledgments

I owe a great debt to all of my colleagues, at Davis and elsewhere, from whom I have learned so much. Foremost among them is Roger Stanier, my Ph.D. supervisor so many years ago, who taught me, by example, how to think as a scientist. Roger died many years ago, but I am confident that he would have appreciated this book, and would have seen his influence in it. My colleagues in the Section of Microbiology at Davis also deserve a great deal of credit for anything of worth here. I have learned more microbiology in 30-plus-years of hallway conversations with them than from any other source. The reviewers, mentioned below, were also a critical part of the process of making this book and deserve a good deal of the credit for its strengths. Also my students, thousands of them, who have been the cheerful guinea pigs of my pedagogical experiments, deserve much credit for helping form my teaching approaches, which in turn led to this book.

The editorial and production staff at Jones and Bartlett Publishers deserve a medal for their patience, persistence, and expertise in guiding me through this project. Brian McKean originally signed me, and his enthusiasm and vision guided me through the difficult initial years. More recently, Cathy Sether has been my editor and conscience. Lou Bruno steered the book through production and was unfailingly professional, even when I probably had him tearing his hair out privately. Jennifer Ryan and her colleagues handled photo research expertly, despite my all-too-frequent vagueness. They, and all the rest of the J&B team, deserve a great deal of credit for this book; I would not have finished it without them.

Many colleagues read part or all of this book in manuscript, and their comments were of immense value. I am grateful to all of them and hope that they will all see their influence in the final product. Of course, any remaining flaws are my responsibility alone. The reviewers were:

Stephen Aley, University of Texas, El Paso
Mary Allen, Hartwick College
James L. Botsford, New Mexico State University
James F. Curran, Wake Forest University
Mark Davis, University of Evansville
Scott Dawson, University of California, Davis
Paul V. Dunlap, University of Michigan
Luti Erbeznik, Albion College
Rebecca Ferrell, Metropolitan State College of Denver
Susan Godfrey, University of Pittsburgh
George D. Hegeman, Indiana University
Joan Henson, Montana State University
Judith Kandel, California State University, Fullerton
E. R. Leadbetter, University of Connecticut
Lee H. Lee, Montclair State University
Lynn Lewis, University of Mary Washington
Bonnie Lustigman, Montclair State University
Elizabeth A. Machunis-Matsuoka, University of Virginia
JaRue Manning, University of California, Davis
Charlotte M. McCarthy, New Mexico State University
Robert J.C. McLean, Southwest Texas State University
Patricia Parker, California State University, Chico
Valley Stewart, University of California, Davis
Rajendra S. Rana, St. John's University
Julie Shaffer, University of Nebraska at Kearney
Gary L. Sloan, University of Alabama
James Strick, Arizona State University
Jill Zeilstra-Ryalls, Oakland University

Finally, my family deserves tremendous credit for their forbearance during the many years I was writing this book. They—my wife Katy, and my children Emily and Ian—endured far too many weekends I was in my office and vacation days dominated by my laptop. I am forever grateful for their support and love.

Mark Wheelis
University of California, Davis
October, 2007

Introduction to Microbiology

1

WELCOME TO MICROBIOLOGY: the study of the great variety of living organisms that are too small for us to see without a microscope—the **microbes,** or **microorganisms.** You will learn, as you read this book, that despite their minute size, these organisms form the basis for all life on earth. Their activities produce the soil in which plants grow and the atmospheric gases that plants and animals both use. Their activities regulate the temperature of earth, preventing it from freezing or baking; subterranean gelatinous masses of them may lubricate the movements of tectonic plates, and their chemical activities recycle gases on which all life on land depends.

For nearly three quarters of the history of earth, a period of about 3 billion years, microbes were the only life on earth. To us the planet would have looked barren and uninhabited, except for the colored scum at the edge of ponds and on intertidal rocks, yet the seas and lakes teemed with great masses of life. Even today, after nearly a billion years of abundant plant and animal life, the earth is fundamentally a microbial planet, to which the macro-organisms are recent and relatively unimportant additions. Life on earth is like an iceberg: only a small portion of it is visible (Figure 1.1).

In this chapter, we define our subject and then briefly survey the evolutionary history of life on earth, the major categories of microbial life, and the chemical composition of microbial cells.

1.1 Most microorganisms are microscopic, and they include all life forms other than the plants and animals

Microorganism is a term that is difficult to define precisely. Operationally, it refers to any organism that is too small to be seen by the unaided eye. For most people, that would be about 0.1 to 0.2 mm in diameter. Anything less than this cannot be seen without a microscope of some kind and is in a loose sense a microorganism.

In practice, however, the term microorganism is often used to include some macroscopic forms that belong to a group that is largely microscopic (e.g., the fungi, most of the algae), and it excludes some microscopic forms (e.g., some microscopic animals that despite their small size are multicellular with differentiated tissues and organ systems).

Our recent ability to sketch confidently the outlines of the evolution of life on earth, discussed later here and in Chapter 13, allows us to now define microorganism in evolutionary terms. Broadly speaking, microorganism in this sense would include everything except the animals and the plants.

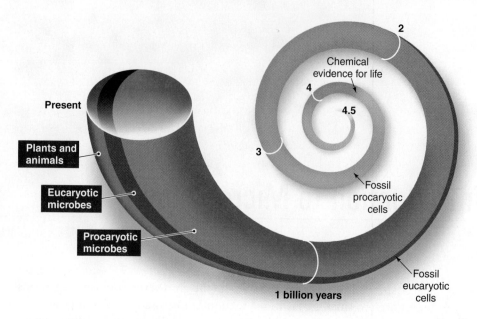

FIGURE 1.1 Time line of life on earth.

1.2 Most microorganisms are unicellular; if multicellular, they lack highly differentiated tissues

The vast majority of microbes are unicellular—that is, the entire organism consists of a single cell. Also fairly common are filaments of cells attached end to end in a row. In some cases, especially in the fungi and the algae (most of which are considered microbes), there are representatives that are multicellular (often macroscopic as well). Even in these forms, however, the cells that make up the organism are not organized into highly differentiated tissues and organs. The differentiation of diverse and very different types of tissues and the organization of these different tissues into organs seem to have been an evolutionary invention of the plants and animals alone (with some rudimentary tissue differentiation in the algae and fungi).

1.3 Microbial life originated shortly after the earth was formed

Cosmologists agree that the earth is about 4.5 billion years old, originating from the coalescence of debris left over after the formation of the sun. It was originally very hot—too hot for liquid water—from the heat of gravitational collapse and from radioactive decay in its core. It was also continually bombarded with large meteorites whose impact released so much energy that nothing living would have been able to survive. It is generally agreed that the earth cooled to habitable temperatures (less than 100°C) about 4.0 billion years ago and that the meteoritic bombardment abated by about 3.8 billion years ago. Coincidentally, this is about the age of the oldest rocks, and these rocks show chemical evidence of microbial life. Shortly thereafter (in geological time), at about 3.5 billion years ago, fossil evidence of microbes exists. Clearly life originated on earth almost immediately after conditions permitted.

The microscope has been as much a tool for geologists as for biologists. Starting in the 19th century, the microscopic organization of mineral grains in rocks was examined by putting very small shards of rock under the microscope and looking through the very thin edges. When the rock is very thin (on the order of 0.1 mm), it is translucent, and light can pass through it, revealing its structure.

It was thus natural that at some point geologists would search shards of rock for fossil microorganisms. The earth's mantle consists of layer on layer of rock, with, unless the rock has been rearranged by tectonic movements, the most recent rock on top and the oldest on the bottom. The top rocks, corresponding to the last quarter of the earth's history, contain macroscopic fossils; the bottom rock, three quarters of the history of the earth, is barren of visible fossils. This discontinuity was recognized even in Darwin's time, and Darwin commented on the apparently sudden appearance of life in the geological era called the Cambrian.

It was an obvious possibility that Precambrian life was microscopic, and thus, geologists were attentive to the possibility of microscopic fossils. They are quite rare, however, and it was not until the 20th century that the first Precambrian microfossils were discovered. In 1918, E. Moore saw fossil cyanobacteria in Precambrian rocks, and similar observations were made by John Gruner in 1923 and Burton Ashley in 1937 (Figure 1B.1). Interest waned, however; the point had been made, and the attention of geologists moved on.

After the Second World War, however, nuclear sciences produced new and precise ways to date rocks (rock dates before this had been little more than educated guesses). This led to the realization that a systematic study of microfossils might reveal when life on earth originated, and there was a resurgence of interest in microbial fossils. These studies have revealed that complex, filamentous organisms similar to modern procaryotes are present in rocks from nearly 3.6 billion years ago (Figure 1B.2). Older rocks exist—up to about 3.8 billion years old. To date, these have not shown unambiguous fossils.

Modern micropaleontology is a complex science. Rocks to be examined are cleaned many times and shattered into small pieces with a hammer, and then thin sections (typically about 100 μm thick) are cut with a fine saw for microscopy. The rest of the rock sample is pulverized and analyzed chemically for the amount of carbon present and for the ratio of ^{13}C to ^{14}C (which can indicate whether the carbon is of biological origin). Sometimes the rock is assayed for specific compounds, like derivatives of chlorophylls or other complex molecules, that indicate the presence of life.

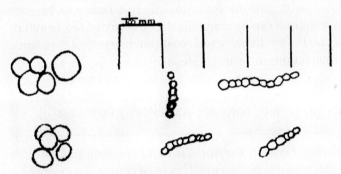

FIGURE 1B.1 Moore's 1918 drawing of microbial fossils.

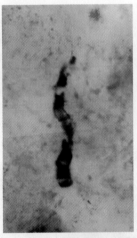

FIGURE 1B.2 Modern micrograph of fossil microbes 3.5 billion years old.

1.4 There are two fundamentally different types of cells: procaryotic and eucaryotic

Careful examination of microbial cells under the microscope reveals that there are two fundamentally different types: relatively large cells with complex interiors and very small and simple cells. Electron microscopic examination confirms that their internal organization is very different. These two types of cells are termed *eucaryotic* for the large, complex ones and *procaryotic* for the small simple ones (Figure 1.2). At the ultrastructural level, eucaryotic cells possess complex internal structures composed of membranes that divide the cytoplasm into a number of different compartments. One of these is the nucleus, which contains the DNA-containing chromosomes. The

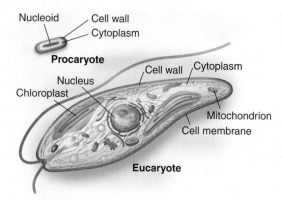

FIGURE 1.2 Procaryotic and eucaryotic cells.

procaryotic cell has a nucleoid (with no membrane around it) instead of a nucleus, and it lacks the internal membrane system. Despite its apparent simplicity at the cellular level, study of the molecular biology of procaryotic cells makes it clear that they too are highly organized structures; however, much of their organization is at the molecular level and is not visible in the microscope. Chapters 4 and 5 discuss microbial cell structure in more detail.

Although there are many exceptions, it is probably fair to think of the procaryotes as specialists at rapid growth at the expense of nutrients dissolved in the water that surrounds them. This has entailed the evolution of small, relatively simple cells with highly effective permease systems to take nutrients from a dilute solution, and a highly efficient, tightly regulated metabolism to make the most of them.

Eucaryotic cells, on the other hand, appear to have initially specialized in predation on the smaller, simpler procaryotes. This move up the trophic scale led to an increase in cell size and complexity and, ultimately, after several billion years, to multicellular organization in the algae, fungi, plants, and animals.

Of course, many exceptions exist to this simple generalization—there are many nonpredatory eucaryotes and many slow-growing procaryotes. Nevertheless, it is a useful way to think of the broad evolutionary strategies that worked to produce the two cell types. Both strategies have proved successful so that today the planet supports an immense variety of life, both eucaryotic and procaryotic, microscopic and macroscopic, unicellular and multicellular. These all interact with each other in an unimaginably complex web of physical and chemical interactions at the microscopic, local, and global levels. This web of interactions has proved to be quite stable for eons: for the first 3 billion years as an exclusively microbial biosphere (containing both procaryotic and eucaryotic microbes) and for the last billion as microscopic and macroscopic organisms commingled. It remains to be seen whether these long-term stabilities can be maintained in the face of the assault of industrial human cultures on the biodiversity and ecological integrity of the planet. We study some of these interactions in Chapters 16–20.

1.5 Microbes, especially procaryotic ones, are unbelievably numerous

The sheer number of microbial cells in the world is mind boggling and vastly in excess of the number of macroscopic organisms. This is, of course, a consequence of their small size; very large numbers of them can fit into a very small space, and very small amounts of nutrients can nourish a very large number of them. For instance, *Escherichia coli*, a very well-studied and reasonably typical procaryote, is a short rod a bit over a micrometer in length and a bit under a micrometer in diameter, weighing about a picogram (10^{-12} g). Although there is much variation in procaryotic cell size, this is an especially common size and can be taken as typical.

A single individual human being is as much an ecosystem as an individual. It has been estimated, very approximately, that there are about 10^{13} human cells that make up the adult human body. This body is, in turn, inhabited by approximately 10^{14} microbial cells—in the intestine, in the mouth, in the vagina, and on the skin—nearly all of them procaryotic. A single milliliter of intestinal contents contains more microbial cells than there are humans on earth (Figure 1.3). Other animals are of course similarly populated. Rich soil is also densely inhabited: 1 gram also contains more microbes than there are humans on earth—so too for lake sediments and for the sediments of the continental shelf. Thus, the total number of microbes on earth is beyond comprehension.

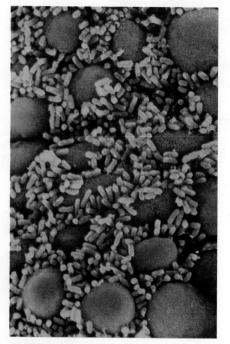

FIGURE 1.3 Intestinal microbes.

Most (90% or more) of these innumerable microbes are procaryotic; the eucaryotic ones, as is usual for larger organisms at a higher trophic level, are vastly outnumbered.

Despite their very small size and mass, the immense numbers of microbes mean that the total microbial biomass (weight of living material) is larger than that of macroscopic life. Indeed, the procaryotic biomass alone is about half of the planetary biomass. Thus, in nearly all quantitative ways, microbes are the most significant life forms on the planet.

Microbes are not just unbelievably numerous; they are genealogically ancient as well. It has been estimated that the average *E. coli* living in the human digestive tract replicates itself about once every 12 hours—or more than 700 times a year. Thus, by the time a human dies at the age of 75 years, the *E. coli* inside him or her have been through more than 50,000 generations since they first colonized his or her intestinal tract. For comparison, there have probably been only about 5,000 generations of humans since anatomically modern humans appeared about 100,000 years ago.

1.6 The tree of life is almost entirely microbial

Considering that the ancestors of today's microbes were actively multiplying and evolving for 3 billion years before the first animals and plants appeared in the fossil record, it is no wonder that the tree of life is almost entirely microbial. There are about 20 or so major lineages of procaryotic cells and a comparable number of eucaryotic microbial lineages, all of which are vastly older than the plants and animals. Plants and animals are two closely related twigs of the tree of life (Figure 1.4).

1.7 There are three major lineages of life on earth

To the best of our current ability to reconstruct the early history of life on earth, it appears that all life on earth shares a single common ancestor that lived about 3.8 billion years ago. It was probably procaryotic in structure, although probably simpler even than the modern procaryotic cell. It was adapted to a world that was very hot

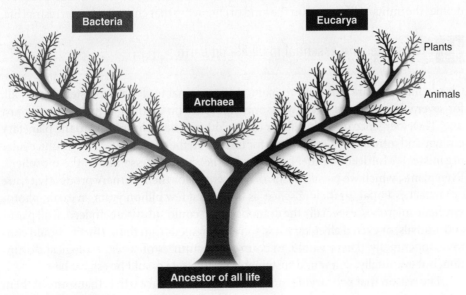

FIGURE 1.4 Phylogenetic tree of life on earth.

and completely lacked oxygen. The descendants of this organism split originally into two major lineages. One of them, called the **bacteria,** produced about a score of known major sublineages, all of them procaryotic.

The other initial branch split again shortly thereafter to produce two lineages, one procaryotic and one that developed onto the eucaryotic branch. These lineages are the **archaea** (procaryotic) and the **eucarya** (eucaryotic). Within the eucaryotic lineage there are another 20 or so known sublineages, plants and animals being two of the most recent newcomers.

Of course when plants and animals first evolved, their bodies provided a new set of habitats for microbes to colonize, with the result that now each individual macroscopic organism is a complex ecosystem inhabited by a large number and wide diversity of microbes. Each of these microbes is specifically adapted to inhabit its host and to take its nutrients from the host. Nutrients may be provided by secretions, such as sebaceous secretions of skin or mucous on mucous membranes or in the case of intestinal bacteria by ingested food that feeds the microbes as well as the host.

Many of the microbes that live in us and on us are benign, even helpful. Others, however, damage their hosts and cause disease. Thus, another reason that microbes are of great interest to humans is the great damage that a few of them cause us—directly by causing human disease or indirectly by causing diseases of our crops or domestic animals.

1.8 Procaryotes inhabit an immense range of habitats

Microbes, particularly procaryotes, inhabit a wide range of habitats, including some that have qualities that would suggest they are completely uninhabitable—for instance, boiling hot springs or highly alkaline or acidic ponds. In general, if a place contains liquid water, a source of energy (light or a reduced organic or inorganic compound), and some dissolved minerals, some procaryotic organism has probably evolved to inhabit it. Procaryotic habitats range in temperature from $-10°C$ to nearly $120°C$ (as long as the water is liquid due to dissolved solutes or hydrostatic pressure) and in pH from less than 1 to more than 10. Procaryotes are found miles under the surface of the earth growing on inorganic minerals or on hydrocarbons, and they are found in all surface soils and waters that are not permanently frozen. This enormous range probably defines the limits of life based on the kind of chemistry that characterizes terrestrial life.

1.9 Procaryotes are essential to all life on earth

Clearly microbes are sufficient to maintain a sustainable global ecosystem—after all, for several billion years, there were only microbes on earth, and the planet did just fine. The emergence of multicellular life introduced new complexity to the planetary ecology and introduced new habitats for microbes (the tissues of the new multicellular organisms). Multicellular life, however, has not become essential to the biosphere. Even plants, which we usually think of as essential for their primary productivity, are not in fact essential. In their absence, as in the first few billion years on earth, photosynthetic microbes, especially the cyanobacteria, could substitute. Indeed, if all plants and animals, or even if all eukaryotes, were eliminated from the earth, life would continue indefinitely. There would, of course, be centuries of severe ecological disruption, but eventually, a new and sustainable steady state would be established.

The reason that procaryotes are essential to all life on earth is that some of them catalyze transformations of chemical compounds that are essential to the sustainability of life. Because no eukaryotes do some of these reactions, these particular procaryotes are essential. We consider some of these transformations in Chapter 16.

1.10 The dry weight of microbial cells consists mainly of macromolecules and lipids

Like all cells, microbes consist mostly of water. *E. coli,* for instance, is about 70% water; only 30% of its weight consists of other chemical compounds. This 30% is called the **dry weight,** and it, in turn, consists principally of a variety of different macromolecules and of lipid (Figure 1.5). **Macromolecules** are very large molecules, typically with molecular weights above 10,000 daltons. They have a polymeric structure, being composed of many similar monomers covalently linked together. Lipids are much smaller, with molecular weights typically under 1,000 daltons; they are not polymeric in structure, and they are hydrophobic.

Macromolecules constitute more than 85% of the dry weight; about 10% is lipid. Thus, all of the small molecules of the cell combined constitute less than 5% of the dry weight. These include all of the various monomeric building blocks from which macromolecules are assembled (amino acids, nucleotides, sugars), all of the various intermediates in biochemical pathways, a number of enzyme cofactors, and a variety of inorganic ions.

1.11 Among the macromolecules, proteins are the largest and most diverse category

Among the macromolecules, there are several major subcategories: proteins, nucleic acids, polysaccharides, and heteropolymers. The last is a category of macromolecule in which two different types of monomers are covalently combined. In the case of bacteria such as *E. coli,* the principal heteropolymer combines many short peptides and a polysaccharide core to form a structure called **murein,** found in the cell wall.

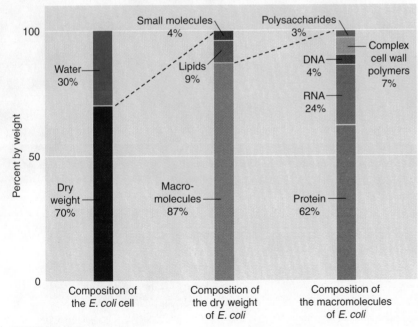

FIGURE 1.5 Cell composition.

Another chemically complex macromolecule found in the *E. coli* cell wall combines polysaccharide with lipid to form **lipopolysaccharide.**

The major class of macromolecule, at least from a quantitative point of view, is protein. Proteins are polymers of amino acids, with molecular weights that vary from somewhat below 10,000 daltons to well over 100,000 daltons. They are not only the most abundant, but also the most diverse of the macromolecules; even such a small cell as *E. coli* has over 2 million protein molecules, of more than 1,000 different types.

Proteins are so diverse because of the many roles they play. Almost every one of the more than 1,000 biochemical reactions of cellular metabolism is catalyzed by a separate enzyme. In addition, there are many permeases in the cell membrane to transport nutrients in and wastes out. There are structural proteins that form organelles within the cell or on its surface, and there are receptor proteins in the membrane that communicate information about the external environment, etc.

1.12 The cytoplasm is a dense suspension of ribosomes

Although proteins are very diverse, with over 1,000 different kinds found in even simple procaryotic cells such as *E. coli,* fewer than 100 proteins make up most of the protein mass of the cell. These are the ribosomal proteins. Ribosomes are organelles composed of protein and RNA; their role is to synthesize protein (in collaboration with mRNA and charged tRNAs). In most procaryotes, there are three molecules of RNA and about 80 protein molecules per ribosome. The number of ribosomes per cell varies; the *E. coli* cell that we are considering here contains about 18,000. Thus, over 1.4 million molecules of ribosomal protein are present in an average *E. coli*, or about three-fourths of all the protein of the cell, and the cytoplasm consists largely of a very concentrated suspension of ribosomes (Figure 1.6).

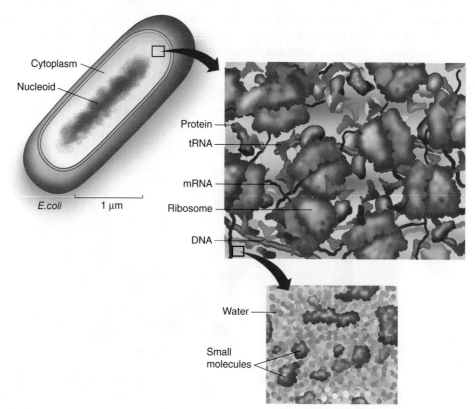

FIGURE 1.6 The procaryotic cytoplasm.

CHAPTER 1 INTRODUCTION TO MICROBIOLOGY

Why are there so many ribosomes? Because about 54% of the dry mass of the cell is protein (62% of 87%), the most significant task during multiplication is to make protein. Furthermore, protein synthesis is slower that that of other macromolecules. Only two replisomes (the complex of enzymes that replicates DNA) can duplicate the *E. coli* chromosome in 40 minutes; a few thousand enzyme molecules can double the mass of RNA, polysaccharide, and heteropolymers in 40 minutes or so. To double the mass of protein in a comparable time, however, takes 10,000 or more ribosomes working at their maximum rate.

1.13 The small molecules of the cell are in constant flux

The small molecules of the cell, despite their minor contribution to the dry weight, are abundant and diverse. How can they be abundant if they are only 5% of the dry weight? Because they are so much smaller than the macromolecules, a very small mass of them consists of a very large number of molecules. If we assume an average molecular weight of about 100, then an average *E. coli* cell would contain approximately 70 million small molecules. Because there are so many different kinds of small molecules in the cell, however, the number of any particular molecule is much smaller—typically 1,000 to 100,000.

Because most of these small molecules are intermediates in biochemical pathways or the monomeric precursors of macromolecules, they are continually removed and transformed into something else (into another intermediate in the case of biochemical intermediates or into macromolecules in the case of monomers). Thus, the collection of molecules of any given particular type, which we call a "pool," is in constant flux. Molecules are continually being removed by being transformed into something else, and these are continually replaced by conversion of molecules from a precursor pool. The rates can be very high; in *E. coli* multiplying at a rapid rate (one generation every 30 to 40 minutes), these small molecule pools turn over entirely (i.e., the number of molecules removed and replaced is equal to the total number in the pool) in a matter of a minute or less. Thus, what we see as a relatively stable chemical composition is in fact a highly dynamic equilibrium.

As an example of the highly dynamic pools, let us calculate an approximate average turnover time for the amino acid pools in *E. coli*. The average *E. coli* cell contains 2 million protein molecules, averaging about 300 amino acids in length. Thus, there are 600 million amino acids polymerized into protein in a single *E. coli* cell. For that cell to double in size and then divide into two *E. coli* cells, it has to double the number of proteins to 4 million before dividing, requiring a further 600 million amino acids to be polymerized. Because there are 20 different amino acids, an average of 30 million molecules of each is polymerized. If we assume an average pool size for amino acids of 100,000 molecules (amino acid pools tend to be larger than most), then the pools turn over 300 times over the course of one cell division. If we take 40 minutes as the time that it takes for this (*E. coli* can multiply faster or slower than this, but this is a common time in the laboratory), then each pool is completely replaced every 10 seconds.

Thus, we see that the procaryotic cell is a highly complex, concentrated mixture of macromolecules and ribosomes in constant frenetic chemical activity. Eucaryotic microbes are even more complex, with their multiple subcellular compartments and intracellular motility and cytoskeletal systems. All cells are highly dynamic entities. Although we necessarily draw them as static, remember that every picture or drawing of a cell is an instantaneous snapshot that freezes what is, in the living cell, a seething mass of chemical activity.

Summary

Microbes are the most abundant organisms on earth by many orders of magnitude, and their biomass is equal to that of all multicellular organisms combined. They are critical to the sustainability of life on earth. They inhabit a wide range of habitats that are lethal for multicellular organisms: temperatures over 100°C, or pH of over 10 or under 2. They are the foundation of all life on earth, and their activities have changed the planet in fundamental ways. They can easily be regarded as the most important life forms on earth.

Study questions

1. Briefly summarize the argument that procaryotes are the most important life form on the planet; or, if you wish, argue that other organisms are more important. If you choose the latter, be sure that you consider and rebut the case for the procaryotes.
2. Summarize the basic features of the procaryotic cell and how it differs from the eucaryotic cell.
3. Draw a time line showing the history of life on earth, showing the approximate times of emergence of procaryotic cells, eucaryotic cells, and multicellular plants and animals. Draw to scale.
4. Assume that the average *E. coli* cell has a volume of 1 μm^3 and contains 8×10^{-15} grams of small organic molecules (the thousand or so building blocks, precursor metabolites, and pathway intermediates). Assume that the average molecular weight of these compounds is 200 g/mole. What is the approximate total concentration of these molecules?
5. Assume that the average *E. coli* cell has a volume of 1 μm^3 and contains 2×10^6 molecules of protein. What is the approximate total concentration of these molecules?
6. Assume that the average *E. coli* cell has a volume of 1 μm^3 and contains ATP at a concentration of 10 mM. How many ATP molecules are there per cell?
7. Assume that the average *E. coli* cell has a volume of 1 μm^3 and its cytoplasm is approximately pH 7. How many protons would you expect to find per cell?
8. Use your answer for question 6 to calculate how fast the ATP pool turns over. Assume that it takes 40 millimoles of ATP to make one gram (dry weight) of cells, that each cell contains 3×10^{-13} grams of dry weight, and that it takes 40 minutes for a cell to grow and divide into two.

History of Microbiology

2

MICROSCOPES WERE INVENTED AROUND 1600 A.D.; however, for decades they were of very poor quality, and although they could reveal considerable detail of fleas, bees legs, and other structures that were barely visible, they did not have the ability to bring most of the microbial world into view. This changed in the late 17th century, when a Dutch amateur scientist named Antoni van Leeuwenhoek (Figure 2.1) began experimenting with simple magnifying glasses, consisting of a single lens that could magnify up to several hundred times. Fascinated with what they revealed, he spent half a century examining an enormous range of material with his various lenses. In September 1674, he communicated a momentous discovery to the secretary of the British Royal Society (a scientific society established to foster the communication of scientific findings):

> *About two hours distant from this Town [Delft, where Leeuwenhoek lived] there lies an inland lake, called the Berkelse Mere, whose bottom in many places is very marshy, or boggy. Its water is in winter very clear, but at the beginning or in the middle of summer it becomes whitish, and there are then little green clouds floating through it; which, according to the saying of the country folk dwelling thereabout, is caused by the dew. . . . Passing just lately over this lake . . . and seeing the water as above described, I took up a little of it in a glass phial; and examining this water next day, I found floating therein . . . very many little animalcules [little animals], whereof some were roundish, while others, a bit bigger, consisted of an oval. . . . And the motion of most of these animalcules in the water was so swift, and so various, upwards, downwards, and round about, that 'twas wonderful to see: and I judge that some of these little creatures were above a thousand times smaller than the smallest ones I have ever yet seen.[1]*

FIGURE 2.1 Antoni van Leeuwenhoek.

With this simple observation, an entire realm of life was brought to human attention (Figure 2.2).

Leeuwenhoek continued his observations on his "little animalcules" over the next several decades, describing many different ones, including both procaryotic

---

[1]All Leeuwenhoek quotations are from Clifford Dobell, *Antony van Leeuwenhoek and his "Little Animals,"* New York: Harcourt, Brace and Company, 1932.

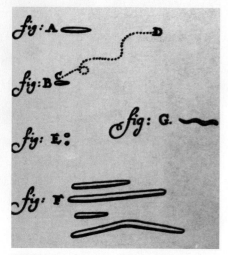

FIGURE 2.2 Leeuwenhoek's drawing of microbes.

and eucaryotic microbes (although he did not use those terms). He showed them to many scientific colleagues and lay people; the reaction was not always enthusiasm:

> *I have had several gentlewomen in my house, who were keen on seeing the little eels in vinegar: but some of 'em were so disgusted at the spectacle, that they vowed they'd ne'er use vinegar again. But what if one should tell such people in future that there are more animals living in the scum on the teeth in a man's mouth, than there are men in a whole kingdom? . . . For my part I judge, from myself (howbeit I clean my mouth like I've already said), that all the people living in our United Netherlands are not as many as the living animals that I carry in my own mouth this very day.*

To this day, many are uncomfortable with the notion that each one of us individuals is an enormously rich and varied ecosystem, teeming with life. Let us hope that as you learn more about the microbial world, you will share Leeuwenhoek's wonder and enthusiasm, rather than the aversion of his genteel friends.

2.1 Microbiology is founded on two basic methods: microscopy and pure culture technique

The essential feature of microbes is that they are so small (especially the dominant ones, procaryotes). Understanding them is thus depended on two particular technologies: *microscopy* to visualize them and *pure culture technique* to allow each individual type of microbe to be studied separate from others.

The first of these is obvious; because microbes are invisible without microscopes, we did not even know of their existence until the development of microscopes capable of bringing them into our visual range. Further understanding of the range of microbial morphology and behavior depended in turn on improvements in microscope design, which allowed better resolution and higher magnification. In the same way that understanding plant and animal biology required dissection to reveal internal structure, so understanding of microbiology required the development of electron microscopy, which allowed the exploration of the internal architecture of the microbial cell (and of plant and animal cells as well).

Microscopy alone, however, only allows a descriptive approach to microbiology: we can observe and describe microbes and to a limited extent experiment with their behavioral responses to various stimuli. If we want to study their biochemistry, genetics, and molecular biology, however, we need additional methods. Most microbes are too small for biochemical study of single cells; rather, populations of identical cells need to be studied. How to get these homogeneous populations is quite a challenge; because microbes are so small and so numerous, it is not easy to get rid of all but one kind of microbe and to keep a population multiplying without it becoming contaminated with other microbes. Pure culture technique consists of a set of interrelated activities, developed in the late 19th century, designed to do this. We return to this shortly.

2.2 Leeuwenhoek discovered microbes because he used a single lens of unusually high quality

At the time Leeuwenhoek discovered the microbial world, microscopes had been in widespread use by scientists for at least a half century. Why did it take so long for microbes to be discovered? Certainly the possibility of microbial life must have been obvious; every naturalist has had the experience of closely observing very tiny animals, such as fleas, and many must have wondered just how small was the smallest animal. In 1508, Alexander Benedictus described tiny worms and mites right at the edge of visual detection. Shortly before Leeuwenhoek discovered true microbes, Robert Hooke in England discovered a fossil foraminiferan (a eucaryotic microbe that makes a tiny shell that fossilizes well). Although a microbial product, the fossil shell was visible with the unaided eye as a speck of sand. With some exaggeration he wrote that he had discovered "a very good instance of the curiosity of Nature in another kind of Animals which are remov'd, by reason of their minuteness, beyond the reach of our eyes." Why did not he, or any one of numerous scientists with microscopes in 17th century Europe, scoop Leeuwenhoek?

The answer undoubtedly lies in the quality of Leeuwenhoek's microscopes. Lenses at that time were made by hand by taking a small piece of glass and laboriously grinding it into shape with a paste of fine abrasive sand and then polishing it with a finer abrasive. The shape of the lens and, consequently, its magnifying power and the severity of its distortion of the image were dependent on the ability of the maker in this very demanding craft. Leeuwenhoek, for whatever reason, was one of the most skilled lens grinders the world has ever seen (Figure 2.3).

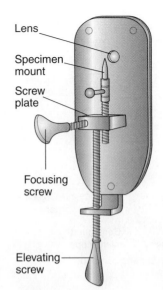

FIGURE 2.3 Leeuwenhoek microscope.

He may also have invented a method of melting glass to form a very thick, but not quite spherical, bead of glass. Nine of his microscopes survive (at his death his estate included about 250 complete microscopes and another 170 lenses, but most have since disappeared). The surviving instruments range in magnification from about 70× to 270×. Eight of them have lenses that were ground; the most powerful one, however, appears to have been made by melting rather than grinding. Exactly how he did this we do not know, but a visitor to Leeuwenhoek's house in 1710 quoted him as saying "that he had succeeded, after ten years speculation, in learning a useful way of blowing lenses which were not round." The photographs in Figure 2.4, both taken through one of his own microscopes (the 270× one), show some red blood cells and a thin slice of cork cut by Leeuwenhoek himself with his shaving razor.

Leeuwenhoek had another advantage over many, although not all, contemporary microscopists: he used a simple microscope rather than a compound one

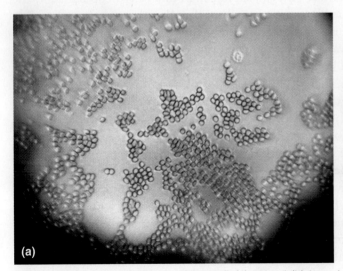
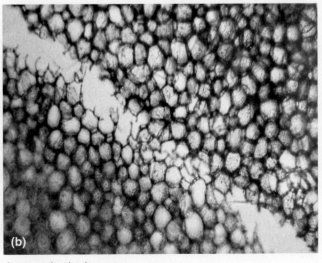

FIGURE 2.4 Photomicrographs of red blood cells (a) and cork (b) through a Leeuwenhoek microscope.

Leeuwenhoek discovered microbes because he used a single lens of unusually high quality

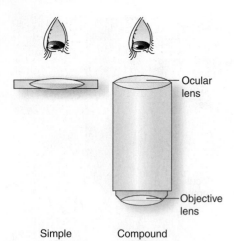

Ocular lens

Objective lens

Simple microscope

Compound microscope

FIGURE 2.5 Simple and compound microscopes.

(Figure 2.5). These are technical terms, referring to the number of lenses. A simple microscope has a single lens; today we call such an instrument a magnifying glass. We are used to thinking of simple microscopes as low-power instruments, but as we have seen Leeuwenhoek routinely made quite high-power ones (of the nine that survive, five are over 100×).

Compound microscopes have two different types of lenses, which magnify in series: one lens produces an image, which is in turn magnified by the second lens. The power of a compound microscope is the product of the powers of its two lenses: a 10× objective lens coupled to a 10× ocular lens gives a total magnification of 100×. Although compound microscopes offer many advantages, which we discuss in the next chapter, they must be made of exceptionally high-quality lenses because any distortions in the first lens are magnified by the second. In the 17th and 18th centuries, optics was not enough of a science to guide the production of high-quality compound microscopes. Simple microscopes offered higher quality images because of their lesser distortion. This gave Leeuwenhoek another advantage over microscopists (such as Robert Hooke) who routinely used compound microscopes.

2.3 Advances in optics in the mid-19th century led to much improved microscopes

After Leeuwenhoek's death, relatively little new was discovered in the microbial world until the 19th century. The reason for this is that few scientists were able to produce simple microscopes of the magnification and quality of his. This is not to say that microbiology was stagnant; several scientists contributed new information, but there were few significant discoveries. There was also substantial improvement in microscope design, mainly in the mechanical aspects of focusing and using the instruments (both simple and compound), but there were no breakthroughs in lens design or manufacture.

In the middle of the 19th century, however, great advances took place in the science of optics, which allowed a clear understanding of the sources of distortion in lenses and, thus, allowed a rational approach to the design of lenses that were free of the kinds of distortion that made earlier compound microscopes so poor. These advances, coupled with lens-grinding machines that could reproducibly manufacture lenses with nearly perfect surfaces, led to compound microscopes that, by the middle of the century, had become so nearly perfect that they compare favorably with the most expensive research microscopes manufactured today.

The principal innovation was to replace each of the lenses of the compound microscope with a set of lenses, each of which was designed to correct a particular type of distortion. Thus, each of the lenses in a primitive compound microscope became a *lens system* in the modern instruments. With this innovation, the compound microscope finally surpassed the best simple microscopes, and the latter are now rarely used except for low-power work in the field.

The final step in the evolution of the modern compound microscope was the development of the **condenser,** a device that produced even illumination of the specimen. When the magnification is high, images tend to be very dim, and very bright illumination is required. Early microscopes used mirrors to direct sunlight or artificial light onto the specimen, but this was generally uneven and insufficiently bright for very high magnification. The development of the condenser, which produced even illumination, in combination with the incandescent light bulb, solved the illumination problem that limited useful magnification until the last half of the 19th century.

These developments led to a great resurgence of interest in microbiology, and throughout the latter half of the 19th century, there was a rapid expansion of knowledge of the diversity of microbes and their habitats (Figure 2.6). In conjunction with the pure culture techniques that were developed in the 1880s and 1890s, these advances in microscopy led to an explosion of knowledge of the microbial world that was so dramatic that the last two decades of the 19th century are sometimes termed the "golden age" of microbiology. The term is accurate in its portrayal of the excitement and dynamism of that period; it is inaccurate in implying that subsequent periods were anticlimactic. In fact, microbiology is a dynamic discipline that for more than a century has been growing and changing at an ever increasing rate.

FIGURE 2.6 Ferdinand Cohn's 1877 drawing of bacterial types.

2.4 Advances in light microscopy in the 20th century were largely confined to improved methods of enhancing contrast

Although the advances in optics of the 19th century pretty much exhausted the potential for improvements in useful magnification or resolving ability of the light microscope, there was still the problem of contrast. Many microbial cells are so small that they are nearly transparent and have almost no contrast with their background. The approach used in the late 19th and early 20th century relied on **staining**—adding chemical dyes that would bind to microbial cells and increase their contrast. These normally kill the cells, however, and thus cannot be used to help visualize living cells.

Two significant innovations in the 20th century have allowed great improvements in this area: **phase-contrast microscopy** and **darkfield microscopy.** Both of these are optical methods for making objects contrast more with their background; phase contrast makes the objects much darker than they would otherwise be, whereas darkfield shows them as bright objects against a black background. Most microbiologists today use phase-contrast microscopy for routine use, but darkfield is used for exceptionally small microbes or to visualize structures, such as bacterial flagella, that cannot be seen in the phase-contrast microscope (Figure 2.7).

2.5 The most significant advance in microscopy in the 20th century was the development of electron microscopy

The ability of a light microscope to visualize very small objects is limited by the physical properties of light, as we discuss in the next chapter. In most practical applications,

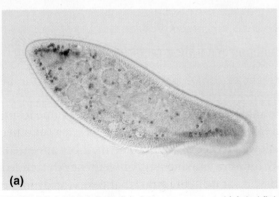
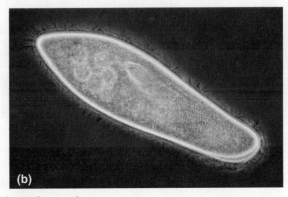

Bacterial flagellum

FIGURE 2.7 (a) Brightfield, (b) phase-contrast, and (c) darkfield photomicrographs.

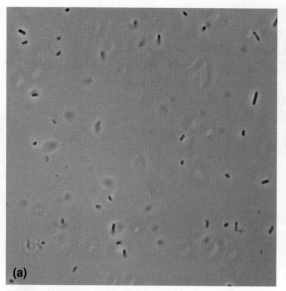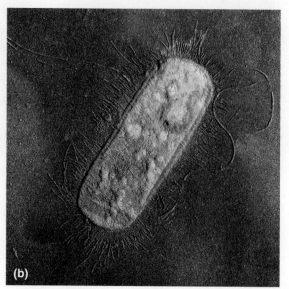

FIGURE 2.8 *E. coli* in the light (a) and electron (b) microscopes.

light microscopes cannot visualize objects less than about 0.1 µm. This is sufficient to visualize easily almost all microbes except viruses; however, it does not allow much detail within microbial cells to be seen. For that we now depend on the *electron microscope,* an instrument whose limit of resolution is at least a thousand times better than the light microscope (Figure 2.8).

The electron microscope was invented in the 1930s, when physicists realized that the wave nature of elementary particles meant that they had a wave length like light but much shorter. Thus, it was theoretically possible to make a microscope that used a beam of electrons instead of a beam of photons (light), and because of the much shorter wavelength, such a microscope would have a much greater ability to resolve very small objects. Early electron microscopes were not very good, but by the 1950s, they had improved markedly and began to be used on biological specimens. Because the electron beam in an electron microscope needs to be in a vacuum, it took a number of years to work out appropriate techniques to preserve biological specimens to be put under vacuum for observation. It was also necessary to develop methods to enhance contrast, as biological specimens are nearly transparent to an electron beam. By the late 1950s, however, electron microscopes were beginning to revolutionize our understanding of cells, and they continue to do so even today. There is no theoretical limit to the resolving ability of the electron microscope, but practical considerations limit most of them to about a nanometer or a bit less.

2.6 The controversy over spontaneous generation required the development of sterilization methods

From the very first, there were controversies over where the microbes came from. No microbes could be seen in fresh well water or rain water, but if some hay or other organic material was put in and allowed to decay, numerous microbes appeared. Were they formed spontaneously from the decaying hay, or were they descended from microbes that came in with the hay or from the air or that were already in the water but simply too few to be seen?

FIGURE 2.9 Louis Joblot in his laboratory (woodcut).

There was a long history of popular belief in spontaneous generation from decaying material; at least since medieval times insects, worms, maggots, and other lower animals were thought to be formed this way. This was pretty much discredited by Francisco Redi, who in the mid-17th century demonstrated quite elegantly that maggots were a stage in the life cycle of flies and that they came from eggs flies had laid. The extremely small and simple microbes that seemed to come from nowhere, however, were still considered by many to be spontaneously formed. The first experiments to address these questions were done by Louis Joblot (pronounced Zhou-blow), a French scientist and contemporary of Leeuwenhoek (Figure 2.9). Joblot showed that microbes were produced by multiplication of microbes brought in on the hay or suspended in the air, not spontaneously (see MicroTopic box on page 18).

Over the next century a number of people repeated Joblot's experiments or did variants of them. In some cases, the results confirmed Joblot's; in other cases, they did not. Furthermore, it was suggested that the air inside a sealed flask became spoiled when the flask was boiled, and it could not support the decay that was required for spontaneous generation.

FIGURE 2.10 Louis Pasteur.

Thus, the picture remained murky until the work in the 1860s of the French scientist Louis Pasteur (Figure 2.10) and the English scientist John Tyndall (Figure 2.11). In an elegant series of experiments, Pasteur convinced most of the scientific community that spontaneous generation did not occur. The most memorable of these was the experiment with "swan-necked flasks." He took flasks with long, drawn-out necks, containing a nutrient medium, and boiled them. He then bent their necks in a burner flame so that even though there was still unimpeded continuity between the outside air and the inside air, any microbes suspended in the air would settle in the neck without reaching the flask. The flasks remained clear, with no decay and no microbes for months. Yet when he snipped the necks off to allow airborne dust to settle into the flasks, their contents began to decay, and within a day, many microbes were seen. This demonstrated conclusively that spontaneous generation did not happen, and it appeared to show what was sufficient to sterilize a liquid infusion (Figure 2.12).

Pasteur was lucky, however; he was working in a fairly clean environment, and he used meat infusions rather than hay infusions. As we now know, hay is often full of the heat-resistant spores of bacteria, some of which are not killed by even an

FIGURE 2.11 John Tyndall.

The controversy over spontaneous generation required the development of sterilization methods

Louis Joblot, a contemporary of Leeuwenhoek's, was the first to address the issue of the origin of microbes in infusions of decaying organic material experimentally. His experimental design was elegant and straightforward and provided a clear prototype for his successors to imitate or modify.

Joblot reasoned that there were three formal possibilities for the microbes that suddenly appeared in his hay infusions: they were the product of multiplication of microbes that came in with the hay (he did not consider the water a likely source because he used clean fresh water in which he could see no microbes); they were the product of multiplication of microbes that settled out of the air; or they were spontaneously generated from the materials of the decaying hay.

On the 4th of October, 1711, I placed a little fresh hay in ordinary cold water in two different vessels. One I stoppered the best I could with well-wetted fine parchment and the other I left open. Two days later, I observed three types of animals which were present in both infusions. This experiment seems very appropriate to convince one that these animals were produced from eggs that other animals had deposited on the hay, and not from those which were dispersed in air [however, these results are also consistent with spontaneous generation] (Figure 2B.1).

On the 13th of October, I boiled similar fresh hay in ordinary water for more than a quarter of an hour. I then placed an equal quantity in two vessels of about the same size. I sealed one right away even before it has cooled down: in the other one which I had left open I saw animals at the end of a few days but not in the infusion which had been sealed. I kept it sealed a considerable amount of time in order to find some living insect, if any should grow, but finding none, I left it open and after a few days I observed some. From this, one should understand that these animals were born from the eggs distributed in the air since those which would have been encountered on the hay had been completely ruined by boiling water (Figure 2B.2).[2]

Joblot interpreted these experiments as convincing disproof of spontaneous generation:

Here it seems to me are experiments in sufficient quantity to show that neither alteration nor corruption nor bad odor are the causes of the generation of animals whatever they may be. Let us then try to establish a hypothesis to explain those most surprising things that may be seen in the infusions of plants. I would propose that in the air close to the earth there flies or swims an innumerable number of small animals of various species, which while fastening themselves to the plants of their choice to rest or nourish themselves, deliver their little ones or deposit their eggs there. And that these same animals also drop young and eggs into the air through which they travel.[2]

Although Joblot's vision of microbes flying through the air dropping eggs and babies is a bit fanciful, he was very close to the truth. The air close to the earth is indeed full of suspended microbes stirred up from the soil, shed from the surfaces of animals, or released from the mouths and noses of animals during breathing, coughing, or sneezing. These suspended microbes settle onto all surfaces, including plants, and can multiply if they find nutrients and water available.

Unfortunately, Joblot's experimental procedure, although it worked in this particular instance, is not robust enough to be easily repeatable, which led numerous imitators to incorrect conclusions. The length of time he boiled his hay infusions was insufficient to reliably sterilize them. He did not use sterile glassware, and his method of stoppering may have been inadequate (particularly because he used nonsterile parchment).

Nevertheless, he deserves a great deal of admiration for his clear experimental design and accurate interpretation so early in the study of the microbial world.

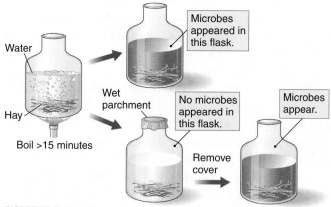

FIGURE 2B.2 Joblot's second experiment.

[2]All Joblot quotations are from Hubert Lechevalier, Louis Joblot and his microscopes, *Bacteriological Reviews* 1976;40:241–258. Copyright American Society for Microbiology.

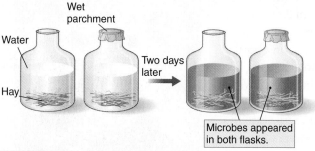

FIGURE 2B.1 Joblot's first experiment.

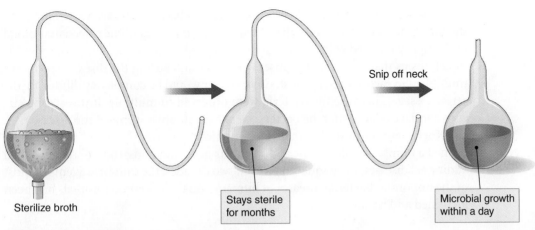

Snip off neck

Sterilize broth

Stays sterile for months

Microbial growth within a day

FIGURE 2.12 Pasteur's swan-necked flasks.

hour's boiling. It was John Tyndall (Figure 2.11) who discovered the heat resistance of spores and worked out a way of reliably killing them by repeated boiling, rather than a single boiling step.

Thus, one of the results of the final disproof of spontaneous generation was the development of methods to sterilize materials and to maintain them sterile. This was part of the technology of pure culture; the other part came from the effort to show that different microbes caused different diseases.

2.7 The germ theory of disease provided an explanation for the phenomenon of contagion

The cause of human, plant, and animal diseases has been an important mystery for millennia. Contagion, the ability of a disease to pass from one person (or plant or animal) to another, was recognized at least as long ago as classical Greek times, but the mechanism of such contagion was unclear, as nothing could be observed passing from the diseased organism to the healthy one. In the 200 years after Leeuwenhoek's discovery, a number of people speculated that microbes might cause disease. Many others, however, argued that organisms so small were unlikely to be able to cause such large effects. The issue remained one primarily for speculation until the late 19th century, when the disease anthrax was studied scientifically. Anthrax is primarily a disease of cattle and other grazing animals, but it is occasionally transmitted to humans. Several different scientists showed that the blood of animals that had died of anthrax was full of bacteria and that the transfer of minute amounts of infected blood could transfer the disease to a healthy animal. Was it the blood itself, however, or the bacteria in the blood, that caused the disease?

Robert Koch, a German doctor (Figure 2.13), was the one who finally showed that it was the bacteria that caused the disease and not something else invisible that was contained in infected blood. Working in a simple home laboratory, he made a series of careful observations that showed the complete life cycle of the causative organism, which he named *Bacillus anthracis*. He showed that the organism made spores that could last for years, explaining how fields could remain contaminated for decades. He also transmitted the disease from one mouse to another by inoculating a bit of infected spleen from a mouse that had died of anthrax under the skin of a healthy mouse. In this way, he transferred anthrax for a series of 20 successive

FIGURE 2.13 Robert Koch.

transmissions. In each of the 20 infected mice, the bacteria that were introduced in the bit of infected tissue multiplied to very large numbers in the spleen and blood of the newly infected animal. Because the infected tissue of the 20th mouse was as infectious as that of the first mouse, whatever was causing the disease must have multiplied in each successive host; otherwise, it would be completely diluted by the serial passages. Because the bacteria were observed to multiply, it was strong circumstantial evidence that the bacteria were the causative agents of the disease.

Koch went on to study other diseases, most importantly tuberculosis, and he trained a number of physicians and scientists in his methods; they, in turn, studied additional diseases. Consequently, within two decades, the causative agents of most of the important bacterial diseases of humans and their domestic animals had been identified and isolated.

2.8 Pure culture technique was necessary to show that different diseases were caused by different microbes

Although Koch's demonstration of the bacterial etiology of anthrax convinced nearly everyone that this particular disease was caused by a bacterium, there was still a great deal of question about how general the phenomenon was. Because each disease is quite different from others, to believe in a microbial origin for each implied that there were a large number of different microbes. This was disputed by many, who looked in the microscope and saw only a few different types. Koch himself listed 10 diseases that seemed to be associated with identical-looking bacteria. Clearly, a method was necessary for obtaining each bacterium associated with a disease in pure culture, separate not only from other material derived from the diseased host, but also free of all other types of bacteria. Only then could a specific microbe be associated with a specific disease. Koch realized this soon after his experiments with anthrax.

The breakthrough came when Koch noticed that there were separate regions of microbial growth on the surface of a potato that had been sliced with a sterile knife and then allowed to sit for a day. Potato slices had been used for several years as a means of growing up masses of a microbe from a tiny original sample, but Koch appears to be the first to have realized that solid media could be used to separate different kinds of microbes. He immediately realized that the microbes must have come from the air, and that each different kind multiplied where it landed on the potato. Because the microbes could not move about on the surface, the different types could not mingle and stayed separate from each other. This provided the key insight: growth on solid surfaces, rather than in liquid, would allow the preparation of pure cultures. A mixed population of bacteria from a natural source, if spread out on the surface of a solid nutrient medium, would produce a series of localized growths, termed **colonies,** each one of which would constitute a homogeneous culture, or **pure culture.**

Potatoes are not convenient for this purpose because they are opaque (making it difficult to see colonies) and because many bacteria do not grow well on them (they are deficient in nutrients that a number of bacteria need). After the key insight that solid growth media were required, however, it was a simple matter to develop ways of solidifying the commonly used liquid media to give a transparent, solid nutrient gel. First, gelatin was used as a solidifying agent. This did not work very well because it would melt at the temperature at which most cultures of disease-causing bacteria were grown (37°C, the normal temperature of the human body) and because many bacteria could break the gelatin down and use it as a nutrient, in the process making the solid medium liquid again. Finally, Fanny Hesse, the American wife of one of Koch's coworkers, suggested the use of agar, a solidifying agent used in cooking. Agar turned out to be ideal; after being melted, it would stay liquid down to temperatures of less than 50°C, thus allowing media to be poured into containers at a temperature that was not scalding. After solidifying, however, it stayed solid up to nearly 100°C, thus allowing incubation at almost any temperature. Finally, as a very complex polysaccharide from marine algae, it was not normally attacked by bacteria from human, plant, animal, or terrestrial sources (Figure 2.14).

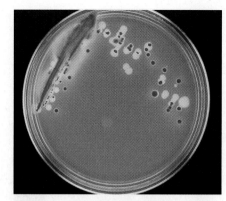

FIGURE 2.14 Colonies on a Petri plate. Two different types are visible.

The development of solid media as a way of isolating the growth of different bacteria from a natural source completed the constellation of techniques needed to allow routine work with pure cultures: (1) techniques to render materials sterile, (2) techniques to keep foreign microbes from entering sterile media or previously inoculated cultures, and (3) techniques for isolating a single type of microbe from all others in a mixed population. Using these techniques, Koch and his collaborators (and imitators) were able to demonstrate a microbial cause for most of the major human and animal diseases within a short period of time (about 10 years at the end of the 19th century).

2.9 Koch's postulates define what is logically necessary and sufficient to prove the microbial etiology of a disease

On the basis of his experiments with anthrax and tuberculosis, Koch articulated four principles that must be satisfied before one could with confidence claim that a disease is caused by a microbe:

1. The microbe must be present whenever the disease is (e.g., all animals suffering from anthrax must show the particular bacterium in its blood or tissues), but absent from healthy organisms.
2. The microbe must be isolated from the diseased host organism and obtained in pure culture.
3. The microbe must be inoculated back into a healthy host, which must then become ill with the same disease.
4. The same microbe must be found in the experimentally infected host.

These four principles are termed *Koch's postulates* and are widely recognized as logically necessary to prove the microbial etiology of a disease. Since Koch's time, however, so many diseases have been shown to have a microbial etiology that we no longer demand that all of Koch's postulates be met before we will accept that a given disease is caused by a microbe. This is particularly true when there are special difficulties, as when a disease affects only humans. In such a case it may be ethically impossible to fulfill the third postulate. The first postulate is also sometimes difficult to meet in some cases, and we now rarely insist on the presence of the microbe in **all** cases; a statistically significant association of the microbe with the disease is generally considered sufficient.

2.10 Enrichment culture allows the isolation of microbes that are present in the source material in very low numbers

The use of solid media for the isolation of pure cultures was an enormous step forward, and it allowed the isolation of many different microbes and the demonstration that many identical-looking microbes were in fact different (they caused different diseases, they were biochemically different, they preferred different temperatures or different media, they required different nutrients, etc.—all things that required pure cultures to determine). The direct isolation of cultures on solid media suffered from one major drawback, however—it was difficult or impossible to isolate organisms that were a very small minority in the source material. For the isolation of very rare organisms with particular biochemical abilities, Martinus Beijerinck (Figure 2.15) developed **enrichment culture**.

In this technique, a small amount of a natural sample (soil, water, blood, tissue, etc.) is placed in a liquid medium that is designed to select for certain types of organisms—for instance, organisms that can use a particular organic compound as their principle food. As the various organisms multiply in this medium, the ones that grow the fastest increase their proportion in the total population, whereas ones that grow slowly or not at all decrease their proportion. After growth, the resulting population, with its altered ratios among the various microbes in the original sample, is plated on solid medium, and the various different major types are isolated.

The only necessity for performing enrichment culture is to be able to devise a liquid medium that selectively favors a particular type of organism. If this can be done, such organisms should be able to be isolated if they occur in nature, despite being quite rare. The application of enrichment culture has allowed the isolation of many different physiologic types of microbe that we would otherwise have never known existed.

FIGURE 2.15 Martinus Beijerinck.

2.11 Several microbes have become model systems for the study of cell biology, molecular genetics, and development

Microbiology in the 20th century has become one of the major disciplines of modern biology. Part of the reason is that beginning after the Second World War a number of scientists chose bacteria or their viruses as a simple model organism in which to study generally important biological phenomena, such as the regulation of gene expression, the biochemistry of the cell cycle, the physical chemistry of photosynthesis, the control of differentiation, etc. The result has been an explosion of knowledge about microbes that has benefited all of biology and brought microbiology squarely into the mainstream as one of the most exciting and important biological sciences.

Among the microbial model systems that all biologists should be familiar with are:

Escherichia coli is a simple intestinal bacterium of mammals that is the best studied organism in the world and has been the organism of choice for most studies of the molecular aspects of gene regulation and for studies of the procaryotic cell cycle. It is also the closest we have to a "typical" bacterium, being a member of the largest and most diverse group of gram-negative bacteria, and having a variety of characteristics that are very common among the bacteria (average cell size, genome size, respiratory or fermentative metabolism, etc.).

Caulobacter crescentus is a bacterial inhabitant of soil and water and has a long stalk at one end of the cell with a sticky *holdfast* at the end to anchor it to a solid surface (commonly another, but larger, cell, from which small amounts of organic compounds leak out and provide nutrients for the *Caulobacter*). *Caulobacter* has become a major model system for studying the regulation of asymmetry in cell morphology, as it divides asymmetrically to produce two different daughter cells (a sessile one with the stalk and the other a motile cell with a flagellum at one end).

Bacillus subtilis is a common soil bacterium that produces a dormant *endospore* when it is starved. It has been very useful in studying the genetics and molecular biology of differentiation.

Saccharomyces cerevisiae is a yeast (a unicellular fungus) and is the most common eucaryotic microbial experimental system. Its major uses have been in studying the eucaryotic cell cycle and in studying the targeting of proteins to various compartments of the eucaryotic cell.

Of course, thousands of other procaryotic and eucaryotic microbes are being actively studied in various laboratories in the world, and their study provides a valuable perspective on the more detailed study of model organisms. The study of a wide variety of microbes is necessary to determine how general are the conclusions from the study of model organisms and to reveal new phenomena that are not encountered in the relatively few microbes that are models. This is especially important because the microbes in general and the procaryotes in particular are by a very large margin the most biochemically and genetically diverse organisms on earth (Figure 2.16).

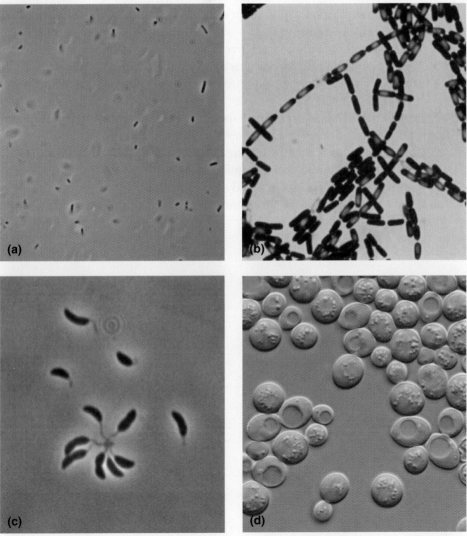

FIGURE 2.16 (a) *E. coli,* (b) *B. subtilis* (vegetative cells and spores), (c) *C. crescentus,* and (d) *S. cerevisiae.*

Model systems for the study of cell biology, molecular genetics, and development

Summary

By its very nature, the microbial world is hidden from our view. It took the careful and detailed work of Leeuwenhoek to bring it to our attention, but to learn anything significant about the organisms that inhabit it had to wait several centuries for the development of techniques that would allow single microbes to be isolated from all others and to multiply into large populations without being contaminated by other microbes. After such pure cultures were available, progress in understanding microbes was rapid. One of the most important advances was the demonstration that infectious disease is caused by microbes, an old idea that required pure cultures to prove. In the century since this early pioneering work, microbiology has become one of the most dynamic of the biological sciences, and a large proportion of our modern insights into the molecular mechanisms of life were first obtained by studying microbes, often one of the small number of intensively studied species that we call model organisms.

Study questions

1. What factors contributed to the fact that Leeuwenhoek discovered the microbial world with simple microscopes and that his observations were not surpassed for over a century.
2. What two technologies characterize the field of microbiology, and why?
3. What advances in microscopy characterized the 19th and 20th centuries?
4. Compare and contrast the experiments of Joblot and of Pasteur regarding spontaneous generation.
5. Explain the technical capabilities necessary for pure-culture technique, and briefly explain how each was achieved.
6. Compare and contrast pure-culture technique using solid media and enrichment culture technique using liquid culture.
7. Summarize Koch's postulates, and explain why each step is necessary.
8. Explain the concept of a "model organism" and summarize the principal microbial model systems.

Methods of Microbiology

3

W E SAW IN THE last chapter that microbiology is a discipline that is unusually dependent on a distinctive set of methods; indeed, it is often defined by these methods. This dependency is a consequence of the very small size of microbes, which means that we cannot see them without microscopy, and the fact that we cannot do experiments with them without pure culture techniques. In this chapter, we examine in some detail these techniques.

These methods constitute a core of techniques common to all microbiology. In addition, almost all microbiologists use a wide variety of other techniques, especially those of biochemistry, genetics, and molecular biology.

Microscopy

Microscopes have improved immensely since Leeuwenhoek first observed microbes with a simple hand lens. Compound microscopes have become highly refined and now approach their theoretical limits. Optical methods of enhancing contrast are now in routine use; fluorescence microscopy allows the detection of specific molecules within cells, and the invention and improvement of electron microscopy have opened an entirely new world of submicroscopic observation.

3.1 Compound microscopes are used for observing microbes

The compound microscope is now routinely used for observing microbial cells. It gives a wider field of view, less distortion, and less eyestrain than the simple microscope, and it can be equipped for phase contrast, fluorescence, or other sophisticated optical refinements.

In the compound microscope, the objective lens forms a magnified **real image** inside the tube of the microscope. This image is real and can be seen if a piece of frosted glass or a piece of paper is placed into the tube at the right spot. The ocular lens then magnifies this image again. The ultimate magnification is the product of the individual magnifications: a 100× objective combined with a 10× ocular gives a total magnification of 1,000×.

Unlike the objective, the ocular lens does not produce an image. In conjunction with the eye, however, it produces the illusion of an image, which we call a **virtual image.** Whether a lens produces a real image or a virtual image depends on the relationship between the specimen and the focal point of the lens; you will learn about this in physics (Figures 3.1 and 3.2).

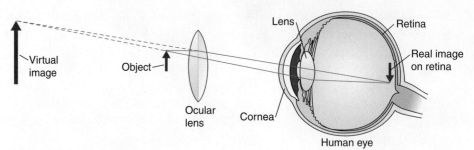

FIGURE 3.1 Virtual image formation.

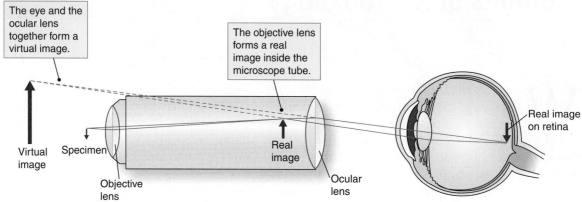

FIGURE 3.2 Image formation in the compound microscope.

3.2 Resolution of the light microscope is normally limited to about 0.2 μm

FIGURE 3.3

FIGURE 3.4

During normal use, fundamental physical properties limit the resolution of a light microscope. Resolution is the ability to distinguish between two closely spaced objects. For example, in Figure 3.3, are we looking at a single dumbbell-shaped object or two spheres very close together as in Figure 3.4?

The limit of resolution (d) is the distance separating the centers of two dots that can just barely be distinguished as separate.

The limit of resolution of a microscope is related to the wavelength of the light being used (λ) and the numerical aperture of the objective lens (NA) by this formula:

$$d = 0.5\lambda/NA$$

The numerical aperture is in turn a function of two physical properties of the objective lens (Figure 3.5): its diameter and its **working distance** (how close to the specimen it has to come to be in focus). This is expressed quantitatively as

$$NA = \eta \sin \alpha$$

Where η is the refractive index of the material in between the lens and the specimen (in air, η = 1.0). Because the physical properties of each lens are fixed and cannot be altered, the only way to improve resolution with a given lens is to increase the refractive index of the material between the lens and the specimen. This is normally done for only very high-power lenses (e.g.,

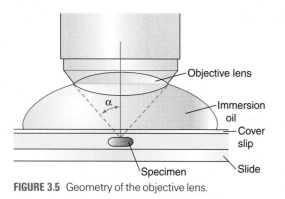

FIGURE 3.5 Geometry of the objective lens.

CHAPTER 3 METHODS OF MICROBIOLOGY

100× objectives), as it is only at total magnifications of about 1,000× that the limit of resolution becomes a problem. Thus, very high-power objective lenses are made to be used with oil ($\eta = 1.5$) and are called **oil immersion** lenses.

Using about 500 nm as an approximation of the wavelength of visible light (actually it includes wavelengths from about 400 nm to about 800 nm) and a figure of about 1.3 for the numerical aperture of a modern, high-quality objective lens using immersion oil between the specimen and the lens, we obtain the following value for the limit of resolution for a modern research-quality microscope:

$$d = (0.5)(500 \text{ nm})/1.3 = 190 \text{ nm} = 0.19 \text{ } \mu\text{m}$$

Although this is the theoretical limit of resolution of the light microscope, it is not the smallest object that one can see; it is actually possible to see an object about half the size of the limit of resolution, but of course without any detail.

This limit is on the order of the size of a procaryotic cell—typically about 1 μm in diameter. Thus, although cells can be easily seen, very little detail inside procaryotic cells is visible with light microscopy; even in the larger eucaryotic cells, intracellular structures are often right at the limit of resolution, and little of their detail can be seen.

This limit of resolution applies during normal use; there are several circumstances under which significantly smaller objects can be seen clearly. One is using darkfield illumination, discussed later; the other is the use of videotaping of a microscopic image and then using sophisticated computer-based image enhancement techniques. With this latter approach, clear images of objects less than 20 nm can be obtained.

3.3 Brightfield microscopy is used for observing stained cells

When used in the traditional fashion, a microscopic specimen is flooded with light from below. After passing through and around the specimen, it enters the objective lens, which forms an image as described previously. The specimen looks darker than the background because it absorbs some light. Because many microbial cells are so small that they absorb very little light, they are often nearly transparent and very difficult to see in this way. Consequently, **brightfield microscopy,** as this traditional way of using the microscope is called, is rarely used anymore for observing living cells.

There is still a function for this type of microscopy, however: to observe stained cells. Originally, stains were developed to enhance contrast in brightfield microscopy, before the invention of the phase contrast microscope. Stains are rarely used this way anymore, as phase contrast is faster and allows living cells to be observed (staining normally kills cells), but **differential staining** is still used. Differential staining uses dyes that will stain some microbes and not others or some intracellular structures and not others. This allows different types of microbes and organelles to be distinguished by their staining reactions, even if they are morphologically indistinguishable. The most common differential stain in use today is the **Gram stain,** named after its inventor, the Danish physician Christian Gram who invented the stain in the 1880s. This stain differentiates one group of bacteria (the "gram-positive bacteria," G+) from all other groups of bacteria (gram-negative, G−), based on the structure of their cell wall. It is not widely used outside the bacteria. We now know that almost all gram-positive bacteria fall into a single major sublineage, and thus, the Gram stain gives us evolutionary information (Figure 3.6).

Because phase-contrast microscopy, the other widely used method of light microscopy, distorts colors, it cannot be used when the object is to determine the

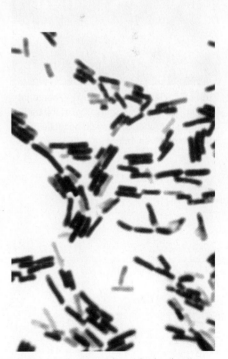

FIGURE 3.6 G+ (purple) and G− (red) cells.

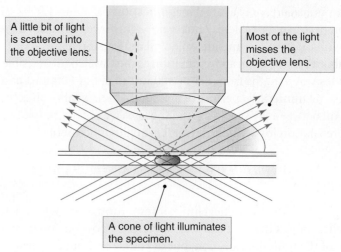

A little bit of light is scattered into the objective lens.

Most of the light misses the objective lens.

A cone of light illuminates the specimen.

FIGURE 3.7 Light path in the darkfield microscope.

color of a stained object (as it is in the Gram stain and other differential stains). Hence, brightfield is routinely used for observing stained preparations.

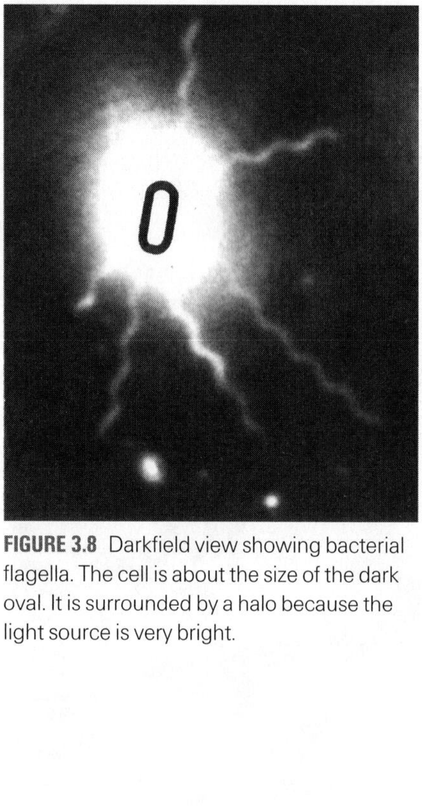

3.4 Darkfield microscopy can resolve very small objects

We have all had the experience of being in a darkened room into which a beam of sunlight shines. In that beam we can see dust motes clearly shining against the dark background, yet when the room is fully illuminated, we see nothing in the air. The reason that we can see the tiny motes (which can be under 100 μm) is that they scatter light, making a bright spot against the black background. The same principle can be used in microscopy. If the condenser is set such that it illuminates the specimen with a cone of light that never enters the objective lens, the field of view will be black because no light is directly entering the lens. If there are objects in the field that scatter light, however, some of that light will enter the lens (Figure 3.7). If there is enough of it to register in the eye, we will see the object as bright against a dark background. With sufficiently intense light, extremely small objects can be seen—for instance, bacterial flagella, which are less than 20 nm thick (Figure 3.8).

Darkfield microscopy is difficult to use routinely, and, thus, it is generally used only by those with a specific need for it, such as microbiologists working with the very smallest microbes. Some spirochetes (e.g., *Treponema*, the causative agent of several human diseases, including syphilis) are less than 0.2 μm in diameter, and they are barely visible in phase contrast. They are routinely observed with darkfield microscopy.

3.5 Phase-contrast is routinely used for observing live microbial cells

Because of their very small size, most microbial cells absorb and scatter little of the light that passes through them; thus, they are nearly transparent and are very difficult to see with transmitted light. One of the great advances in microscopy in the 20th century was the development of sophisticated optical means of enhancing contrast by manipulating the light inside the microscope. The most common type of contrast instrument is the **phase-contrast microscope;** considerably less common, but equally effective, is the **interference-contrast microscope.** Most microbiologists now routinely use one of these means for observing live microbes suspended

FIGURE 3.8 Darkfield view showing bacterial flagella. The cell is about the size of the dark oval. It is surrounded by a halo because the light source is very bright.

in water. We describe the phase-contrast microscope here, as it is the most widely used; interference contrast operates on the same basic physical principles, although the detailed mechanisms are different, and hence, so is the image.

3.6 Phase-contrast is achieved by modifications to the condenser and the objective lens

In the brightfield microscope, the condenser functions to illuminate the specimen evenly and brightly, and the objective lens produces a magnified image. In a phase-contrast microscope, both of these elements are modified—the condenser by the addition of an annular diaphragm and the objective by the addition of a phase plate (Figure 3.9).

The phase-contrast microscope, like the darkfield microscope, illuminates the specimen with a cone of light, rather than an even flood of light. This is because the condenser has an **annular diaphragm**—a transparent ring in an otherwise opaque plate. Thus, a ring of light enters the condenser lens and is then bent into a hollow cone, with the specimen at its apex. Unlike the darkfield microscope, however, in

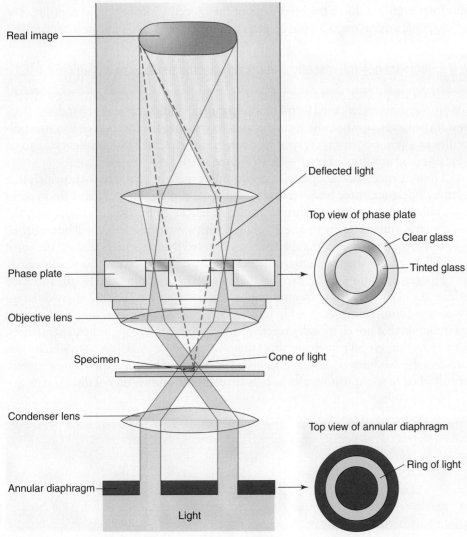

FIGURE 3.9 The light path in the phase-contrast microscope.

the phase-contrast microscope, the cone of light is angled such that it enters the objective lens.

As this light passes through the specimen, some of it gets deflected, particularly from the edges of cells and organelles. Some of it gets absorbed. The great majority of it, however, passes straight through. Ultimately, the lens systems of the microscope will bring the deflected rays and the transmitted rays to a focus at the same point. The image is thus made up of light that came straight through the specimen and light that was deflected by the specimen.

After passing through the specimen, the cone of undeviated light enters the objective lens and passes through a ring of tinted glass on the **phase plate.** This **phase ring** is sized and positioned such that all of the light in the illuminating cone goes through it. Because the ring is tinted, like dark sunglasses, the light that passes directly through the specimen is dimmed by more than 70%. Of course, the background light is dimmed the same amount because it too goes through the phase ring.

The deflected light, although a minor component of the light leaving the specimen, is not dimmed because it passes randomly through any portion of the phase plate; most of it thus misses the tinted phase ring. Even though it is not tinted, however, the phase plate other than the ring is made of thicker glass; it thus shifts the phase of the deflected light. The total phase shift is one half wavelength. The deflected light is now 180° out of synchrony with the direct light, and when they combine in the image plane, destructive interference reduces the brightness of the object, contributing to contrast. The MicroTopic box on pages 32 and 33 gives a more detailed view of what is happening.

3.7 Fluorescence microscopy can locate specific molecules within cells

One of the most widely used forms of microscopy in biological research today is fluorescence microscopy because it can be used to visualize the location of specific molecules in a cell. For instance, specific proteins, or specific DNA sequences, can be visualized by this kind of microscopy (Figure 3.10).

Fluorescence microscopes are very complex instruments, but their principle is simple. There are three basic steps to fluorescence microscopy. First, a fluorescent molecule (a **fluorophore**) is linked to the molecule to be visualized. Second, the specimen is illuminated with a beam of light with a wavelength that will be absorbed by the fluorophore. Third, the specimen is observed through filters that cut out most of the light except the wavelengths that are emitted by the excited fluorophore.

Fluorescent compounds are ones that absorb light of one wavelength and then release the absorbed energy by emitting light of a longer wavelength. Many different fluorescent compounds are used in fluorescence microscopy; some are small organic compounds that are chemically reactive and can bind to certain cellular structures. Others are fluorescent proteins or short single-stranded DNA sequences to which a fluorescent dye has been covalently attached. The details of how specific fluorescent tags are attached to specific molecules in cells are complex, and we do not discuss it here.

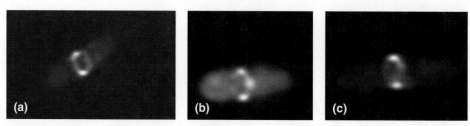

FIGURE 3.10 Fluorescence micrograph of *E. coli* showing location of a specific protein in an *E. coli* cell.

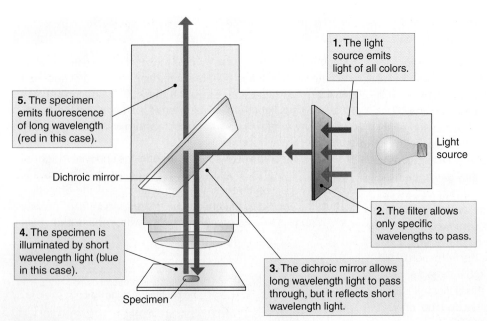

5. The specimen emits fluorescence of long wavelength (red in this case).

1. The light source emits light of all colors.

Light source

Dichroic mirror

4. The specimen is illuminated by short wavelength light (blue in this case).

2. The filter allows only specific wavelengths to pass.

Specimen

3. The dichroic mirror allows long wavelength light to pass through, but it reflects short wavelength light.

FIGURE 3.11 Light path in the fluorescence microscope.

Most fluorescence microscopes are constructed around a set of filters and a **dichroic mirror,** a piece of glass that acts as a mirror for some wavelengths, but is transparent to others. The illuminating light is passed through a filter that filters out all wavelengths except those in the region that will excite the fluorophore. The filters can be changed to allow different fluorophores to be used. The dichroic mirror then reflects the exciting light to the specimen. Fluorescence emitted by the specimen passes back through the dichroic mirror and to the image plane (Figure 3.11).

The image formed by a fluorescence microscope can be pretty fuzzy. This happens when the fluorescent molecules are distributed in a specimen through a wide zone above and below the plane on which the microscope is focused (which can be as narrow as 100 nm, one tenth of the thickness of a typical procaryotic cell). Because all of the fluorophores are fluorescing, but only some are in focus, the image is fuzzy. A modification of the fluorescence microscope, termed the **confocal microscope,** removes light that is out of focus. The optical mechanism by which this is done is too complex for us to describe here, but it is important to know of the technique, as it is becoming widely used.

3.8 The electron microscope uses electromagnetic lenses to bend a beam of electrons

One of the insights of the quantum mechanical revolution in physics in the first third of the 20th century was the understanding of the dual nature of elementary particles. Thus, visible light, normally considered to be composed of waves, was realized to have many properties of a stream of particles (termed *photons*). The same was true of electrons—normally considered elementary particles, they have many of the properties of an electromagnetic wave. Thus, the fundamental nature of these elementary particles was understood to be different from both a pure wave and a pure stream of particles but, rather, to have properties of both.

One of the practical applications of this understanding was that if a way could be found to bend a beam of electrons, similar to the way glass lenses bend a beam

When light rays pass through a wet mount containing a suspension of bacterial cells, we can think about three different sets of rays (Figure 3B.1). One set goes through the water alone—it will be brought to a focus in the image plane and will form the bright background. We will call these rays number 1. Second, some light goes through the cells—some of this light is absorbed, but most goes right through the very small cells and is brought to a focus to form the image of the cells. This image is slightly darker than the background because of the light that was absorbed. We will call these rays number 2. Finally, some light is deflected by the cells. Much of the deflected light is lost; however, some is deflected at a small enough angle that it still enters the objective lens, and it too is brought to a focus and contributes to the formation of the image of the cells. We will call these rays number three. For physical reasons that we do not discuss here, this deflected light is retarded by one fourth of a wavelength, but that does not prevent it from contributing to image formation (Figure 3B.2).

In the brightfield microscope, rays 2 and 3 are added together to form the image of the cells. In the phase-contrast microscope, however, rays 3 are *subtracted* from rays 2, thus making the image darker. This by itself would make very little difference, as so little light is deflected that subtracting it instead of adding it makes a negligible difference in the image. Thus, a second manipulation of the light reduces the intensity of rays 1 and 2 dramatically; now, when rays 3 are subtracted from what is left of rays 2, the difference is dramatic.

These manipulations are the result of the undeviated light all passing through the objective as a ring. The deviated light, rays 3, however, passes through the entire surface of the objective. Thus, the two types of light can be treated separately. The way this is done is to construct the objective with a **phase plate** built

FIGURE 3B.1 Light rays in the phase-contrast microscope.

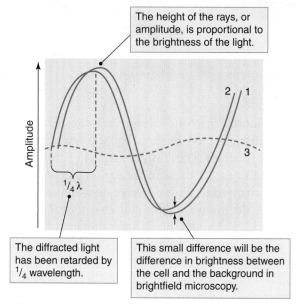

The height of the rays, or amplitude, is proportional to the brightness of the light.

The diffracted light has been retarded by $^1/_4$ wavelength.

This small difference will be the difference in brightness between the cell and the background in brightfield microscopy.

FIGURE 3B.2 Amplitude of light rays entering the objective lens.

of photons, then a microscope could be constructed using an electron beam instead of a light beam, and it would have vastly improved resolving power because the wavelength would be so much smaller. Work to construct such an instrument began in the 1930s, but it was not until the 1950s that they were sufficiently well developed to become useful in biology. Since then, however, their use has revolutionized our understanding of the architecture of the cell. This ability to resolve the fine details of internal cell structure has been coupled to the techniques of biochemistry and molecular biology to study the chemical composition of subcellular structures so that we now have an immensely detailed understanding of the structure and function of cells. This is one of the great achievements of modern biology.

In an electron microscope, the illuminating electron beam is generated by an **electron gun.** This produces electrons by passing a current through a thin filament, which ejects electrons into the vacuum within the gun. The electron gun generates a powerful electrical field that accelerates the electrons toward the positive electrode. Some pass through a small hole in the positive electrode to form a beam directed into the condenser lens.

into it. The phase plate is a relatively thick piece of glass with a thinner ring etched into it. The ring, or **conjugate area** of the phase plate, is positioned so that all of the undeviated light, rays 1 and 2, pass through it. It is also tinted very dark, and thus, it reduces the intensity of rays 1 and 2 significantly.

The rest of the phase plate, or the **complementary area,** is where the deviated light, rays 3, passes through. It is not tinted, and thus, these rays are not reduced in intensity. It is thicker than the conjugate area, however, and thus, the light that passes through it is slowed down briefly. The thickness is set so that light is retarded by one fourth of the wavelength. Because the deviated light is already retarded by one fourth of the wavelength, the total retardation is one half of wavelength. The deviated light rays are thus exactly out of phase with the undeviated light (Figure 3B.3).

When these three sets of light rays reach the focal plane, rays 1 form the background. The image of the cells is the result of rays 2 and 3 combined. Because rays 2 and 3 are 180° out of phase, however, they interfere destructively with each other. The result is that rays 3 are subtracted from rays 2, making the image of the cells darker than it would otherwise have been (Figure 3B.4).

The phase-contrast microscope was a brilliant invention that transformed the practice of microscopy. For the first time scientists could clearly observe living microbial cells, with good contrast without using stains. The invention was the work of the Dutch physicist Frits Zernicke in the 1930s, for which he received the Nobel Prize in physics in 1953.

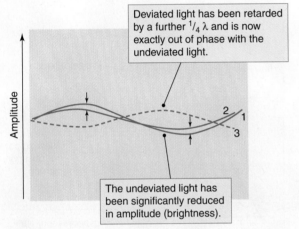

FIGURE 3B.3 Amplitude of light rays passing through the phase plate.

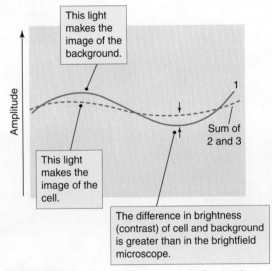

FIGURE 3B.4 Amplitude of light rays on the image plane.

To bend a beam of electrons, necessary to form an image, electromagnetic lenses are used. These consist of a powerful magnetic field shaped like a lens. This field is created by passing an electric current through a fine wire wound hundreds of times around a central hole. The current creates a magnetic field in the hole, whose strength is proportional to the current that creates it. Thus, unlike the glass lenses of a light microscope, the power of an electromagnetic lens can be varied by just turning a knob (Figure 3.12).

Because the human eye is not sensitive to electrons, the electron microscope has a **projector lens** instead of an ocular lens. The projector lens produces a second real image on a screen, which has a phosphorescent coating that glows when it is hit by the beam. We thus see the image on the screen.

3.9 The resolution of the electron microscope is a thousand-fold better than the light microscope

The same basic equation that predicts the limit of resolution for the light microscope works for the electron microscope, and thus, we can predict the resolution of

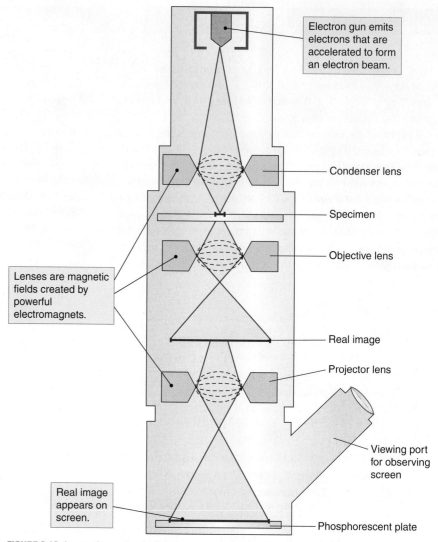

Electron gun emits electrons that are accelerated to form an electron beam.

Condenser lens

Specimen

Objective lens

Lenses are magnetic fields created by powerful electromagnets.

Real image

Projector lens

Viewing port for observing screen

Real image appears on screen.

Phosphorescent plate

FIGURE 3.12 Image formation in the electron microscope.

the latter by knowing the wavelength of the electron beam and the geometry of the objective lens.

Physical theory indicates that an electron's wavelength is a function of its speed—the faster it goes, the lower the wavelength (and hence the better the resolution). Speed, in turn, is a function of the accelerating voltage (V). Skipping a lot of algebra and using known values of physical constants, we can calculate that the wavelength of a beam of electrons is

$$\lambda \approx 1.2/\sqrt{V} \text{ nm}$$

If we substitute this into the equation for resolution, we get

$$d \approx 0.7/\sin \alpha \sqrt{V} \text{ nm}$$

If we substitute values for a typical electron microscope ($\sin \alpha \approx 10^{-2}$, $V \approx 10^5$ volts), we get a theoretical resolution of close to 0.2 nm (2 Å), a thousand times smaller than the light microscope, easily able to resolve even the finest details of intracellular structure.

CHAPTER 3 METHODS OF MICROBIOLOGY

3.10 Specimens in the electron microscope need to be dehydrated

Although the beam of electrons in an electron microscope is very intense and has high velocity, the electrons are so light that they are easily deflected by collision with air molecules. Thus, the interior of the electron microscope is kept under vacuum. This is a problem for biological specimens, however, which can be severely distorted if simply placed in a vacuum.

The distortion is due to the evaporation of water and dissolved gases from cells; because this is a major fraction of the volume of the cell, its loss can result in collapse, shrinkage, and distortion of structures. Consequently, one of the first challenges to face electron microscopists was to develop a way of **fixing** the structures of the cell and then slowly dehydrating it before putting it into the microscope.

Typically, fixation involves the use of chemical agents that **cross-link** the macromolecules of the cell, covalently linking them to their neighbors. Thus, even when the water is removed, they retain their relative positions. Nevertheless, the preparation of specimens for the electron microscope remains a real art.

3.11 Thin sectioning slices through cells shows interior structures

Probably the most common form of sample preparation for electron microscopy is **thin sectioning**, in which the specimen is embedded in a small block of plastic after fixation and dehydration and then cut into very thin slices (typically about 100 nm, or approximately one-tenth of the thickness of a typical procaryotic cell). This allows a view of the interior of the cell, and a large amount of our knowledge of the fine structure of cells comes from thin sectioning (Figure 3.13).

Thin sections have to be stained before viewing in the electron microscope because organic material is nearly transparent to an electron beam. Some fixatives also act as stains; osmium tetroxide (OsO_4), for example, acts both as a fixative by virtue of its four reactive oxygens and as a stain because its heavy metal atom easily deflects the electron beam. It is wise to remember that what we see in a thin section is not the cellular material itself, but the stain bound to it. If the binding is not equal to all cell components, we can be misled about the nature of cell structure. Much of the history of electron microscopy has been occupied with trying to determine what is real and what is an artifact due to distortions caused by fixation, drying, or staining.

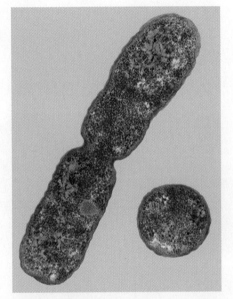

FIGURE 3.13 Thin section of a bacterial cell.

3.12 Negative staining reveals the texture of the outside of cells and isolated subcellular organelles

One of the earliest techniques of electron microscopy, still widely used for very small objects, like virus particles and protein molecules, is **negative staining.** The fixed but still wet specimen is placed in a drop of stain (usually a heavy metal salt), which then precipitates on and around the specimen. More stain accumulates at the edges of objects and in surface depressions. This technique thus is particularly useful for showing surface structures (Figures 3.14 and 3.15).

3.13 Shadow casting shows the surface of very small isolated organelles

Shadow casting also uses the deposition of a heavy metal around a specimen to show surface structure; however, in this case, the fixed, dried specimen is placed in a vacuum chamber, and the metal is evaporated from an electrode. This results in deposition of

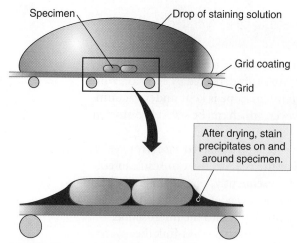

FIGURE 3.14 Negative staining technique.

Specimen

Drop of staining solution

Grid coating

Grid

After drying, stain precipitates on and around specimen.

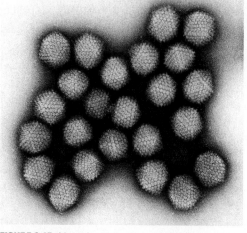

FIGURE 3.15 Negative stain of virus particles.

metal on the side toward the electrode, with little stain deposited on the side away from it. When such a preparation is examined in the electron microscope, it looks as if it is being illuminated from a low angle—hence the name (Figures 3.16 and 3.17).

3.14 Freeze etching does not require fixation or dehydration

The one method of sample preparation for electron microscopy that does not require the fixation and dehydration steps is **freeze etching.** It has thus been of great importance in evaluating whether things seen by thin sectioning are real or are artifacts caused by the stresses of fixing and drying.

In freeze etching, the specimen is frozen very rapidly, usually by immersing it in liquid nitrogen at −190°C (rapid freezing prevents the formation of large ice crystals during the freezing process, which can cause a lot of damage to cell structures). The

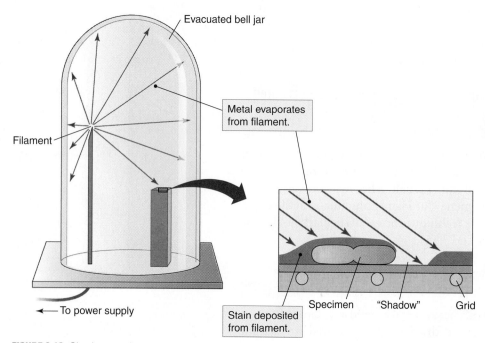

Evacuated bell jar

Metal evaporates from filament.

Filament

To power supply

Stain deposited from filament.

Specimen   "Shadow"   Grid

FIGURE 3.16 Shadow casting technique.

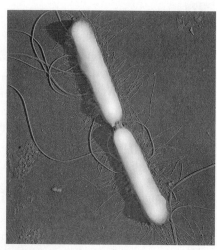

FIGURE 3.17 Shadow-cast electron micrograph of a bacterial cell.

CHAPTER 3 METHODS OF MICROBIOLOGY

frozen specimen is then fractured with a sharp, cold knife, which acts as a wedge. The plane along which the specimen splits ahead of the knife is the plane of least resistance. With biological specimens, this is often the interface between a membrane and the ice or down the middle of the membrane (because the hydrophobic forces holding membranes together are weaker than the hydrogen bonds holding the water molecules together in ice).

The specimen is then usually put under vacuum to allow some of the ice to sublimate. This lowers the surface level of the ice, revealing some of the structure that was hidden before etching and giving a more three dimensional view. The fractured, etched specimen is then shadowed just as in shadow casting, and a layer of carbon is deposited on top of the shadow. This **replica** is then lifted off the specimen, taking the stain with it, and observed in the electron microscope (Figures 3.18 and 3.19).

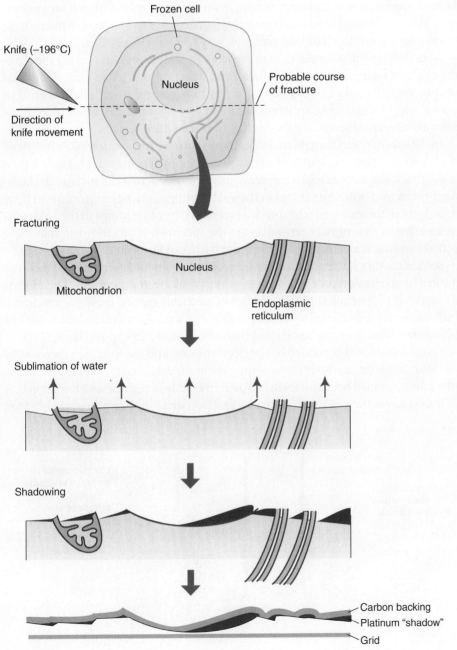

FIGURE 3.18 Freeze etching technique.

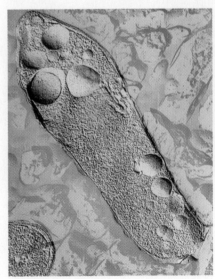

FIGURE 3.19 Freeze etch electron micrograph.

Freeze etching does not require fixation or dehydration

Freeze etching is of particular value because live cells are quickly frozen and thus protected from the distorting effects of fixing and dehydrating. Thus, freeze etching is often used to help decide whether structures seen in other techniques, like thin sectioning, are real or artifacts. If real, they should also be seen in freeze-etched preparation; if an artifact, they should not.

Freeze etching is also an excellent way to visualize membranes. It is especially good for detecting membrane proteins because when a membrane fractures down the middle, the proteins either stay with the specimen and are seen as bumps or are torn away, leaving a visible hole. In either case, their number and distribution is obvious.

3.15 Scanning electron microscopy shows surfaces with great depth of field

The scanning electron microscope is actually an electron probe, rather than a microscope, but it can be used to form an image. Images in the scanning electron microscope typically have great depth of field and look very realistic (Figure 3.20).

As in the transmission electron microscope, a sample preparation for the scanning electron microscope requires fixing and dehydration. There is an additional step, however: the sample is coated with a very thin (several atoms thick) coating of gold by using a shadowing apparatus adapted such that the specimen is continuously turned, evenly spreading the coating over the entire surface.

In the scanning electron microscope, the electron beam is squeezed by magnetic lenses down to a very tiny spot only a few square nanometers in area. This spot is then scanned back and forth across the specimen; as it moves across the surface, the high-intensity electron beam hits atoms in the gold coating, ejecting electrons from their orbits. Most of these ejected electrons are recaptured by other atoms in the gold coating, and they are eventually carried away (the specimen is grounded so that these electrons, as well as electrons from the beam itself, can flow out of the specimen).

Some electrons, however, are completely ejected from the specimen, and these are captured by an electron collector, forming a very weak electrical current. This electrical current is proportional to the number of electrons ejected from the specimen, which is in turn proportional to the shape of the surface. When the electron beam hits a high point, more electrons are ejected than when it hits a low spot (Figure 3.21).

The current that is generated by the electron beam and the specimen then is used to form an image on a cathode ray tube (just like in a television set). The electron beam in the tube is scanned back and forth in a pattern exactly in register with the beam that is scanning across the specimen (Figure 3.22). Thus, when the beam passes over a high

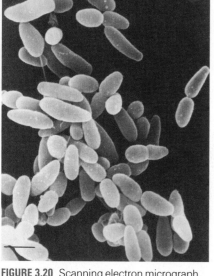

FIGURE 3.20 Scanning electron micrograph.

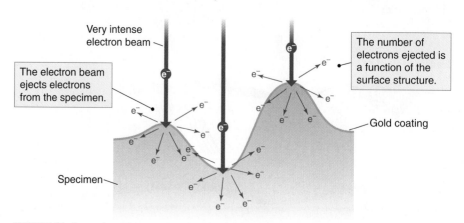

FIGURE 3.21 Secondary electrons in the electron microscope.

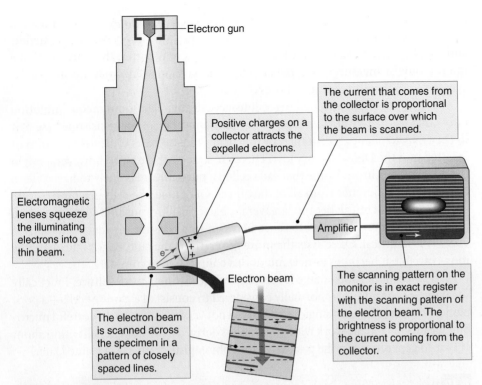

Electron gun

Electromagnetic lenses squeeze the illuminating electrons into a thin beam.

Positive charges on a collector attracts the expelled electrons.

The current that comes from the collector is proportional to the surface over which the beam is scanned.

Amplifier

Electron beam

The electron beam is scanned across the specimen in a pattern of closely spaced lines.

The scanning pattern on the monitor is in exact register with the scanning pattern of the electron beam. The brightness is proportional to the current coming from the collector.

FIGURE 3.22 Image formation in the electron microscope.

spot in the specimen, more electrons than average are ejected, and the beam in the tube makes a bright spot on the screen; when a depression in the specimen is encountered, few electrons are ejected, and the screen shows a dark spot. This shading produces an image on the screen with a pronounced three-dimensional feeling.

Magnification in the scanning electron microscope is a simple matter of the ratio of the size of the pattern that the electron beam scans and the size of the pattern on the screen. For instance, if the screen is 10 cm on a side and the microscope beam is scanning a square that is 100 μm on a side, the magnification is 100,000 μm/100 μm, or 1,000×. Magnification is controlled simply with a dial, which controls magnets that scan the beam back and forth.

Pure Culture Technique

Although individual microbes may be observed in the microscope and their morphology and behavior observed, the study of their biochemistry and genetics requires that populations (called **cultures**) be studied rather than individuals. Obviously, for experimental results to mean anything, all of the cells in the population must be essentially the same; such populations are called **pure cultures.** A set of techniques, mostly developed in the late 19th century by Koch, Pasteur, and their collaborators, allows the routine isolation, cultivation, and study of pure cultures. Probably more than anything else, it is these techniques that define microbiology.

3.16 Pure cultures do not consist of identical cells

Unfortunately, the concept of pure cultures is more complex that it appears. Pure cultures do not consist of identical cells for several reasons. One is that an actively multiplying pure culture consists of a mixture of cells in different stages of the cell cycle: small "baby" cells that were just "born," large "mother" cells about to divide, and everything in between. In most cases this does not matter; the activities that a microbiologist measures (i.e., the activity of a particular enzyme) are simply the average of all the different size classes of cells.

Pure cultures are also heterogeneous genetically. The spontaneous mutation rate—that is, the rate at which spontaneous chemical reactions cause one base pair in the DNA to change to another, or one or more base pairs to be deleted from or inserted into the DNA—is very low. There are so many base pairs, however, even in the procaryotic chromosome that each cell in a pure culture is likely to have at least one base pair difference from all of the others. Of course, the cells in a pure culture are still nearly identical; because an average bacterial chromosome contains several million base pairs, a single difference between two cells means that their DNA is 99.9999% identical. Of course, the more generations the culture goes through, the more variable it becomes as new mutations continue to accumulate.

These complications make it very difficult to define a pure culture. Practically speaking, a pure culture is normally considered to consist of a **clone** of cells—a population all derived from a single ancestral cell not very far in the past. A careful microbiologist will normally work with populations derived from a single cell within about 100 or so generations in the past. How microbiologists do this is explained later.

3.17 Pure culture technique consists of three interrelated techniques

Pure culture technique requires reliable methods for (1) sterilizing growth media and glassware, (2) introducing desired cells into sterile growth media or removing samples from pure cultures without accidentally introducing other contaminating microbes, and (3) isolating single cells, or their progeny, to obtain pure cultures.

Sterility means the absence of living organisms, and it is in principle an all or nothing phenomenon—something is either sterile or it is not; nothing is "90% sterile." However, sterility can be expressed as a probability—something may have a 90% probability of being sterile after a certain treatment, such as boiling for 15 minutes. This would mean that if you boiled 100 samples for 15 minutes, you would expect about 90 of them to be sterile, and the rest would not be sterile; each individual sample would either be sterile or not.

3.18 The most common method of sterilizing is autoclaving

As Tyndall discovered, even prolonged heating at 100°C is insufficient to kill some bacterial spores. Thus, sterilizing by heat is routinely done at higher temperatures; 121°C is standard. Because this is over the boiling point of water at atmospheric pressure, it is done in a fancy pressure cooker called an **autoclave.** The autoclave maintains a pressure of 15 lb/in^2 (1.06 kg/cm^2), at which pressure the boiling point of water is 121°C. To prevent boiling of liquids, the atmosphere in an autoclave is saturated with water vapor; autoclaving is sometimes termed "steam sterilizing" for this reason.

3.19 Heat-sensitive solutions are sterilized by filtration

Many chemical compounds that microbiologists sometimes need to place in media are heat sensitive and decompose at the temperature of the autoclave. Thus, media containing these compounds have to be sterilized by alternative means. The most common is to pass them through a filter in which the pore size is so small that most cells cannot go through. Although smaller pore size filters sterilize more effectively, they are slower, and they clog more rapidly than larger pore size filters. Hence, we usually use the largest pore size filter that experience indicates is usually effective. In most cases, a pore size of 0.4 μm is sufficient (Figure 3.23).

Filtration is less reliable than autoclaving as a means of sterilization, as some bacterial cells are very small and can pass through even very small-pore filters, and of course, most viruses can pass these filters. Fortunately, in most applications, viruses and very small bacteria are rarely encountered. One case in which this can be a serious problem is in the filter sterilization of media for the culturing of animal cells (called **tissue culture**). Tissue culture media often contain animal serum (blood from which the cells have been removed); this cannot be autoclaved because the proteins in serum coagulate when heated. Thus, tissue culture media are routinely filtered to sterilize them. Animals from which the serum is derived (usually cattle), however, may be infected with bacteria of the mycoplasma group, whose cells are very small (as small as 0.2 μm in diameter), and are very plastic because they lack a rigid cell wall. They can therefore pass easily through filters of 0.4 or 0.2 μm pore size, or even smaller. Mycoplasma contamination is known to have ruined many tissue culture experiments; antibiotics are now routinely included in the medium to kill any mycoplasmas that get through the filter.

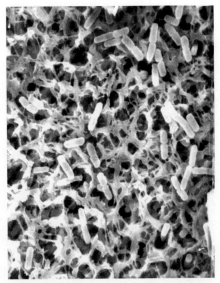

FIGURE 3.23 Bacterial cells on the surface of a filter.

3.20 Glassware is sterilized by dry heat

Reusable glassware, such as flasks and pipets, is usually sterilized in an oven, usually set to about 180°C. This temperature easily kills even the hardiest bacterial spores; however, laboratories increasingly are using plastic, disposable labware to replace reusable glass. This material is produced commercially at low cost, and it saves the effort of washing, drying, and sterilizing glassware. Sterile plastic ware is sterilized at the factory, either by exposure to gamma irradiation or to the toxic gas ethylene oxide.

3.21 Bunsen burner flames help to prevent contamination during transfer into or out of containers

Sterile media in flasks and other containers are kept sterile by stoppering the container so as to prevent any access by microbes to the interior. This can be by making the cap overlap the container by a significant amount so that any microbes would have to travel a convoluted route to enter. The same principle keeps solid media in a Petri plate sterile. Alternatively, the container may be plugged with a fibrous material such as cotton or foam rubber, which will trap particles in the mesh of fibers. In both cases, air can enter the container, but suspended particles, such as microbial cells, get trapped in the meshwork of cotton or rubber fibers.

Of course, the principal opportunity for unwanted microbes in air to gain access to a container is when it is opened to introduce a few cells of a pure culture, to withdraw a sample, or for any other reason. Thus, microbiologists very early developed techniques to allow them to maintain the container free of unwanted airborne organisms while they manipulated it. The principal tool here is a simple Bunsen burner flame. Whenever a flask is opened, its mouth is passed quickly through the flame. Even though the exposure to the flame is momentary (too brief to heat the glass), it is sufficient for the heat and the ionized gases of the flame to kill microbes on the surface, thus preventing them from falling into the container and contaminating it. Similarly, any instrument, such as a pipet, that is going to be used to add or remove material from the flask is also first flamed before being used. The mouths of containers are flamed again before the stopper is reinserted.

This simple technique is quite effective, and routine flaming of glassware as it is used helps keep contamination to a minimum.

3.22 Pure cultures are isolated by streaking on solid media

Solid media are prepared by adding agar, a complex polysaccharide derived from marine algae, to liquid media. The agar dissolves at the high temperature of the autoclave and remains liquid as it cools down to a temperature of about 45°C. Below that it solidifies into a firm, transparent gel, which will not melt again until the temperature is increased to nearly 100°C. It is resistant to hydrolysis by most bacteria.

Solid media are normally poured into *Petri plates,* whose overlapping halves work just like the caps on flasks to prevent airborne contaminants from reaching the agar after it has been poured into the plate and is cooling. After the agar has solidified, however, Petri plates are normally incubated upside down to prevent any condensation that might form on the lid from dripping onto the agar. In this position, although contaminants can reach the interior of the plate, they settle on the lid, rather than on the agar.

To obtain pure cultures, microbes are normally **streaked** onto solid media. A small amount of **inoculum** (material containing microbes to be placed onto solid media or into liquid media) is picked up on a sterile **inoculating loop,** made of wire. The loop is then drawn repeatedly across the surface of the plate, depositing microbes as it moves. The plate is then incubated at the desired temperature, and some time later (typically 24 hours, but sometime more for slowly growing organisms), **colonies** are visible wherever a microbial cell capable of growth on the particular medium was deposited on the agar surface. A colony is a mass of cells derived by the multiplication of a single cell; normally colonies contain about 10^7 to 10^8 cells (Figure 3.24).

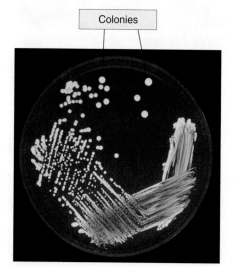

Colonies

FIGURE 3.24 A streaked plate.

3.23 Different media are necessary for different microbes

All cells have to find in their medium all nutrients required for growth, or they cannot multiply. Different cells, however, have widely varying requirements, and thus, there is no single medium that satisfies all; every medium is to some extent selective.

Of course, some media will support the growth of more types of microbes than others. What we call **rich media,** or **complex media,** typically have a mixture of many different organic compounds, including all of the amino acids, purines, pyrimidines, vitamins (enzyme cofactors), etc. They are usually made by hydrolyzing natural products, such as meat, with enzymes to break down macromolecules and release the monomers (peptone and tryptone are examples), or they are made by extracting the small molecules from other cells (yeast extract is an example). Rich media will usually support growth of many different types, including ones that require **growth factors** (specific compounds required for growth, such as vitamins, amino acids, or other compounds). There are microbes that do not grow in rich media, however; they have more simple nutritional needs and find one or more components of rich media to be toxic.

Other media, called **minimal media** (sometimes **mineral media**), contain mineral salts of major bioelements, such as sulfur, nitrogen, and phosphorous. This is supplemented, when necessary, with a single organic compound that serves as both carbon and energy source. If we are culturing a photosynthetic microbe, we would not need such a supplement. Minimal media are much more restrictive, as no organism with a growth factor requirement can grow unless that factor is added separately, and only cells that have the enzymes to metabolize the particular carbon and energy source can grow.

3.24 Cultures are incubated under different conditions according to their relations with oxygen

Microbes fall into three major categories based on their relations to oxygen. Some are **obligate aerobes** and require oxygen to grow. These organisms use aerobic respiration as their only mode of metabolism and, hence, cannot grow when their terminal electron acceptor (oxygen) is absent. Others are **obligate anaerobes,** for which oxygen is toxic. These organisms are restricted to habitats where oxygen is absent and to modes of metabolism that do not require it (anaerobic respiration, fermentation, certain types of photosynthesis). Finally, some organisms are **facultative anaerobes** (the term facultative aerobes means the same thing, but is rarely used). These organisms can grow in the presence or absence of oxygen. Most facultative anaerobes switch between aerobic respiration and fermentation, according to the availability of oxygen.

Obviously, different methods of cultivation are needed for anaerobes and aerobes. Aerobes are often simply cultivated under a normal atmosphere of air, with no special measures taken to insure adequate supplies of oxygen (which is about 20% of air). In liquid cultures, however, microbes can multiply to huge numbers (in excess of 10^9 per ml), and they can use dissolved oxygen faster than it can diffuse into the medium unless special measures are taken. Most frequently, liquid cultures are shaken vigorously to insure that the medium is continuously exposed to the atmosphere.

On a Petri plate, aerobes can also use oxygen so rapidly that the interior of a colony becomes anaerobic. Thus, for obligate aerobes, colonies consist of an outer layer of growing cells and an inner core of nongrowing cells that are oxygen starved (Figure 3.25).

It is much more challenging to culture microbes under anaerobic conditions. This requires air to be excluded, which is a difficult task, especially for very sensitive anaerobes, which can be killed by even momentary exposure to oxygen. The most straightforward way of cultivating these organisms is to use an anaerobic glove box—a sealed box in which the atmosphere is composed so that there is no oxygen (i.e., 95% N_2, 5% CO_2). In such an anaerobic chamber, materials are introduced and removed through an airlock so that the interior is never exposed to air, and materials are manipulated by gloves that are hermetically sealed to the chamber (Figure 3.26).

For less fastidious anaerobes, which can tolerate exposure to air but which cannot grow in its presence, and for anaerobic culturing of facultative anaerobes, simpler techniques suffice. Petri plates can be streaked, liquid cultures inoculated, and other manipulations performed under air, and then the cultures placed under anaerobic conditions. For Petri plates, this usually means that the plates, after being streaked, are placed in an **anaerobic jar** in which a commercially available "gas pack" has been placed just before sealing the jar (Figure 3.27). When water is added to the gas pack, it evolves H_2 gas, which then combines with the O_2 in the air on a platinum catalyst built into the lid of the jar. This chemically removes most of the oxygen from the air and allows many anaerobes to grow well. Liquid cultures of these aerotolerant anaerobes can be grown in sealed tubes or flasks that are filled to the brim to exclude air.

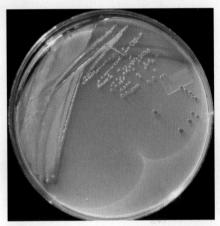

FIGURE 3.25 Colonies of *Rhodobacter;* the red centers indicate anaerobic conditions.

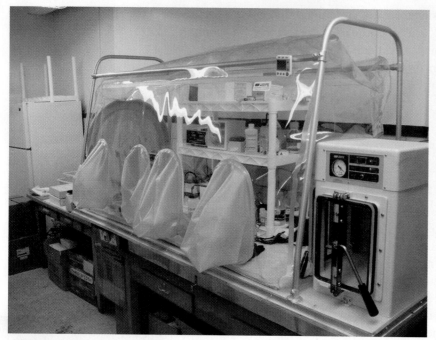

FIGURE 3.26 Anaerobic chamber.

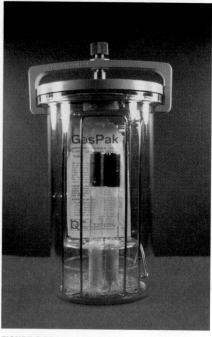

FIGURE 3.27 Anaerobic jar.

3.25 Some pathogenic microbes require special containment facilities

Pure culture techniques were designed to prevent airborne microbes from contaminating microbial cultures; that is, they were designed to keep microbes out—not in. When the microbes being cultured are **pathogens** (microbes that cause disease), however, additional techniques may be necessary to prevent them from infecting the investigator or others in the vicinity. Of course, such accidental infections are an unavoidable hazard of working with pathogens. Accidents happen and cannot be completely prevented; dropped flasks, a slip with a hypodermic needle, or any of dozens of other accidents can infect laboratory workers. Such accidents can be minimized (e.g., unbreakable plastic flasks), but not completely prevented. Accidental infection is thus a fact of life for medical microbiologists, and dozens have died of laboratory infections during the 120-year history of microbiology.

In addition to accident, laboratory infections can occur for some microbes by the aerosol route. Certain laboratory manipulations create **aerosols**—suspensions of fine droplets or particles in air—and these can cause infection when breathed into the lungs. Not all pathogens can cause infection by the aerosol route, but for those that can, normal laboratory manipulations can cause dangerous and invisible aerosols. For these organisms, special containment techniques or equipment is used to reduce the chance of aerosol infection.

In normal practice, any aerosols produced by laboratory manipulation are very low concentration; that is, the number of particles or droplets released is low. Thus, the hazard is only to laboratory workers in the immediate vicinity; normally, there is no danger to anyone outside the laboratory. Thus, special containment techniques are normally needed only to protect laboratory workers rather than the general public, which is not at risk. If high concentration aerosols are deliberately created (e.g., to study the process of airborne infection), however, then those outside may possibly be at risk.

Generally, four categories of hazard are recognized, usually referred to as **biosafety levels (BSL)** (Table 3.1). Biosafety level 1 (BSL 1) is for nonpathogenic microbes.

BSL	Agents	Practices	Safety Equipment (Primary Barriers)	Facilities (Secondary Barriers)
Table 3.1	**Summary of Recommended Biosafety Levels for Infectious Agents**			
1	Not known to consistently cause disease in healthy adults	Standard microbiological practices	None required	Open bench top sink required
2	Associated with human disease, hazard = percutaneous injury, ingestion, mucous membrane exposure	BSL-1 practice plus: Limited access Biohazard warning signs "Sharps" precautions Biosafety manual defining any needed waste decontamination or medical surveillance policies	Primary barriers = class I or class II BSCs or other physical containment devices used for all manipulations of agents that cause splashes or aerosols of infectious materials; PPEs: laboratory coats; gloves; face protection as needed	BSL-1 plus: Autoclave available
3	Indigenous or exotic agents with potential for aerosol transmission; disease may have serious or lethal consequences	BSL-2 practice plus: Controlled access Decontamination of all waste Decontamination of lab clothing before laundering Baseline serum	Primary barriers = class I or class II BCSs or other physical containment devices used for all open manipulations of agents; PPEs: protective lab clothing; gloves; respiratory protection as needed	BSL-2 plus: Physical separation from access corridors Self-closing, double-door access Exhausted air not recirculated Negative airflow into laboratory
4	Dangerous/exotic agents which pose high risk of life-threatening disease, aerosol-transmitted lab infections; or related agents with unknown risk of transmission	BSL-3 practices plus: Clothing change before entering Shower on exit All material decontaminated on exit from facility	Primary barriers = All procedures conducted in class III BSCs or class I or class II BSCs in combination with full-body, air-supplied, positive-pressure personnel suit	BSL-3 plus: Separate building or isolated zone Dedicated supply and exhaust, vacuum and decon systems

"Sharps precautions" refers to procedures designed to reduce accidents with hypodermic needles or broken glass. BSC stands for biological safety cabinet. PPE stands for personal protective equipment (such as lab coats and gloves). "Baseline serum" refers to the taking of a blood sample from each worker before they enter the facility for the first time to be used as a reference in case of suspected laboratory infection. "Negative airflow into laboratory" means that the laboratory is maintained at a lower pressure than the outside air so that any air leakage will be into the facility, not out of it.

Source: Centers for Disease Control and Prevention. *Biosafety in Microbiological and Biomedical Laboratories,* 4th ed. Washington, DC: U.S. Government Printing Office, 1999.

BSL 2 is for those pathogens that do not readily transmit by the aerosol route. BSL 3 is for pathogens that transfer readily by the aerosol route and that pose a serious health threat. The highest category, BSL 4, is for pathogens that transmit readily by the aerosol route and that cause fatal diseases for which there is no cure.

One of the primary pieces of protective equipment in most BSL 2, BSL 3, and BSL 4 laboratories is the **biological safety cabinet (BSC),** a chamber that is constructed so as to minimize the chance of any aerosol particles escaping into the laboratory air.

There are several types. Class III BSCs are completely enclosed, and workers access the interior through gloves sealed to the front panel. Air is exhausted through very fine filters termed **HEPA filters** (for high-efficiency particulate air filters). Class I and II BSCs have an opening in the front through which workers insert their hands. Fans create air currents that enter the BSC through this opening, preventing particles inside the cabinet from leaving this way. Exhaust air is filtered through HEPA filters (Figure 3.28).

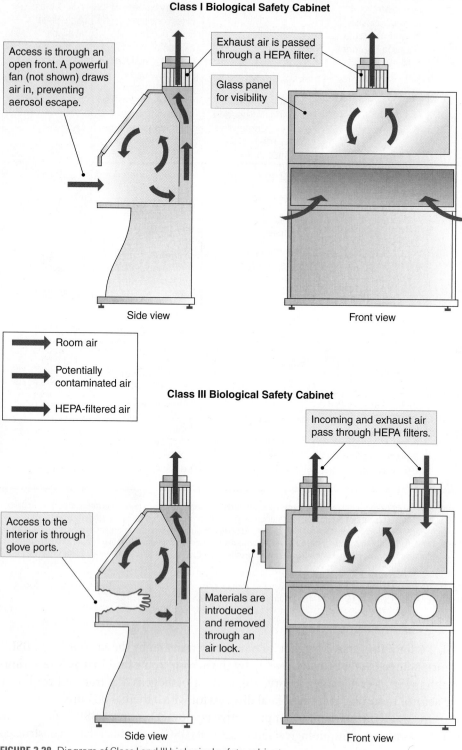

Class I Biological Safety Cabinet

Access is through an open front. A powerful fan (not shown) draws air in, preventing aerosol escape.

Exhaust air is passed through a HEPA filter.

Glass panel for visibility

Side view

Front view

→ Room air

→ Potentially contaminated air

→ HEPA-filtered air

Class III Biological Safety Cabinet

Access to the interior is through glove ports.

Incoming and exhaust air pass through HEPA filters.

Materials are introduced and removed through an air lock.

Side view

Front view

FIGURE 3.28 Diagram of Class I and III biological safety cabinets.

CHAPTER 3 **METHODS OF MICROBIOLOGY**

Even with these features, laboratory infections still occur, although at much reduced rates compared with the period before the introduction of formal biosafety systems. For instance, in 2003, the virus that causes the serious disease SARS (see section 20.8) infected laboratory workers in high-containment laboratories in three different countries (fortunately, none of the three infected workers started epidemics of SARS, which they easily could have done).

Summary

Microbiology as a field is characterized by the special techniques that are required to study organisms too small to be seen—microscopy to visualize them and pure culture techniques to allow large populations of essentially identical organisms to be prepared free of other microbes. The basic elements of both of these technologies were established in the late 19th century. In the 20th century, the principal innovations were the development of electron microscopy and the application of biochemistry and genetic techniques (developed for the study of plants and animals) to the study of microbes. Now, at the start of the 21st century, genomics and associated techniques are again enriching microbiology. Microbiology continues to evolve in intellectually exciting and practically important ways; nevertheless, it remains ultimately dependent on our ability to see microbes and to obtain pure cultures of specific ones.

Study questions

1. Compare and contrast image formation in the light and electron microscopes.
2. Describe the contemporary uses of brightfield, phase-contrast, and darkfield light microscopy.
3. Calculate the limit of resolution in micrometers of a light microscope using green light (wavelength about 400 nm) and an objective lens with a numerical aperture of 1.2.
4. What would be the maximum useful magnification of such a microscope, assuming that your eye has a limit of resolution of 0.2 mm? (Maximum useful magnification is the magnification at which the maximum amount of information is obtained; further magnification enlarges the image but yields no more information.)
5. If you wished to design a phase-contrast microscope that made the specimen brighter than the background, rather than darker, what modifications would you make?
6. Summarize the principal techniques of specimen preparation for the electron microscope and the principal uses of each.
7. Discuss the concept of a pure culture.
8. Why do microbiologists commonly use cultures rather than individual cells?
9. What are the principal elements of pure-culture technique?
10. How are media and materials sterilized?
11. Discuss the chemical composition and uses of different kinds of media.
12. Summarize the principal features of the biosafety level system—the kinds of organisms worked with at each level and the kinds of containment measures applied.

4

Procaryotic Cell Structure and Function

P ROCARYOTIC ORGANISMS ARE in very many ways the dominant life on earth—in terms of total protoplasmic volume, they make up about 90%. Every blade of grass, bush, shrub, tree, bird, insect, cat, dog, fish, human, alga, protozoan, and mushroom altogether constitute only 10% of the biomass on earth. We are like the tip of an iceberg, maintained by the much greater mass of unseen material below the surface. In terms of numbers of individual organisms, procaryotic life forms outnumber eucaryotic ones by many orders of magnitude; there are probably at least a thousand procaryotic organisms for every eucaryotic one (including eucaryotic microbes), and there are probably many billions of procaryotic organisms for every one macroscopic eucaryotic life form.

Much of this evolutionary success is due to the very simple cell structure of procaryotes that we explore in this chapter.

4.1 The procaryotic cell is small and structurally simple

Of course, there is a large size range of procaryotic cells; nevertheless, the great majority of procaryotic organisms consist of a single small cell—typically with a volume of about 1 μm^3 (10^{-15} L)—and a cell envelope, a cell membrane, a nucleoid, the cytoplasm, and little more (Figure 4.1). Most of the rest of the procaryotes are similar, with only a few additional simple structures besides those already mentioned. Such a simple cell can replicate itself in a shorter period of time than the more complex eucaryotic cell and thus competes exceptionally well for dissolved nutrients.

Structural simplicity does not mean molecular simplicity, however; the absence of many separate compartments and different membrane systems as in the eucaryotic cell means that the few compartments and membranes that are present must perform multiple functions. Thus, the molecular complexity of the procaryotic cell is substantial.

Although most procaryotic cells are small, with diameters between 0.7 and 2.0 μm^3, there is a large range of procaryotic sizes. At the small end, there are some members of the mycoplasma group that have spherical cells as small as 0.25 μm in diameter, with a volume of 0.01 μm^3. At the large end, some filamentous forms discovered near deep-sea hot springs have disk-shaped cells that are about 120 μm in diameter by about 25 μm thick, thus averaging about 270,000 μm^3. However, thin sections of these organisms reveal that they are largely empty space: most of their cells consist of a large vacuole, or space, filled with an aqueous solution. Thus, the actual cytoplasmic vol-

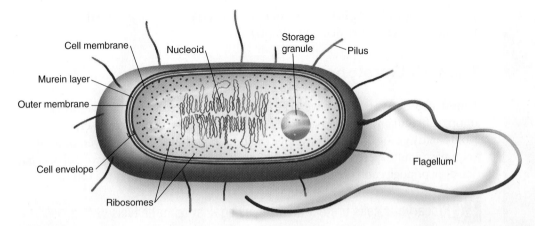

FIGURE 4.1 Diagram of a typical procaryotic cell.

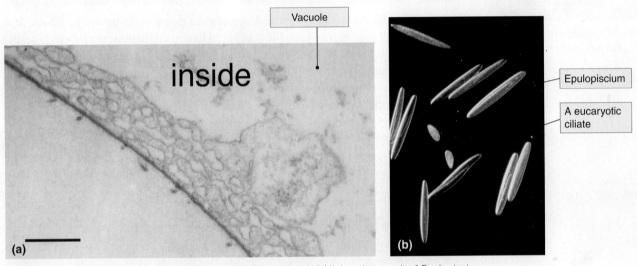

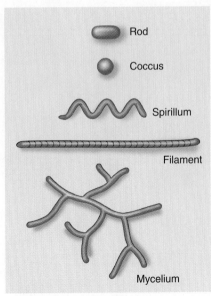

FIGURE 4.2 (a) Thin section of a marine *Thiomargarita* (bar = 1 μm); (b) light micrograph of *Epulopiscium*.

ume is only 20% of the cell volume, or about 50,000 μm³ (Figure 4.2a). Another recently discovered organism, *Epulopiscium,* is a true giant (Figure 4.2b). It can be so large as to be visible without a microscope (up to 60 μm in diameter and over 500 μm long, over 1 million cubic micrometers!). It was assumed since its discovery to be eucaryotic; however, recent thin sections revealed it to have the characteristic structure of procaryotic cells. It does not have the large vacuole seen in the very large deep-sea filaments. Presumably, more such oddities await discovery.

4.2 Most procaryotes are unicellular

Most procaryotic organisms are unicellular—that is, the entire organism consists of a single cell. Cell shape is normally cylindrical (**rods**), spherical (**cocci**—singular **coccus**), or helical (**spirilla**—singular **spirillum**) (Figure 4.3). A few are multicellular, most commonly **filamentous** (a chain of identical cells attached end to end). A few are **mycelial**—branching tubes of cytoplasm contained within walls. Even when they are multicellular, however, there is rarely any differentiation of individual cells; all cells in the organism are usually identical.

Occasionally, a filament or a mycelium will produce differentiated cells with particular function—commonly spores, but occasionally with a different metabolism,

FIGURE 4.3 Microbial shapes.

like the nitrogen-fixing heterocysts of cyanobacteria (Figure 4.4). We discuss spores and heterocysts in more detail in Chapter 11.

The Cell Envelope

Almost all procaryotic cells have a cell envelope, but its structure and chemistry differ among different groups of procaryotes. These differences characterize a series of broad evolutionary groups. The cell envelope can be a single homogeneous layer, or it can be a complex, multilayered structure. If the envelope contains a rigid layer of polysaccharide or peptidoglycan (which we describe shortly), we call that layer the **cell wall.**

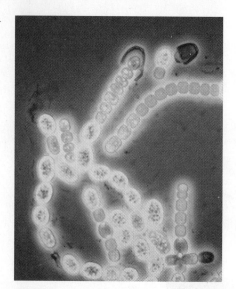

FIGURE 4.4 Light micrograph of cyanobacterium showing differentiated cell types.

4.3 Procaryotic cells protect themselves against osmotic lysis with a cell envelope

Procaryotic cells have a concentrated cytoplasm, dense with proteins, RNAs, and small molecules (see Figure 1.6). Most of them live in environments in which solute concentration is less than their cytoplasm; only a few procaryotes that live in high-salt lakes have an environment that is close to isotonic. Consequently, the osmotic pressure (often termed **turgor pressure**) of most procaryotic cells is very high, often several atmospheres; those with very thick walls (the gram-positive bacteria) can have turgor pressures of well over 10 atmospheres. Even the lower pressures in most procaryotes are well above the ability of the cell membrane alone to contain. Thus, all procaryotic cells need a mechanism to prevent osmotic lysis. In almost all cases, the strategy is like that of plant cells: to contain the pressure with a layer external to the cell membrane that resists expansion of the cell volume.

4.4 There are four major types of cell envelope among the bacteria; the most common is the gram-negative envelope

Most bacteria have a cell envelope that consists of a thin rigid layer (the cell wall), overlain with a membrane, termed the **outer membrane** (Figure 4.5). The rigid layer consists of a single giant molecule termed **murein,** which is a form of **peptidoglycan,** because it is a heteropolymer containing both peptide chains and glycan (polysac-

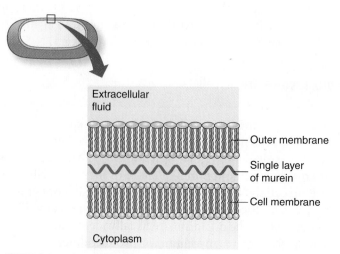

FIGURE 4.5 The gram-negative envelope.

CHAPTER 4 PROCARYOTIC CELL STRUCTURE AND FUNCTION

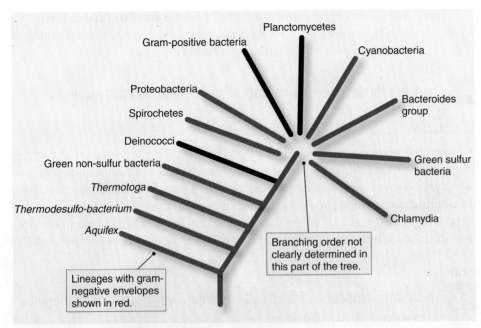

FIGURE 4.6 Phylogenetic distribution of the gram-negative envelope.

In the phylogenetic tree:
Planctomycetes, Gram-positive bacteria, Cyanobacteria, Proteobacteria, Spirochetes, Bacteroides group, Deinococci, Green non-sulfur bacteria, Green sulfur bacteria, Thermotoga, Thermodesulfo-bacterium, Chlamydia, Aquifex

Branching order not clearly determined in this part of the tree.

Lineages with gram-negative envelopes shown in red.

charide) chains. This type of envelope is termed **gram negative** because cells with this type of envelope do not retain the Gram stain (see Section 3.3) when washed with alcohol (Figure 4.6).

A second type of envelope is termed **gram positive** because cells with this kind of envelope retain the Gram stain when washed with alcohol. The gram-positive envelope is a thick cell wall made largely of murein, with no outer membrane (Figure 4.7).

One group of bacteria, the deinococci, has an envelope that appears to be a combination of gram-negative and -positive structures: a thick murein layer, overlain with an outer membrane (Figure 4.8). The deinococcal outer membrane, however, is chemically different from that of gram-negative bacteria, and it is not clear whether they are homologous structures. These bacteria stain gram positive because of their thick murein wall that retains the Gram stain.

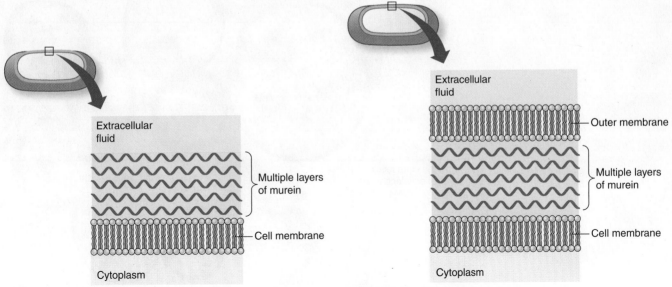

Figure 4.7 (left):
Extracellular fluid
Multiple layers of murein
Cell membrane
Cytoplasm

Figure 4.8 (right):
Extracellular fluid
Outer membrane
Multiple layers of murein
Cell membrane
Cytoplasm

FIGURE 4.7 The gram-positive envelope.

FIGURE 4.8 The deinococcal envelope.

There are four major types of cell envelope among the bacteria

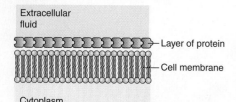

Extracellular fluid

Layer of protein

Cell membrane

Cytoplasm

FIGURE 4.9 The planctomyces envelope.

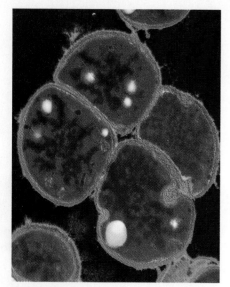

FIGURE 4.10 An archaeal cell wall composed of polysaccharide.

Finally, the planctomyces group of bacteria has an envelope that consists of a layer of protein (Figure 4.9). These bacteria stain gram negative because the thin layer of protein is unable to prevent the Gram stain from rinsing out.

4.5 There are three types of archaeal cell envelopes; the most common is a layer of protein

Archaeal cell envelopes are similarly diverse. The most common is a layer of protein, like the planctomycetes among the bacteria. Some archaea (all of them members of the methanogen group) have a rigid wall made of a murein-like compound termed **pseudomurein;** however, no archaea are known to have murein itself. Finally, some archaea have walls that look very much like bacterial gram-positive walls in the electron microscope but are chemically different. They consist of complex polysaccharides (Figure 4.10).

4.6 Procaryotes without a defined cell envelope layer have their membranes strengthened by glycolipids

One subgroup of the gram-positive bacteria (the **mycoplasma** group) has lost the ability to make a murein wall, and there are several different archaea that similarly cannot make a defined cell envelope. Many of the mycoplasmas live as parasites within the animal body, where the osmolarity is higher than outside, and their osmotic stress is less. Others, however, are free living, as are the archaea that lack cell envelopes.

These envelope-less procaryotes all seem to have large quantities of **glycolipid** in their membranes (glycolipids are molecules that have a polysaccharide portion covalently linked to a lipid). The lipid portion is embedded in the cell membrane and the polysaccharide portion lies on the outer surface of the membrane. Presumably hydrogen bonding among the polysaccharide chains helps to strengthen the membrane (Figure 4.11).

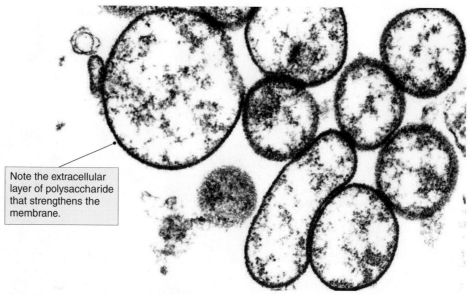

Note the extracellular layer of polysaccharide that strengthens the membrane.

FIGURE 4.11 Thin section of a mycoplasma.

CHAPTER 4 PROCARYOTIC CELL STRUCTURE AND FUNCTION

4.7 Murein is a form of peptidoglycan in which the individual glycan strands are cross-linked to each other by the peptides

Murein is the component of both gram-negative and gram-positive bacterial envelopes that is responsible for the wall's principal function: containing turgor pressure. It consists of a strand of chemically modified polysaccharide: alternating N-acetyl-glucosamine (usually abbreviated NAG) and N-acetylmuramic acid (NAM). Each of the muramic acid residues has a short peptide chain linked by its amino terminus to the carboxyl group of the muramic acid. Thus, the repeating unit of murein is a disaccharide (NAM and NAG), with a peptide attached to the NAM (Figure 4.12). The combined nature of murein—a polysaccharide (glycan) backbone and peptide side chains—makes it a member of a class of heteropolymers termed **peptidoglycans,** which also includes the archaeal pseudomurein.

Notably, the amino acids alternate L and D configuration. This is one of very few instances in the biological world in which D amino acids are found. It is thought that the alternation of L and D amino acids makes the murein more resistant to degradation by enzymes in the environment that hydrolyze peptide bonds, as most protease enzymes have active sites that are specific for the peptide bond between two L amino acids.

Strands of murein average about 30 monomers in length (total of about 60 sugar residues). Because a single monomer is about a nanometer across the two sugars, one would expect such a strand to be about 30 nm long. However, like many polysaccharides, murein coils into a helix, held together by hydrogen bonding among the sugars (Figure 4.13). Thus, the actual length of a murein strand is considerably less than its extended length of 30 nm. It takes several hundred individual murein strands to encircle an average bacterial cell of around 1 μm in diameter.

The spacing of the sugars in the murein helix is such that each peptide chain is oriented 90° to the previous and following chains. Thus, there are four monomers per turn of the helix, and the peptide chains of these four monomers point up, right, down, and left (looking down the axis of the chain). It is these peptide chains that allow separate murein strands to be cross-linked into a single huge molecule.

The cross-linking reaction is termed **transpeptidation** (Figure 4.14). As synthesized, the monomer has a peptide chain of five amino acids. This pentapeptide can

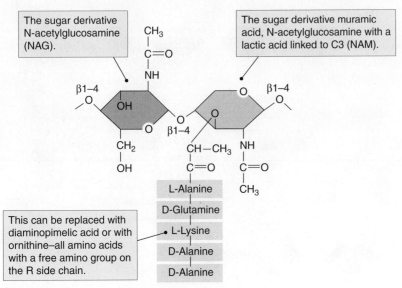

FIGURE 4.12 The repeating unit of murein.

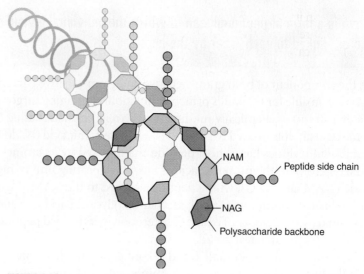

FIGURE 4.13 Helical murein chain.

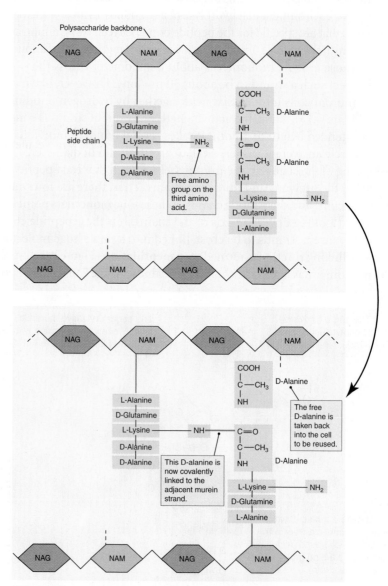

FIGURE 4.14 The transpeptidation reaction.

CHAPTER 4 PROCARYOTIC CELL STRUCTURE AND FUNCTION

vary considerably in its amino acid sequence in different bacteria, but in all cases, the last two residues are both D-alanine, and the third amino acid has a free amino group. An enzyme catalyzes the transfer of the bond between the two D-alanines to the diaminoacid on an adjacent murein strand, splitting off the terminal D-alanine in the process.

4.8 Murein consists of a single giant molecule, a single layer thick in the gram-negative bacteria, many layers thick in the gram-positive bacteria

Because the peptides of murein project from the polysaccharide backbone in alternating directions, the individual murein strands can be linked into a sheet, with each strand linked only to those on either side of it, or they can be linked to strands above and below also. In the former case, characteristic of the gram-negative bacteria, the murein forms a single layer surrounding the cell. In the latter case, found in the gram-positive bacteria, the murein forms a three-dimensional mesh, with multiple layers of murein linked to each other.

In both cases, however, every individual strand is covalently linked to those that surround it so that the entire murein layer forms a single giant molecule. The molecular weight obviously varies according to the thickness of the layer and the size of the cell, but it is always huge. In an average-sized gram-negative cell, with its single layer, the molecular weight is more than 10^{10}; in a large gram-positive bacterium, it is likely to be 10^{11} to 10^{12} or more. In filamentous and mycelial bacteria, in which the murein layer is continuous over the entire organism, the molecular weight can be many times greater even than this. Murein, even the smaller molecules of it, is the largest known molecule in the world.

4.9 Archaeal pseudomurein is a peptidoglycan similar to bacterial murein

Like murein, pseudomurein is a polymer of disaccharides with short peptides attached to one of the sugars. It probably assumes a similar helical structure and functions in much the same way to confer rigidity to cells; however, there are numerous differences of detail (Figure 4.15).

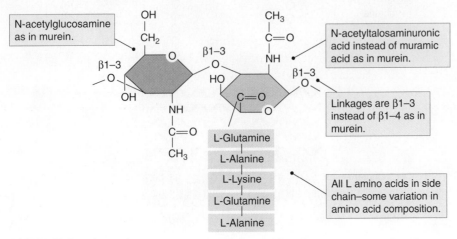

FIGURE 4.15 Pseudomurein monomer structure.

Both structures contain NAG as one of the sugars of the backbone; however, pseudomurein contains N-acetylalosaminuronic acid instead of muramic acid. The amino acid composition of the peptide side chain is also different, and none of the amino acids has the D configuration. Sugars in the backbone are linked β1–3 instead of β1–4.

Pseudomurein is cross-linked in a similar fashion to murein. The carboxyl group on the glutamic acid side chain of one peptide chain reacts with the amino group on the side chain of the lysine of another peptide side chain.

4.10 Gram-positive bacterial walls contain teichoic acids

In addition to murein, the gram-positive wall contains other compounds, most notably **teichoic acids.** These are polymers of sugar alcohol phosphates, usually glycerol phosphate or ribitol phosphate. Some of these teichoic acids are covalently bound to a lipid at one of their ends, forming **lipoteichoic acids.** Their lipid portion is embedded in the cell membrane, with the polar teichoic acid portion enmeshed with the murein. Other teichoic acids, termed **wall teichoic acids,** are covalently attached to murein strands (Figures 4.16 and 4.17).

The function of teichoic acids is not known, but they are presumed to be important because they are ubiquitous among gram-positive bacteria.

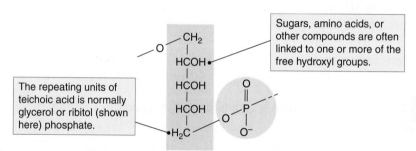

Sugars, amino acids, or other compounds are often linked to one or more of the free hydroxyl groups.

The repeating units of teichoic acid is normally glycerol or ribitol (shown here) phosphate.

FIGURE 4.16 Structure of ribitol teichoic acid.

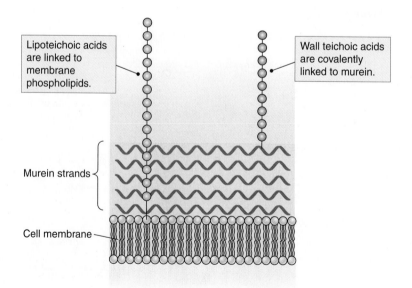

Lipoteichoic acids are linked to membrane phospholipids.

Wall teichoic acids are covalently linked to murein.

Murein strands

Cell membrane

FIGURE 4.17 Schematic of the gram-positive wall.

4.11 The gram-negative outer membrane protects the murein

There are many enzymes secreted by bacteria or by fungi that attack murein, either by hydrolyzing the glycosidic bonds of the glycan backbone or the peptide bonds of the cross-linking side chains. A particularly common enzyme, found also in animal body fluids, is **lysozyme,** which breaks the glycosidic bond that links the monomers of the murein.

The gram-positive bacteria have such a thick layer of murein that some loss on the outer surface is tolerable. The gram-negative bacteria, however, with their single layer of murein, are very vulnerable to lysis by enzymes that hydrolyze murein. One of the functions of the outer membrane is to protect the murein, as this membrane provides a physical barrier to large molecules such as proteins and makes it impossible for them to gain access to the murein layer.

4.12 The outer membrane is highly asymmetric

The outer membrane is a lipid bilayer, like most biological membranes. Unlike most other membranes, however, it is highly asymmetric, with the inner and outer leaves made of different lipids. The inner leaf is composed of phospholipids, like the cell membrane. The outer leaf, however, is composed almost entirely of a unique molecule termed **lipopolysaccharide.** This molecule has a lipid core, to which is covalently attached a long polysaccharide chain. The lipid portion forms the outer leaf of the membrane, and the polysaccharide forms a dense mesh on the outer surface of the cell (Figure 4.18).

The lipid portion, termed **lipid A,** is composed of a phosphorylated glucosamine dimer to which are attached six to eight fatty acids. The polysaccharide consists of a

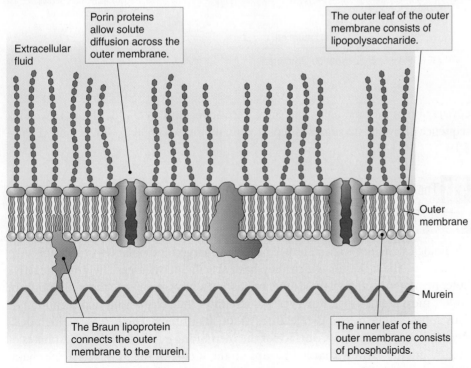

Porin proteins allow solute diffusion across the outer membrane.

The outer leaf of the outer membrane consists of lipopolysaccharide.

Extracellular fluid

Outer membrane

Murein

The Braun lipoprotein connects the outer membrane to the murein.

The inner leaf of the outer membrane consists of phospholipids.

FIGURE 4.18 Structure of the gram-negative outer membrane.

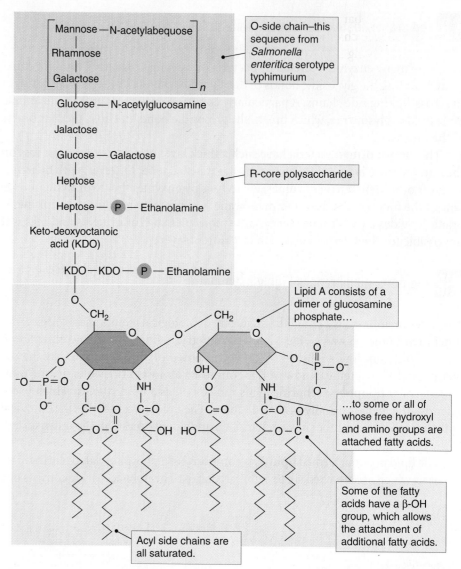

O-side chain–this sequence from *Salmonella enteritica* serotype typhimurium

R-core polysaccharide

Lipid A consists of a dimer of glucosamine phosphate…

…to some or all of whose free hydroxyl and amino groups are attached fatty acids.

Some of the fatty acids have a β-OH group, which allows the attachment of additional fatty acids.

Acyl side chains are all saturated.

FIGURE 4.19 Lipopolysaccharide structure.

core region of various sugars and sugar derivatives, and an **O-side chain,** in which a sequence of five to six sugars is repeated many times to make a long chain (Figure 4.19).

4.13 The outer membrane is impermeable to both hydrophobic and hydrophilic compounds

All biological membranes are impermeable to charged and polar (hydrophilic) compounds, except those for which they have specific permeases. The gram-negative outer membrane is, like all these other membranes, impermeable to these compounds. Unlike most membranes, however, it is also impermeable to nonpolar compounds. This appears to be due to two features of the membrane. The first is the high density of charges at the outer surface of the membrane because of the phosphates of lipid A and the various charged groups on the R core of the polysaccharide chain. This high density of charges forms many hydrogen bonds to water and to other side chains, forming a barrier to the diffusion of hydrophobic compounds. Second, the hydrocarbon side chains of the lipid A are saturated, which allows them to pack very tightly, leaving little room for hydrophobic compounds to slip in between lipid molecules.

4.14 Dissolved compounds enter the gram-negative periplasm via porins

Because the outer membrane is impermeable to both polar and nonpolar compounds, gram-negative cells need a mechanism to allow nutrients and waste products to cross the outer membrane into or out of the space between the two membranes, termed the **periplasm.** Most compounds diffuse passively through pores in the membrane made by special proteins termed **porins.** These membrane proteins have a pore through their center that allows the diffusion of compounds smaller than the size of the pore. Most gram-negative cells have porins that form holes approximately 1 nm in diameter, which allows the free diffusion of compounds of molecular weights of about 1,000 or less. Larger molecules (especially proteins, such as lysozyme) cannot fit through the porins and are excluded from the periplasm.

The necessity for all nutrients and waste products to diffuse through the porins has an interesting consequence: the effective surface area of a gram-negative cell is only a few percent of its geometric surface area. For instance, if we take *E. coli* as a typical cell, it is a rod approximately 2 μm long and 0.8 μm in diameter. Its total surface area is approximately 6 μm^2 (6×10^6 nm^2). The surface area of each 1 nm diameter pore is about 0.8 nm^2. There are about 10^5 porins per cell, and thus, each cell has an effective surface area for diffusion of about 8×10^4 nm^2, or about 1% to 2% of the total surface area. This would seem a high cost to pay for an organism that survives by taking up dissolved nutrients; however, the ubiquity of the gram-negative type of wall suggests that this cost is minor compared with the advantages of this type of envelope.

4.15 The outer membrane is physically linked to the murein

The outer membrane is linked to the murein through a lipoprotein, sometimes called the **Braun lipoprotein** after its discoverer. This protein has several fatty acids added to its N-terminal end and these are embedded into the inner leaf of the outer membrane. The carboxyl end of the protein is covalently joined to free amino groups in the murein by enzymes in the periplasm. An average *E. coli* cell has about 10^5 molecules of this lipoprotein, making it (along with porins) one of the most abundant proteins of the cell.

4.16 The periplasm is a dense solution of protein

Because the gram-negative cell has two separate membranes, it has two discrete compartments: the cytoplasm, enclosed by the cell membrane, and the **periplasm** between the cell membrane and the outer membrane. The periplasm is a concentrated solution of proteins, including proteins that bind nutrients and transfer them to permeases in the cell membrane. This keeps the concentration of free nutrients in the periplasm very low, which favors nutrient diffusion through the porins into the periplasm. Other periplasmic proteins include enzymes involved in the final stages of murein synthesis and cross-linking and chaperone proteins that conduct LPS and outer membrane proteins across the periplasm.

The Cell Membrane

Like all cells, procaryotes have a **cell membrane** (sometimes called the cytoplasmic membrane, or plasmalemma, or plasma membrane). This membrane separates the cytoplasm from the exterior environment and allows the cytoplasm to have a chemical composition different from the environment. Without this compartmentation, cellular life would be impossible.

4.17 The procaryotic cell membrane contains hopanoids in place of sterols

Most eucaryotes have sterols in their membranes, which help to fill the spaces caused by unsaturated fatty acid chains and also help to maintain the correct fluidity of the membrane over a wide temperature range. Very few procaryotes can make sterols, although a few that grow in the animal body will incorporate host sterols into their membranes. Instead of sterols, most procaryotes synthesize compounds termed **hopanoids.** Hopanoids are compounds that can have a shape similar to that of sterols and that are made from a sterol precursor, squalene (Figure 4.20). Thus, the pathway of squalene synthesis appears to be nearly universal among cells; procaryotes and eucaryotes differ in the subsequent fate of the squalene.

4.18 The cell membrane is a two-dimensional fluid

Like all biological membranes, the forces that hold individual phospholipids to other phospholipids, or to membrane proteins, are weak. They consist largely of van der Waals interactions at very close range. The aggregate of all of these attractions is very strong and contributes to the stability of the membrane. For any individual molecule, however, they are weak and readily broken. Thus, the individual molecules in a membrane can move readily in the plane of the membrane, although they almost never leave the plane of the membrane (to do so they would have to interact with the highly polar water, which they cannot by virtue of their hydrophobicity).

A consequence of this is that phospholipids and proteins diffuse freely within the membrane, and thus, they achieve a random distribution. When the cell needs specific molecules to be located at specific places, for instance at the pole of a rod-shaped cell or at the cell's midpoint, then these molecules have to be anchored in place by being attached to molecules in the cell envelope or in the cytoplasm.

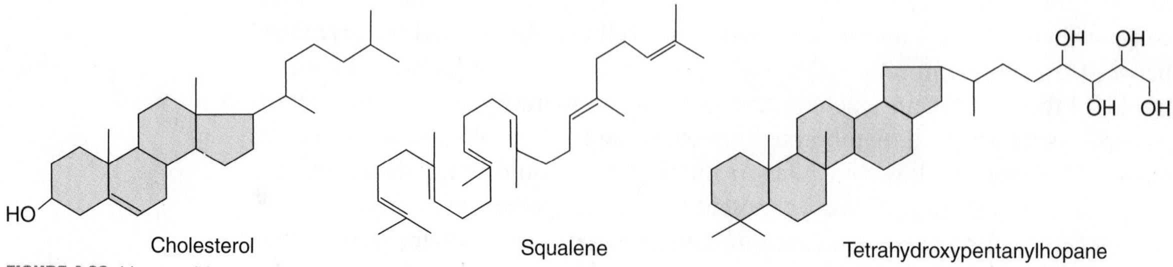

FIGURE 4.20 Hopanoid structure.

Cholesterol   Squalene   Tetrahydroxypentanylhopane

CHAPTER 4 PROCARYOTIC CELL STRUCTURE AND FUNCTION

4.19 The procaryotic cell membrane has more protein and more different proteins than most membranes

Procaryotic cell membranes are typically about 25% lipid (by weight) and 75% protein. Because lipids have an average molecular weight of around 700 daltons, however, whereas proteins have average molecular weights of nearly 100 times this, lipids outnumber proteins in the membrane by about 30 to 1. This large amount of protein is unusual and reflects the fact that the procaryotic cell membrane has a number of functions absent from the eucaryotic cell membrane. This is because the eucaryotic cell has a number of other membrane systems for specialized functions: the Golgi, the endoplasmic reticulum, the nuclear envelope, mitochondria and chloroplasts, endosomes, etc. All of these are absent from procaryotic cells, and their functions are performed by the cell membrane. Thus, the proteins that are responsible for all these different functions are all in the cell membrane, rather than dispersed among many different membranes. *E. coli,* a typical procaryote, has over 700 different membrane proteins.

This membrane in all cells has the basic function of controlling the internal chemical composition. Remember that the lipid portion of membranes is impermeable to polar compounds; however, most of the nutrients that procaryotes need to take in and some of the waste products that they need to excrete are polar compounds. For this reason, about a third of the proteins in the membrane are **permeases,** proteins that allow polar molecules to cross the membrane. There are many types of permeases, but all of them have the function of facilitating molecular movements across the membrane.

Some procaryotes get their energy by fermentation; however, most use some form of respiration or photophosphorylation and thus have membrane-embedded electron transport proteins. These are almost all in the cell membrane (the principal exception is the cyanobacteria, which have separate thylakoid membranes) (Figure 4.21).

The cell membrane is the site of synthesis of cell membrane proteins, outer membrane proteins, periplasmic proteins, and secreted proteins. Ribosomes bound to the inner surface translate mRNA for these categories of protein. Cell membrane proteins then enter the cell membrane directly as they are translated. Periplasmic or outer membrane proteins pass through the cell membrane to the other side, where outer membrane proteins are bound by chaperone proteins that conduct them across the periplasm to the outer membrane. Secreted proteins use a special secretion apparatus to pass directly from their site of synthesis to the outside.

Another principal function of the procaryotic cell membrane is lipid and cell wall synthesis. All of the enzymes that make the various lipids of the cell membrane and the outer membrane of gram-negative cells are made by enzymes embedded in the cell membrane. Similarly, many of the enzymes that assemble the cell wall are in the cell membrane.

Finally, as in all cells, the cell membrane includes sensory proteins that allow the cell to determine certain aspects of the chemical or physical conditions of the environment so that they can respond in an appropriate way.

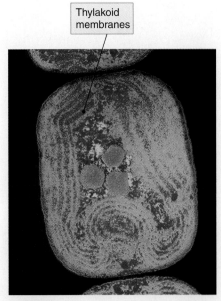

Thylakoid membranes

FIGURE 4.21 Thin section of a cyanobacterium.

4.20 The cell membrane may invaginate into the cell to provide extra surface area

Some bacteria with high respiratory or photosynthetic rates have such a high concentration of electron transport proteins and other proteins involved with energy generation that the surface area of the cell is insufficient. These bacteria synthesize

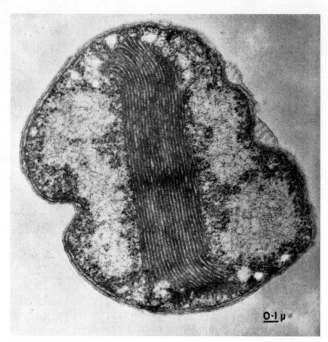

FIGURE 4.22 A procaryotic cell showing invaginations of the cell membrane.

extra membrane, which invaginates into the cytoplasm as a folded stack of membranes, or as membranous tubules. In such cases, the cell membrane is physically continuous, even though it may be functionally specialized, with the electron transport functions concentrated in the intracytoplasmic portion and transport and synthesis functions concentrated in the surface portion (Figure 4.22).

4.21 Some procaryotes have intracellular membrane-bound organelles

Procaryotic cells typically lack all intracellular membrane-bound organelles; however, this is not a universal feature of the procaryotic cell. There are some procaryotes that have membrane-bound organelles in their cytoplasm. These organelles include the following: the thylakoids of cyanobacteria, the acidocalcisomes (see Section 4.32) that store phosphate, and anammoxisomes (see Section 14.10), where ammonia and nitrate are oxidized to nitrogen gas. Thylakoids are restricted to the cyanobacteria, and anammoxisomes are found only in some planktomycetes. Acidocalcisomes, however, are found in a number of bacteria from several different lineages, as well as in eucaryotes.

4.22 The archaeal cell membrane is a bilayer or monolayer of ether lipids

Bacterial and eucaryotic cell membranes are always bilayers of phospholipids (glycerol–fatty acid diesters, with a phosphate esterified to the third glycerol carbon). Indeed, this structure was until recently assumed to be universal, shared by all life on earth; however, archaeal membranes are now known to be different. They are usually bilayers consisting of a different type of phospholipid: **glycerol-diphytane-diethers.** These lipids have the same amphipathic character as the more traditional lipids (i.e., they are molecules with both hydrophobic and hydrophilic portions), but they differ in two principal ways: the lipids are branched, usually saturated, hydrocarbon chains; and the hydrocarbon side chains are linked to the glycerol by ether bonds rather than ester bonds (Figure 4.23).

In some groups of archaea, particularly the hyperthermophilic forms that grow at temperatures near boiling, two diethers can be linked covalently tail to tail to form a **diglycerol–dibiphytane–tetraether.** These lipids span the membrane, with the two polar ends on opposite surfaces. A few groups of hyperthermophiles have membranes made up principally of such tetraethers; in such a case, the membrane is a monolayer, rather than the bilayer structure found in all the rest of the living world (Figure 4.24).

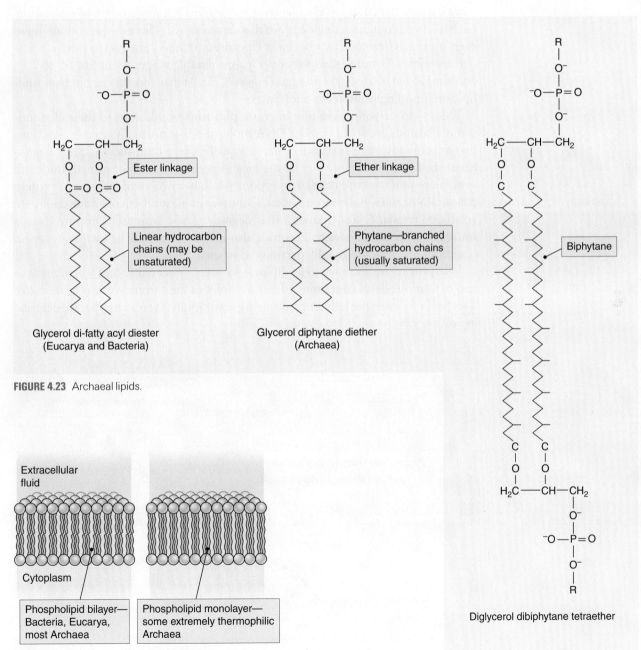

FIGURE 4.23 Archaeal lipids.

Glycerol di-fatty acyl diester
(Eucarya and Bacteria)

Ester linkage

Linear hydrocarbon chains (may be unsaturated)

Glycerol diphytane diether
(Archaea)

Ether linkage

Phytane—branched hydrocarbon chains (usually saturated)

Biphytane

Diglycerol dibiphytane tetraether

Extracellular fluid

Cytoplasm

Phospholipid bilayer—Bacteria, Eucarya, most Archaea

Phospholipid monolayer—some extremely thermophilic Archaea

FIGURE 4.24 Archaeal membrane structures.

The archaeal cell membrane is a bilayer or monolayer of ether lipids

The Nucleoid

Special structures in all cells contain the chromosomes, which encode all the cells' proteins and RNAs and which have sequences that regulate the expression of these genes. These structures are thus of central importance, as it is these proteins and RNAs that determine all cellular structures and behaviors and that make each different species unique. In contrast with eucaryotic cells' nucleus, procaryotic cells have no nuclear envelope, and thus, we use a different term for the structure: the **nucleoid.**

4.23 The nucleoid is haploid and usually contains a single circular chromosome

Most bacteria have a single, circular chromosome, with a single origin of replication. Each nucleoid contains a single copy of this chromosome. There are exceptions, however: some procaryotes have one linear chromosome, and some two or three circular chromosomes. Genome sizes vary over a considerable range: about 5×10^8 to 10^{10}, encoding about 500 to 8,000 proteins (Figure 4.25). Archaea are similar but have multiple origins of replication like eucaryotes.

Many procaryotes have one or more **plasmids** in addition to their chromosome. Plasmids are small circular DNA molecules that normally encode functions that are only advantageous under certain, usually rare, circumstances. They are distinguished from chromosomes by the lack of genes essential for growth under all conditions and for core cellular functions (such as rRNA and ribosomal protein genes, tRNA genes, polymerases, etc.). Examples of plasmid-encoded functions are antibiotic resistance and enzymes that degrade exotic aromatic compounds as a source of carbon and energy. Plasmids also usually encode their own segregation system so that they are equally partitioned when a cell divides.

No procaryote is known to be diploid; however, procaryotic cells will often have two to four nucleoids containing identical copies of the chromosome (Figure 4.26). This happens when the cells are multiplying very rapidly, for reasons that we describe in Section 9.9.

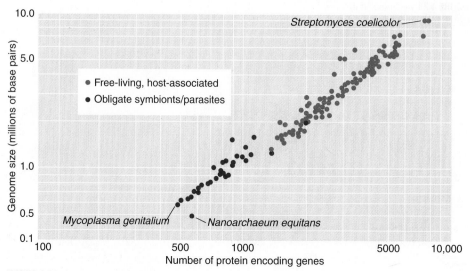

FIGURE 4.25 Genome size and gene content in bacteria.

CHAPTER 4 PROCARYOTIC CELL STRUCTURE AND FUNCTION

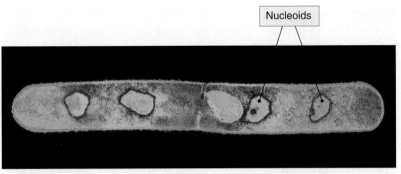

FIGURE 4.26 Multiple nucleoids in a rapidly growing cell.

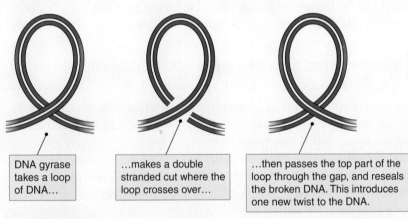

DNA gyrase takes a loop of DNA…

…makes a double stranded cut where the loop crosses over…

…then passes the top part of the loop through the gap, and reseals the broken DNA. This introduces one new twist to the DNA.

FIGURE 4.27 DNA gyrase action.

4.24 The procaryotic chromosome is supercoiled

The DNA of the chromosome is **supercoiled;** that is, the double helix itself is twisted. The twisting is such that the double helix is under stress to unwind, but the unwinding tension is opposed by the hydrogen bonds between bases. The result is that the chromosome twists like a kinked garden hose, or a twisted rubber band. The supercoiling is actively introduced into the DNA by a special enzyme called **DNA gyrase** (Figure 4.27). Another category of proteins, homologous to eucaryotic histones and called either **histones** or **histone-like proteins,** assists in maintaining the proper amount of supercoiling. Supercoiling is essential to proper transcription and DNA repair, and cells that are unable to maintain the proper degree of supercoiling will die.

4.25 The nucleoid is a highly ordered structure

The nucleoid is not simply a concentrated mass of DNA in the center of the cytoplasm. It has a substantial amount of order to it, as we are beginning to learn. We still do not have a complete picture of this structure, but the last few years have revealed the basic outlines. The average procaryotic cell contains a chromosome that would be about 1,000 times the length of the cell when fully spread out—as if an adult human swallowed a mile of vermicelli (angle-hair pasta). Clearly, the nucleoid must have significant organization to both pack that much DNA into the cell and to make sure that it is all available for transcription, replication, and repair.

The nucleoid is arranged with the origin of replication at one end and the terminus at the other. In between, the chromosome is condensed into approximately 500 loops by special condensing proteins. The boundaries of the loops are chosen at random and are highly dynamic—they are continually breaking down and reforming. This is probably how the DNA continually interchanges between the surface and the interior.

The central portions of the nucleoid are very densely packed, and it would seem that these regions would be difficult to transcribe. This is consistent with measurements of the location of transcription (see MicroTopic on page 67), which show that transcription occurs on the surface of the nucleoid (where the densely packed core transitions to a more loosely packed region where the longer loops extend into the cytoplasm). As the longer loops may extend some distance into the cytoplasm, transcription extends some distance away from the visible surface of the nucleoid.

4.26 Transcription and translation are coupled in procaryotes

Because transcription occurs in contact with the cytoplasm, in which there is a dense suspension of ribosomes, translation of mRNA can begin as soon as the ribosome-binding site is exposed. Thus, translation begins less than a second after transcription, and ribosomes move down the mRNA as it is made, keeping up with the RNA polymerase. Transcription and translation in procaryotes thus occur simultaneously and are said to be **coupled** (Figure 4.28). Coupling in protein synthesis is important because it is the basis for a mechanism of gene regulation termed attenuation, which we discuss in Section 10.18.

4.27 Compaction of DNA into the nucleoid requires neutralization of DNA charges by polyamines

At neutral pH, as in cells, the phosphates of DNA carry a negative charge. It is thus difficult to pack DNA tightly. In eucaryotic cells, most of the charges are neutralized by the basic histone proteins that are an integral part of chromatin; however, procaryotic

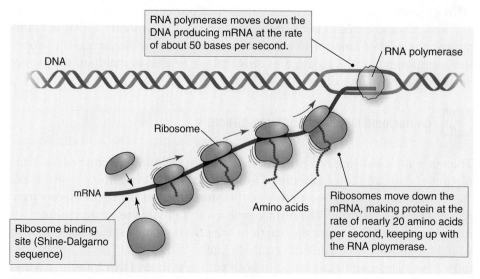

FIGURE 4.28 Coupling of transcription and translation.

CHAPTER 4 PROCARYOTIC CELL STRUCTURE AND FUNCTION

The localization of transcription in cells can be visualized by several different techniques. One of the most common uses radioactive labeling. Remember that uracil is a base that is used only in RNA synthesis, and thus, cells grown in the presence of radioactive uracil will make radioactive RNA. If the labeling period is short (a few seconds), no other cellular constituent will be radioactive. Thus, the strategy to localize transcription in procaryotic cells was to give cells a brief pulse of radioactive uracil (a few seconds only) and then remove the excess with a brief (a few seconds) wash and then kill the cells. Most radioactivity in the cells will thus be RNA still in the process of being formed (called **nascent RNA**). This is because the very brief period between initial exposure to radioactive uracil and killing the cells is not sufficient for most RNA molecules to be completed. Thus, most of the RNA will still be attached to its DNA template and to RNA polymerase at the site where its synthesis was arrested when the cells were killed. So most of the radioactivity will be where transcription occurs.

Visualizing the location of radioactivity at the submicrometer scale is a challenging task. The most common method is called **autoradiography.** In this technique, the killed cells are prepared for electron microscopy by making thin sections. The sections then have a thin layer of photographic emulsion placed on top of them. They are then placed in the dark for a few weeks, during which time radioactive decay of the uracil exposes the film. The exposure results in silver grains being deposited in the emulsion where a decay event occurred. The film is then developed (which removes silver atoms not precipitated by exposure), and the thin section is put in the electron microscope. The observer sees the thin section of cells, with a scattering of silver grains near where radioactive decay events occurred (Figure 4B.1).

At each site of transcription there may be only a few molecules of radioactive uracil, and not all of those will decay before the film is developed. Furthermore, not all decay events result in a silver grain because the particle released by the decay event leaves in a random direction, and less than half will enter the emulsion. Once in the emulsion, many will pass through without hitting a silver atom. Finally, particles traveling obliquely through the emulsion may cause a silver grain to form at some distance from the site of the decay event. Thus, we cannot simply look at the electron micrograph and see where transcription occurs; we have to treat the data statistically. We plot the location of each silver grain with respect to the nucleoid of the cell, visible under the silver grains (Figure 4B.2).

When this is done, we obtain clear results. Silver grains are rare over the core of the nucleoid, common at the interface of the nucleoid and the cytoplasm and decline in frequency with distance from the nucleoid. The decline as distance from the surface of the nucleoid increases is due in part to the fact that there is a random distribution of the length of cytoplasmic loops. So more of the DNA in loops is close to the surface than distant from it.

The conclusion thus is that transcription is largely confined to the surface of the nucleoid and to longer loops that extend into the cytoplasm (which cannot be seen in the electron microscope because they are obscured by the dense concentration of ribosomes).

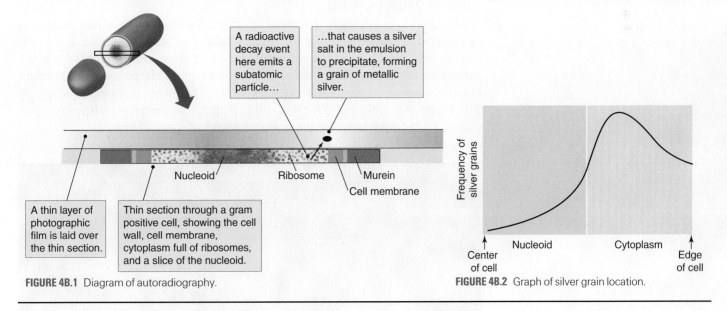

FIGURE 4B.1 Diagram of autoradiography.

FIGURE 4B.2 Graph of silver grain location.

Compaction of DNA into the nucleoid requires neutralization of DNA charges by polyamines

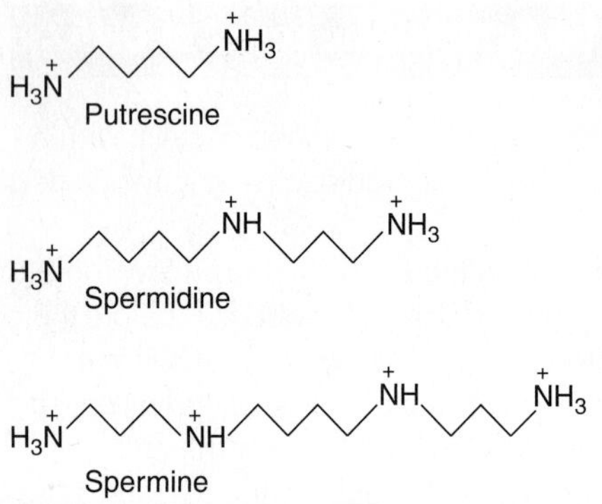

Putrescine

Spermidine

Spermine

FIGURE 4.29 Polyamine structures.

histones appear to be confined to the surface of the nucleoid and do not help to neutralize charges in the interior of the tightly packed nucleoid.

Instead, procaryotic nucleoids contain large amounts of one or more **polyamines.** These are small organic compounds with two, three, or four amino groups spaced approximately as far apart as the phosphates in DNA. Because the amino groups are protonated (and thus positively charged) at neutral pH, they can effectively neutralize the charges on the DNA and allow it to pack tightly (Figure 4.29).

The Bacterial Cytoskeleton

For a long time, procaryotes were thought to lack anything equivalent to the eucaryotic cytoskeleton (see Chapter 5). It is now clear, however, that they have proteins that are homologous to the proteins of the eucaryotic cytoskeleton and that these proteins are part of a cytoskeleton that is functionally similar to that of eucaryotes, although the details differ significantly.

The two principal proteins of the eucaryotic cytoskeleton are actin and tubulin, components of microfilaments and microtubules, respectively. Homologs of both are found in procaryotes, and both form filaments like eucaryotic microfilaments (no microtubule equivalents are found in procaryotes).

4.28 An actin-like cytoskeleton maintains cell shape and forms the procaryotic mitotic apparatus

Rod-shaped and spiral bacteria have a helical band of filaments made of an actin-like protein called **MreB** that underlies their cell membrane. The protein subunits have little identity with eucaryotic actin in terms of their amino acid sequence (about 10%), but their three-dimensional structure is nearly identical. Disruption of these filaments leads to spherical cells, and thus, it is thought that they determine cell shape (Figures 4.30 and 4.31).

In addition to helping to determine cell shape, the MreB filaments are thought to be involved in intracellular motility. In particular, they seem to constitute the procaryotic equivalent of a mitotic apparatus. Soon after chromosome replication is initiated, specific sequences near the duplicated origin regions, the procaryotic equivalent of centromeres, appear to become attached to MreB filaments, and are moved rapidly to opposite ends of the cell. After replication is complete, the termini also move apart (Figure 4.32). So far, however, no motor proteins have been identified.

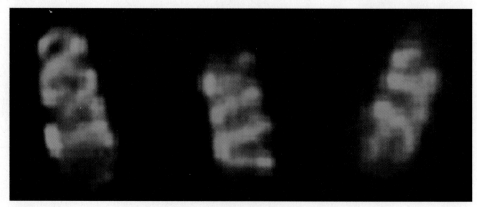

FIGURE 4.30 Fluorescent light micrograph showing MreB helices.

CHAPTER 4 PROCARYOTIC CELL STRUCTURE AND FUNCTION

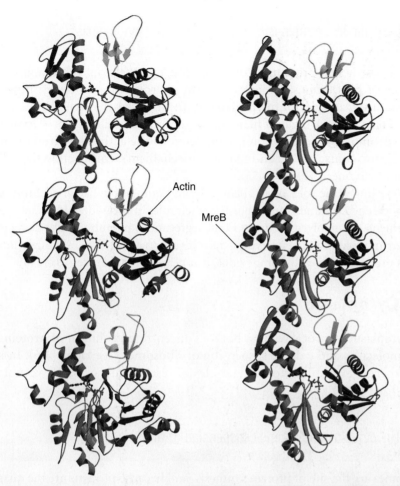

Actin

MreB

FIGURE 4.31 Filaments of actin and MreB compared. Reprinted with permission from the *Annual review of Biophysics and Biomolecular Structure*, Volume 33 © 2004 by Annual Reviews. www.annualreviews.org. Figure provided by Jan Löwe, Ph.D., The Medical Research Council, Laboratory of Molecular Biology.

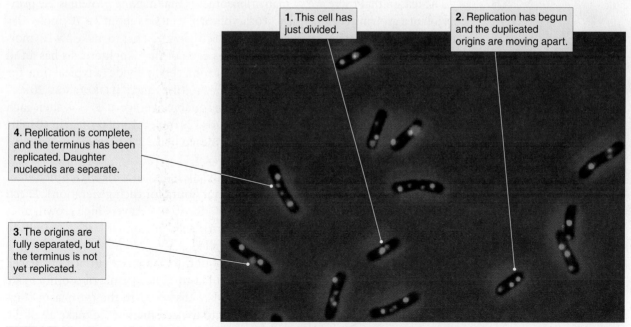

1. This cell has just divided.

2. Replication has begun and the duplicated origins are moving apart.

4. Replication is complete, and the terminus has been replicated. Daughter nucleoids are separate.

3. The origins are fully separated, but the terminus is not yet replicated.

FIGURE 4.32 Visualization of chromosomal origins (green) and termini (red) by fluorescence microscopy.

An actin-like cytoskeleton maintains cell shape and forms the procaryotic mitotic apparatus

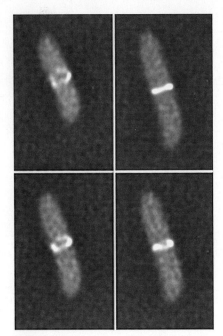

FIGURE 4.33 Fluorescent light micrographs showing FtsZ.

4.29 A tubulin-like protein is involved in bacterial cytokinesis

The other principal protein of the eucaryotic cytoskeleton, tubulin, also has a homolog in procaryotes. It does not form microtubules, however, but rather forms another category of microfilament. In this case, the tubulin-like protein (called **FtsZ**) forms a band of filaments around the middle of a procaryotic cell. When chromosome segregation is completed by the actin-like MreB filaments, the FtsZ ring contracts (by an unknown mechanism), constricting the cell and initiating cytokinesis (Figure 4.33).

Interestingly, the roles of actin and tubulin homologs in bacterial chromosome segregation and cytokinesis are opposite of their roles in eucaryotes. In bacteria, actin-like filaments are involved in chromosome segregation, and tubulin-like filaments are involved in cytokinesis. In eucaryotes, the roles are reversed. Thus, although the component proteins are clearly homologous, it appears that the processes are not.

The Cytoplasm

The cytoplasm of procaryotes is both a concentrated solution of proteins and small molecules and a dense suspension of ribosomes (see Figure 1.6). In many cases, there may also be storage granules of reserve nutrients, gas vacuoles, or other structures.

4.30 The cytoplasm is a dense suspension of ribosomes

Ribosomes are the site of protein synthesis, and because proteins are the quantitatively most prominent component of cells, there are typically tens of thousands of ribosomes per procaryotic cell. This immense number of ribosomes gives the procaryotic cytoplasm a granular appearance in thin sections in the electron microscope.

The reason that there are so many ribosomes is that making protein is the principal job of a growing cell. A typical procaryotic cell contains about 2×10^6 molecules of protein (over half of its dry weight). For a cell to divide, it has to make 2×10^6 molecules of new protein so that when it divides each of the daughter cells has a full complement of protein. If it takes 40 minutes for a cell to divide (a typical time for *E. coli* at 37°C in minimal medium in laboratory culture) and if it takes about 20 seconds for a ribosome to make a protein (a reasonable average at 37°C), then each ribosome can make three proteins per minute, or 120 molecules of protein in 40 minutes. Thus, it would take about 16,000 ribosomes to make 2×10^6 molecules of protein in 40 minutes.

As you would expect, the number of ribosomes per cell varies as a function of growth rate. At very slow growth rates (several hours for each generation), *E. coli* cells are quite small and contain as few as 1,000 ribosomes. At very high growth rates (20 minutes per generation in rich medium), cells are much larger, mainly to accommodate the 80,000 or so ribosomes needed.

In growing cells, many ribosomes are bound to the inner surface of the cell membrane, instead of floating free in the cytoplasm. Cytoplasmic ribosomes make the soluble proteins of the cell—those that are dissolved in the cytoplasm (Figure 4.34). In contrast, the ribosomes bound to the cell membrane make all of the membrane proteins of the cell, as well as periplasmic and secretory proteins.

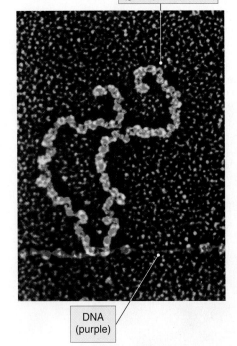

mRNA (red) with attached ribosomes (green)

DNA (purple)

FIGURE 4.34 High-resolution electron micrograph showing ribosomes. Note the coupling of transcription and translation.

4.31 Storage granules are polymeric reserves of nutrients

Many procaryotic cells store some form of nutrient reserve. Such reserves are stored as polymers, which prevents the turgor pressure from going up too much (100,000 molecules of sugar in a procaryotic cell would have a large impact on the osmolarity of the cytoplasm; 100 molecules of polysaccharide, each 1000 sugar residues long, would contain the same amount of sugar but have essentially no impact on osmolarity). These polymeric reserve materials usually form granules in the cytoplasm.

In many cases, some form of carbon is stored as an energy reserve. The most common are polymers of glucose, forming **glycogen,** and polymers of β-hydroxybutyric acid, forming **poly-β-hydroxybutyrate** (PHB). Glycogen granules are usually quite small, invisible in the light microscope; however, PHB granules are often very large and easily visible (Figures 4.35 and 4.36).

Other cells may store other types of compounds. After carbon, sulfur is probably the most often encountered in storage granules, as many procaryotes use reduced sulfur compounds in their energy metabolism (see Chapters 7 and 8). Many of these organisms store sulfur as elementary sulfur (S_n) or as polysulfide (S_n^{-2}). These *sulfur globules* are more liquid than solid and are usually deposited in the periplasm between the cell membrane and cell wall; thus, they are not strictly speaking cytoplasmic structures, even though they appear in the light microscope to be in the cytoplasm (Figures 4.37 and 4.38).

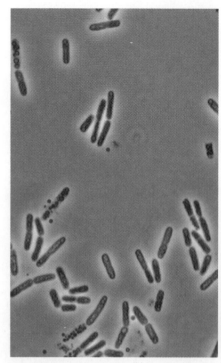

FIGURE 4.35 PHB granules.

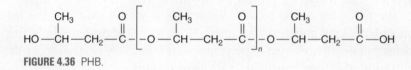

FIGURE 4.36 PHB.

FIGURE 4.37 Light micrograph showing sulfur globules.

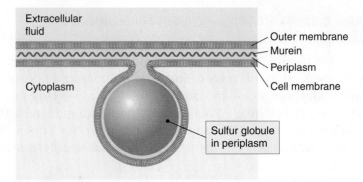

FIGURE 4.38 Periplasmic location of sulfur globules.

4.32 Phosphate is stored in membrane-bound vesicles termed acidocalcisomes

Phosphate is found in very low concentrations in many environments, and a number of bacteria make phosphate storage structures termed **volutin granules** (originally observed in the bacterium *Spirillum volutans*). Recently, it has been recognized that these are identical to the organelles termed **acidocalcisomes,** found in a number of protists and even in human cells (Figures 4.39 and 4.40). They thus are thought to have evolved prior to the last common ancestor of all life on earth.

Acidocalcisomes are surrounded by a membrane that contains transporters for a variety of ions, including proton pumps that make the interior highly acidic (hence the first part of the name). Phosphate is stored inside in the form of **polyphosphate,** long polymers of phosphate up to 700 phosphates in length. The high density of negative charges in polyphosphate is partially neutralized by the high concentration of H^+ and partially by divalent cations, especially Ca^{2+} (hence the middle part of the name).

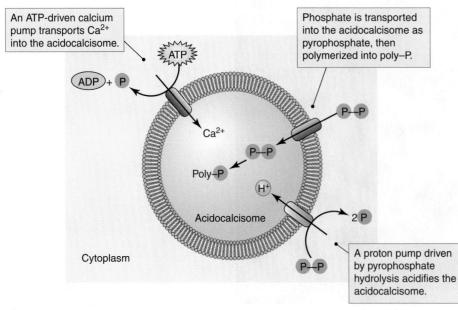

FIGURE 4.39 The acidocalcisome.

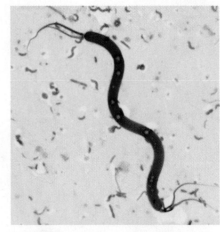

FIGURE 4.40 LM of bacterium with volutin granules.

CHAPTER 4 PROCARYOTIC CELL STRUCTURE AND FUNCTION

4.33 Gas vacuoles provide buoyancy to aquatic cells

Procaryotic cells have a density slightly greater than that of water, and thus, aquatic procaryotes sink unless kept afloat by active motility or some other mechanism. There are many aquatic procaryotes that are not motile and that thus need some other mechanism of achieving buoyancy. A widespread solution, in both bacteria and archaea, is the **gas vacuole** (Figure 4.41).

Gas vacuoles are composed of many individual **gas vesicles**: hollow tubular structures of protein, with conical ends (Figure 4.42). The proteins that make up the gas vesicle have a highly hydrophobic interior surface, so as the vesicles are assembled water is excluded from the interior because it cannot cross the hydrophobic vesicle wall. Instead, the vesicle fills with whatever gas is in solution in the cytoplasm. The result is a rigid, hollow structure filled with gas rather than water, reducing the overall density of the cell and allowing it to float.

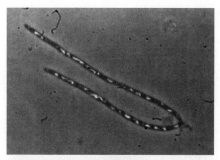

FIGURE 4.41 Gas vacuolate cells.

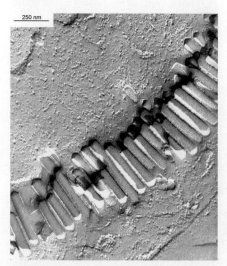

FIGURE 4.42 Electron micrograph of a gas vesicle.

4.34 Magnetosomes allow procaryotic cells to distinguish north from south

Some motile aquatic and marine bacteria have a cytoplasmic structure termed a **magnetosome** (Figure 4.43). This consists of a row of granules of a magnetic mineral (usually magnetite—Fe_3O_4), enclosed in a membranous invagination of the cell membrane (magnetosomes are thus, like sulfur globules, periplasmic). The row of magnetite granules is organized by cytoskeletal filaments of the actin-like protein MreB that lie just inside the membrane, paralleling the row of magnetic granules.

The magnetosome acts like a compass needle and allows these bacteria to determine up and down, because in the latitudes in which they live the earth's magnetic field is tilted relative to the earth's surface; lines of force thus run up and down as well as north and south (Figure 4.44). Most bacteria with magnetosomes are

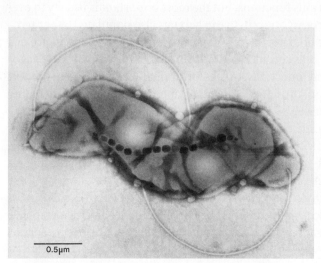

FIGURE 4.43 Transmission electron micrograph of magnetotactic bacterium; negative stain of magnetosome.

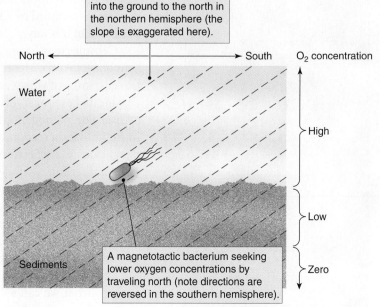

FIGURE 4.44 The orientation of lines of magnetic force and oxygen gradients in sediments.

Lines of magnetic force slope into the ground to the north in the northern hemisphere (the slope is exaggerated here).

North ← → South O_2 concentration

Water

High

Low

Zero

Sediments

A magnetotactic bacterium seeking lower oxygen concentrations by traveling north (note directions are reversed in the southern hemisphere).

microaerophiles, and they use the magnetosomes to find the right concentration of oxygen. In the northern hemisphere, most magnetotactic bacteria will swim north if the oxygen concentration is too high and south if it is too low. These directions are reversed in southern hemisphere magnetotactic bacteria.

Surface Layers and Appendages

Many bacteria have various structures external to their envelope. Some of these are thin appendages that protrude from the cell surface; others are layers of several kinds. The most common are a surface layer of protein, termed an **S-layer,** and a thick gelatinous layer termed a **capsule.**

4.35 Many procaryotes have a surface layer of protein external to the envelope

Many bacteria, both gram negative and gram positive, as well as some archaea, have a layer of protein on the outer surface of the envelope. This S-layer is held together at least partly by quaternary interactions among the protein molecules, which form a two dimensional crystalline array on the cell surface. The function of these layers is not clear; perhaps it provides some protection from viruses or other predators (Figure 4.45).

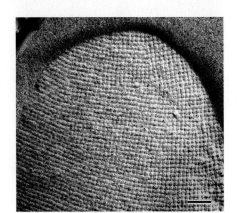

FIGURE 4.45 Negative stain of a bacterial S-layer.

4.36 Capsules are gelatinous layers of polysaccharide external to the wall

Many procaryotes have a layer of variable thickness external to the wall (and S-layer if there is one) termed a **capsule.** Capsules are almost always composed of polysaccharide, although a few examples of polypeptide capsules are known. Capsules are gelatinous in consistency—a loose mesh of polysaccharide strands with most of the space and weight being water. Capsules can be visualized in the light microscope, at least if they are thick enough, by suspending cells in India ink. The fibrous mesh of the capsule prevents the India ink particles from getting close to the cells so that cells look as if they are surrounded by a halo (Figure 4.46). Thin capsules, however, may not be able to be seen in this way.

Capsules may have various functions, but the most important is usually to prevent phagocytosis. Most procaryotic cells live in habitats also populated by eucaryotic cells, some of which are predatory on the smaller procaryotes. The presence of a capsule makes procaryotic cells considerably more difficult for eucaryotes to phagocytose. For pathogens that infect the tissues of animals, capsules can prevent immune system recognition, thus preventing a number of host defenses including phagocytosis (see Section 19.9).

Another function of some capsules is adherence. Many cells attach themselves to a solid substratum by means of capsular material. For instance, bacteria in the human mouth must attach to tooth enamel or epithelial cells or live in the gingiva (the crack between the teeth and gums) in order to avoid being swallowed. Those bacteria that attach to teeth do so by means of their capsules; indeed, a single mass of capsular slime may enclose many different bacteria. As these bacteria metabolize nutrients, especially sugars, they produce acids that dissolve the underlying enamel to form what we call a cavity, or **dental caries.** We brush and floss our teeth to remove these adherent bacteria, not to remove food particles (which by themselves are relatively harmless).

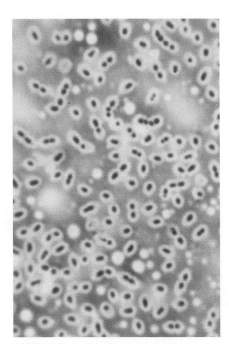

FIGURE 4.46 The bacterial capsule (blue halo).

Sometimes the capsular material is produced in large amounts that are sloughed off, rather than adhering to the cell. These are often termed **slime layers** rather than capsules.

4.37 Procaryotic cells often are found as biofilms

The aggregates of bacteria on teeth that cause caries are an example of a **biofilm**—an aggregation of cells held together in a common matrix of capsular material. Biofilms were long thought to be accidental accumulations of cells; however, recent work suggests that they are actively formed by cells and remodeled as they mature (Figure 4.47). Biofilms may contain a single kind of microbe, or they may be mixed (as in the mouth).

4.38 Pili mediate the specific attachment of cells to other cells

Many procaryotes have fine, hair-like appendages protruding from their cell surface. These are termed **pili** (singular **pilus**), sometimes called **fimbriae** (singular **fimbria**). They are fibers made up of many molecules of globular protein, held together by quaternary forces. They are assembled by the addition of new **pilin** molecules (the individual protein monomers of which the pilus is made) at the base, where they are anchored in the outer membrane in gram-negative cells. Pili may be distributed randomly on the cell surface, or they may be clustered at one or both poles of a rod-shaped cell (Figures 4.48 and 4.49).

The proteins at the very tip of the pilus are often different from the pilins that make up the bulk of the pilus. These molecules at the tip are termed **adhesins,** and they have specific binding sites that allow the pilus to attach to specific structures. For instance, many pathogens have pili with adhesins that bind specifically to glycoproteins on the surface of animal cells. This allows the pathogen to attach to host cells and resist being swept out of the body (by the flow of saliva, urine, intestinal contents, or even the slow movement of fluid past the cells in tissues).

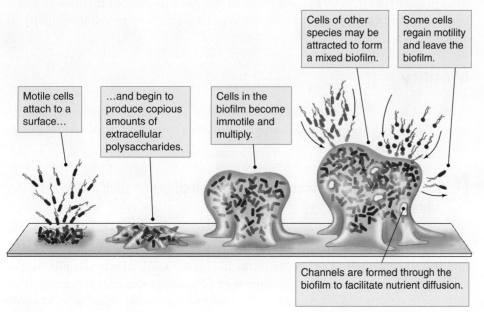

Motile cells attach to a surface…

…and begin to produce copious amounts of extracellular polysaccharides.

Cells in the biofilm become immotile and multiply.

Cells of other species may be attracted to form a mixed biofilm.

Some cells regain motility and leave the biofilm.

Channels are formed through the biofilm to facilitate nutrient diffusion.

FIGURE 4.47 Maturation of a biofilm.

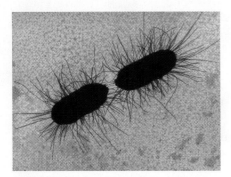

FIGURE 4.48 Electron micrograph showing a piliated cell.

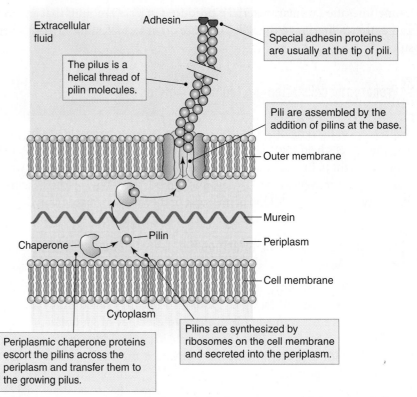

Extracellular fluid

Adhesin

Special adhesin proteins are usually at the tip of pili.

The pilus is a helical thread of pilin molecules.

Pili are assembled by the addition of pilins at the base.

Outer membrane

Murein

Chaperone

Pilin

Periplasm

Cell membrane

Cytoplasm

Periplasmic chaperone proteins escort the pilins across the periplasm and transfer them to the growing pilus.

Pilins are synthesized by ribosomes on the cell membrane and secreted into the periplasm.

FIGURE 4.49 Structure and growth of a type I pilus (a common type). Other types of pili are assembled by different mechanisms.

Thus, the function of most pili is attachment to other cells. Sometimes these are cells of the same kind; sometimes they are different (as in the case of pathogenic bacteria attaching to host cells). In any case, the specificity of attachment is determined by the structure of the adhesins at the tip of the pilus.

In some cases, pili mediate less specific attachment. This is due to nonspecific forces (i.e., electrostatic attractions of positive and negative charged surfaces) and usually involves attractive forces between the pilin molecules themselves and a surface. In addition to attachment, some pili are involved in cell movement, as described later.

Motility

Many procaryotic cells are motile. Most move by swimming, using organelles called flagella. Others can glide along a solid surface, using one of several different mechanisms.

4.39 Flagella are rigid, helical organelles in which the rotation moves cells through liquid

Flagella (singular, **flagellum**) are hollow tubes composed of globular proteins; however, they are helical rather than straight (Figure 4.50). Also unlike pili, the flagellar filament is assembled by addition of new **flagellin** molecules at the tip, rather than the base. Newly synthesized flagellin molecules move through the hollow core of the flagellum until they reach the tip, where they are to be added.

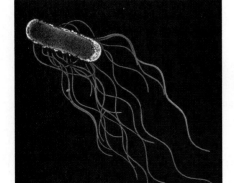

FIGURE 4.50 Electron micrograph of a flagellated cell.

CHAPTER 4 PROCARYOTIC CELL STRUCTURE AND FUNCTION

Flagella are often nonrandomly distributed on the cell surface. Frequently, one or several flagella are found at the tip of rod-shaped or spiral cells; this is termed **polar** flagellation. Alternatively, flagella may be distributed over the entire surface, a pattern termed **peritrichous** flagellation. The patterns of flagellar insertion are normally constant for procaryotic species; they are thus useful for identifying procaryotes (Figure 4.51).

Flagella typically have three distinct structural regions. Most of the organelle consists of the helical **filament,** whose rotation provides the motive force for swimming. This is attached to a flexible region at the surface of the cell termed the **hook.** The function of the hook is to act as flexible coupling to allow the flagellar filament to point in different directions. Finally, the hook is connected to the **basal body,** which both embeds the flagellum firmly in the cell envelope and membrane and acts as microscopic rotary motor to rotate the hook and filament. The entire structure is composed of protein. Typically, the filament consists of thousands of copies of a single type of protein (flagellin), although there are cases in which two or more proteins form the filament. Several different proteins compose the hook, and there are many different proteins in the basal body (Figure 4.52).

Flagella function by being rotated. Because they are helical, their rotation exerts force on the cell, much like the helical threads on a bolt or screw can exert force when rotated. In most cases, procaryotic flagella *push* the cell; that is, cells move by rotating their flagella in only one direction, such that the cell is pushed ahead by the rotating flagella at its rear. Rotation in the opposite direction does not normally result in movement.

For polarly flagellated cells, this means that the pole with the flagella is usually the trailing end, or back. For peritrichously flagellated cells, either end may be the front, and the flagella all trail toward the other end (possible because of the flexible hook) (Figure 4.53).

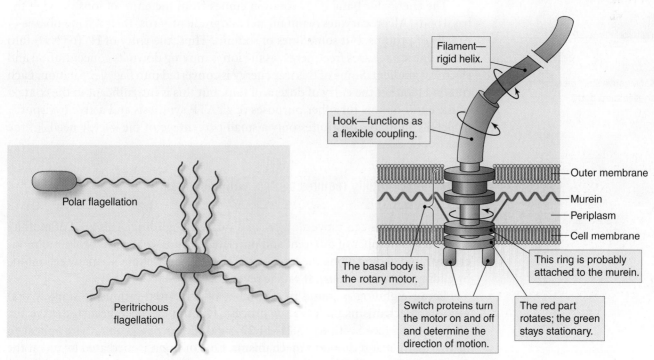

FIGURE 4.51 Flagellar insertion patterns.

FIGURE 4.52 Flagellar structure.

Flagella are rigid, helical organelles in which the rotation moves cells through liquid

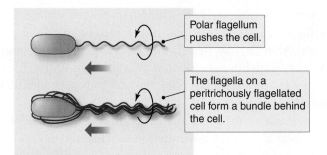

FIGURE 4.53 Motility of polarly and peritrichously flagellated cells.

Polar flagellum pushes the cell.

The flagella on a peritrichously flagellated cell form a bundle behind the cell.

4.40 Flagella are rotated by the entry of ions through the basal body

Unlike eucaryotic flagella, whose movements are driven by ATP hydrolysis within the flagella themselves, procaryotic flagellar filaments have no active catalytic activity. Their movement is passive, and it is the basal body, not the filament, that is responsible for the movement. The basal body thus functions as a tiny rotary motor fixed in the cell wall and membrane.

The basal body contains two disks, the bottom one of which is attached to a rod that penetrates through a hole in the other disk and attaches to the hook. This bottom disk is thus rigidly attached to the flagellum. The other disk is attached to the murein. Thus, when the two disks are rotated against each other, the lower disk rotates (and thus the rod, hook, and filament as well); the other disk remains stationary.

In gram-negative bacteria, there are usually two additional rings and a collar in which the rod passes through the murein and outer membrane (Figure 4.54). These are assumed to act as bearings for the rotating rod. In gram-positive bacteria, there appears to be a larger collar, but there are no additional rings.

The energy for basal body rotation comes from the entry of ions through the basal body. All procaryotes maintain an ionic gradient across their cell membrane—usually of protons, but sometimes of sodium. Thus, the entry of H^+ (or Na^+) into the cytoplasm releases free energy, as the ion is moving down its concentration and electrical gradient. Some of this free energy is converted into flagellar rotation. Each rotation requires the entry of dozens of ions, but this is insignificant in the context of the entry of ions for other purposes (e.g., ATP synthesis and active transport). Generally motility constitutes only a small percentage of the energy needs of the procaryotic cell.

4.41 Gliding motility requires contact with a surface

Some procaryotes can move along a solid surface by **gliding.** This kind of motility is found in a number of different and unrelated groups. It consists of slow to moderately rapid gliding along the long axis of the cell (or trichome—many cell motile by gliding are filamentous). It can be reversed.

Because gliding is found in a number of unrelated groups, it is likely that multiple mechanisms underlie the process. Gliding has been best studied in the myxobacteria (see Sections 11.21–11.27), and in this one group, there appear to be two distinct and different mechanisms. One of them is mediated by pili at the front end of the cell. The pili are extruded, attach to the surface (probably non-

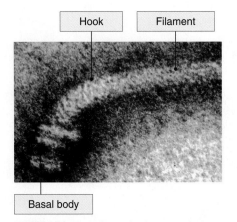

Hook   Filament

Basal body

FIGURE 4.54 Negative stain electron micrograph of flagellar basal body.

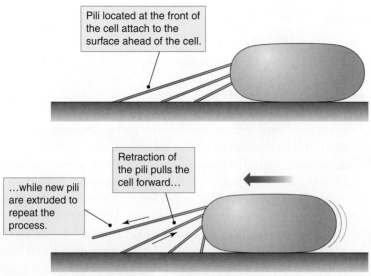

Pili located at the front of the cell attach to the surface ahead of the cell.

...while new pili are extruded to repeat the process.

Retraction of the pili pulls the cell forward...

FIGURE 4.55 Pili-mediated gliding.

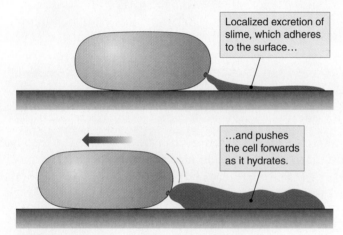

Localized excretion of slime, which adheres to the surface...

...and pushes the cell forwards as it hydrates.

FIGURE 4.56 Gliding by polysaccharide extrusion.

specifically by van der Waals interactions), and then are retracted, pulling the cell forward (Figure 4.55).

The other type of gliding done by myxobacteria appears to depend on the extrusion of polysaccharide from the trailing end of the cell. This slime attaches to the surface, and as it is hydrated, it expands and pushes the cell forward (Figure 4.56).

Summary

The procaryotic cell is superficially simple. In the light microscope, internal structures are rarely seen, and even in the electron microscope, most procaryotes seem to be devoid of organelles. Nevertheless, they are very complex at the molecular level, and we still are far from a complete understanding of even the best studied procaryotes. They are arguably the most successful life form on the planet, as they outnumber eucaryotes by a factor of probably a billion and constitute 90% of the biomass on earth. Clearly, their simplicity is not primitive, but is rather a highly advanced strategy for competing effectively in the endless struggle for survival and reproduction.

Study questions

1. Describe the structure and chemical composition of the various types of bacterial and archaeal cell envelopes.
2. For all types of procaryotic walls, and for those that lack a defined cell envelope, identify the structures that resist osmotic lysis.
3. Compare and contrast the gram-negative and gram-positive cell envelopes.
4. Compare and contrast murein and pseudomurein.
5. Describe the permeability properties of the gram-negative outer membrane.
6. Calculate the surface to volume ratio (in $\mu m^2/\mu m^3$) through which nutrients can diffuse for a gram-positive rod of length 2 μm and width 1 μm.
7. Calculate the surface-to-volume ratio (in $\mu m^2/\mu m^3$) through which nutrients can diffuse for a gram-negative rod of length 2 μm and width 1 μm. Assume that there are 2×10^5 porins per cell, with a pore diameter of 0.8 μm.
8. Compare and contrast the cell membranes of bacteria and archaea.
9. Describe the principal features of the nucleoid.
10. Compare and contrast flagellar and gliding motility.

Cell Structure and Function in Protists

5

PROCARYOTIC MICROBES ARE SPECIALISTS at rapid growth at the expense of dissolved nutrients as a consequence of their small size and simple cell structure; eucaryotic microbes, or **protists,** are larger, more complex, and grow more slowly, often at the expense of particulate food. Procaryotes cannot ingest particulate material and thus can only digest it by secreting hydrolytic enzymes. This is inefficient because most of the hydrolytic products diffuse away; only a small proportion of the material is absorbed by the cell that produces the enzymes. The early evolution of the eucaryotic cell type is thought to be a result of the evolution of several structural features that allowed phagocytosis and hence allowed the proto-eucaryote to become predators. This move up the trophic ladder freed them from competition with the more numerous and more efficient procaryotes and opened the way for a rapid diversification of cell structure and an increase in cell size. Today, probably at least 2 billion years after the eucaryotic microbes first appeared on earth, they are an immensely diverse and complex group of organisms.

5.1 Most eucaryotes are microbes

Although they are nowhere near as numerous as procaryotic microbes, eucaryotic microbes are still orders of magnitude more abundant than macro-organisms. Not only are eucaryotic microbes much more numerous than macro-organisms, but most lineages of eucaryote are microbes (Figure 5.1). Macroscopic organisms are a relatively recent evolutionary innovation and did not arise until probably at least a billion years after the first eucaryotic microbes. The eucaryotic microbes, thus, had more than a billion years to diversify before any of the macroscopic lineages first arose.

The recent realization of the great antiquity and evolutionary diversity of eucaryotic microbes has raised some very interesting classification issues. For instance, if we consider that the plants, animals, and fungi each constitute a separate kingdom of life, how should each of the microbial lineages be classified? Should they each constitute a kingdom, resulting in at least a score of kingdoms of eucaryotic life (and presumably an equal number of procaryotic kingdoms)? This has a great deal of logic to it, but it would be a major departure from tradition, in which all eucaryotic microbes were assigned to a one kingdom and all procaryotes to another. It will take years for the implications of our new understanding of microbial evolution to be fully assimilated into our classification schemes.

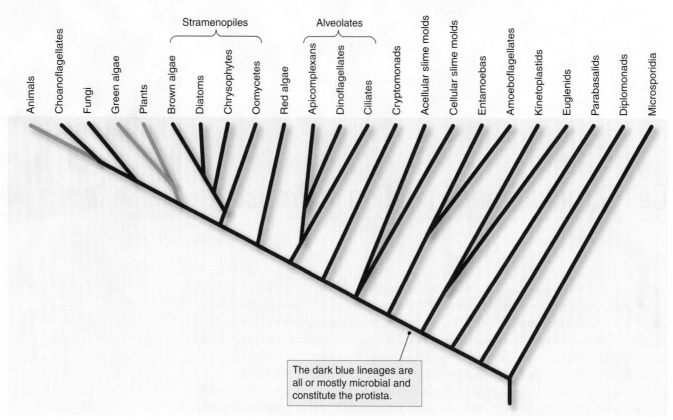

Stramenopiles   Alveolates

Animals
Choanoflagellates
Fungi
Green algae
Plants
Brown algae
Diatoms
Chrysophytes
Oomycetes
Red algae
Apicomplexans
Dinoflagellates
Ciliates
Cryptomonads
Acellular slime molds
Cellular slime molds
Entamoebas
Amoeboflagellates
Kinetoplastids
Euglenids
Parabasalids
Diplomonads
Microsporidia

The dark blue lineages are all or mostly microbial and constitute the protista.

FIGURE 5.1 Eucaryotic phylogeny.

5.2 Eucaryotic cells are characterized by an endomembrane system and a cytoskeletal system

All eucaryotic cells have a number of separate membrane systems, such as the cell membrane, the endoplasmic reticulum (ER), the Golgi body, the endosome, and others. Although physically separate, these membranes all exchange materials with each other, and thus, they are not functionally separate; rather, they are all part of the same dynamic system. These membranes are collectively termed the **endomembrane system,** and they constitute a universal feature of the eucaryotic cell (Figure 5.2).

The other universal feature of the eucaryotic cell is the possession of a **cytoskeletal system,** consisting of microtubules and microfilaments. These two classes of fibrous elements in the eucaryotic cell are responsible for most intracellular movement, and they may provide structural support to maintain cell shape. The cytoskeletal system thus functions not only like a cytoskeleton, but like the muscles of the cell as well.

These two systems seem to be the basis for the evolutionary success of the eucaryotic cell because they are the basis for the ability to phagocytose particulate food. The process of phagocytosis and intracellular digestion requires the participation of the cytoskeletal system to move membrane vesicles to the right places for the different membranes to interact. Thus, it appears that the crucial first events in the evolution of the eucaryotic cell were the development of a rudimentary cytoskeletal system and the ability to manage the flow of material from one membrane to another.

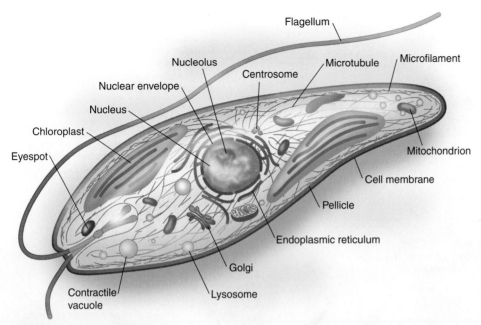

FIGURE 5.2 A typical eucaryotic cell.

5.3 Special signal sequences on proteins target them to particular places in the eucaryotic cell

The endomembrane system, plus mitochondria and chloroplasts if present, partitions the eucaryotic cell into multiple membrane-bounded compartments (the nucleus, the ER, the Golgi, lysosomes, vacuole, mitochondria, chloroplasts, etc.). Each of these has its own distinctive proteins in both the membranes and in the interior spaces. Thus, there is a significant problem of protein targeting and sorting. Proteins, which are made on ribosomes in only a few locations in the cell, must contain some features that allow them to be specifically transported to their proper destination; without such specific targeting mechanisms, the eucaryotic cell could not exist.

The information for proper targeting of most proteins is all contained within the primary structure of each protein. That is, each protein that must be transported somewhere other than its site of synthesis has a short sequence of amino acids somewhere in its primary sequence that is a specific **signal sequence** indicating its ultimate destination. Such signal sequences are often not sequences of specific amino acids but rather are regions of particular chemical properties; for instance, the targeting to the ER involves a sequence of approximately 20 to 30 residues at the amino terminus of a protein in which a couple of positively charged residues are followed by a long run (at least 16) of hydrophobic residues.

5.4 There are two types of targeting mechanisms

There are two fundamentally different approaches to protein targeting: most proteins whose targeting involves crossing one of the membranes of the endomembrane system have a signal sequence of amino acids that, when translated, causes a block to further translation (Figure 5.3). Only when the ribosome is bound to the

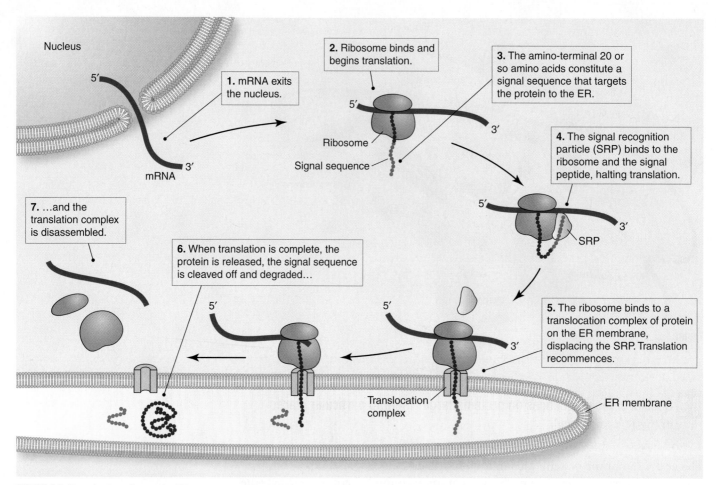

Nucleus

1. mRNA exits the nucleus.

5′

mRNA

3′

2. Ribosome binds and begins translation.

5′

Ribosome

Signal sequence

3′

3. The amino-terminal 20 or so amino acids constitute a signal sequence that targets the protein to the ER.

4. The signal recognition particle (SRP) binds to the ribosome and the signal peptide, halting translation.

5′

3′

SRP

7. …and the translation complex is disassembled.

6. When translation is complete, the protein is released, the signal sequence is cleaved off and degraded…

5′

5′

3′

5. The ribosome binds to a translocation complex of protein on the ER membrane, displacing the SRP. Translation recommences.

Translocation complex

ER membrane

FIGURE 5.3 Protein targeting to the ER.

ER does translation recommence. The protein product then enters the ER membrane or the interior of the ER **co-translationally** (simultaneously with translation).

A **posttranslational** mechanism operates for proteins targeted to mitochondria and chloroplasts (Figure 5.4). In this case, as the protein is being translated by ribosomes in the cytoplasm, it binds to **chaperone** proteins. These proteins bind to other unfolded proteins and can maintain the bound proteins in this unfolded state. The complex of chaperone and organelle protein then diffuses to its target site, where receptors for the signal sequence bind it to the target membrane. The chaperones dissociate and the protein is translocated and only then folds into its mature conformation. The signal sequence is normally cleaved from the protein and degraded to amino acids, which are reused. Proteins of the nucleus are also transported posttranslationally, but through the nuclear pores (see Section 5.8).

5.5 Proteins that cross two membranes have two signal sequences

Posttranslational targeting of proteins sometimes requires proteins to cross two membranes. Both mitochondrial and chloroplast proteins have a primary signal sequence that directs them to the matrix or stroma, crossing both the outer and inner membranes at once. Thus, proteins that are destined for the intermembrane space of either organelle, as well as proteins targeted to the thylakoid space, have a

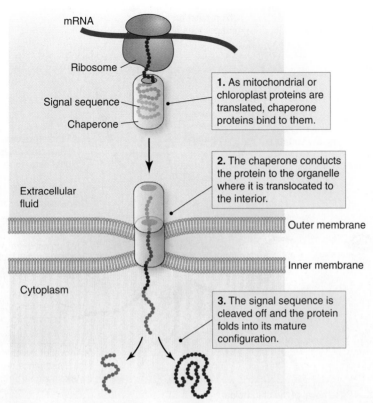

1. As mitochondrial or chloroplast proteins are translated, chaperone proteins bind to them.

2. The chaperone conducts the protein to the organelle where it is translocated to the interior.

3. The signal sequence is cleaved off and the protein folds into its mature configuration.

mRNA

Ribosome

Signal sequence

Chaperone

Extracellular fluid

Outer membrane

Inner membrane

Cytoplasm

FIGURE 5.4 Posttranslational protein targeting.

second membrane to cross. These proteins have two signal sequences—one directs them to the mitochondrion or chloroplast, and it is cleaved off as the protein enters the organelle. As the protein enters, it is bound by organelle chaperone proteins to keep it unfolded. These chaperones then conduct the protein to a second translocation complex, where the second signal sequence directs their transfer across the second membrane (Figure 5.5).

The Endomembrane System

The endomembrane system consists of all of the membranes of the cell that exchange material with each other by way of transport vesicles. Usually, this includes the cell membrane, the nuclear envelope, the ER, the Golgi body, the endosome, and lysosomes or vacuoles. Most eucaryotic cells also contain mitochondria, and many contain chloroplasts as well; these organelles, although membranous, are not part of the endomembrane system.

5.6 The endomembrane system exchanges materials by the budding and fusion of membrane vesicles

The material flow through the endomembrane system is mediated by membranous vesicles. These are produced by a process of **budding,** in which specific proteins of the cytoskeletal system draw out a bit of membrane to create a bleb, which is then pinched off to form a separate vesicle (Figure 5.6).

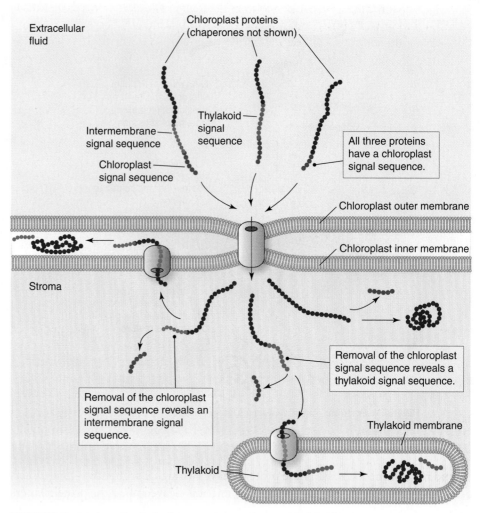

FIGURE 5.5 Protein targeting to the intermembrane space and thylakoids.

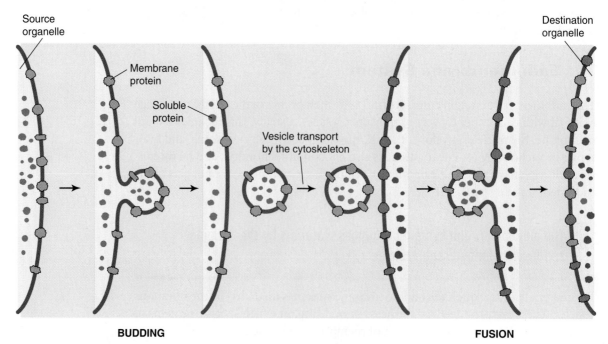

BUDDING

FUSION

FIGURE 5.6 Membrane budding and fusion.

Vesicle **fusion** is the opposite of budding. Normally, vesicles are moved from their site of formation to their destination by the cytoskeletal system. Once there, fusion proteins fuse the vesicle and the destination membrane. The result is that whatever membrane proteins were embedded in the vesicle membrane (originally in the membrane from which the vesicle budded) become part of the destination membrane. Similarly, anything inside the vesicle (e.g., soluble proteins) winds up inside the destination organelle. Thus, vesicle budding and fusion is a method both for membrane material to be exchanged among membranous organelles of the cell, and it is a method of transferring material from the inside of one organelle to the inside of another.

5.7 The nuclear envelope encloses the chromosomes

The **nuclear envelope** is a specialized region of the ER (or vice versa) that consists of two concentric membranes (Figure 5.7). The outer surface is, like the ER to which it is connected, studded with ribosomes.

The nuclear envelope surrounds the **nucleoplasm**—the material within the nuclear envelope. The nucleoplasm is chemically very different from the cytoplasm, a consequence of the specificity of the nuclear pore complexes that we discuss in the next section. The nucleoplasm is especially rich in RNA, both as a consequence of active transcription of the DNA and because there are a number of specifically nuclear RNAs (i.e., RNAs that are retained within the nucleus). The protein composition of the nucleoplasm is also very different from that of the cytoplasm, nuclear proteins being targeted to the nucleus as we described in Section 5.4.

The nucleoplasm also contains the chromosomes, of which there are almost always more than one (dozens is common), and the chromosomes are linear, unlike the circular chromosomes of most procaryotes. There are multiple origins of replication on each chromosome like the archaea but unlike bacteria.

Chromosomes consist of **chromatin**—a complex of DNA and histone proteins. In the interphase nucleus in most organisms, the chromatin is in an extended conformation unlike the condensed chromatin usually seen during mitosis (Figure 5.8). This

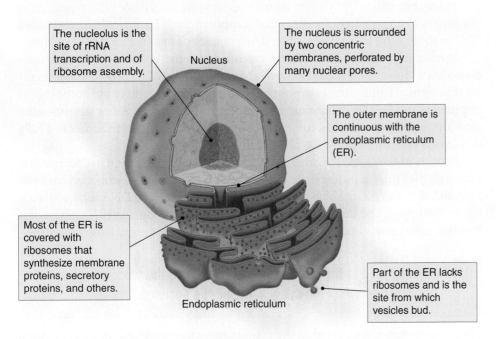

The nucleolus is the site of rRNA transcription and of ribosome assembly.

Nucleus

The nucleus is surrounded by two concentric membranes, perforated by many nuclear pores.

The outer membrane is continuous with the endoplasmic reticulum (ER).

Most of the ER is covered with ribosomes that synthesize membrane proteins, secretory proteins, and others.

Endoplasmic reticulum

Part of the ER lacks ribosomes and is the site from which vesicles bud.

FIGURE 5.7 The eucaryotic nucleus and ER.

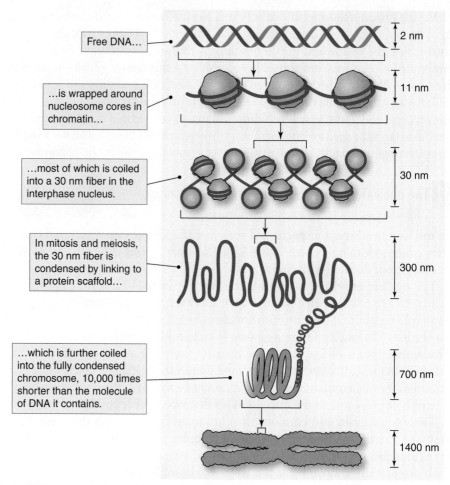

Free DNA...

...is wrapped around nucleosome cores in chromatin...

...most of which is coiled into a 30 nm fiber in the interphase nucleus.

In mitosis and meiosis, the 30 nm fiber is condensed by linking to a protein scaffold...

...which is further coiled into the fully condensed chromosome, 10,000 times shorter than the molecule of DNA it contains.

2 nm

11 nm

30 nm

300 nm

700 nm

1400 nm

FIGURE 5.8 Chromatin organization.

consists of chromatin (DNA coiled around nucleosome cores made of histones), most of which is, in turn, coiled into a fiber that is approximately 30 nm in diameter (naked DNA is about 2 nm in diameter, and a string of nucleosomes is 11 nm in diameter).

During mitosis and meiosis, the extended chromatin (30-nm fiber) is condensed in a sequence of successive levels of condensation. The proteins that mediate this, termed **condensins,** are poorly defined, but some of them are homologous to proteins involved in maintaining the procaryotic nucleoid. A few protists—the dinoflagellates and euglenids, for instance—maintain their chromosomes in a fully condensed state throughout the cell cycle.

Because of multi-level organization of chromatin, transcription in eucaryotes is more complex than in procaryotes. Before chromatin can be transcribed, it has to be **remodeled,** a process in which the 30 nm fiber uncoils and the DNA unwraps from the nucleosomes. The details of this process remain to be elucidated, but part of it involves acetylation of the histone proteins that composes the nucleosomes cores.

5.8 Pore complexes in the nuclear envelope regulate the passage of materials between the cytoplasm and the nucleoplasm

The nucleoplasm is connected to the cytoplasm by numerous **pores** that penetrate both the inner and outer membrane of the nuclear envelope. These pores are partially

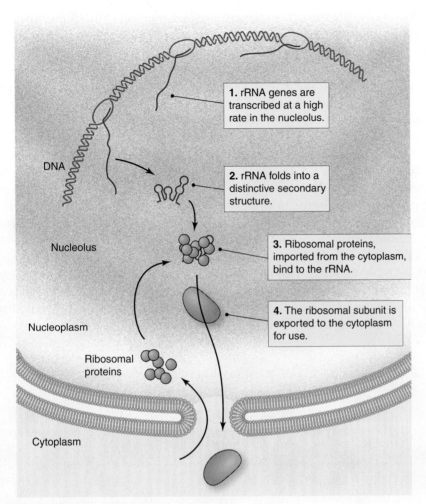

1. rRNA genes are transcribed at a high rate in the nucleolus.

2. rRNA folds into a distinctive secondary structure.

3. Ribosomal proteins, imported from the cytoplasm, bind to the rRNA.

4. The ribosomal subunit is exported to the cytoplasm for use.

DNA

Nucleolus

Nucleoplasm

Ribosomal proteins

Cytoplasm

FIGURE 5.9 Ribosome assembly and export.

blocked, however, by a group of proteins termed the **pore complex.** Thus, although there is free flow of water and small solutes such as ions between the two compartments and of small proteins (less than about 50,000 molecular weight), larger molecules cannot cross unless they are recognized by the pore complex and allowed to pass. The principal movements of macromolecules through the core complex include mRNA and tRNA exiting the nucleus for use in protein synthesis in the cytoplasm and nucleoplasmic proteins entering the nucleus after synthesis in the cytoplasm.

Another category of material that passes the nuclear pore complex consists of partially assembled ribosomes. Ribosomes are assembled to their nearly complete form in the **nucleolus**—the region of the nucleus where rRNA genes are being transcribed at high rate. Here ribosomal proteins, imported from the cytoplasm, bind to rRNA (probably simultaneously with transcription) to form the small and large subunits, which are then individually exported to the cytoplasm where they function in translation (Figure 5.9).

5.9 mRNA molecules are capped and tailed to mark them for export

One of the characteristic features of the eucaryotic cell is the segregation of transcription and translation into different compartments. Transcription to produce mRNA occurs within the nucleus, of course (that is where the DNA is), whereas translation of mRNA occurs in the cytoplasm. Obviously, mRNA must be one of the types of molecule that the nuclear pore complexes allow to pass. Complicating things is the fact that there are many different types of RNA, some of which need to be exported (e.g., tRNA and mRNA) and some of which need to be retained in the nucleus (e.g., spliceosome RNAs). Obviously, there must be structural features or special sequences associated with different RNA classes that allow the nuclear pore complexes to distinguish which RNAs to export and which to retain.

For mRNAs the principle signals are a specially modified **cap** at the 5′ end of the message and a **tail** of multiple adenine nucleotides. The cap is a modified guanosine nucleotide that is added 5′-to-5′, forming a distinctive structure (Figure 5.10). This guanosine residue is often methylated, and the first one or two residues of the message proper may also be methylated. The enzymes that cap mRNA are located in the nucleus, and they cap the message as it is being transcribed—thus, in eucaryotes transcription and capping **coupled.**

Tailing involves a targeted cleavage of nascent pre-mRNA at a special sequence (AAUAAA) that is recognized by the **tailing enzyme.** This enzyme cleaves the pre-mRNA downstream a few bases from the recognition sequence and then adds adenine residues one at a time to form the **poly-A tail** (Figure 5.11). The number of adenine residues is variable, usually between 50 and 200.

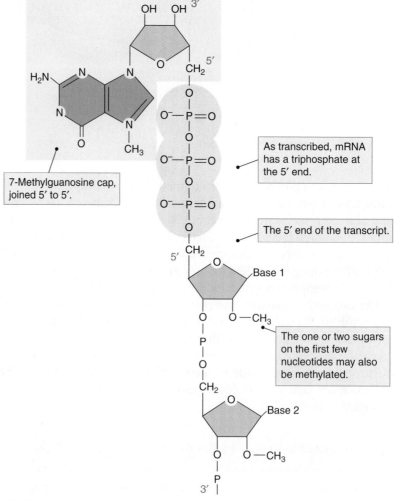

FIGURE 5.10 The mRNA cap.

CHAPTER 5 CELL STRUCTURE AND FUNCTION IN PROTISTS

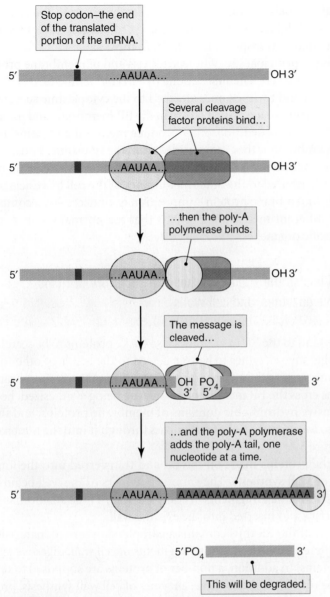

FIGURE 5.11 mRNA tailing.

Both capping and tailing have purposes in addition to constituting signals to the nuclear pore complexes that this is an RNA to export. Both probably make the RNA more slow to be degraded by enzymes (called **exonucleases**) that break down RNAs by clipping off one nucleotide at a time from the ends. In addition, the 5′ cap serves as part of the ribosome binding site for the initiation of translation.

In addition to capping and tailing, mRNAs that contain introns have to be **spliced** before they can be exported to the cytoplasm. Splicing of some introns occurs spontaneously as the result of an intramolecular chemical reaction. Most introns, however, are spliced by a complex termed the *spliceosome,* a complex of RNA and protein.

5.10 The endoplasmic reticulum is responsible for membrane synthesis

The ER is an extension of the nuclear envelope (see Figure 5.7). Most of its surface is covered with ribosomes attached on the cytoplasmic surface; this ER is often termed **rough ER** because of its appearance in the electron microscope. A small region near the Golgi is usually free of attached ribosomes; this is the region from which vesicles bud to transport ER material to the Golgi.

The ER is the principal site of lipid synthesis and of membrane protein synthesis in the eucaryotic cell. The first is done by enzymes embedded in the ER membrane and the second by ribosomes attached to the cytoplasmic surfaces of the ER. As phospholipids are made, they dissolve in the ER membrane and increase its area. As the membrane-attached ribosomes produce membrane proteins, they dissolve into the ER membrane as they are extruded from the ribosome. Thus, both the lipid and protein components of the entire endomembrane system are all made by the ER and then distributed to the other membranes of the cell by vesicle budding and fusion. The ER membrane protein composition is, of necessity, a complex mixture of the many different membrane proteins that are normally found in the membranes of specific organelles.

5.11 The ER is also the site of synthesis of secretory proteins, digestive enzymes, and cell walls

Ribosomes bound to the ER membranes also make proteins to be secreted from the cell and the digestive enzymes to be transferred to the inside of other organelles of the endomembrane, such as lysosomes or vacuoles. As with membrane proteins, these proteins cross the ER membrane as they are being synthesized; however, they lack the extensive hydrophobic domains of membrane proteins, and thus, they do not insert into the membrane but rather pass through it into the lumen of the ER as shown in Figure 5.3.

Among the proteins made on the ER and transferred into the lumen are the enzymes of cell wall synthesis. The various polymers of the wall begin to be made and linked up into small fragments of complete wall. These are ultimately exported to the cell surface and inserted into the existing wall.

The lumen of the ER thus contains a complex mixture of materials, each destined for a different final destination: fragments of cell wall, digestive enzymes, and secretory proteins. In addition, a number of proteins are supposed to remain in the ER and not be transported elsewhere: enzymes of cell wall synthesis, protein glycosylating enzymes (which we will discuss later).

5.12 The Golgi sorts the mixed contents of ER membranes and lumen into different vesicles

We have seen that the ER membrane contains a mixture of proteins that are normal ER membrane proteins, mixed together with the membrane proteins of the cell membrane, the vacuole, the Golgi, and all of the other components of the endomembrane system. Similarly, the lumen contains fragments of cell wall, digestive enzymes, secretory proteins, etc. Thus, before these membrane proteins and luminal proteins can be transported to their ultimate destinations, they have to be sorted to separate them from the other materials. This sorting is done by the Golgi.

Materials are transported from the ER to the Golgi in membrane vesicles. Once at the Golgi, the vesicles fuse with the cisterna on one face of the Golgi (called the **cis** face). Materials are transferred from one cisterna to the next through the stack by vesicle budding and fusion. Eventually, vesicles are produced at the other face

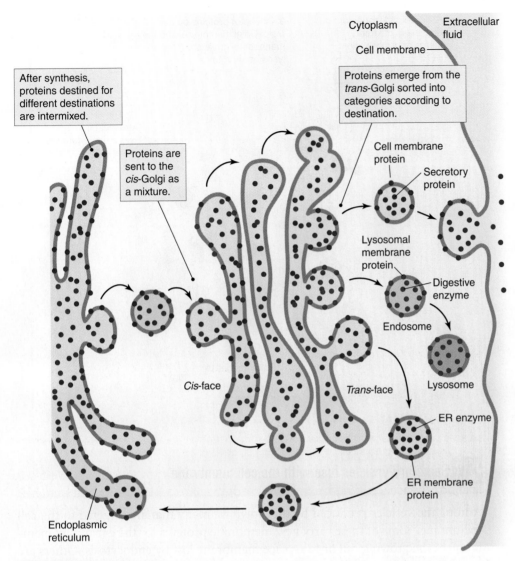

After synthesis, proteins destined for different destinations are intermixed.

Proteins are sent to the *cis*-Golgi as a mixture.

Proteins emerge from the *trans*-Golgi sorted into categories according to destination.

Cytoplasm

Extracellular fluid

Cell membrane

Cell membrane protein

Secretory protein

Lysosomal membrane protein

Digestive enzyme

Endosome

Lysosome

Cis-face

Trans-face

ER enzyme

ER membrane protein

Endoplasmic reticulum

FIGURE 5.12 Protein sorting.

(the **trans** face) that are homogeneous, with contents and membrane proteins appropriate to their destination (Figure 5.12).

The underlying mechanisms by which the Golgi groups related proteins together is very complex and is not yet completely understood. Some of it has to do with patterns of glycosylation. Enzymes in the ER lumen and in the lumen of the Golgi cisternae add various oligosaccharides to various spots on the surface of proteins in the ER (including membrane proteins with some of their surfaces exposed to the lumen). The patterns of glycosylation are different for different categories of protein, and in at least some cases, it is their different patterns of glycosylation that allows the Golgi to recognize the different proteins. For instance, lysosomal proteins characteristically are marked with oligosaccharides containing mannose phosphate. Within the Golgi, receptor proteins in the membrane bind to the mannose phosphate, and the related receptor proteins are brought together as vesicles are formed at the trans face of the Golgi (Figure 5.13). Thus, various different lysosomal proteins, all with the same pattern of glycosylation, are brought together in the same vesicles at the trans-Golgi surface.

The Golgi sorts the mixed contents of ER membranes and lumen into different vesicles

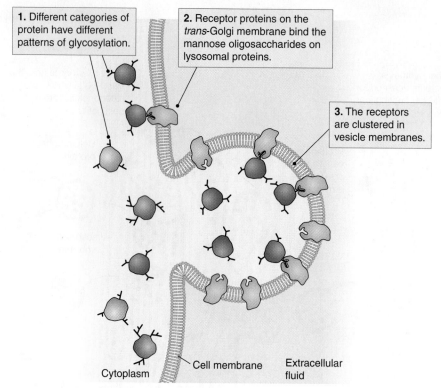

1. Different categories of protein have different patterns of glycosylation.

2. Receptor proteins on the *trans*-Golgi membrane bind the mannose oligosaccharides on lysosomal proteins.

3. The receptors are clustered in vesicle membranes.

Cytoplasm

Cell membrane

Extracellular fluid

FIGURE 5.13 Oligosaccharides and protein targeting.

5.13 Secretory vesicles fuse with the cell membrane

Among the vesicles produced by the Golgi are many that are targeted to the cell membrane. Some of these carry new membrane proteins for the expanding membrane of the growing cell or to replace membrane lost by endocytosis. Others are **secretory vesicles,** which carry inside them materials to be exported from the cell, such as proteins or pieces of cell wall. These vesicles move from the Golgi to the cell membrane, where they fuse, inserting their membrane constituents into the cell membrane, and ejecting their contents to the outside, as shown in Figure 5.12. The general term for this process, in which vesicles fuse with the cell membrane to release their contents to the outside, is **exocytosis.**

5.14 Lysosomal vesicles fuse with the endosome, which then targets proteins to the lysosome

Digestive enzymes targeted to the lysosomes or vacuole of the cell are transported by vesicles to second sorting organelle, the **endosome.** The endosome is mainly involved in sorting proteins taken in from the cell membrane during the process of endocytosis, but it also routes lysosomal enzymes to the lysosome.

One of the major functions of the endosome is to separate materials from their receptors (Figure 5.14). The lysosomal enzymes are transported to the endosome bound to a membrane receptor that recognizes the lysosome-specific mannose-6-P bound to the proteins. Within the endosome, however, the enzymes dissociate from

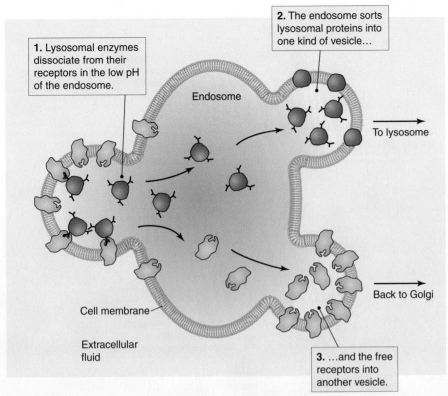

1. Lysosomal enzymes dissociate from their receptors in the low pH of the endosome.

2. The endosome sorts lysosomal proteins into one kind of vesicle…

Endosome

To lysosome

Back to Golgi

Cell membrane

Extracellular fluid

3. …and the free receptors into another vesicle.

FIGURE 5.14 Sorting of lysosomal enzymes and receptors in the endosome.

their receptors because the pH of the endosome is low (about 5.5, compared with about 6.5 for the trans Golgi).

The endosome, by mechanisms not yet understood, then sorts the membrane receptor and the lysosomal enzymes into different transport vesicles. The former are sent back to the trans Golgi and the latter to the lysosome.

5.15 Endocytosis is the first step in intracellular digestion and in recycling surface receptors

The process of endocytosis is the opposite of exocytosis: vesicles bud from the cell membrane into the cytoplasm, enclosing a bit of the external medium inside a vesicle. This process is fundamental to all eucaryotic cells and serves a number of functions. Most notably, it is the means by which many eucaryotic cells obtain their food, as we discuss shortly. Endocytosis is also the first step in recycling surface receptors and in recovering proteins that do not belong permanently in the cell membrane.

Surface receptors sense the chemical conditions of the extracellular environment by binding specifically to particular small molecules (called **ligands**). When the receptor binds its particular ligand, it undergoes a conformational change that activates a domain of the protein on the inside to become active (Figure 5.15). This activity constitutes the cell's signal that the environment contains the particular chemical.

Of course, it is necessary to turn the signal off as well. If the ligand binds so tightly that it does not readily dissociate, it can be stripped off by a more complex process that begins by endocytosing it.

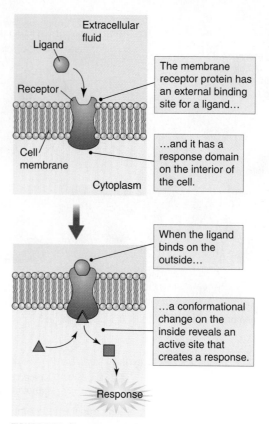

Extracellular fluid

Ligand

Receptor

The membrane receptor protein has an external binding site for a ligand…

Cell membrane

…and it has a response domain on the interior of the cell.

Cytoplasm

When the ligand binds on the outside…

…a conformational change on the inside reveals an active site that creates a response.

Response

FIGURE 5.15 Signal transduction by a surface receptor.

5.16 The endosome recycles membrane proteins

After endocytosis of vesicles containing surface receptors, the vesicles are transported to the endosome, where they fuse with the endosome membrane. The acidic pH normally causes the dissociation of the ligand from the receptor, and the two then part company. The ligand is normally transported to the lysosome for digestion, and the receptor is transported back to the cell surface (Figure 5.16).

5.17 Vesicles containing material to be digested fuse with lysosomes

Many protists ingest particulate materials, such as bacterial cells, which are then digested inside their vacuoles in the process termed **intracellular digestion.** Endocytic vesicles that contain particulate matter do not normally fuse with the endosome; rather, they fuse directly with a lysosome. In this new hybrid vesicle (called a **phagolysosome**), the macromolecular constituents of the food material are broken down by the battery of digestive enzymes from the lysosome. These include enzymes that break down all major types of macromolecule, including complex ones like murein and lipopolysaccharide. The monomeric products (sugars, amino acids, nucleotides, etc.) are then transported into the cytoplasm.

5.18 Exocytosis eliminates indigestible residue

Although there are many different lysosomal hydrolases, targeting most types of macromolecules, there is often an indigestible residue left in the phagolysosome

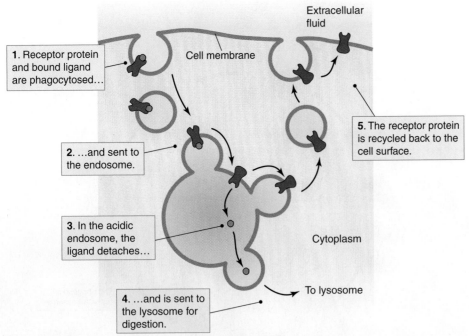

1. Receptor protein and bound ligand are phagocytosed…

Extracellular fluid

Cell membrane

5. The receptor protein is recycled back to the cell surface.

2. …and sent to the endosome.

3. In the acidic endosome, the ligand detaches…

Cytoplasm

To lysosome

4. …and is sent to the lysosome for digestion.

FIGURE 5.16 Recycling of surface receptors via the endosome.

of macromolecules that resist digestion. This residue is expelled to the outside by exocytosis.

Some protists, notably the ciliates, have a defined pathway that phagolysosomes take (Figure 5.17). In these organisms, there is a morphologically defined spot on the cell surface where phagocytosis takes place, often at the base of a groove or pit termed a **cytostome.** The phagolysosome then follows a course around the periphery of the cell to another defined spot where exocytosis occurs: the **cytoproct** (Figure 5.18).

5.19 Many protists are armed with extrusomes

Many predatory protists—those that ingest other living microbes—make complex vesicular structures that aid in capturing prey. Others make similar structures with defensive function. These structures are called **extrusomes.** They are sacs that lie just beneath the cell surface, which contain a coiled or compressed filament ending in a sharp dart-like structure or other mechanism for attaching to prey. When activated, the extrusome membrane fuses with the cell membrane, and the dart and filament are explosively discharged to the exterior, impaling the prey. Often the darts inject a toxin that immobilizes the prey as well. The synthesis of extrusomes is poorly understood but probably involves the ER and the Golgi, as do secretory vesicles, which are also transported to the cell periphery and which also fuse with the cell membrane.

One category of extrusome, **trichocysts,** contains a proteinaceous rod that is thought to lengthen many times when it is hydrated. Thus, when the containing sac ruptures and allows water in, the rod suddenly lengthens, driving the tip forcefully into prey or predator. Other extrusomes contain a long, hollow, coiled fiber that is fired by everting it with osmotic pressure. After firing, extrusomes are apparently discarded and replaced with new ones (Figure 5.19).

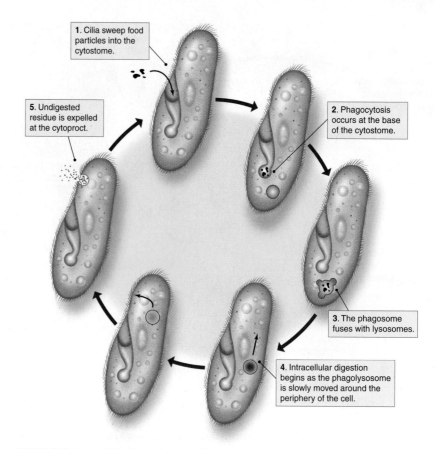

1. Cilia sweep food particles into the cytostome.

2. Phagocytosis occurs at the base of the cytostome.

3. The phagosome fuses with lysosomes.

4. Intracellular digestion begins as the phagolysosome is slowly moved around the periphery of the cell.

5. Undigested residue is expelled at the cytoproct.

FIGURE 5.17 Intracellular digestion in a ciliate.

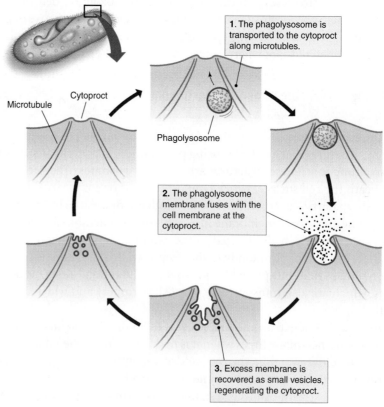

1. The phagolysosome is transported to the cytoproct along microtubles.

Microtubule   Cytoproct

Phagolysosome

2. The phagolysosome membrane fuses with the cell membrane at the cytoproct.

3. Excess membrane is recovered as small vesicles, regenerating the cytoproct.

FIGURE 5.18 The cytoproct.

CHAPTER 5 **CELL STRUCTURE AND FUNCTION IN PROTISTS**

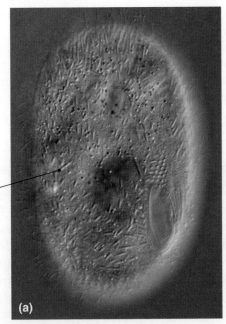

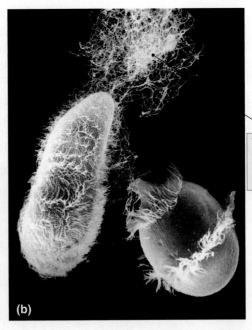

Trichocysts lie just beneath the cell membrane.

The ciliate has fired dozens of trichocysts in an attempt to avoid predation.

FIGURE 5.19 (a) Unfired and (b) fired trichocysts.

Mitochondria and Chloroplasts

Most eucaryotes—although not all—have mitochondria, which allow them to respire (without mitochondria they must ferment to generate ATP). Many have in addition to their mitochondria, chloroplasts as well, which allows them to be photosynthetic during the daytime. Both organelles are surrounded by two concentric membranes; typically, the mitochondrial inner membrane is extensively invaginated by folds or tubules, whereas the chloroplast inner membrane is smooth. The chloroplast, however, has an additional set of membranes internal to the inner and outer ones; these are the **thylakoid membranes** and are the location of the photosynthetic light harvesting machinery.

Many anaerobic protists have **hydrogenosomes** instead of mitochondria. Like mitochondria, these oxidize pyruvate produced by fermenting sugars. Unlike mitochondria, the oxidation is only partial, and one of the products is hydrogen gas.

5.20 Mitochondria and chloroplasts are not part of the endomembrane system

Both mitochondria and chloroplasts are complex organelles with multiple membrane systems; however, they are not part of the endomembrane system because they are not linked by vesicle fusion and budding to any other membrane systems of the cell.

5.21 Mitochondria and chloroplasts have their own chromosomes and their own protein-synthesizing system

Mitochondria and chloroplasts both have their own chromosomes and their own ribosomes, tRNAs, RNA polymerase, amino acid–activating enzymes, and all of the rest of the machinery for making protein. Most of the components and enzymes needed for the translation, transcription, and replication of these chromosomes are encoded in the organelle chromosome. Many of the inner membrane proteins are also encoded there as well.

Chromosome organization and ribosome structure in these organelles are often quite different from that of the nuclear chromosomes and cytoplasmic ribosomes. Frequently, the organelle structures are similar to those of procaryotes; for instance, many have circular chromosomes, like the majority of procaryotes, and their ribosomes are typically procaryotic in size (70S instead of the more typical 80S of many eucaryotes).

5.22 Many mitochondrial and chloroplast proteins, and some lipids, are imported from the cytoplasm

Although mitochondria and chloroplasts have their own chromosome and protein synthetic system, they do not encode all their own proteins. Indeed, in most cases a majority of the organelle-specific proteins are made on cytoplasmic ribosomes (free, not bound to the ER) and are then transported into mitochondria and chloroplasts as described in Section 5.4.

Both mitochondria and chloroplasts typically synthesize some of their own lipids; however, much of their membrane lipid may come from synthesis at the ER. This is possible because the cell has **phospholipid transfer proteins** that extract specific phospholipids from the ER and transfer them to the mitochondria and chloroplasts.

5.23 Mitochondria and chloroplasts exchange small molecules with the cytoplasm via permeases in their inner membrane

Mitochondrial and chloroplast functioning requires that they exchange small molecules with the cytoplasm—for instance, metabolites such as pyruvate, ATP, ADP, PO_4^{2-}, glyceraldehyde-P (the principal product of photosynthesis), or the amino acids and nucleotides needed for transcription, translation, and DNA replication. Both organelles are surrounded by two concentric membranes, and thus, these small polar molecules have to be able to cross both membranes; the mechanisms are different for the inner and outer membranes.

The outer membrane has **porins** in it, very like the outer membrane of the Gram-negative bacterial wall (no surprise, as the outer membrane of mitochondria and chloroplasts is probably evolutionarily derived from the outer membrane of their procaryotic ancestors). Thus, the intermembrane space of mitochondria and chloroplasts is similar in its small-molecule chemical composition to the cytoplasm, as there is free diffusion through the porins. The macromolecular composition is very different, however, as cytoplasmic proteins cannot pass through the porins.

The inner membrane does not have porins in it, and solutes have to cross by diffusion or via permeases. Some small molecules, such as water, CO_2, and O_2, cross by diffusion. Other molecules can only cross if there are specific permeases for them.

5.24 Mitochondria and chloroplasts have an evolutionary origin different than that of the nucleus

These two organelles are thought to have an evolutionary origin different from that of the rest of the cell around them; they are thought to be the degenerate relics of bacteria that formerly lived **endosymbiotically** within the cytoplasm of an ancient eucaryote (Figure 5.20). Mitochondria are thought to have been transferred early,

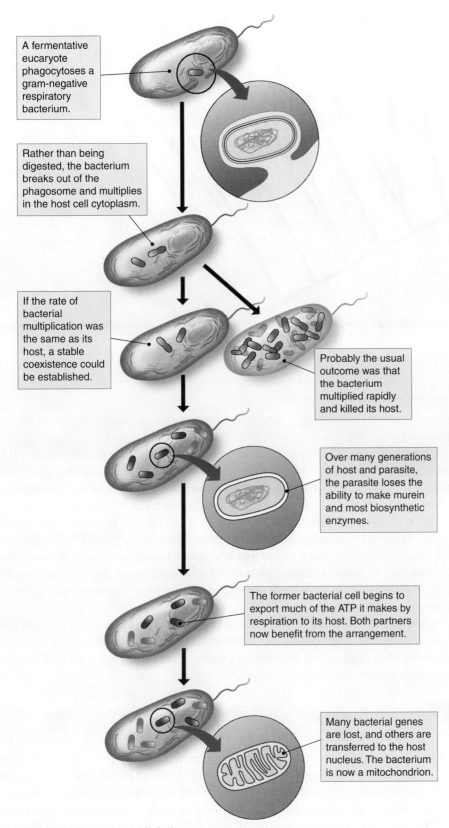

A fermentative eucaryote phagocytoses a gram-negative respiratory bacterium.

Rather than being digested, the bacterium breaks out of the phagosome and multiplies in the host cell cytoplasm.

If the rate of bacterial multiplication was the same as its host, a stable coexistence could be established.

Probably the usual outcome was that the bacterium multiplied rapidly and killed its host.

Over many generations of host and parasite, the parasite loses the ability to make murein and most biosynthetic enzymes.

The former bacterial cell begins to export much of the ATP it makes by respiration to its host. Both partners now benefit from the arrangement.

Many bacterial genes are lost, and others are transferred to the host nucleus. The bacterium is now a mitochondrion.

FIGURE 5.20 The endosymbiotic theory.

Mitochondria and chloroplasts have an evolutionary origin different than that of the nucleus

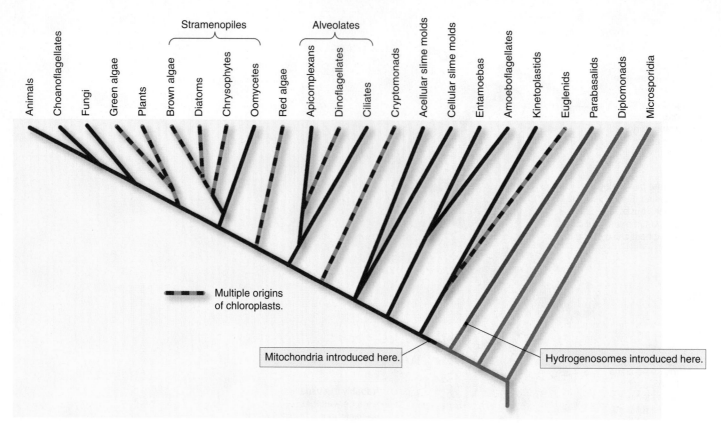

FIGURE 5.21 Eucaryotic acquisition of mitochondria, hydrogenosomes, and chloroplasts.

before many lineages had split from the main trunk of eucaryotic evolution. It is thought that an early respiratory bacterium (whose closest relatives among modern bacteria include *E. coli*) came to inhabit a fermentative eucaryotic cell. It obtained shelter from predators, a constant chemical environment, and continuous provision of food. The host obtained an advantage in making ATP (we look at the details in Section 7.24). Eventually, the endosymbiotic bacterium lost the DNA sequences that coded for unnecessary structures like a cell wall, much of its biosynthetic enzymes, etc., to become integral parts of the cell. Many genes also were transferred to the nucleus, after the ability was developed to transfer proteins made in the cytoplasm into mitochondria.

Chloroplasts are thought to have evolved similarly, except that it is thought that there have been several separate endosymbiotic events, with different cyanobacteria inhabiting the cytoplasm of different respiratory host cells (Figure 5.21). These led to the different groups of photosynthetic eucaryotes.

5.25 Hydrogenosomes are another relict of symbiosis

Many anaerobic protists contain **hydrogenosomes.** These are about the same size as mitochondria, but they lack the convolutions of the inner membrane. They appear to be, like mitochondria, the relicts of ancient endosymbioses between a fermentative protist and a bacterium. Unlike mitochondria, hydrogenosomes are descended from fermentative ancestors, and they thus lack the TCA (Krebs) cycle and electron transport chains. They appear to oxidize pyruvate by an enzyme system that is found in many strictly anaerobic bacteria. Instead of producing NADH as pyruvate is oxi-

CHAPTER 5 CELL STRUCTURE AND FUNCTION IN PROTISTS

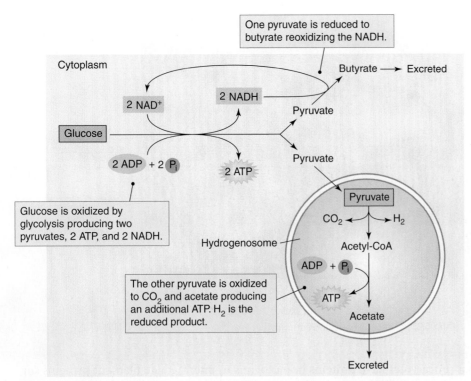

One pyruvate is reduced to butyrate reoxidizing the NADH.

Glucose is oxidized by glycolysis producing two pyruvates, 2 ATP, and 2 NADH.

The other pyruvate is oxidized to CO_2 and acetate producing an additional ATP. H_2 is the reduced product.

FIGURE 5.22 The hydrogenosome.

dized to acetyl-CoA (as is done by aerobes), this system produces hydrogen gas (Figure 5.22). Many of these anaerobic protists produce butyrate and acetate as their fermentation end products. Producing butyrate from pyruvate oxidizes two NADH, thus balancing the NADH produced in glycolysis. Thus, half of the pyruvate can go to acetate, CO_2, and H_2, producing an extra ATP.

5.26 Some chloroplasts appear to be more recent endosymbioses

Many unicellular algae contain chloroplasts with the typical structure of inner, outer, and thylakoid membranes. Many others, however, have more complex chloroplasts, with a third bounding membrane, and they may also contain a functional nucleus (a nucleomorph) between the two outermost membranes (Figure 5.23). These structures appear to be the relics of a relatively recent endosymbiosis between the host chemotroph and a unicellular eucaryotic alga. In many protistan groups, there is a mixture of phototrophic and chemotrophic representatives, with the phototrophs containing such reduced eucaryotic endosymbionts, suggesting that such endosymbioses are frequent.

The Cytoskeletal System

The eucaryotic cytoskeletal system is composed of two principal kinds of protein-aceous structures: microtubules and microfilaments. In addition, animals have an extensive system of intermediate filaments, but there is little evidence that protists have these. It has long been thought that procaryotes completely lacked cytoskeletal elements; however, as we saw in Chapter 4, it is now known that they have proteins that are functionally equivalent and evolutionarily homologous to the proteins of microtubules, microfilaments, and even intermediate filaments. Indeed, the finding

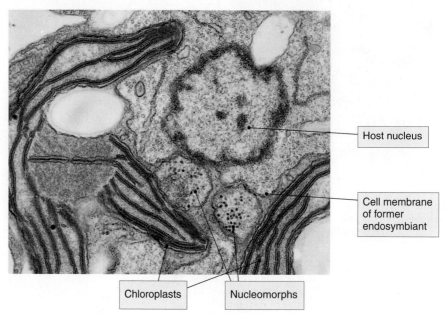

Host nucleus

Cell membrane of former endosymbiant

Chloroplasts

Nucleomorphs

FIGURE 5.23 Chloroplast and nucleomorphs in a protist.

that procaryotes have proteins homologous to intermediate filament proteins suggests that we will find them in protists too, although they may have different functions than in animal cells.

Microtubules and microfilaments are involved in intracellular movements, both in conjunction with **motor proteins** that bind to the microtubule or microfilament and then move along it. The details of cytoskeletal organization can vary greatly among organisms; however, the presence of a cytoskeletal system composed of these two elements is universal among eucaryotes.

5.27 Microtubules are hollow tubes composed of thousands of molecules of tubulin

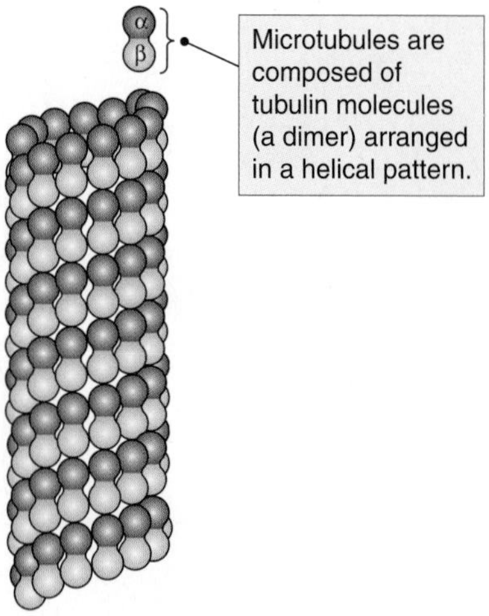

Microtubules, as the name implies, are small hollow tubes, about 24 nm in diameter. They are composed of thousands of copies of the protein **tubulin** (Figure 5.24). Tubulin is itself a dimer, consisting of one α-tubulin and one β-tubulin. These associate into a cylinder that is 13 tubulin molecules around.

Because the asymmetric tubulin dimer is oriented along the length of the microtubule, the microtubule as a whole is asymmetric, and its two ends are not identical. One end (called the [+] end) is the preferential site of addition or removal of tubulin subunits during growth or shrinkage of microtubules. The other (−) end can also gain or lose tubulin subunits, but usually does so more slowly than the (+) end. In the living cell, microtubules are quite dynamic. There is a continual process of elongation and shrinkage. Individual microtubules may shrink to nothing, and new ones form and grow.

5.28 Centrosomes organize the cell's microtubule network

Most of the microtubules in the interphase eucaryotic cell radiate from a central point, termed a **microtubule organizing center,** or **centrosome** (Figure 5.25). The centrosome often appears to have little or no organized structure, consisting of an

Microtubules are composed of tubulin molecules (a dimer) arranged in a helical pattern.

α
β

FIGURE 5.24 Microtubule structure.

CHAPTER 5 CELL STRUCTURE AND FUNCTION IN PROTISTS

amorphous mass of dense material. Alternatively, a pair of centrioles at right angles to each other may be found in the middle of the centrosome. Centrioles are short cylinders composed of nine triplets of fused microtubules. Centrioles are found in many protists, but are lacking in many others.

Centrosomes organize the network of microtubules in the cytoplasm of the interphase cell, such that most of the microtubules radiate from it. These microtubules have their (−) end at the centrosome and their (+) end toward the periphery.

5.29 Microtubules serve as tracks for endomembrane vesicles to slide on

One major function of microtubules is to provide a set of tracks throughout the cell along which vesicles of the endomembrane system can be transported. This motility is mediated by a set of motor proteins that have at one end binding sites for microtubules, structured such that not only can the protein bind to microtubules, but in the presence of ATP, they can move along it. The other end of the motor protein has a binding site for the membrane vesicle. Thus, these motor proteins attach vesicles to microtubules and then move along the microtubule to their destination in an energy-requiring process (Figure 5.26). There are two principal types of microtubular motor proteins: **kinesins** move toward the (+) end of microtubules, whereas **dyneins** move toward the (−) end.

The cytoskeletal system and the endomembrane system are thus functionally connected. The eucaryotic cell is simply too big for diffusion to be an effective mechanism for moving large objects like membrane vesicles around, and thus, it depends on the cytoskeletal system to transport vesicles.

5.30 Microfilaments are chains of actin monomers

Microfilaments are long, thin, thread-like structures that are around 9 nm thick (Figure 5.27). They are composed of thousands of monomers of the protein **actin**, the most abundant protein in the eucaryotic cell. Each microfilament has a helical

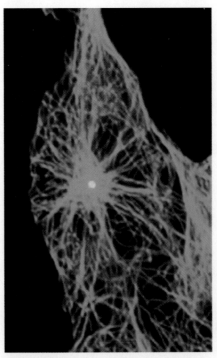

FIGURE 5.25 Micrograph of cell stained to show microtubules and centrosome.

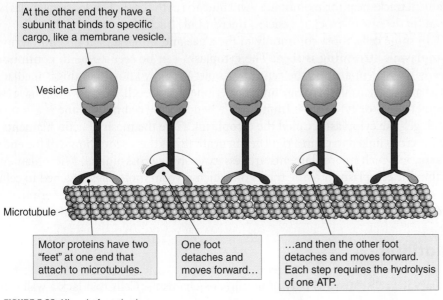

At the other end they have a subunit that binds to specific cargo, like a membrane vesicle.

Vesicle

Microtubule

Motor proteins have two "feet" at one end that attach to microtubules.

One foot detaches and moves forward…

…and then the other foot detaches and moves forward. Each step requires the hydrolysis of one ATP.

FIGURE 5.26 Kinesin functioning.

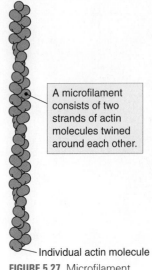

A microfilament consists of two strands of actin molecules twined around each other.

Individual actin molecule

FIGURE 5.27 Microfilament structure.

substructure, and it has a polarity like a microtubule (with a [+] end and a [−] end). Unlike microtubules, however, there is no correlation between the (+) or (−) end of microfilaments and the cell center or periphery; microfilaments form a dense mesh in the cytoplasm, with some having their (+) end toward the periphery, some their (−) ends, and some running at an angle to the cell surface.

5.31 Microfilaments maintain cell shape and stabilize the membrane

Microfilaments form a dense mesh within the cytoplasm. Where individual microfilaments cross in the mesh, they are joined to each other by cross-linking proteins. The mesh is also linked to the cell membrane by binding of the filament to any of a variety of membrane proteins. This meshwork of microfilaments maintains cell shape and provides stability to the cell membrane. It can also anchor membrane proteins in particular places, preventing them from moving in the fluid membrane.

Although the cytoplasmic mesh of microfilaments appears to be very stable, in fact it can be rapidly rearranged, particularly in amoeboid cells that continuously change their shape.

5.32 Microfilaments also provide tracks for membrane vesicles to slide along

Like microtubules, microfilaments can be tracks for vesicles of the endomembrane system to slide along. This sliding is mediated by microfilament motor proteins, most notably a class of proteins termed **myosins.** There are a number of different forms of myosin in every cell, differing mainly in the domain that attaches to cargo, such as a vesicle. The other end, which attaches to the microfilament, is quite similar in all myosins.

The role of microfilaments in vesicle transport is most important at the cell periphery because the density of microtubules is low there. Because microtubules are radially arranged, with their origin at the centrosome near the center of the cell, microtubule density is greatest at the center of the cell and least at the periphery. Thus, directed movements of vesicles at the cell periphery would be difficult using microtubular tracks alone. Endocytic and exocytic vesicles thus probably move along microfilament tracks near the membrane, switching to (in the case of endocytic vesicles) or from (in the case of exocytic vesicles) microtubule tracks deeper inside the cell.

In some cells, most commonly in the algae and fungi, a phenomenon termed **cytoplasmic streaming** is seen. The cytoplasm can be seen to be in continuous steady movement around the central vacuole. We now know that this is mediated by a band of microfilaments around the periphery of the cell. Many protists have two distinct zones of cytoplasm. Immediately beneath the cell membrane is a zone of rigid, gel-like cytoplasm, called the **ectoplasm.** Here the mesh of actin filaments is highly cross-linked to form a rigid gel. Beneath this is the viscous liquid of the **endoplasm,** in which microfilaments are less extensively cross-linked. Microfilaments within the ectoplasm provide tracks on which myosin molecules attached to cellular organelles in the endoplasm move in a circular motion around the periphery of the cell. These organelles in turn drag the rest of the endoplasm with them.

Motility

Eucaryotic cells have a variety of motility mechanisms. Cells that lack a wall often move across a solid surface by **amoeboid movement,** in which changes in cell shape mediate movement. Alternatively, cells may have one or more *flagella* or *cilia*—organelles that are composed of bundles of microtubules that slide against each other.

5.33 Amoeboid movement is mediated by microfilaments and myosin

The details of the motility mechanism are not yet clear. We do know that this type of motility is based on microfilaments and myosin, and it is thought that two separate mechanisms are involved (Figure 5.28). The first is the polymerization of actin to form microfilaments. This is thought to occur vigorously at the leading edge of the cell, leading to a protrusion being extended. This protrusion is called a **pseudopod** (Latin for "false foot") and can be a thin spike or a broad, flat protrusion.

The second type of mechanism also involves microfilaments. Pseudopod extension is driven by hydraulic forces that are created by the contraction of actin and myosin filaments in the ectoplasm at the rear of the cell, creating pressure at the anterior end of the cell. Simultaneously, a localized thinning of the ectoplasm at the front allows a pseudopod to be pushed forward.

5.34 Flagella and cilia contain a bundle of microtubules that slide against each other

Flagella and cilia are, like procaryotic flagella, appendages protruding from the cell surface that propel the cell through the medium with a swimming motion. They are completely different in structure and in the mechanism of movement, however, from the organelles of the same name in procaryotes. Procaryotic and eucaryotic flagella are not homologous, despite their having the same name.

Flagella and cilia have identical structure; they differ in the way they move and in their length (flagella are generally longer). Their most striking feature is a highly structured bundle of microtubules that runs down the center of each (Figures 5.29 and 5.30). Most commonly, this bundle, the **axoneme,** consists of

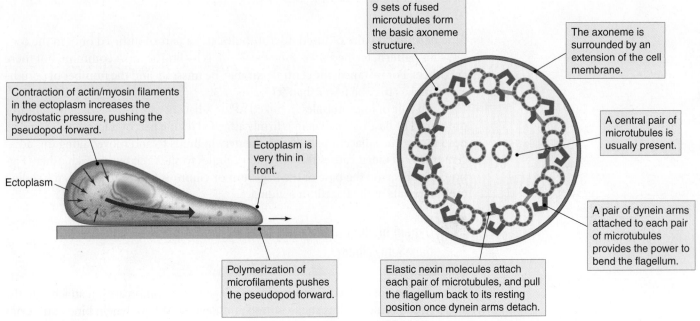

Contraction of actin/myosin filaments in the ectoplasm increases the hydrostatic pressure, pushing the pseudopod forward.

Ectoplasm is very thin in front.

Ectoplasm

Polymerization of microfilaments pushes the pseudopod forward.

9 sets of fused microtubules form the basic axoneme structure.

The axoneme is surrounded by an extension of the cell membrane.

A central pair of microtubules is usually present.

A pair of dynein arms attached to each pair of microtubules provides the power to bend the flagellum.

Elastic nexin molecules attach each pair of microtubules, and pull the flagellum back to its resting position once dynein arms detach.

FIGURE 5.28 Mechanism of amoeboid movement.

FIGURE 5.29 Diagram of a flagellar axoneme.

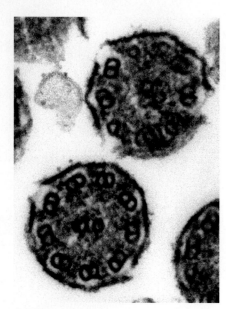

FIGURE 5.30 Cross-section showing the structure of cilia and flagella.

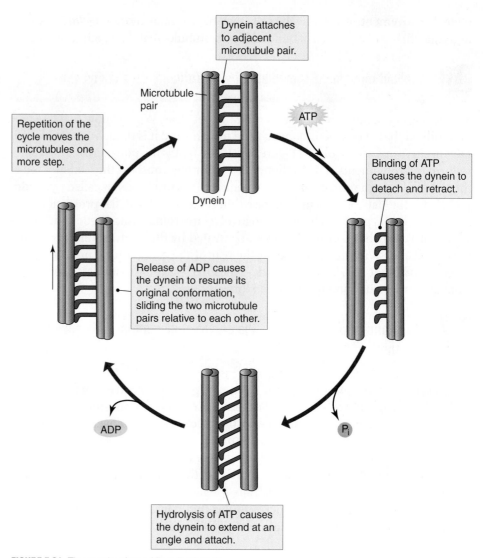

Dynein attaches to adjacent microtubule pair.

Microtubule pair

Dynein

ATP

Binding of ATP causes the dynein to detach and retract.

Repetition of the cycle moves the microtubules one more step.

Release of ADP causes the dynein to resume its original conformation, sliding the two microtubule pairs relative to each other.

ADP

P_i

Hydrolysis of ATP causes the dynein to extend at an angle and attach.

FIGURE 5.31 The mechanism of flagellar bending.

nine peripheral pairs of fused microtubules and a pair of unfused ones in the center. This pattern, termed a *9 + 2 arrangement,* is by far the most common, but there are others. For instance, the central pair may be missing, and the number of peripheral doublets may be fewer than 9 (e.g., 6 or 3).

Each pair of microtubules in linked to the adjacent one by a set of dynein motor proteins. The base of the dynein is firmly attached to one pair of microtubules, and its head is near the adjacent pair. When the dynein binds to and moves along this adjacent microtubule, it causes the two microtubules to slide relative to each other (Figure 5.31). Because the base of the flagellum or cilium is firmly anchored, the sliding motion results in the flagellum bending.

5.35 Dynein binding to flagellar microtubules controls the state of the flagellum/cilium

When the dynein molecules in a region of a flagellum or cilium are not attached to the adjacent molecule, the organelle is limp and flexible. When dynein binds but is not actively moving, the flagellum/cilium is stiff and rigid, and when the dyneins on one side move, while the dyneins on the other side are unattached, the organelle bends. Coordinated, but distinct, regulation of the dynein cycle on each set of microtubules is obviously critical to the ability of the organelle to function effectively, but it is still not well understood. Ca^{2+} concentration, the Ca^{2+} binding protein **calmodulin,** and cAMP all appear to be involved.

Adjacent pairs of microtubules are permanently linked by flexible proteins called **nexin,** and thus, there is a limit to how far they can slide against each other before these linkers are stretched to the maximum. Nexin in the resting state is about 30 nm long; in its fully stretched state, it is about 140 nm. When the dynein stops moving and detaches, the stretched nexin molecules relax and the flagellum/cilium returns to its initial state.

5.36 Flagella and cilia originate in a centriole-like basal body

Not only is the structure of both flagella and cilia identical, but they also have identical anchoring structures, termed **basal bodies.** These are very similar or identical to centrioles and are found at the base of all flagella or cilia even in cells whose centrosome does not include a centriole. Remember that centrioles consist of small cylinders composed of nine sets of three fused microtubules. When this structure acts as a basal body, two of each set of three fused microtubules continue past the basal body to become the nine fused pairs that make up the core of the flagellum or cilium.

In ciliates, which have dozens to many hundreds of cilia, the basal bodies are all interconnected by bundles of microtubules and microfilaments (Figure 5.32). These

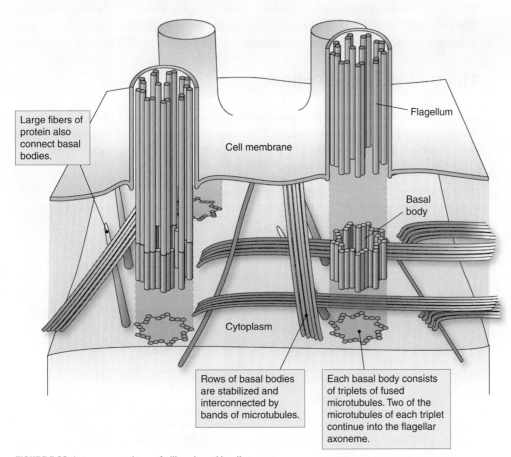

Large fibers of protein also connect basal bodies.

Cell membrane

Flagellum

Basal body

Cytoplasm

Rows of basal bodies are stabilized and interconnected by bands of microtubules.

Each basal body consists of triplets of fused microtubules. Two of the microtubules of each triplet continue into the flagellar axoneme.

FIGURE 5.32 Interconnections of ciliary basal bodies.

are presumed to help coordinate the beat pattern of adjacent cilia to prevent them from interfering with each other.

5.37 Flagella and cilia differ in their beat pattern and in their length

Although flagella and cilia have identical structure, they are usually distinguished by their length (flagella usually being quite a bit longer than cilia) and their beat pattern. Typically, flagella beat by propagating a bend down the length of the filament, beginning at the base (Figure 5.33). As the bend moves down the length of the filament, it presses against the medium and moves the cell forward. Because the bends alternate in direction, the result looks like a sine wave moving along the filament. Although flagella typically propagate waves away from the cell, thus generating a pushing force, some protists can propagate a wave toward the cell, generating a pulling force. Flagella are also capable of much more complex patterns of movement, but these make little contribution to motility.

Cilia, on the other hand, typically have a pattern of movement that has two distinct phases: a power stroke and a recovery stroke (Figure 5.34). During the power stroke, most of the filament is rigid, and bending occurs only near the base. This sweeps the cilium through the medium like an oar. During the recovery stroke, the cilium is mostly limp and is pulled back to its original position by a bend (opposite the bend of the power stroke) that propagates from the base to the tip.

It is not always easy to distinguish cilia from flagella, however. Of course, length is a continuum, and it is impossible to say exactly where a cilium ends and a flagellum begins. Similarly, many protists have flagella (cilia) that are capable of a multitude of different movements, including those typically thought of as flagellar, and those thought of as ciliar, and many other motions as well. This, plus the identical structure, suggests that the distinction between the two organelles is not always clear.

The Cell Wall and Pellicle

Many protists have a rigid polysaccharide cell wall that resists the cell's internal osmotic pressure (**turgor pressure**). There are a wide variety of different types of wall, but none has the chemical or structural complexity of the procaryotic wall.

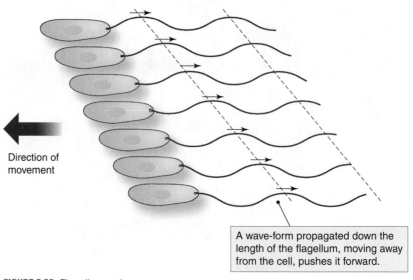

Direction of movement

A wave-form propagated down the length of the flagellum, moving away from the cell, pushes it forward.

FIGURE 5.33 Flagellar motion.

CHAPTER 5 CELL STRUCTURE AND FUNCTION IN PROTISTS

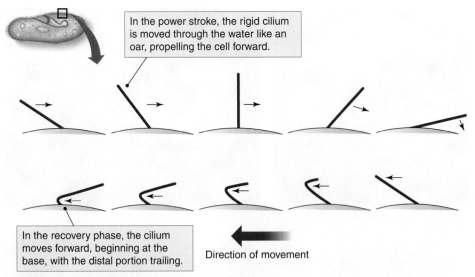

In the power stroke, the rigid cilium is moved through the water like an oar, propelling the cell forward.

In the recovery phase, the cilium moves forward, beginning at the base, with the distal portion trailing.

Direction of movement

FIGURE 5.34 Ciliar beat pattern.

Many other protists have a **pellicle** instead of a cell wall. This structure lies immediately beneath the membrane and is attached to it.

Some protists, notably some of the amoebas, make extracellular shells, or **tests,** that protect against predation rather that osmotic stress.

5.38 The eucaryotic cell wall is usually composed of polysaccharide

Many groups of protists have a rigid cell wall, including many of the photosynthetic groups (algae), the fungi, and several others. Usually these consist of a thick layer of simple polysaccharide, such as cellulose, pectin, or other polymers of glucose. Polymers of mannose are also occasionally seen, as are polymers of N-acetylglucosamine (termed **chitin**). Sometimes the wall has a layered appearance, and the layers may be different polysaccharides (Figure 5.35).

Occasionally, the polysaccharide wall material is organized into a series of discrete plates that interlock or overlap to form the wall. Best known for this type of wall are the dinoflagellates (Figure 3.36).

5.39 Some protists make walls that are heavily impregnated with inorganic salts

Some groups of protists have walls composed principally of inorganic salts, such as silicon oxides or calcium carbonate. Best known of these are the diatoms, an algal group that makes elaborately sculptured siliceous walls. Another group is the coccolithophorids, another algal group (Figure 5.37). These organisms make walls of overlapping plates of calcium carbonate.

5.40 The pellicle is a complex structure that includes the cell membrane and an underlying layer of protein or polysaccharide

Instead of a wall, many protists have a **pellicle,** a term that refers to a variety of different structures, all characterized by having the layer that strengthens the cell

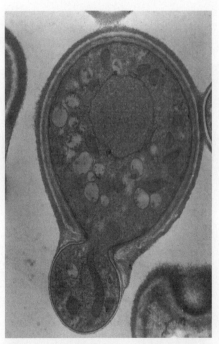

FIGURE 5.35 Thin section of a yeast cell.

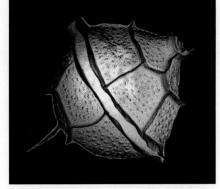

FIGURE 5.36 A dinoflagellate.

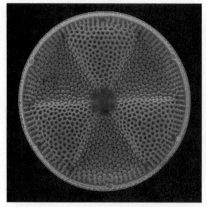
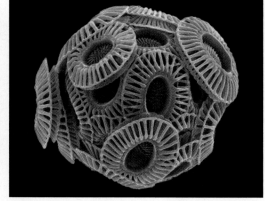

FIGURE 5.37 Diatom and coccolithophorid.

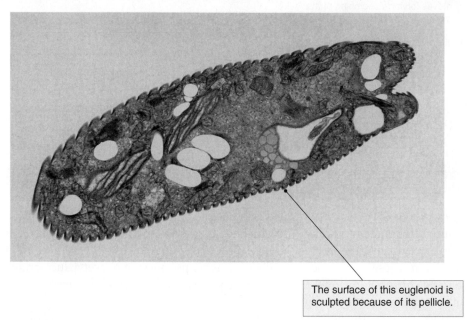

The surface of this euglenoid is sculpted because of its pellicle.

FIGURE 5.38 Protistan pellicle.

membrane in the cytoplasm, rather than on the exterior. Generally, the term pellicle includes the cell membrane, as well as the strengthening layer immediately below it. Pellicles in different groups of protists probably represent a number of independent evolutionary inventions.

Structurally, pellicles usually consist of a layer of plates immediately beneath the cell membrane, composed of protein or polysaccharide. This structure is sometimes nearly as rigid as a thick wall but usually is considerably more flexible, allowing a limited range of change of cell shape. It may also be folded in a precise pattern, giving the outside of the cell a sculpted appearance (Figure 5.38).

5.41 Some amoebas make shells for protection against predation

Most amoebas have no defined cell wall or a pellicle; they presumably protect themselves against osmotic lysis by effective osmoregulation, an active contractile vacuole (see next section), and cytoskeletal support for the cell membrane. Some of them, however, make an extracellular structure for protection against predation

(Figure 5.39). These shells, or **tests,** can be very elaborate. Tests are normally made of minute grains of sand or other minerals, stuck together with a secreted mucous that hardens like cement. Testate amoebas typically extend their pseudopods from the opening of the shell and move and phagocytose just like naked amoebas.

Similar in principle are the foraminifera (Figure 5.40), a group of protists that make a shell similar in construction to the testate amoebas, but whose pseudopods are typically very thin and fuse and branch to form a pseudopodal network outside the shell. These fine pseudopods may be extended from the main opening in the shell, or they may be extend through fine pores that often penetrate the shell at numerous spots.

The Contractile Vacuole

A highly specialized form of exocytosis is characteristic of protists that lack a rigid wall and inhabit fresh water or other hypotonic environments: water excretion via an elaborate system of membranes termed the **contractile vacuole complex.**

5.42 The contractile vacuole collects water from the cytoplasm and expels it to the outside by exocytosis

The broad outlines of contractile vacuole function are common to all protists that possess them. The vacuole enlarges as it accumulates water from the cytoplasm, becoming visible in the light microscope, and then disappears as it contracts and expels its contents to the outside (Figure 5.41). A complete cycle of filling and discharge may take anywhere from a few seconds to nearly an hour, depending on the species and the external osmolarity.

When the contractile vacuole is full, it fuses with the cell membrane and expels its contained water to the outside. It is still unclear if the expulsion of water is by

FIGURE 5.39 A testate amoeba. © Dr. Richard Kessel and Dr. Gene Shih. Visuals Unlimited.

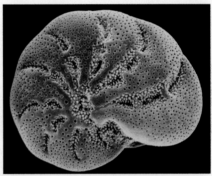

FIGURE 5.40 Foraminiferan.

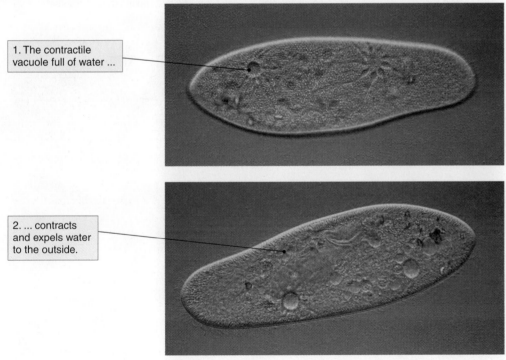

1. The contractile vacuole full of water ...

2. ... contracts and expels water to the outside.

FIGURE 5.41 The contractile vacuole.

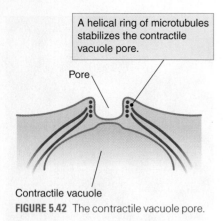

A helical ring of microtubules stabilizes the contractile vacuole pore.

Pore

Contractile vacuole

FIGURE 5.42 The contractile vacuole pore.

active contraction of the vacuole or if it is simply due to the turgor pressure of the cell collapsing the vacuole once its lumen is continuous with the outside.

Most ciliates and flagellates have a distinctive structure on the cell surface, termed the **contractile vacuole pore,** through which this exocytosis occurs. The contractile vacuole pore is a localized indentation of the cell membrane into a short cylinder that penetrates through the pellicle (Figure 5.42). It is stabilized by microtubules that wrap around it and that radiate from the pore into the cytoplasm. In other protists, the discharge site is not recognizable as a distinctive structure and may move on the cell surface.

5.43 The contractile vacuole collects water through a system of tubules or vesicles

The contractile vacuole membrane system has two principal components: the **contractile vacuole** itself, a membrane-bound vesicle that accumulates water for expulsion, and the **spongiome,** an adjacent system of membranous vesicles or tubules that conveys water to the contractile vacuole.

Two broad categories of contractile vacuole complex may be recognized: that characteristic of the amoebas, in which the contractile vacuole is formed by the coalescence of small vesicles, and fragments again into vesicles during discharge; and that characteristic of the ciliates and flagellates, in which a permanent contractile vacuole alternately inflates and collapses (Figure 5.43). In the amoebas, the spon-

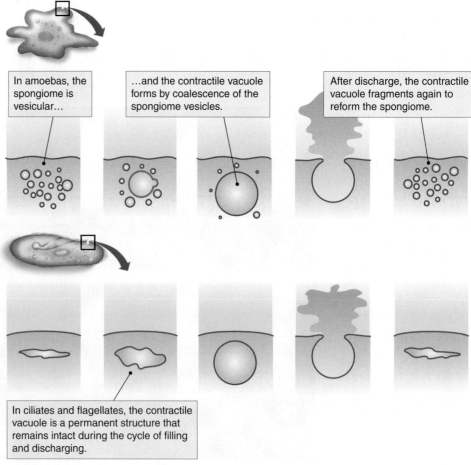

In amoebas, the spongiome is vesicular…

…and the contractile vacuole forms by coalescence of the spongiome vesicles.

After discharge, the contractile vacuole fragments again to reform the spongiome.

In ciliates and flagellates, the contractile vacuole is a permanent structure that remains intact during the cycle of filling and discharging.

FIGURE 5.43 The contractile vacuole cycle in amoebas and ciliates.

CHAPTER 5 CELL STRUCTURE AND FUNCTION IN PROTISTS

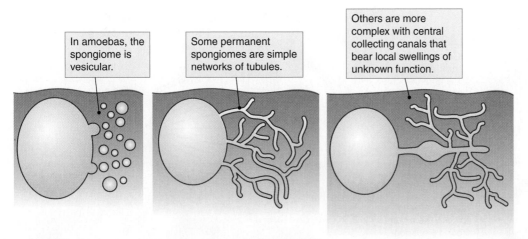

In amoebas, the spongiome is vesicular.

Some permanent spongiomes are simple networks of tubules.

Others are more complex with central collecting canals that bear local swellings of unknown function.

FIGURE 5.44 Spongiome types.

giome is typically vesicular, as it consists of vesicles that have yet to fuse with the enlarging contractile vacuole. In flagellates and ciliates, the spongiome consists of a system of tubules that empty into the contractile vacuole (Figure 5.44).

The mechanism of discharge of the contractile vacuole is unclear. Microfilaments may contract, squeezing the vesicle and expelling the contents, or the contents may simply be discharged by the osmotic pressure within the cell once the vesicle fuses with the cell membrane.

Reproduction in Protists

Like other eucaryotes, cellular reproduction in protists is by some form of mitosis and cytokinesis. There are many variants of these processes, however, unlike the relatively homogeneous processes of plants and animals.

5.44 Closed mitosis is common in protists

Mitosis in plant and animal cells is characterized by the fragmentation of the nuclear envelope into thousands of tiny membrane vesicles in prophase or prometaphase. This allows the microtubules of the developing mitotic spindle to penetrate the nucleoplasm and attach to the condensed chromosomes. This form of mitosis is termed **open mitosis.** In most protists, however, one or another form of **closed mitosis** occurs. In closed mitosis, the nuclear envelope remains intact, or nearly so, throughout the process.

There are at least three different types of closed mitosis (Figure 5.45). Sometimes the mitotic spindle forms entirely within the nucleus. There are microtubule organizing centers embedded in the nuclear envelope that are responsible for organizing the spindle. In telophase, the nuclear envelope is pinched in the middle, and the nucleus ultimately divides in two.

In other cases, there is a partial fragmentation of the nuclear envelope, such that it develops gaps through which the cytoplasmic spindle fibers can penetrate the nucleoplasm and attach to the chromosomes. In telophase, the gaps are repaired, and the nucleus divides.

Probably the most divergent form of closed mitosis is found in the dinoflagellates. In this group, the nuclear envelope remains intact, and the spindle forms in

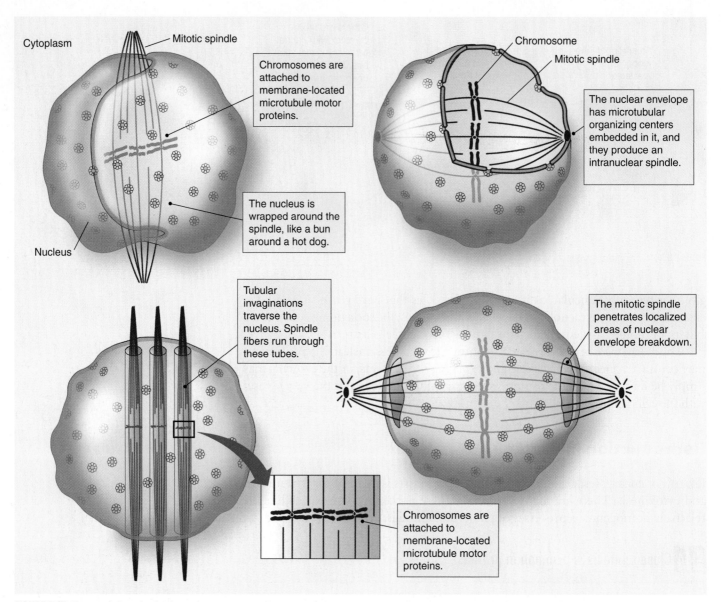

Cytoplasm — Mitotic spindle

Chromosomes are attached to membrane-located microtubule motor proteins.

The nucleus is wrapped around the spindle, like a bun around a hot dog.

Nucleus

Chromosome

Mitotic spindle

The nuclear envelope has microtubular organizing centers embedded in it, and they produce an intranuclear spindle.

Tubular invaginations traverse the nucleus. Spindle fibers run through these tubes.

The mitotic spindle penetrates localized areas of nuclear envelope breakdown.

Chromosomes are attached to membrane-located microtubule motor proteins.

FIGURE 5.45 Types of closed mitosis.

the cytoplasm. The nucleus wraps around the spindle, or it develops membrane-bounded channels through the nucleus. Chromosomes attach to nuclear envelope proteins, which in turn attach to the spindle.

In addition to closed mitosis, other variations on the norm in plants and animals occur. For instance, in several groups, such as the euglenids and dinoflagellates, chromosomes remain mostly condensed throughout interphase.

5.45 Sex is occasional in protists; reproduction is normally asexual

Sexual reproduction in protists does occur in almost all groups that have been well studied, but all of the evidence suggests that it is quite rare in nature. Reproduction in protists is almost universally asexual by mitosis. Nevertheless, sexual reproduction probably plays a large role in maintaining genetic diversity in the population.

Summary

The protists combine the large populations and rapid growth capacity of unicellular organisms with the complexity of the eucaryotic cell. They probably originated as predators of the smaller, simpler procaryotes, but now they have diversified to an extraordinary extent, including multicellular forms, parasitic forms, and various types of metabolism, habitats, and life cycles. The complexity of some unicellular forms rivals the complexity of small multicellular organisms. This is truly a fascinating and very successful group of organisms. We return to them in Chapter 15.

Study questions

1. Summarize the mechanisms by which proteins are targeted to their proper locations in the eucaryotic cell.
2. Compare and contrast the bacterial nucleoid and the eucaryotic nucleus.
3. Compare and contrast mitochondria, chloroplasts, and hydrogenosomes.
4. Compare and contrast microtubules and microfilaments.
5. Describe the importance and role of endocytosis in the protistan cell.
6. How do protists living in a hypotonic environment cope with osmotic pressure?
7. The contractile vacuole presumably requires energy in some form for its functioning. Which steps might be energy requiring, and why?
8. Compare and contrast eucaryotic flagella and cilia.
9. Compare and contrast procaryotic and eucaryotic flagella.
10. Describe the relationship between the mitotic spindle and the nuclear envelope in closed and open mitosis.

6 Viruses and Other Acellular Entities

It is a common truism that all life consists of cells, or at the simplest, of a single cell. This insight was a product of the systematic application of the microscope to studying the structure of macro-organisms in the 19th century. What then are we to make of viruses? These share certain properties that are otherwise characteristic of living systems: ability to replicate, ability to mutate, ability to recombine, a genealogical history, and so on. At least during part of the time, however, they consist of a structure much smaller than a cell (Figure 6.1): a bit of nucleic acid in a protein coat (sometimes with a membrane around it).

Viruses have a life cycle in which the acellular form alternates with a cellular stage—the infected cell (Figure 6.2). It is this stage—the infected cell—that shows the characteristic features of life. The acellular stage is biochemically and genetically inert. Questions about whether viruses are alive are puzzling if only the acellular stage is considered; if the infected cell is taken into account, viruses are unquestionably alive. Because they have an absolute requirement for a host cell to support their multiplication, however, they are obligate parasites.

6.1 Viruses have an acellular stage in their life cycle

Viruses alternate between two states in their life cycle: the actively replicating state, in which viral replication takes place inside an infected host cell, and the **virion,** an acellular, biologically inert structure. The virion may have an extremely simple structure, or it may be quite complex. It may be very tiny—only slightly larger than the largest protein molecules—or as large as a small cell—barely visible in the light microscope. At its simplest, the virion consists of a **capsid** of protein molecules arranged into a geometric shell and a bit of nucleic acid contained within. Most viruses whose hosts are animal cells have in addition a membrane surrounding the capsid, termed an envelope (Figure 6.3).

6.2 Virions contain one or more chromosomes of either DNA or RNA, double-stranded or single-stranded

The genome of a virus consists of one to several pieces of nucleic acid, short compared with the chromosomes of cells. Most viruses have several dozen genes, but there are some with as few as three and some with several hundred. Most of the time all of the genes are on a single nucleic acid molecule, but sometimes the genome is

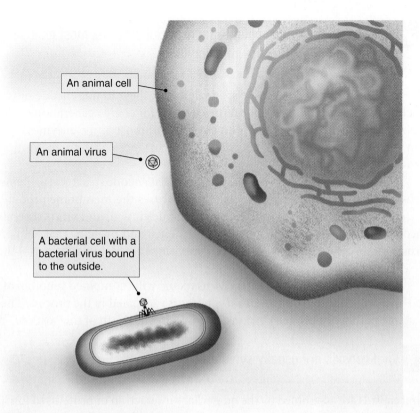

FIGURE 6.1 Relative size of cells and viruses.

An animal cell

An animal virus

A bacterial cell with a bacterial virus bound to the outside.

divided among several separate molecules to form a **segmented genome.** Viral chromosomes consist of either DNA or RNA, which may be either single or double stranded in the virion.

The basic structure of the genome of a given virus is constant—the number of chromosomes, DNA or RNA, and single or double stranded. Thus viruses collectively are the most genetically diverse entities on earth. Indeed, they provide the only known instances in which the hereditary information is encoded in something other than double-stranded DNA.

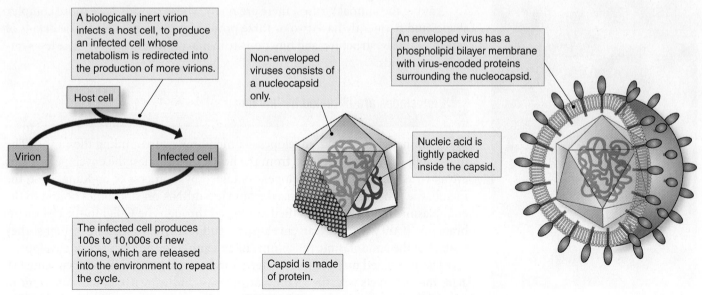

A biologically inert virion infects a host cell, to produce an infected cell whose metabolism is redirected into the production of more virions.

Host cell

Virion

Infected cell

The infected cell produces 100s to 10,000s of new virions, which are released into the environment to repeat the cycle.

Non-enveloped viruses consists of a nucleocapsid only.

An enveloped virus has a phospholipid bilayer membrane with virus-encoded proteins surrounding the nucleocapsid.

Nucleic acid is tightly packed inside the capsid.

Capsid is made of protein.

FIGURE 6.2 The viral life cycle.

FIGURE 6.3 Structure of a simple virus and an enveloped virus.

Virions contain one or more chromosomes of either DNA or RNA

6.3 Viral genomes are incomplete—the host cell provides most genetic information needed by the infecting virus

The reason viral genomes can be so small is that they do not encode most of the machinery necessary for multiplication: enzymes for energy generation; ribosomes, tRNAs, and other machinery of protein synthesis; most of the enzymes of replication and transcription; and of course all of the enzymes necessary to synthesize the virus's structures.

The few genes virions do possess are those that encode structural proteins of the virus (the capsid proteins and in enveloped viruses the membrane proteins), any proteins necessary for the replication of viral nucleic acid and virion assembly not encoded in the host genome, regulatory proteins, and proteins that help progeny viruses escape from the host cell. This is why virions are biologically inert—they lack the enzymes necessary for metabolism.

Thus, the replicating form of the virus requires the combined genomes of virion and host cell, although only the viral genome is replicated in the process. The result is a large number of progeny virions, and usually the death of the host cell.

6.4 Viral capsids are usually symmetrical

Viral capsids are assembled by the quaternary interaction of proteins to form a hollow shell. There may be only one protein forming the capsid, or several. In most cases, the patterns of aggregation lead to a capsid that has one or more axes of symmetry. Two symmetrical forms are common: a **cylindrical,** or tubular capsid, and a nearly spherical **icosahedral** capsid, with 20 triangular faces (Figure 6.4).

A third form of capsid combines the two forms of symmetry. An icosahedral capsid containing the nucleic acid is joined to a cylindrical tail that functions in injecting the nucleic acid into the host. These cylindrical **tails** often have additional structures at their tip—base plates, tail fibers, and so on—that function to attach the virus to its host cell. These are called **binal** capsids and are found only in some groups of viruses whose host cells are procaryotic (termed **bacteriophages,** often shortened to **phages**) (Figure 6.5).

Among the animal viruses, there are many whose capsids have more complex shapes and frequently have two or three protein layers (Figure 6.6). The details of capsid assembly, structure, and function are in most cases unclear for these less symmetrical capsids.

6.5 Envelopes are obtained by budding

Many animal viruses have envelopes—a membrane surrounding the capsid. The phospholipid bilayer is derived from the host's membranes; the envelope proteins are all virus encoded. The envelope is obtained by a process of budding. First, the membrane proteins are synthesized from viral mRNA on ribosomes bound to the endoplasmic reticulum. They then are sorted through the Golgi to the cell membrane, or if they are nuclear envelope or endoplasmic reticulum proteins, they remain in the endoplasmic reticulum (ER) membrane and the nuclear envelope.

The assembled **nucleocapsid** (a term for the capsid containing the viral nucleic acid) then interacts with the membrane proteins, inducing a bud, the membrane of which becomes the envelope (Figure 6.7).

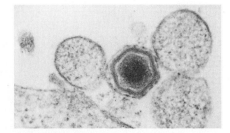
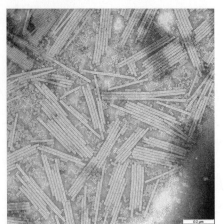

FIGURE 6.4 Icosahedral virus (top) and cylindrical viruses (bottom).

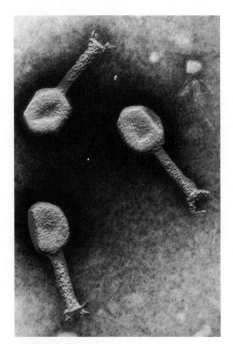

FIGURE 6.5 Binal viruses.

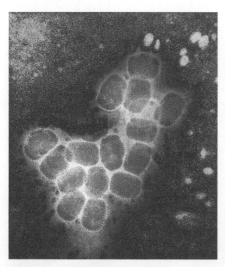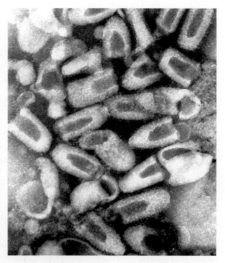

FIGURE 6.6 Irregular-shaped viruses.

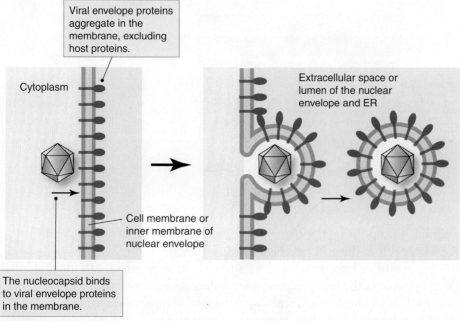

Viral envelope proteins aggregate in the membrane, excluding host proteins.

Cytoplasm

Extracellular space or lumen of the nuclear envelope and ER

Cell membrane or inner membrane of nuclear envelope

The nucleocapsid binds to viral envelope proteins in the membrane.

FIGURE 6.7 Viral budding.

This process not only provides the envelope, but it simultaneously helps the newly assembled virions to exit the cell. In the case of budding from the cell membrane, the process itself results in virion release. In the case of budding through the nuclear membrane or the endoplasmic reticulum, the virion winds up in the lumen of the ER, from whence it can access the outside through the normal secretory pathway. Presumably this happens because the viral membrane proteins have signal sequences that mimic those of normal host secretory proteins.

6.6 Virus multiplication is a five-stage process

Virus multiplication requires that (1) the virion encounter and attach to a host cell, (2) its nucleic acid reach the site of replication and expression (cytoplasm or nucleoplasm), (3) the nucleic acid be replicated and expressed, (4) the progeny virions be

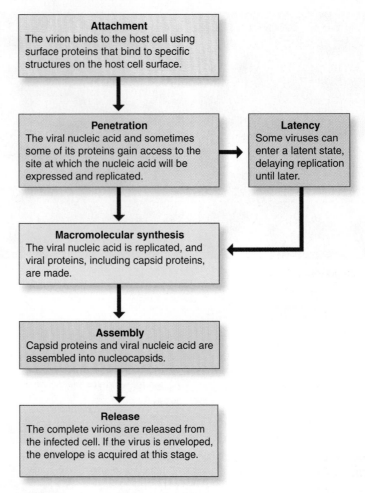

Attachment
The virion binds to the host cell using surface proteins that bind to specific structures on the host cell surface.

Penetration
The viral nucleic acid and sometimes some of its proteins gain access to the site at which the nucleic acid will be expressed and replicated.

Latency
Some viruses can enter a latent state, delaying replication until later.

Macromolecular synthesis
The viral nucleic acid is replicated, and viral proteins, including capsid proteins, are made.

Assembly
Capsid proteins and viral nucleic acid are assembled into nucleocapsids.

Release
The complete virions are released from the infected cell. If the virus is enveloped, the envelope is acquired at this stage.

FIGURE 6.8 Scheme of viral multiplication.

assembled from the component pieces (proteins and nucleic acids), and (5) they then escape from the host cell (Figure 6.8).

Some viruses have developed an alternate strategy—they can integrate their DNA into one of the host chromosomes, where it becomes latent. Of course, every time the host chromosome is replicated, the viral sequences are replicated as part of the chromosome. We discuss latency in a later section.

Attachment and Penetration

Because virus multiplication requires the combined efforts of the host and viral genomes, the first steps are to get these two genomes together, at least in the same cell, sometimes in the same subcellular compartment. It is here that most of the host–cell specificity of viruses is imposed. Many viruses infect only a single species; the polio virus, for instance, infects only humans. Other viruses are less specific and can infect a number of closely related species (e.g., yellow fever virus can infect humans and several other primates). Finally, there are some viruses that can infect widely different hosts—the rabies virus, for example, can infect a wide range of mammals, including humans, bats, skunks, foxes, and dogs. Whether the host range of a virus is highly specific or broad depends mainly on the ability of the virion to attach to a potential host cell, which in turn depends on the attachment proteins of the virion and the receptor molecules on the host cell.

6.7 Attachment of virion to host cell is highly specific

Virions, of course, are not actively motile. They are so small, however, that they can be moved easily by the slightest currents in air or liquid. When they by chance collide with a host cell, they attach by virtue of noncovalent interactions between one of their capsid or envelope proteins and a protein or carbohydrate on the surface of the host cell. Because such interactions require a close match of shape and charge, they can be highly specific—even closely related host cells may have a surface structure different enough that they do not attach to the same viruses.

Of course, because attachment is a chance event, most virions never encounter a host cell with the right chemical structure on its surface, and they eventually are inactivated in the environment. This occurs because their proteins and nucleic acid are digested by proteases and nucleases in the environment or because they accumulate lethal mutations in their nucleic acid (mutations can continue to occur slowly in the inert virion).

6.8 Bacteriophage penetration is by nucleic acid injection

Viruses of bacteria and archaea face the problem of getting their nucleic acid across the thick or multilayered cell envelope and the cell membrane. They have developed a variety of mechanisms for doing so, leaving the capsid entirely outside the cell (Figure 6.9).

The binal viruses typically have a contractile tail, with an inner core that penetrates through the cell wall when the tail sheath contracts. The nucleic acid then moves through the hollow core into the cytoplasm (Figure 6.10).

The injection mechanism of the binal viruses is the most complicated known. Most bacterial or archaeal viruses have simple icosahedral or filamentous capsids. Like the binal viruses, after attachment, their DNA is injected into the host cell, leaving the capsid on the outside. The mechanisms by which they inject their DNA are poorly understood. Most unenveloped animal viruses also appear to inject their

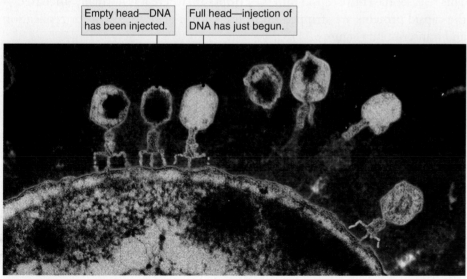

Empty head—DNA has been injected.

Full head—injection of DNA has just begun.

FIGURE 6.9 Virions attached to *E. coli*.

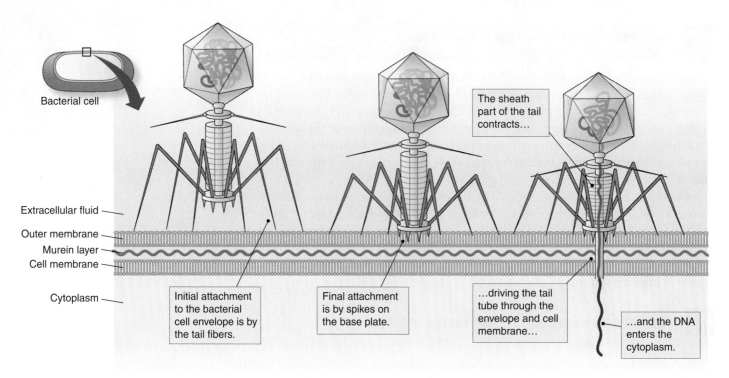

Bacterial cell

Extracellular fluid

Outer membrane

Murein layer

Cell membrane

Cytoplasm

The sheath part of the tail contracts…

Initial attachment to the bacterial cell envelope is by the tail fibers.

Final attachment is by spikes on the base plate.

…driving the tail tube through the envelope and cell membrane…

…and the DNA enters the cytoplasm.

FIGURE 6.10 The stages of bacteriophage DNA.

nucleic acid, leaving the capsid on the outside. The mechanism is unknown, but probably involves virion proteins forming a pore through the host cell membrane.

6.9 Penetration is by membrane fusion in enveloped viruses, followed by uncoating

Animal cells have no cell wall, and thus, the only barrier to viral access is the cell membrane. Because many animal viruses are enveloped, they can gain entry by catalyzing membrane fusion between their envelope and the host cell membrane. This releases the entire capsid into the cytoplasm, where it is normally **uncoated**—the capsid proteins are stripped off of the nucleic acid, freeing it for replication and transcription (Figure 6.11).

In many cases, the membrane fusion is delayed and does not occur until after the virion is endocytosed. The acidic pH of the endosome, where the endocytosed virion winds up, causes a conformational change in viral envelope proteins, leading to fusion of the viral envelope and the endosome membrane.

Unenveloped animal viruses, like polio virus, are generally taken up by endocytosis, and the acidic pH of the endosome then causes conformational changes in one or more capsid proteins, which then appear to form a pore through the endosome membrane. The details of these processes are not well understood.

6.10 Plant viruses penetrate via wounds, or they are injected by insects

Like bacteriophages, plant viruses have to breach a thick cell wall to gain entry to their host cells. Unlike the bacteriophages, however, plant viruses do not have injection mechanisms to accomplish this, and thus, they generally depend on mechanical

In the 1950s, a number of biologists began to study several of the bacteriophages with binal structure as simple model systems in which features generally characteristic of living systems could be studied. One of the first questions was the role of the DNA and of the protein of the phages. This question was addressed experimentally before Watson and Crick published their model for the structure of DNA but after Oswald Avery and his collaborators had demonstrated that DNA was the material of which genes were made. There were still many who questioned Avery's conclusion, however, and thought that genes might be made of protein. A set of experiments conducted by Alfred Hershey and Martha Chase is widely viewed as confirming that genes were DNA. The reality is not so simple.

At the time, it was known that phages were composed of DNA and protein and that the DNA was contained in the head. It was also known that these phages adsorbed to their host cells by their tails and that the capsid and tail structures remained attached to the outside throughout the infection. Obviously, however, phage genes had to enter the cell to direct the synthesis of new phages. Hershey and Chase's experiments were designed to determine whether it was DNA or protein that entered.

To address this, they prepared phages propagated on host cells in the presence of ^{35}S, which becomes incorporated into protein (because the amino acids cysteine and methionine contain sulfur), and they prepared a different batch of phage propagated in the presence of ^{32}P, which becomes incorporated into DNA. Because protein contains very little phosphorous and DNA contains very little sulfur, these radioactive labels should be specific to the two compounds.

They then allowed phages to absorb to new host cells (in two different experiments—one with the ^{35}S-labeled phage and one with the ^{32}P-labeled phage). The mixtures were then put in a blender and blended at high speed for varying lengths of time. This treatment does not harm the cells, as they are too small for the blender blades to cut them, but the treatment creates intense currents that Hershey and Chase reasoned would strip the phage coats off the bacteria. They could then separate the cells from the phage coats by centrifuging the mixture, with the cells being sedimented into the pellet at the bottom of the centrifuge tube and the much smaller phage coats remaining in suspension. They could then simply measure the radioactivity in the pellet and the supernatant and determine how much protein and how much DNA entered the cells and how much was left outside to be stripped off. Unfortunately, the results were quite ambiguous (Figure 6B.1).

Certainly the DNA and protein behaved differently. About 80% of the protein was separated from the cells by blending, but only 30% of the DNA was. They interpreted this to mean that most of the DNA entered the cell, whereas most of the protein remained outside; however, the 20% of protein that remained attached could have entered the cells and played a role in phage multiplication. (We now know that it is mainly phage tail fibers, base plates, and broken tails that remained attached.) Thus, this experiment was not as decisive as it is usually portrayed.

Hershey and Chase were very cautious in interpreting their results: "We have shown that . . . most of the phage DNA enters the cell, and a residue containing at least 80 per cent of the sulfur-containing protein . . . remains at the surface. This residue consists of the material forming the protective membrane of the . . . phage particle, and it plays no further role in infection. . . . Our experiments show clearly that a physical separation of the phage T2 into genetic and non-genetic parts is possible. . . . The chemical identification of the genetic part must wait, however. . . ."

Others, however, were not so cautious, and this experiment was hailed by many scientists as proving that the genetic material was DNA, and it is this view that has made its way into most textbooks. Why were such ambiguous results so overinterpreted? Probably the principal reason was that by 1952, when Hershey and Chase published, there was already widespread, although far from universal, acceptance that DNA was the genetic material, based on the very careful work from Avery's laboratory. Although controversial when it was published in 1944, with the passage of years, it became more generally accepted. Furthermore, Hershey and Chase published their paper nearly simultaneously with the publication of the Watson and Crick model for the structure of DNA, which suggested an elegant mechanism for its replication (one of the major features of the genetic material). Thus, the impact of the Hershey and Chase paper was probably more due to its timing than to its intrinsic merit.

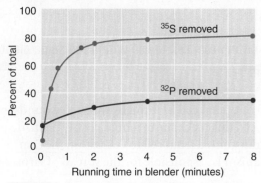

FIGURE 6B.1 Results of the Hershey-Chase experiment.

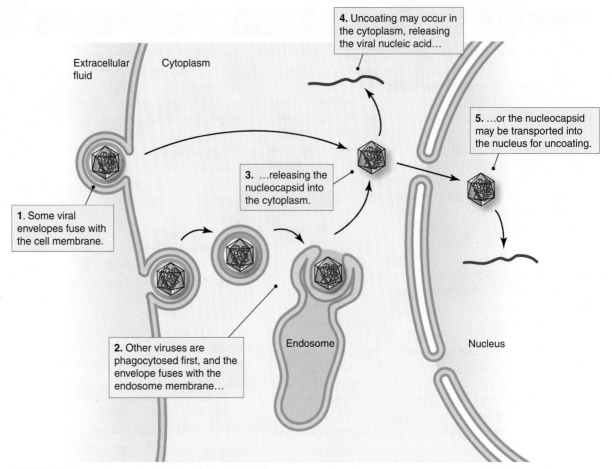

4. Uncoating may occur in the cytoplasm, releasing the viral nucleic acid…

5. …or the nucleocapsid may be transported into the nucleus for uncoating.

3. …releasing the nucleocapsid into the cytoplasm.

1. Some viral envelopes fuse with the cell membrane.

2. Other viruses are phagocytosed first, and the envelope fuses with the endosome membrane…

Extracellular fluid

Cytoplasm

Endosome

Nucleus

FIGURE 6.11 Penetration and uncoating of an animal cell.

damage to the wall to initially gain entry. Sometimes this is random damage, from abrasion or other trauma; most commonly, however, plant viruses are transmitted from plant to plant by insects that bite or suck on plant leaves. The insect mouth parts become contaminated with virions when the insect feeds on an infected plant and are then able to infect the next plant the insect feeds on when it damages cell walls during feeding.

6.11 Plant viruses are transmitted from cell to cell within the same plant through plasmodesmata

Although the initial infection of a plant cell is normally through wounds inflicted by random trauma or by insects, this mechanism would clearly be inadequate to insure that the viruses were transmitted from cell to cell within the same infected plant. Thus, a separate mechanism is responsible for the spread of viruses within a plant.

Adjacent cells in plants and algae are connected by thin cytoplasmic connections termed **plasmodesmata.** Although small, these are large enough for viral nucleic acid to move from cell to cell, assisted by special viral proteins that actively move the nucleic acid through the plasmodesmata, powered by ATP hydrolysis. Thus, a single-penetration event, infecting a single plant cell, may be sufficient to infect all the cells within a local area, such as a leaf.

Longer distance movement through a plant occurs when the virus enters the plant's vascular system. It can then be transferred systemically to all the plant tissues. How release into the vascular system and penetration from the vascular system into cells are mediated is not well understood. It appears, however, to generally require virions, rather than naked nucleic acid.

Macromolecular Synthesis

After the viral nucleic acid is in the host cell and in the case of eucaryotes in the appropriate compartment (usually the cytoplasm or nucleus), it can be replicated to produce more of itself and its genes expressed to produce viral proteins. Frequently, host cell DNA replication and protein synthesis are shut down; all necessary host gene expression has occurred before viral penetration (e.g., producing the enzymes to produce ATP and the various amino acids and nucleotides necessary for viral protein and nucleic acid synthesis). Thus, continued expression of host genes is not needed and could compete with viral replication, transcription, and translation.

The mechanisms of replication may be quite ordinary, for some double-stranded DNA viruses, or it may require novel, virally encoded enzymes (e.g., to replicate RNA). The mechanisms of viral nucleic acid replication are extremely varied and complex; we only discuss a representative selection of mechanisms here.

6.12 DNA viruses of eucaryotic cells usually replicate in the nucleus and RNA viruses in the cytoplasm

In eucaryotic cells, the enzymes of DNA replication are located in the nucleus (and in chloroplasts and mitochondria, but these are not sites of viral replication). Because some or all of the enzymes of viral DNA replication are usually host enzymes, DNA viruses must replicate in the nucleus. There are a few exceptions, and these all have special mechanisms for DNA replication. For instance, the pox viruses carry enzymes of DNA replication in the virion, and these enzymes accompany the DNA into the cytoplasm where they function.

Of course, when viral DNA replication and transcription occur in the nucleus, viral mRNA is exported through the nuclear pores for translation, and viral proteins synthesized in the cytoplasm are imported back into the nucleus.

Most RNA viruses of eucaryotes replicate in the cytoplasm, presumably for the efficiency of having their RNA replication and protein synthesis in the same cellular compartment. Exceptions include influenza viruses and retroviruses, which replicate their RNA largely in the nucleus but which assemble in the cytoplasm.

6.13 DNA phages often replicate by a "rolling circle" mechanism

Many DNA viruses of procaryotes, both double and single stranded, replicate by a mechanism termed **rolling circle replication.** We use the widely studied bacterial virus λ (which infects *E. coli*) as an example (shown in Figure 6.5).

The λ chromosome in the virion is a linear double-stranded molecule with short single-stranded regions at the ends. These single-stranded regions involve

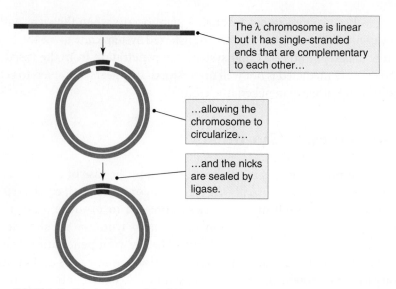

The λ chromosome is linear but it has single-stranded ends that are complementary to each other…

…allowing the chromosome to circularize…

…and the nicks are sealed by ligase.

FIGURE 6.12 Circularization of the λ chromosome.

both DNA strands, one at one end and the other at the other end. If the single strand at the left end is an extension of the Watson strand, then the region at the right end is an extension of the Crick strand. Furthermore, the two ends are complementary to each other, and thus, they can hydrogen bond, making the linear chromosome into a circular one with two single-stranded nicks in it (Figure 6.12). This happens immediately on introduction of the DNA into its host cell. This is followed by sealing the two nicks by a host cell enzyme (DNA ligase) that is part of the host DNA replication system.

After circularization, the replication process begins. First, normal bidirectional replication produces several new circular chromosomes as a prelude to rolling circle replication. This bidirectional replication is essentially the same process used to replicate the host DNA, and it uses mainly host enzymes.

Then a nick is made in one strand only of each of these circular chromosomes, at the origin of replication (Figure 6.13). The 3′-OH that is generated by the nicking is used as the primer for DNA polymerase to begin leading strand synthesis, using the other strand as a template. This process pushes the old strand with a 5′-P end out of the way. As the process continues around the circle, the displaced single strand becomes longer, and eventually lagging strand synthesis begins using it as the template. Because the template for leading strand synthesis is a circle, the polymerase can keep going around indefinitely, producing an increasingly long tail that consists of many λ chromosomes covalently joined end to end, a structure termed a **concatemer** (Figure 6.13).

The final step in λ DNA replication is for a specific viral enzyme to cleave the concatemer into individual chromosomes by making staggered nicks in the specific sequences that mark the ends of the linear chromosome.

The process of rolling circle replication is not found in all DNA phages, but it is very common. There is a good deal of variation in the process of initially generating a circular template, and there is also quite a bit of variation in how the concatemer is cleaved. The rest of the process is pretty similar for all viruses that use this mechanism. The principal difference in single-stranded DNA viruses is that lagging strand synthesis is suppressed, and thus, the concatemer remains single stranded.

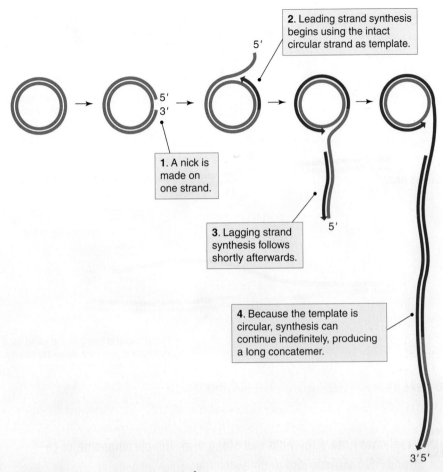

2. Leading strand synthesis begins using the intact circular strand as template.

5′

1. A nick is made on one strand.

5′
3′

3. Lagging strand synthesis follows shortly afterwards.

5′

4. Because the template is circular, synthesis can continue indefinitely, producing a long concatemer.

3′ 5′

FIGURE 6.13 Rolling circle replication in λ.

A few DNA viruses of eucaryotes replicate by a rolling circle mechanism, for example, the herpes viruses, but it is much less common than in the bacteriophages.

6.14 The chromosome of (+) strand RNA viruses acts as mRNA to produce a replicase

Single-stranded RNA viruses are conventionally divided into two types. **Positive (+) strand viruses** have a chromosome that has the same base sequence as viral mRNA, whereas **negative (−) strand viruses** have a chromosome whose sequence is complementary to the mRNA. Their replication strategies are fundamentally different; however, they both face the same basic problem: their host cells do not have enzymes for replicating RNA, and thus, they have to provide this function themselves. In the case of (+) strand viruses, when the chromosome is released into the cytoplasm, it is recognized by ribosomes as an mRNA because it has a ribosome binding site. This allows the chromosome to act as an mRNA and be translated (Figure 6.14). One product is an enzyme, termed an **RNA replicase,** that will replicate the RNA.

The initial product of the RNA replicase is a number of RNA strands complementary to the chromosome. These (−) strands are then used as templates to produce the progeny (+) strand chromosomes.

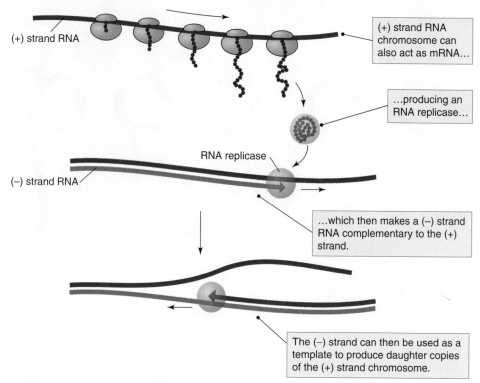

FIGURE 6.14 Steps in the replication of a (+) strand RNA virus.

The boxes in the figure read:

(+) strand RNA chromosome can also act as mRNA...

...producing an RNA replicase...

...which then makes a (–) strand RNA complementary to the (+) strand.

The (–) strand can then be used as a template to produce daughter copies of the (+) strand chromosome.

Labels: (+) strand RNA; RNA replicase; (–) strand RNA

6.15 A replicase enters the host cell along with the chromosome of (—) strand RNA viruses and double-stranded RNA viruses

The (–) strand and double-stranded RNA viruses face a particularly acute problem in getting replication started. Not only do their host cells not have any RNA replicase, but their chromosomes cannot act directly as an mRNA to direct the synthesis of one. In both cases, the only solution seems to be for the virion to carry one or more copies of the RNA replicase enzyme, which accompanies the chromosome into the cell (Figure 6.15). The replicase then uses the (–) strand as a template to produce a series of (+) strands, which can act as mRNA to produce more replicase (as well as other viral proteins).

These (+) strands not only act as mRNA, but they also act as templates for the production of progeny (–) strand chromosomes, or of progeny double-stranded ones, depending on the virus.

6.16 Retroviruses copy their ss-RNA chromosome into DNA using reverse transcriptase

One group of single-stranded RNA virus, the retroviruses, has a unique method of replication. Their RNA is (+) strand, but instead of acting as mRNA as in other (+) strand RNA viruses, theirs is converted to double-stranded DNA by the enzyme **reverse transcriptase.** The reverse transcriptase is a virally encoded enzyme, and a couple of molecules of it are carried in the virion, entering the host cell along with the RNA (Figure 6.16). This enzyme uses single-stranded RNA as a template to produce a complementary DNA molecule, simultaneously degrading the RNA. The single-

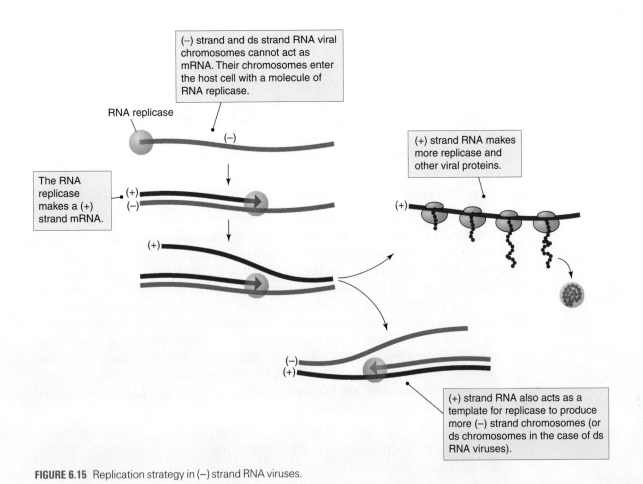

(−) strand and ds strand RNA viral chromosomes cannot act as mRNA. Their chromosomes enter the host cell with a molecule of RNA replicase.

RNA replicase

(−)

The RNA replicase makes a (+) strand mRNA.

(+)
(−)

(+)

(−)
(+)

(+) strand RNA makes more replicase and other viral proteins.

(+)

(+) strand RNA also acts as a template for replicase to produce more (−) strand chromosomes (or ds chromosomes in the case of ds RNA viruses).

FIGURE 6.15 Replication strategy in (−) strand RNA viruses.

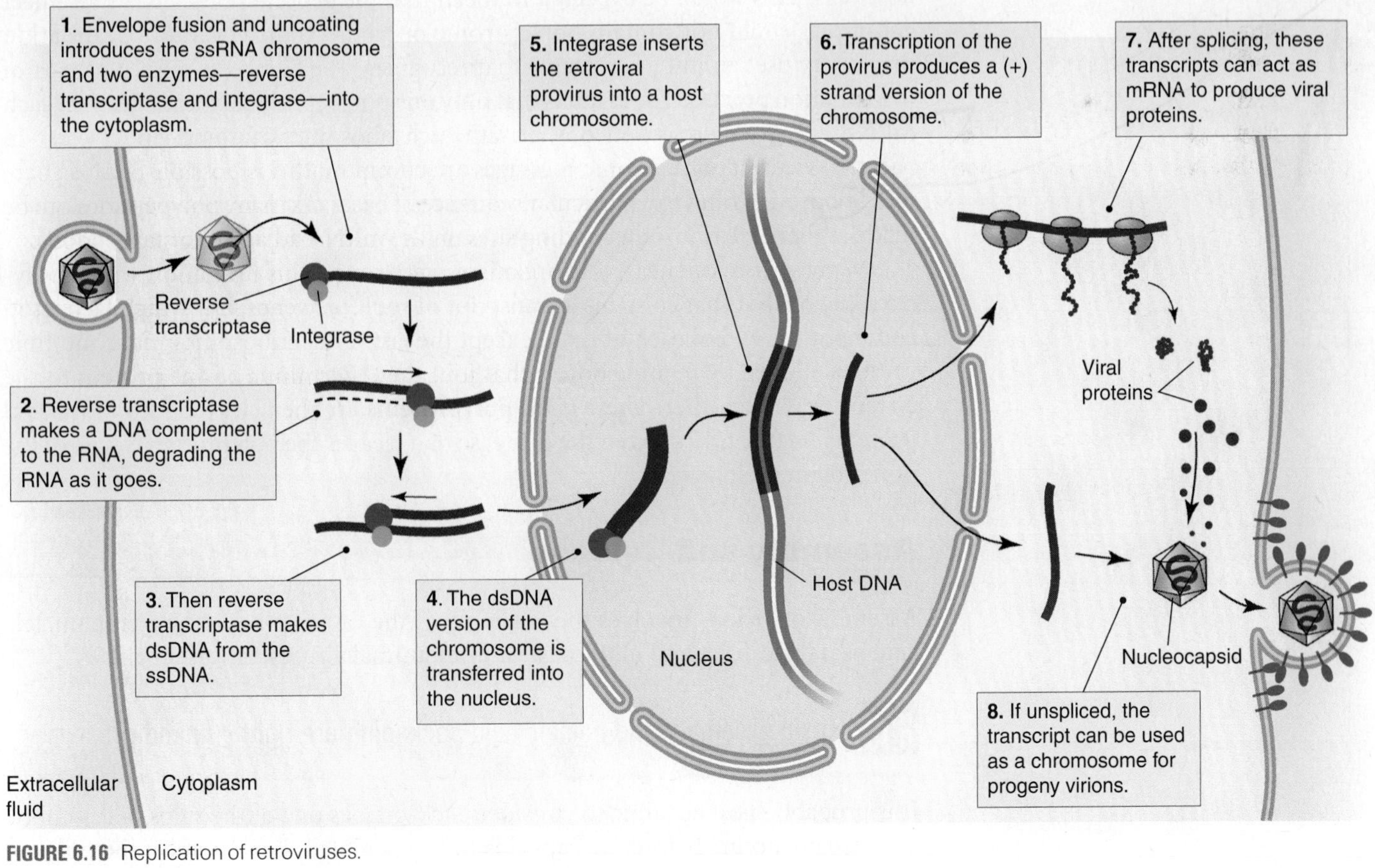

1. Envelope fusion and uncoating introduces the ssRNA chromosome and two enzymes—reverse transcriptase and integrase—into the cytoplasm.

5. Integrase inserts the retroviral provirus into a host chromosome.

6. Transcription of the provirus produces a (+) strand version of the chromosome.

7. After splicing, these transcripts can act as mRNA to produce viral proteins.

Reverse transcriptase   Integrase

2. Reverse transcriptase makes a DNA complement to the RNA, degrading the RNA as it goes.

3. Then reverse transcriptase makes dsDNA from the ssDNA.

4. The dsDNA version of the chromosome is transferred into the nucleus.

Host DNA

Nucleus

Viral proteins

Nucleocapsid

8. If unspliced, the transcript can be used as a chromosome for progeny virions.

Extracellular fluid   Cytoplasm

FIGURE 6.16 Replication of retroviruses.

Retroviruses copy their ss-RNA chromosome into DNA using reverse transcriptase stranded DNA is then used in turn as a template forming a double-stranded DNA molecule. The double-stranded DNA viral chromosome is then transported into the nucleus, where it is integrated into the host chromosome at a random site, a process catalyzed by another protein, **integrase,** that was also carried in the virion. The host RNA polymerase then transcribes the integrated **provirus** to produce single-stranded viral RNA chromosomes and mRNA for viral proteins.

6.17 Viral protein synthesis is regulated

In addition to replicating the nucleic acid, multiplication requires the synthesis of viral proteins. Proteins may be required for a variety of functions, and these functions may have different timing. For instance, nucleic acid replication typically occurs in the early part of the infection, whereas capsid synthesis occurs later. This obviously requires careful regulation, and most viruses make several proteins whose only function is to regulate the timing of synthesis of other proteins. Obviously the extent of this will vary with the complexity of the viral genome. When it is simple, consisting of only a few genes, regulation is rudimentary. The more complex viruses, with scores or hundreds of genes, however, have extremely complex regulatory strategies. Several novel mechanisms of regulation have been first discovered and their details elucidated in viruses.

6.18 Eucaryotic viruses often make polyproteins

Viruses are generally constrained by the limitations of the molecular biology of their host cells. This would be expected to mean that the genomes of viruses that infect eucaryotes could not contain polycistronic operons, as the eucaryotic translation machinery uses an initiation factor that recognizes the 5′ cap on mRNA as part of the initiation process. The result is that only one polypeptide can be made from each mRNA because there is only one site at which ribosomes can bind (the 5′ cap). In bacterial viruses, polycistronic messages are common; this is possible because bacterial ribosomes bind to a particular sequence of bases. As many polypeptides can be made as there are ribosome binding sites on the mRNA adjacent to start codons.

Many viruses of eucaryotic organisms get around this limitation with a polycistronic mRNA that contains a transcript of multiple genes, but which lacks stop codons at the end of each of them except the last. Thus, ribosomes make multiple proteins, linked by peptide bonds that join the C-terminus of one protein to the N-terminus of another. These giant **polyproteins** are then cleaved into individual proteins (Figure 6.17). Often the protease that cleaves the polyprotein is part of the polyprotein itself.

Assembly and Release

Assembly of virions involves the assembly of the capsid, packaging of the nucleic acid into the capsid, and in the case of most animal viruses, envelopment.

6.19 Capsid assembly and nucleic acid packaging are tightly linked processes

It is probably most accurate to consider packaging as one of the steps in assembly, as it usually occurs before the capsid is fully assembled. For instance, the phage λ

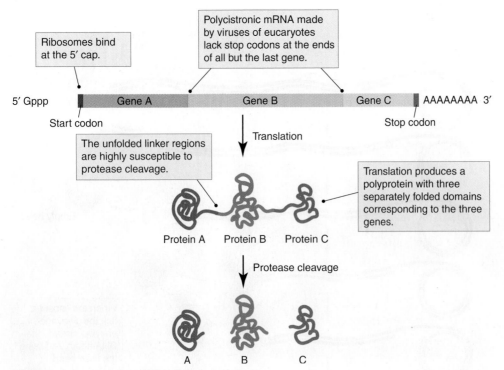

Ribosomes bind at the 5' cap.

Polycistronic mRNA made by viruses of eucaryotes lack stop codons at the ends of all but the last gene.

5' Gppp   Gene A   Gene B   Gene C   AAAAAAAA 3'

Start codon

Stop codon

Translation

The unfolded linker regions are highly susceptible to protease cleavage.

Translation produces a polyprotein with three separately folded domains corresponding to the three genes.

Protein A   Protein B   Protein C

Protease cleavage

A   B   C

FIGURE 6.17 Synthesis and cleavage of a polyprotein.

discussed previously as an example of rolling circle replication, tightly couples assembly, packaging, and replication. Assembly begins nearly simultaneously with replication, and the cleavage of the concatemer of DNA occurs as the DNA enters the capsid.

In some other phages, packaging is by a **headful mechanism.** In this type of packaging, phage DNA from a concatemer (generated by rolling circle replication usually) enters the empty head until the head is full (Figure 6.18). The DNA is then cleaved and head assembly completed. Because it is important that each virion get a complete chromosome, phages that use a headful packaging mechanism typically have heads a bit larger than necessary. Thus, the chromosomes in the virions have duplicate sequences at their ends. These do not get perpetuated, however, because when the chromosome is injected into the next host, recombination between the duplicate sequences eliminates the redundancy and generates the circular template for replication (Figure 6.18).

6.20 Assembly of simple viruses occurs spontaneously; more complex viruses have regulated assembly pathways

In the case of the simpler icosahedral and filamentous viruses, assembly appears to be a spontaneous reaction among the viral proteins and nucleic acid, and it can be made to happen in a test tube if the concentration of proteins and nucleic acid is high (as it would be inside the infected cell). In the more complicated viruses, however, especially the binal phages, the assembly process is extremely complex and requires a number of proteins whose role is to mediate various of the steps.

The assembly process leads to the accumulation of virions in the cytoplasm or nucleoplasm of the host cell, except for enveloped viruses, whose virions are not complete until they obtain their envelope during the release process. The number of virions can be immense—hundreds, thousands, or even tens of thousands of virions per host cell.

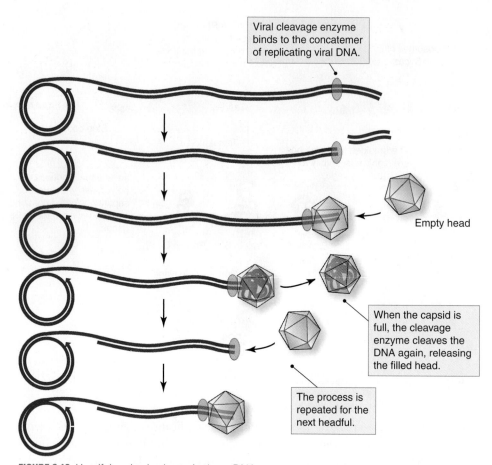

Viral cleavage enzyme binds to the concatemer of replicating viral DNA.

Empty head

When the capsid is full, the cleavage enzyme cleaves the DNA again, releasing the filled head.

The process is repeated for the next headful.

FIGURE 6.18 Headful packaging bacteriophage DNA.

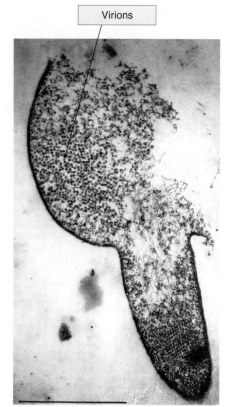

Virions

FIGURE 6.19 A burst infected bacterial cell.

6.21 Release of unenveloped virions is normally by the lysis of the host cell

The last step in the multiplication cycle for unenveloped viruses is the release of the virions, a process that normally requires the lysis of the host cell to release the accumulated virions (Figure 6.19). One or more of the proteins produced late in the infective cycle is responsible for lysing the cell.

6.22 Release of enveloped virions is a continuous process that often does little damage to the host cell

In the case of enveloped virions, release involves budding from the cell or nuclear membrane. This does little damage to the host cell, unless the rate of budding is so high that the cell cannot replace its membrane fast enough. Ultimately, however, the host cell usually dies as a result of cumulative damage from the diversion of its metabolism to the production of virions.

During the synthesis stage of viral multiplication, virus-encoded membrane proteins are made and embedded in the appropriate membrane (cell membrane, nuclear envelope, or endoplasmic reticulum). Nucleocapsid proteins then establish quaternary interactions with the membrane proteins, forming the bud. This process was discussed earlier in this chapter (see Figure 6.7).

Lysogeny and Latency

A number of viruses are capable of becoming latent in their host cells. These viruses thus face a "choice" soon after their DNA enters the host: they can proceed with active replication, as described previously, or they can enter the latent state. The molecular details of how this decision is made have been worked out for a few bacteriophages, but it is poorly understood in animal viruses.

Latency usually involves the integration of the viral chromosome into the host chromosome; thus, viruses capable of latency are generally DNA viruses. In this integrated state, they do not actively replicate themselves, but rather are passively replicated every time the host chromosome is replicated. A number of different bacteriophages can become latent, although the term *latency* is not normally used for them. Instead, they are referred to as **temperate phages** (because they do not always kill their hosts), and the integrated viral chromosome is termed a **prophage.** A host cell that carries a phage in the latent state is termed **lysogenic.** In animal viruses, latency also occurs but has been less thoroughly studied, and its mechanisms are not yet clear.

6.23 Prophages make their host cells immune to subsequent infection by another virion of the same kind

When a cell carries a latent virus, it is generally resistant to subsequent infection by another virion of the same virus or sometimes by related viruses as well. This is due to a repressor protein that is made by the provirus. This repressor prevents the expression of the provirus genes required for lytic growth. Of course, if another copy of the same viral chromosome is injected into the cell, the repressor protein can also bind to it and prevent the expression of its genes and thus prevent it from entering the lytic growth pathway.

6.24 Proviruses can be induced to reenter active multiplication

The integrated latent form of a virus can be triggered to reenter the actively replicating state. Often the trigger for this process (termed **induction** for phage, or **reactivation** for animal viruses) is stress to the host. By some mechanism (well understood for phage, but not for animal viruses), certain types of stress can lead to the destruction of the repressor protein, thus allowing the provirus to enter active multiplication.

Thus, when these viruses enter a population of host organisms, some of the hosts produce virions immediately, which can spread throughout the population. Other hosts become latently infected; later, when stressed, they or their descendants will produce virions.

6.25 Latency is a strategy for intergenerational transmission

In animal viruses, latency is clearly a mechanism for transmission of the virus from one generation of host to another. One of the problems that animal viruses face is that when they infect a host, they often stimulate the immune system, with the result that the host may soon become immune and eliminate the virus (see Chapter 18). Thus, a virus that gains entry into a small host population will soon cause all members of the population to die or become immune. At that point, the virus dies out unless it can get to a different population or unless it can become latent for a period of time sufficient for new members to be born into the population to act as fresh hosts. Latency, and then reactivation, is thus an evolutionary mechanism for delaying virion production in some infected hosts to ensure transmission among generations.

The same is probably true of bacteriophage as well. If a temperate phage gains access to a population of host cells, for instance, the *E. coli* in the mammalian gut, lytic multiplication serves to spread the phage within the population, killing many of the host cells. Some become lysogenic, however, and these can replenish the population of *E. coli*, and serve as well as a long-term reservoir of the phage.

Viral Taxonomy

Virus taxonomy is **artificial;** that is, it does not attempt to group together organisms related by evolutionary descent. This is because too little is known of the origins and evolution of viruses. We can confidently trace the evolutionary relatedness of closely similar viruses—for instance, the relationship between the various types of HIV (the virus that causes AIDS in humans) and the closely related SIV (which causes a similar disease of other primates). Similarly, we know that measles virus is related to the virus that causes the disease rinderpest of cattle; however, we cannot trace viral evolution back much further than this, unlike cellular organisms (see Chapter 13).

6.26 Viruses are classified into artificial groups on the basis of their molecular biology, life cycles, and host organisms

Because we have no idea of the broader relationships among viruses, we classify on the basis of various features of their virions (enveloped or not, capsid symmetry, etc.), the structure of their nucleic acid (DNA or RNA, single or double stranded, segmented genome or not), their replication and transcription strategies, and their hosts (procaryotes, animals, plants, fungi). Although this classification scheme gives no information about relatedness, it does give a great deal of information about the virion structure and molecular biology, and thus, it is in practice a useful approach.

Viroids and Prions

The simplest viruses seem to be about as simple as it is possible to get—they encode only a few proteins and depend on the host for everything else; however, there are several other acellular entities that have evolved to exploit cells for their replication: viroids and prions. Both of these consist of a single type of macromolecule that is replicated by host cells. These entities certainly press the boundaries of life; whether they are considered alive is an interesting semantic question that results from the fact that there is no generally accepted definition of life.

6.27 Viroid RNA is not translated; it is replicated by a rolling circle mechanism

Viroids consist of unencapsulated, or naked, RNA that infect several species of plant. The RNA is a circular single strand; however, because there is a great deal of internal

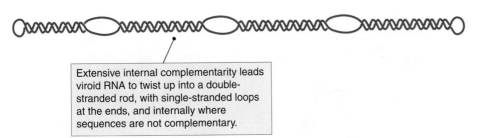

Extensive internal complementarity leads viroid RNA to twist up into a double-stranded rod, with single-stranded loops at the ends, and internally where sequences are not complementary.

FIGURE 6.20 Viroid structure.

complementarity, the single strand hydrogen bonds with itself to form a rigid rod (Figure 6.20). Viroids differ somewhat in size, but all are in the 250 to 400 base range. This size could in principle encode one or two small proteins; however, it is clear that viroid RNA does not encode any product, and thus, viroid replication has to be accomplished entirely with host enzymes.

How viroids are replicated is not entirely clear, but it appears that host machinery uses a rolling circle mechanism to produce a concatemer of RNA complementary to the viroid RNA. This concatemer is then cleaved into monomeric lengths of RNA, which then circularize. These circular complementary RNA molecules then act as template for a second round of rolling circle synthesis, concatemer cleavage, and circularization to produce the progeny viroids. Obviously, viroid replication raises many questions about what these host enzymes do normally within the plant cell.

Like many plant viruses, viroids are transmitted from plant to plant by insects and through plasmodesmata within a plant. How they cross the cell membrane and are transported to the nucleus where they are replicated is unclear.

6.28 Prions are aberrant conformations of a normal mammalian brain protein that replicates by catalyzing a conformational change in the normal proteins

A few years ago it was realized that several degenerative neurological diseases of mammals are contagious and are transmitted by the consumption of infected tissue. The best known of these is BSE (for bovine spongiform encephalopathy), or "mad cow disease." This disease appears spontaneously in a very low percentage of animals; however, modern agricultural systems supplemented cattle feed with ground up offal from slaughterhouses, a practice now banned in most countries. The result was that for a period of time cattle routinely consumed the remains of other cattle. This apparently established the conditions that allowed this disease to become contagious. A similar disease of humans was known in New Guinea, transmitted by ritual cannibalism, and it now appears that some human cases of disease are connected to eating beef contaminated with the agent of BSE. These are called **new variant Creutzfeldt-Jakob disease,** or **nvCJD,** to distinguish it from the spontaneous CJD that occurs at low frequency. A number of other mammals are also known to suffer from similar diseases, either spontaneous, low-level disease, or contagious disease spread by other mechanisms than human contamination of the food chain.

A great deal of work ultimately led to the startling conclusion that the agents of these diseases, termed *prions,* are protein molecules that are normally found in high amounts in normal brain and nerve tissue. The protein is a normal constituent of the membrane of nerve cells, but its function is unknown. The prion

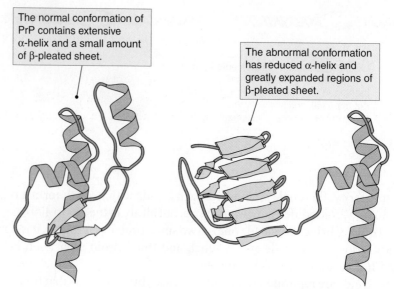

The normal conformation of PrP contains extensive α-helix and a small amount of β-pleated sheet.

The abnormal conformation has reduced α-helix and greatly expanded regions of β-pleated sheet.

FIGURE 6.21 Normal and abnormal conformations of the prion protein.

version of the protein has the same amino acid sequence, but it folds into a different three-dimensional form (Figure 6.21). In this variant conformation, it catalyzes the switch of all the normal protein into the variant conformation. The variant conformation aggregates into long rods, which are released when the cell dies, and the abnormal protein can then perpetuate the process in adjacent cells. Disease, ultimately fatal in all cases, results from the progressive destruction of brain cells (Figure 6.21).

If another animal consumes brain or nervous tissue from an infected animal, the abnormal protein can apparently be taken up intact from the digestive tract into brain tissue, where it begins again the process of converting the normal proteins. This was the basis for the outbreak of mad cow disease in the United Kingdom and also of the ongoing outbreak of nvCJD in the United Kingdom among consumers of infected beef. Fortunately, prions do not appear to be highly contagious, and although millions of people undoubtedly consumed infected beef during the BSE outbreak, to date there have been fewer than 200 cases of nvCJD.

Summary

Viruses are the ultimate parasite, dependent on their hosts for nearly everything, including synthesis of all monomers, nearly all of the components of macromolecular synthesis, and the provision of the energy to fuel all of this. They multiply by a unique process of separate synthesis of virion components followed by assembly and release from the host cell, a process very different from the growth and division of cells.

In addition to their ability to take over a host cell and divert its metabolism to the production of new virions (sometimes more than 10,000 per infected cell), some viruses can also enter a latent state. Latently infected host cells can then serve as a reservoir of the virus, capable of infecting a future generation.

Although almost all virions will eventually decay without ever initiating an infection, viruses are clearly a major evolutionary success. How they evolved is unclear, but there is probably no group of cellular life that is not parasitized by many different viruses; thus, the biological diversity of the viral world is very great, and they constitute one of the most fascinating groups of microbes.

Study questions

1. Why are viruses obligate intracellular parasites?
2. Calculate the ratio of the volume of a typical icosahedral virus with a diameter of 40 nm to that of a typical coccoidal bacterial cell with a diameter of 1 μm and to that of a typical yeast cell with a diameter of 10 μm. For the purposes of calculation, you can assume all three are spheres.
3. Briefly describe the life cycle of a typical bacteriophage with binal symmetry (e.g., λ).
4. Briefly describe the life cycle of a typical plant virus.
5. Briefly describe the life cycle of a typical enveloped animal virus.
6. How does the type of nucleic acid within a virion affect the replication strategy?
7. Discuss the role of latency in viral multiplication.
8. Explain how and why viral classification differs from that of cellular organisms.
9. How do viroids differ from viruses? How are they similar?
10. How do prions differ from viroids? How are they similar?

7

Microbial Metabolism: Fermentation and Respiration

METABOLISM IS THE sum total of all of the chemical reactions of the cell. Conventionally, metabolism is divided into two major categories: biosynthesis and energy metabolism; however, some reactions take part in both types of metabolism, and thus, the distinction is not perfect. In this chapter, we have a short introduction to metabolism and then focus on fermentation and respiration—the major types of energy metabolism of heterotrophs. In the next chapter, we consider photosynthesis and autotrophic metabolism.

Microbial Nutrition, Central Metabolism, and Biosynthesis

We focus mainly on energy metabolism, as that is where the microbes are much more diverse than plants and animals. Of course, the microbes have been evolving for four times as long, and, thus, it is no wonder that they have perfected a number of biochemical strategies unknown in macro-organisms. First, however, we need to review a bit about nutrition and biosynthesis.

7.1 Microbes are divided into four basic nutritional categories

Microbiologists conventionally divide microbes into categories based on their preferred carbon source (organic or CO_2) and their preferred energy source (chemical compounds or light). **Chemotrophs** use chemical compounds as their energy source; **phototrophs** use light. **Autotrophs** use CO_2 as their carbon source; **heterotrophs** use organic compounds. Thus, there are four different combinations:

Chemoheterotrophs oxidize chemical compounds as their source of energy, and they use organic compounds as their source of carbon. Usually a single organic compound can serve as both source of energy and carbon (some of it being oxidized for energy, some being taken up as carbon source). Most microbes fall into this category.

Chemoautotrophs oxidize inorganic chemical compounds as their source of energy, and they use CO_2 as their source of carbon. Only some bacteria and archaea are in this category.

Photoautotrophs use light as their source of energy and CO_2 as their source of carbon. These photosynthetic organisms include only bacteria and those eucaryotes (plants and algae) that contain chloroplasts, which are descendants of bacteria.

Photoheterotrophs use light as their source of energy and use organic compounds as their carbon source. Some bacteria and some algae are in this category.

7.2 Microbial metabolism is also categorized on the basis of its relationship to oxygen

Molecular oxygen (O_2) is a potent oxidizing agent and can be seriously harmful if organisms do not have adequate protective mechanisms (see Section 9.21); however, the same oxidizing ability makes oxygen a highly effective electron acceptor in respiration, and hence, many organisms have adapted to growth in the presence of air (which is about 20% oxygen). These organisms are termed **aerobes,** and they use aerobic respiration as their principal form of energy generation (Table 7.1). If the compounds they oxidize are organic, they are chemoheterotrophs; if inorganic, they are chemoautotrophs. In either case, however, they use oxygen as the terminal electron acceptor.

The aerobes also include all phototrophs that use oxygenic photosynthesis. **Oxygenic photosynthesis** uses water as electron donor and produces oxygen. Because oxygen is produced in this process, all phototrophs that use it have to have the necessary protective mechanisms. Although they may not require oxygen, they have to be able to live in its presence.

Other organisms have never developed the necessary protective mechanisms against the toxic effects of oxygen, and they must live in habitats from which oxygen is excluded (we discuss how in Section 16.3)—mainly the intestinal tracts of animals and the nutrient-rich sediments of lakes and oceans. If these **anaerobes** are chemoheterotrophs, they normally use fermentation as their energy-generating metabolism; however, few of them can use **anaerobic respiration,** in which a compound other than oxygen serves as the electron acceptor at the end of the electron transport chain.

Many bacterial phototrophs use a form of photosynthesis termed **anoxygenic photosynthesis,** in which the electron donor is something other than water, and no oxygen is produced. Most of these organisms are anaerobes, but some perform anoxygenic photosynthesis under aerobic conditions.

7.3 Many microbes can switch from one category to another, depending on the conditions

We tend to think that a particular organism will be characterized by only a single type of metabolism; animals and fungi, for instance, are chemoheterotrophs, and plants are chemoautotrophs. Among the procaryotes, however, many organisms can change their metabolism according to their environment. Most photoheterotrophs can switch to photoautotrophic growth if they run out of organic compounds or to

Table 7.1 Types of Aerobic and Anaerobic Metabolism	
Relationship to O_2	**Type of Metabolism**
Aerobic	Aerobic respiration Oxygenic photosynthesis Anoxygenic photosynthesis
Anaerobic	Fermentation Anaerobic respiration Anoxygenic photosynthesis

chemoheterotrophic growth in the dark. Some can even switch to chemoautotrophic growth if light and organic compounds are both absent.

Most chemoheterotrophs, however, are restricted to that mode of growth, although they may be capable of switching among many different organic compounds as their principal carbon and energy source.

As with the principal metabolic types, many microbes can switch between aerobic and anaerobic metabolism as necessary. Most commonly, many chemoheterotrophs use aerobic respiration when oxygen is present and fermentation (or in some cases, anaerobic respiration) when it is absent. Similarly, some bacteria can grow anaerobically in the light, using anoxygenic photosynthesis, or aerobically as chemoheterotrophs using aerobic respiration. This kind of metabolic and environmental flexibility has been an important advantage to microbes in changeable environments.

The term **facultative** is often used to designate those organisms that can use two or more modes of metabolism. Thus, one speaks of **facultative anaerobes** to indicate organisms that can grow aerobically or anaerobically or **facultative chemoautotrophs** to designate an organism that can grow chemoautotrophically, but that has other alternatives available to it. The opposite of facultative is **obligate;** thus, we speak of obligate anaerobes or obligate chemoautotrophs.

7.4 Central metabolism interconverts a small number of small organic compounds needed for biosynthesis

All of the monomers necessary for macromolecule synthesis and all of the building blocks of lipid synthesis are made from a small number of organic compounds. These **precursor metabolites** can be interconverted, and thus, in most cases, if a cell is provided with any compound that it can use to make any one of the precursor metabolites, it can make all of the others. The biochemical pathways of interconversion are collectively termed **central metabolism** and are present in all cells (Figure 7.1).

Central metabolism obviously serves a biosynthetic role in all cells, but it also may serve to provide ATP as well. The most obvious example is the glycolytic pathway, taught in all basic biology classes as a route of ATP generation from glucose; however, it also provides several of the precursor metabolites for biosynthesis, and thus, it serves biosynthesis as well. When chemoheterotrophs grow on compounds other than glucose (e.g., pyruvate, which many cells can use as their carbon and energy source), the glycolytic pathway may serve only biosynthesis by operating "backward." Glycolysis also has an exclusively biosynthetic role in chemoautotrophs and in phototrophs.

7.5 The pathways of central metabolism, biosynthesis, and macromolecule assembly are nearly identical in all organisms

All organisms have pretty much the same sets of reactions of central metabolism. Furthermore, in nearly all organisms, the pathways by which the precursor metabolites are converted to the amino acids, purine and pyrimidine nucleotides, and enzyme cofactors are also the same. Even the reactions by which macromolecules are assembled from their monomers are nearly identical in all organisms. This commonality, often termed the *unity of biochemistry,* is powerful evidence for the descent of all life on earth from a single common ancestor.

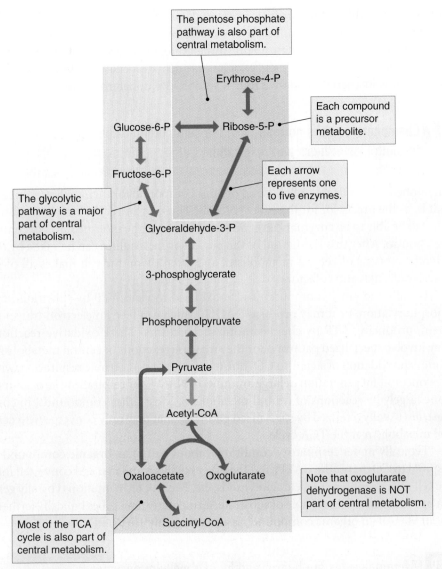

The pentose phosphate pathway is also part of central metabolism.

Each compound is a precursor metabolite.

Each arrow represents one to five enzymes.

The glycolytic pathway is a major part of central metabolism.

Note that oxoglutarate dehydrogenase is NOT part of central metabolism.

Most of the TCA cycle is also part of central metabolism.

Erythrose-4-P

Glucose-6-P

Ribose-5-P

Fructose-6-P

Glyceraldehyde-3-P

3-phosphoglycerate

Phosphoenolpyruvate

Pyruvate

Acetyl-CoA

Oxaloacetate

Oxoglutarate

Succinyl-CoA

FIGURE 7.1 Pathways of central metabolism.

7.6 Oxoglutarate dehydrogenase is found principally in aerobic chemoheterotrophs

The enzyme of the tricarboxylic cycle (TCA cycle; also known as the Krebs cycle and the citric acid cycle) that convert oxoglutarate to succinyl-CoA, **oxoglutarate dehydrogenase,** is not part of central metabolism. Indeed, many organisms lack this enzyme entirely and thus do not have a TCA cycle. In them, the enzymes of the TCA cycle are organized into two separate branches, leading to oxoglutarate and succinyl-CoA respectively. In these organisms, the TCA cycle enzymes have a biosynthetic function only. Organisms that usually lack oxoglutarate dehydrogenase include obligate anaerobes, facultative anaerobes when they are growing under anaerobic conditions, obligate chemoautotrophs, and obligate photoautotrophs.

Nearly all aerobes have the enzyme, including facultative anaerobes when they are growing under aerobic conditions, and facultative chemoautotrophs. This pattern of distribution has led to the conclusion that the TCA cycle evolved first as a branched biosynthetic pathway, as is still seen in the central metabolism of anaerobes. Only later, when oxygen became abundant in the atmosphere and aerobic respiration became a metabolic possibility, did oxoglutarate dehydrogenase evolve and the TCA cycle assume its current oxidative role in aerobic chemoheterotrophs.

7.7 Chemoheterotrophs normally use the same organic compound for both biosynthesis and respiration

Chemoheterotrophic microbes use an organic compound for their carbon source—that is, as the precursor for their biosynthesis. Thus, whatever their carbon source, it has to be able to be enzymatically converted to one of the intermediates of central metabolism. After this is done, all of the precursor metabolites can be made by the interconversions of central metabolism, and from them, the cell makes all of its macromolecules and cofactors.

Usually the same compound is also oxidized to yield ATP by substrate-level phosphorylation, or it may produce NADH, which can feed an electron transport chain, producing ATP by chemiosmotic mechanisms. These oxidative reactions may involve specialized pathways, or they may use reactions of central metabolism. Complete oxidation of sugars to CO_2 during aerobic and anaerobic respiration, with the concomitant generation of large amounts of NADH for respiration, is, of course, done largely by reactions of central metabolism. Most other compounds, in contrast, are usually oxidized by specialized pathways that may only converge with central metabolism at the TCA cycle.

Typically under respiratory conditions, about half of an organic compound is respired to CO_2 to produce ATP, and the other half is assimilated and converted into the cell's own proteins, nucleic acids, lipids, etc. Because fermentation typically generates much less ATP per mole of substrate, fermentative microbes typically ferment about 90% of an organic compound, assimilating only 10% of it.

7.8 Macromolecules are hydrolyzed by extracellular enzymes

Because most of the organic material in the biosphere is in the form of macromolecules, the first step in the degradation of most organic matter is hydrolysis to monomers. Eucaryotic microbes do this within phagolysosomes during intracellular digestion, and the monomers are taken into the cytosol by permeases in the phagolysosome membrane.

Procaryotes, in contrast, are unable to ingest large food molecules, and thus, they hydrolyze macromolecules by secreting extracellular hydrolases and then taking up by active transport the monomers produced. This is a wasteful process, as many of the monomers produced diffuse off and are never recovered by the cell secreting the hydrolases. Indeed, many procaryotes do not themselves produce extracellular hydrolases; rather, they depend on others to secrete the hydrolases, and they then compete for the products.

Probably in order to reduce such competition, some microbes have **exocellular** hydrolases—these are located on the external surface of the cell, but they are not released into the environment. Thus, these cells grow physically attached to their macromolecular substrate, and this close proximity gains them as large a share of the products as possible.

7.9 Most solute uptake by procaryotes is by active transport

Procaryotes typically live in environments that are chronically very low in nutrients. Competition for the traces of nutrients is fierce. This forces them to use active transport by high-affinity permeases to concentrate most of the compounds they take up. Indeed, it is the use of these permeases that maintains habitats so deficient in nutrients. In almost all habitats, almost all of the available soluble nutrients have been assimilated into microbial cells or oxidized to CO_2.

7.10 In many bacteria, sugar uptake is by a phosphotransferase system

All sugars are readily converted to intermediates of central metabolism, and thus, they are widely used as sources of carbon and energy. There is thus intense competition for sugars among microbes, and their environmental concentration is typically low. Thus, like other nutrients, their uptake is by high-affinity permeases and is driven by energy expenditure. This is, however, often technically not active transport. Active transport describes systems that can generate an internal solute concentration greater than the external concentration. Many bacteria use a special type of transport system for sugars, called a **phosphotransferase system,** that phosphorylates the sugar as it is transported. Thus, the system does not actually generate a concentration gradient because what accumulates internally is not the sugar itself, but rather the sugar phosphate (Figure 7.2).

Energy Metabolism

Because the reactions of biosynthesis and macromolecule assembly in microbes are essentially the same as in other organisms, we do not dwell on them here. Instead, in this chapter and the next, we focus on the reactions in which microbes have much greater diversity than macro-organisms: their mechanisms of energy conservation.

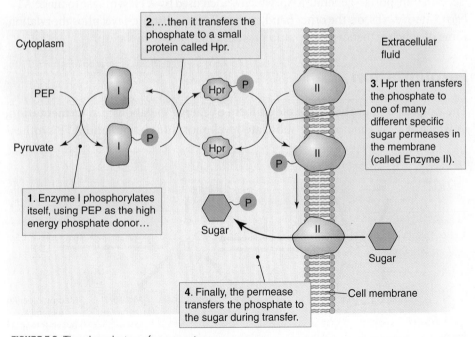

FIGURE 7.2 The phosphotransferase system.

All cells need chemical energy in the form of ATP, GTP, UTP, phosphoenolpyruvate, and a few others. Of these, ATP is used in most energy-requiring reactions of biosynthesis; however, GTP and UTP are frequently used in macromolecular assembly reactions, and PEP is used in many transport reactions. Not only is ATP used in most reactions, but also, it is the phosphate donor to replenish the others when they are depleted. For instance, a number of reactions use GTP, forming GDP; the GDP is converted back to GTP by transferring a phosphate from ATP. Thus, ATP can be considered the general "energy currency" of the cell, and the point of energy metabolism is to replenish ATP pools by driving the phosphorylation of ADP.

There is another class of reactions that require an input of energy, however, but for which the energy is not chemical, but rather electrical and osmotic. Many active-transport reactions, for instance, and the rotation of the procaryotic flagellum are driven by the movement of ions down a concentration gradient. The ion is usually, but not always, H^+, and the chemiosmotic potential that releases energy during ionic movement is due to the exterior of the cell being more acidic and more positively charged relative to the interior of the cell. This membrane potential is symbolized in several different ways; we use the term **protonic potential,** symbolized as **Δp.**

A chemiosmotic potential can be made in two ways (Figure 7.3). In respiratory or phototrophic cells, electron transport systems act as proton pumps to transfer protons across the cell membrane (in procaryotes) or the inner membrane of mitochondria and the thylakoid membranes in chloroplasts (of eucaryotes). Alternatively, the enzyme **ATP synthase** (also called ATPase) can pump protons across the membrane using ATP as a source of energy.

ATP synthase can also be used in the opposite direction, to make ATP from ADP and phosphate, driven by the entry of protons. Thus, ATP synthase is capable of interconverting the two forms of energy. Respiratory and phototrophic cells use electron transport to generate a Δp, which is then used by ATP synthase to make ATP. Fermentative cells, on the other hand, make ATP by substrate-level phosphorylation, some of which is then used to pump protons out to produce the Δp.

Fermentation

The simplest and probably the oldest form of energy metabolism is **fermentation,** in which ATP is generated by substrate-level phosphorylation, and ATP synthase

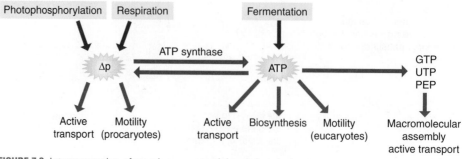

FIGURE 7.3 Interconversion of membrane potential and chemical energy.

CHAPTER 7 MICROBIAL METABOLISM: FERMENTATION AND RESPIRATION

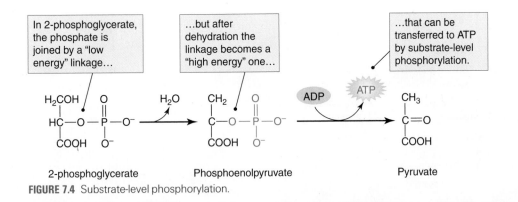

In 2-phosphoglycerate, the phosphate is joined by a "low energy" linkage…

…but after dehydration the linkage becomes a "high energy" one…

…that can be transferred to ATP by substrate-level phosphorylation.

2-phosphoglycerate   Phosphoenolpyruvate   Pyruvate

FIGURE 7.4 Substrate-level phosphorylation.

functions in the direction of ATP hydrolysis to generate a Δp. There are fermentative members of each of the domains, archaea, bacteria, and eucarya, and it looks as if this mode of life is the ancestral one.

7.12 Fermentation is a mode of chemotrophic energy metabolism in which most or all of the ATP is made by substrate-level phosphorylation

In fermentation, organic chemical compounds are converted to organic end products, in the process of which one or more substrate-level phosphorylations occur. The basic strategy of a substrate-level phosphorylation is to phosphorylate an organic compound with inorganic phosphate and then transfer the phosphate to ADP to form ATP. In order to do this, the phosphate has to be added to the organic compound as a "low-energy" phosphate and then converted to a "high-energy" phosphate by rearranging the compound. For instance, in glycolysis, 2-phosphoglycerate, whose phosphate is linked with a "low-energy" bond, is dehydrated to produce phosphoenol pyruvate, in which the bond is "high energy" (Figure 7.4). The phosphate is then transferred to ADP.

7.13 Fermentation produces end products at the same average redox level as the substrates

In fermentations, no exogenous electron acceptor such as oxygen or nitrate is available; hence, there can be no net oxidation of the carbon atoms of the substrate. Because there is nowhere to place electrons, organic carbon compounds are simply rearranged. During this rearrangement, oxidation reactions yield free energy that can be coupled to substrate-level phosphorylation. Three different strategies are used to incorporate oxidation reactions into fermentation pathways (Figure 7.5): (1) part of the substrate molecule is oxidized, and part of it is reduced; (2) the substrate is first oxidized, and then the oxidized intermediate is reduced; and (3) two different substrates are used, one being oxidized and the other being reduced.

7.14 Some fermentations do not involve redox reactions

The great majority of fermentations involve redox reactions. There are a few that do not, however, in which the chemistry is such that a substrate-level phosphorylation

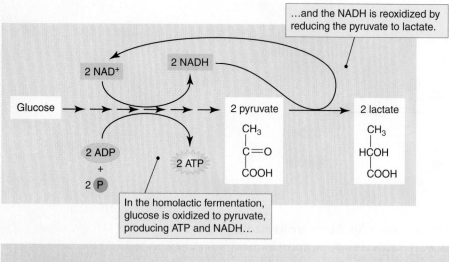

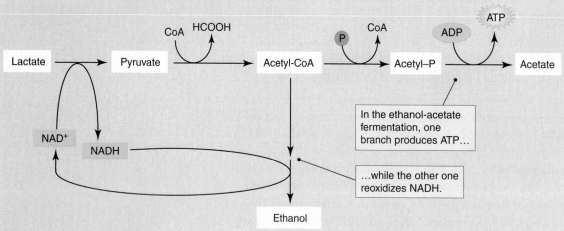

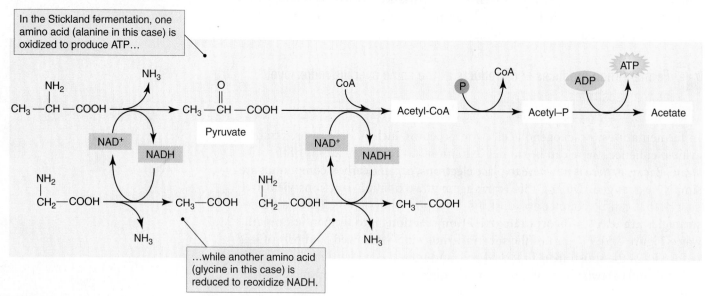

FIGURE 7.5 The homolactic, ethanol-acetate, and Stickland fermentations.

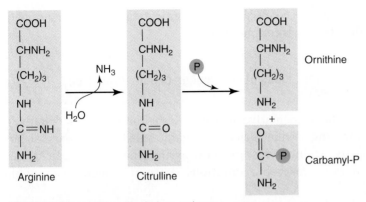

FIGURE 7.6 The arginine dihydrolase pathway.

can occur without any oxidation of the organic substrate. For instance, a common fermentation of the amino acid arginine (Figure 7.6) involves non-oxidative or reductive rearrangements and cleavage to produce ornithine and the high-energy compound carbamoyl phosphate, which can transfer its phosphate to ADP.

7.15 Many fermentations involve a minor component of chemiosmotic energy generation

Many fermentations involve reactions that are catalyzed by membrane enzymes and that pump protons or other cations. Usually the number of ions pumped is small, and thus, the metabolism remains fermentative, with the bulk of the ATP generation being by substrate-level phosphorylation; however, the ion pumping saves some ATP that would otherwise have to be used to maintain the Δp. The most common process is called **fumarate respiration** (Figure 7.7): the reduction of fumarate to succinate, catalyzed by **fumarate reductase,** a membrane enzyme that pumps two

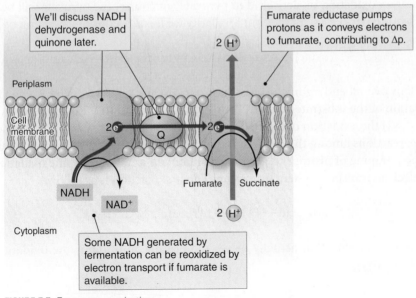

FIGURE 7.7 Fumarate respiration.

Many fermentations involve a minor component of chemiosmotic energy generation

H⁺ per fumarate reduced. Because many fermentations produce succinate as an end product, this strategy is common.

7.16 Many different kinds of compounds can be fermented

Collectively, the fermentative microbes (either obligate or facultative) are capable of fermenting a wide variety of organic substrates: sugars, sugar alcohols, amino sugars, amino acids, purines, pyrimidines, organic acids, and others. Thus, even under anaerobic conditions, substantial degradation of biomass can occur.

Individual fermentative organisms are much more limited, however, each specializing in the fermentation of one or a few different types of compound.

Chemoheterotrophic Respiration

Respiration is a much more efficient mode of energy metabolism than fermentation because it couples the oxidation of organic compounds to the reduction of a separate electron acceptor (usually oxygen). The presence of an electron acceptor separate from the organic substrate allows the latter to be oxidized completely, thus liberating more energy than if significant amounts of it have to be excreted as waste. Furthermore, respiration involves an additional and highly effective mechanism of energy conservation, beyond the substrate-level phosphorylation characteristic of fermentation: chemiosmotic energy conservation.

7.17 Respiration is a type of chemotrophic energy metabolism in which most or all of the ATP is made by chemiosmotic means

Respiration is a mode of energy metabolism in which the oxidation of a chemical compound provides the energy, which is captured by the cell in the form of a Δp. In principle, what happens in respiration is that the oxidizing half-reaction is physically and temporarily separated from the reducing half-reaction. Take, for instance, the respiratory oxidation of lactic acid to pyruvate (of course, the pyruvate will be further oxidized by the TCA cycle, but for clarity we focus on this first reaction only):

$$\text{lactic acid} + 1/2\ O_2 \rightarrow \text{pyruvate} + H_2O$$

This overall endergonic process has four distinct phases (Figure 7.8): (1) the oxidation of the substrate (lactate) coupled to the reduction of the electron carrier NAD⁺; (2) the oxidation of NADH by the electron transport system; (3) a series of redox reactions among the carriers of the electron transport system, accompanied by the pumping of protons to generate or maintain a Δp; and (4) the oxidation of the electron transport system by oxygen. The first reaction is as follows:

$$\text{lactic acid} + NAD^+ \rightarrow \text{pyruvate} + NADH + H^+$$

The redox reaction that actually drives the proton pumping is the oxidation of NADH by oxygen:

$$NADH + H^+ + 1/2\ O_2 \rightarrow NAD^+ + H_2O$$

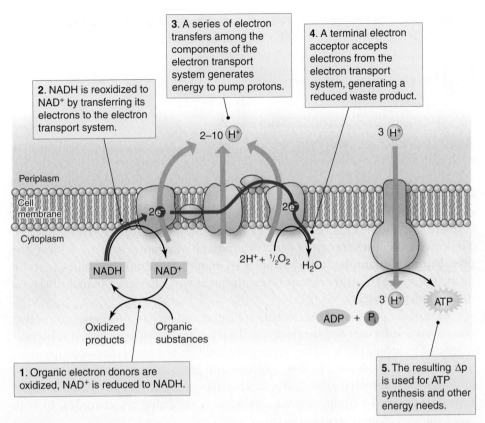

3. A series of electron transfers among the components of the electron transport system generates energy to pump protons.

4. A terminal electron acceptor accepts electrons from the electron transport system, generating a reduced waste product.

2. NADH is reoxidized to NAD$^+$ by transferring its electrons to the electron transport system.

2–10 (H$^+$)

3 (H$^+$)

Periplasm

Cell membrane

Cytoplasm

2e$^-$

2e$^-$

NADH

NAD$^+$

2H$^+$ + $^1/_2$O$_2$

H$_2$O

3 (H$^+$)

3 (H$^+$)

ATP

ADP + P$_i$

Oxidized products

Organic substances

1. Organic electron donors are oxidized, NAD$^+$ is reduced to NADH.

5. The resulting Δp is used for ATP synthesis and other energy needs.

FIGURE 7.8 Outline of respiration.

The two half-reactions of this overall reaction take place separately from each other, however. First, NADH oxidation is coupled to reduction of the first component of the electron transport system:

$$NADH \rightarrow NAD^+ + H^+ + 2e^-$$

The electrons then make their way through the electron transport system in a series of electron transfers, some of which are coupled to the pumping of protons across the membrane. Finally, the last carrier in the electron transport chain donates the electrons to oxygen, producing water:

$$2e^- + 2H^+ + 1/2\ O_2 \rightarrow 2H_2O$$

Thus, the overall strategy is to interpose the electron transport chain between the oxidizing and reducing half reactions, using the free energy that the combined reactions produce to generate the Δp.

7.18 Most oxidations of organic compounds reduce NAD$^+$

Every oxidation must be accompanied by a simultaneous reduction, as electrons removed from one compound (oxidation) must be added to another (reduction)—electrons cannot float freely in solution. Most of the soluble oxidation reactions of metabolism use the same electron acceptor: NAD$^+$. This is advantageous because a single reduced electron carrier (NADH), derived from hundreds of different chemical oxidations, can be reoxidized by the same component of the electron transport system (NADH dehydrogenase, called complex I in mitochondria).

7.19 The electron transport system consists of two or three separate complexes of protein

Respiratory electron transport systems consist of two or three complexes of protein, each embedded in the same membrane (Figure 7.9). For procaryotes, this is the cell membrane; for mitochondria, it is the inner membrane (derived evolutionarily from the procaryotic cell membrane). One of these complexes (the **dehydrogenase complex**) oxidizes a soluble organic compound—normally this is NADH. Another (the **oxidase complex,** called complex IV in mitochondria) reduces oxygen to water. These two complexes are components of almost all respiratory systems. A third (the **cytochrome b/c complex,** called complex III in mitochondria) constitutes a proton pump (although the other two are often pumps as well); however, in many bacteria this complex is absent.

Each of the proteins of the electron transport system has prosthetic groups that are responsible for carrying the electrons. That is, these groups can be reversibly oxidized and reduced. The most common prosthetic groups are **hemes**—tetrapyrroles with an iron atom bound in their center—and **iron-sulfur centers**—iron atoms bound by the sulfhydryl groups of several cysteine residues (Figure 7.10). Electron transport proteins with heme prosthetic groups are called **cytochromes.** In both cases, the iron atoms carry the electrons.

Most of the carrier proteins have several prosthetic groups, spaced closely enough together to allow electrons to hop from one to another through the protein. In most cases, as the electrons pass through the protein, they cause conformational

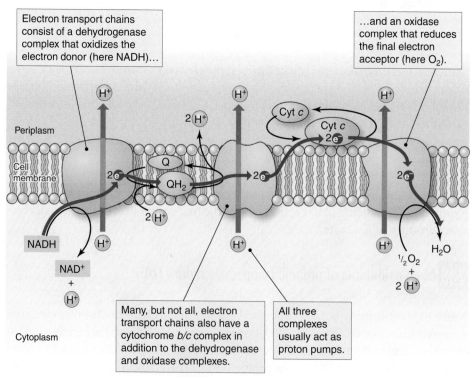

FIGURE 7.9 Components of electron transport systems.

CHAPTER 7 MICROBIAL METABOLISM: FERMENTATION AND RESPIRATION

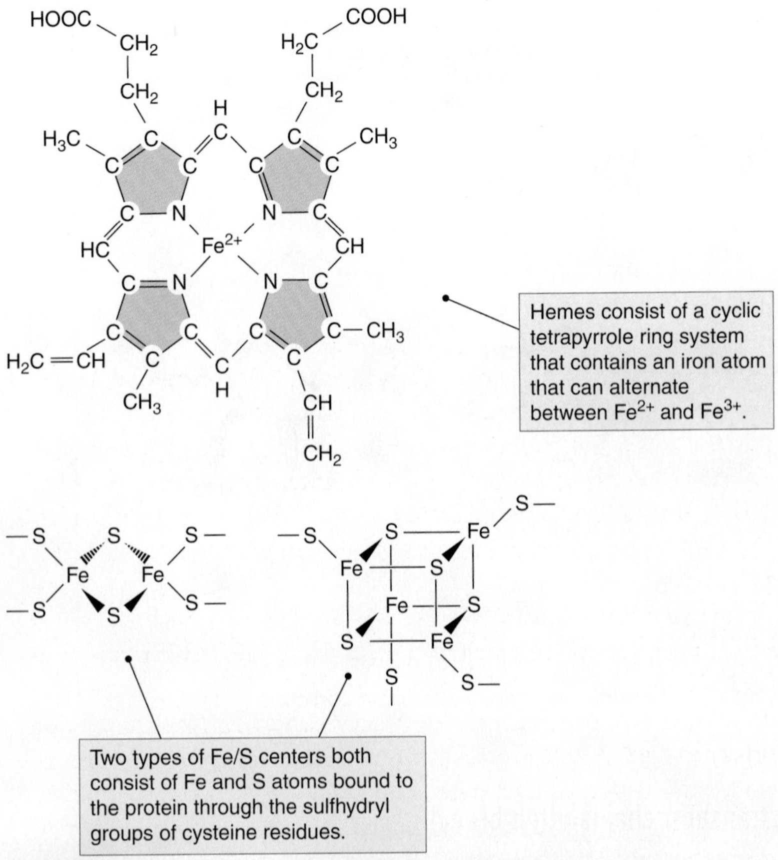

Hemes consist of a cyclic tetrapyrrole ring system that contains an iron atom that can alternate between Fe^{2+} and Fe^{3+}.

Two types of Fe/S centers both consist of Fe and S atoms bound to the protein through the sulfhydryl groups of cysteine residues.

FIGURE 7.10 Heme and iron-sulfur centers.

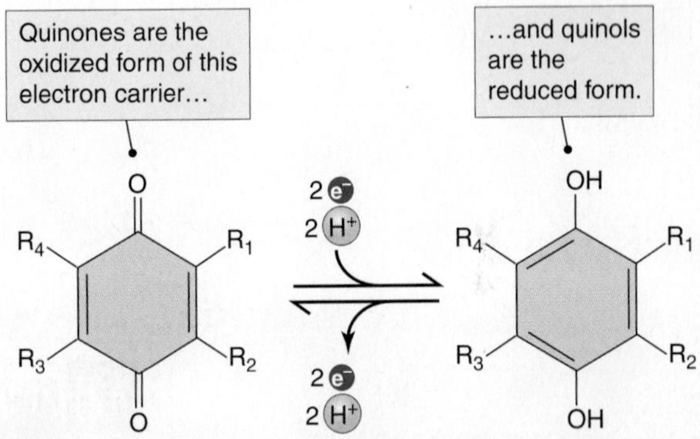

Quinones are the oxidized form of this electron carrier...

...and quinols are the reduced form.

FIGURE 7.11 The oxidized and reduced forms of quinone.

changes that result in protons being released on the outside of the membrane and then replaced from the inside. They thus act as redox-driven proton pumps.

Transfer of electrons among these two or three components is by small, mobile carriers (the protein complexes are quite large, and although they do diffuse within the plane of the membrane, their diffusion rates are slow). One or another quinone (Figure 7.11) mediates transfer from the dehydrogenase complex to the cytochrome b/c complex, and a small, soluble cytochrome c normally mediates transfer of electrons from the cytochrome b/c complex to the oxidase. In procaryotes, this soluble cytochrome c is in the periplasm; in mitochondria, it is in the equivalent space between the two membranes.

7.20 Organisms that lack the cytochrome b/c complex transfer electrons directly from quinol to the oxidase complex

Many respiratory procaryotes lack the cytochrome b/c complex, as well as the periplasmic cytochrome c (Figure 7.12). These organisms can be readily identified by the **oxidase test,** a simple test for the presence of the periplasmic cytochrome c. Oxidase-positive organisms have the cytochrome b/c complex and the soluble cytochrome c; oxidase-negative organisms lack both. In these organisms, the terminal oxidase is called a **quinol oxidase,** as it oxidizes quinol. The oxidase complex found in oxidase-positive organisms is called a **cytochrome oxidase,** as it oxidizes cytochrome c.

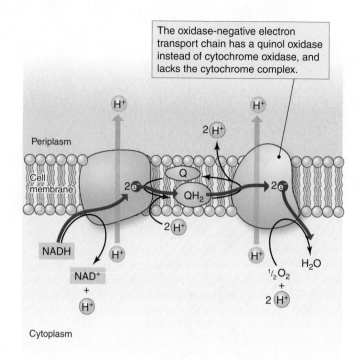

The oxidase-negative electron transport chain has a quinol oxidase instead of cytochrome oxidase, and lacks the cytochrome complex.

FIGURE 7.12 The oxidase-negative electron transport chain.

7.21 Electron transport chains often branch

Many organisms that possess the cytochrome b/c complex, periplasmic cytochrome c, and cytochrome oxidase can also make quinol oxidase (Figure 7.13). Thus, they have a branched electron transport chain. Where the electrons go from the quinol depends on what the quinol encounters first: cytochrome b/c complex, or quinol oxidase. The ratios of the various carriers in a branched electron transport system are often modulated in response to environmental conditions. Usually in well-aerated conditions the amount of quinol oxidase is low, and the amounts of cytochrome b/c complex and cytochrome oxidase are high. Under limited oxygen supply the situation is reversed. The quinol oxidase has a greater affinity for oxygen than cytochrome oxidase and is therefore more efficient at low O_2 concentrations. At high-oxygen concentrations, this is less important, and the greater H^+ pumping of the oxidase positive electron transport system makes it preferable.

7.22 Alternate dehydrogenase complexes are used for different compounds

Although the oxidation of most organic compounds is coupled to the reduction of NAD^+ and therefore couples to the electron transport system via the NADH-dehydrogenase, there are several organic compounds that have their own dehydrogenases and thus reduce quinone directly. Succinate is the best known example, oxidized by mitochondria as well as by many procaryotes. The enzyme in the TCA cycle that oxidizes succinate, called **succinate dehydrogenase** (complex II in mitochondria), is embedded in the cell membrane (inner membrane of mitochondria), and it couples the oxidation of succinate to the reduction of quinone (Figure 7.14). (Many textbooks say that the electron acceptor for this enzyme is FAD, but this is incorrect; FAD is a prosthetic group on the enzyme and is only

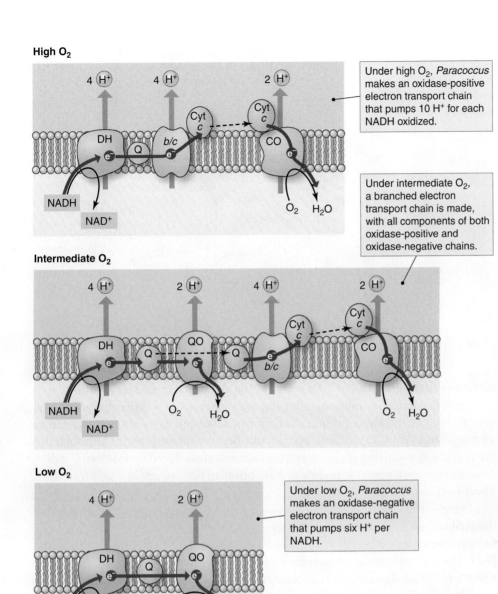

High O₂

4 H^+ 4 H^+ 2 H^+

DH   Q   b/c   Cyt c   Cyt c   CO

NADH

NAD⁺

O₂   H₂O

Under high O₂, *Paracoccus* makes an oxidase-positive electron transport chain that pumps 10 H^+ for each NADH oxidized.

Intermediate O₂

4 H^+ 2 H^+ 4 H^+ 2 H^+

DH   Q   QO   Q   b/c   Cyt c   Cyt c   CO

NADH

NAD⁺

O₂   H₂O

O₂   H₂O

Under intermediate O₂, a branched electron transport chain is made, with all components of both oxidase-positive and oxidase-negative chains.

Low O₂

4 H^+ 2 H^+

DH   Q   QO

NADH

NAD⁺

O₂   H₂O

Under low O₂, *Paracoccus* makes an oxidase-negative electron transport chain that pumps six H^+ per NADH.

FIGURE 7.13 *Paracoccus* electron transport system.

Succinate dehydrogenase is an alternate dehydrogenase complex that reduces quinone.

Periplasm

Cell membrane

Q

2 e^-

QH₂

Cytoplasm

2 H^+

Succinate

Fumarate + 2 H^+

FIGURE 7.14 Succinate dehydrogenase.

temporarily reduced in the course of passing electrons from succinate to the quinone.)

A number of other compounds are oxidized by their own specific, membrane-embedded dehydrogenase enzymes that reduce quinone, including formic acid, glycerol-3-P, and H₂ gas.

7.23 Alternate oxidases are used for different electron acceptors

The most common electron acceptor in respiration is oxygen; however, many procaryotes are able to use nitrate or nitrite in the absence of oxygen. This process is called **nitrate respiration** when the nitrate is reduced to nitrite or ammonia. Nitrate respiration depends on the possession of an alternate oxidase (confusingly called

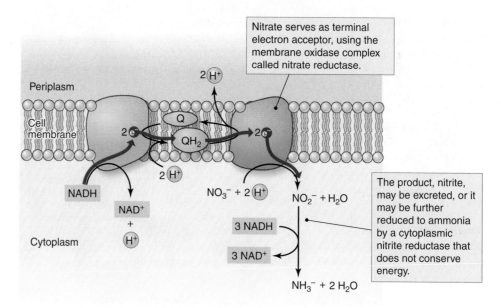

FIGURE 7.15 Nitrate respiration and ammonification.

nitrate reductase). This complex replaces the cytochrome oxidase or quinol oxidase under anaerobic conditions, allowing normal respiration in the absence of oxygen, so long as nitrate is present. The end-product nitrite is often converted to ammonia in a cytosolic reaction, presumably because nitrite is toxic. Although this is a reduction, requiring six electrons for each nitrate reduced to ammonia, it does not conserve any energy, as no protons are pumped nor is there any substrate-level phosphorylation (Figure 7.15).

When the nitrate is reduced to the gaseous end-product N_2, the process is called **denitrification.** This ability requires the simultaneous presence of three alternate terminal oxidases (although again called reductases): one each for NO_3^-, NO_2^-, and N_2O. These complexes serve to transfer electrons from quinol or cytochrome c to nitrate, nitrite, and so forth, but they are not usually proton pumping (Figure 7.16).

Thus, for nitrate respiration and denitrification, the anaerobic electron transport chain is a variant of the aerobic one, and the organisms that do this kind of anaerobic respiration are facultative anaerobes. Under aerobic conditions, they use

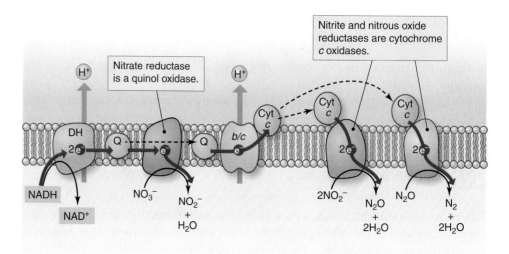

FIGURE 7.16 Denitrification in *Paracoccus.*

Table 7.2	Electron Acceptors for Anaerobic Respirations
Electron Acceptor	**Reduced Product**
NO_3^-	NO_2^-, N_2O, or N_2
SO_4^{2-}	H_2S
CO_2	CH_4
Fe^{3+} (ferric)	Fe^{2+} (ferrous)
Mn^{4+} (manganic)	Mn^{2+} (manganous)
SeO_4^{2-} (selenate)	SeO_3^{2-} (selenite)
Fumarate	Succinate
$(CH_3)_2$ SO (dimethylsulfoxide)	$(CH_3)_2$ S (dimethylsulfide)
$(CH_3)_3$ NO (trimethylamine-N-oxide)	$(CH_3)_2$ N (trimethylamine)

cytochrome oxidase or quinol oxidase to reduce oxygen as electron acceptor; nitrate reductase and other reductases are repressed. Under anaerobic conditions nitrate reductase, nitrite reductase, and so forth, are derepressed and the organisms switch to anaerobic respiration with nitrate or its products as electron acceptor.

Many other compounds can serve as terminal electron acceptors in anaerobic respirations (Table 7.2). Some anaerobic respirations, like fumarate respiration, denitrification, and nitrate respiration (discussed previously), and sulfur- and sulfate-respiration, use an electron transport system identical in general principles to the electron transport system of aerobic respiration. Many of these organisms are facultative anaerobes. For other anaerobic respirations, however, the components of the electron transport systems are different from those of the aerobic system, and the organisms that do these anaerobic respirations are mostly obligate anaerobes.

7.24 ATP yields of microbial respiration differ greatly

Microbial respiratory chains differ greatly in their efficiency, from less than one ATP/NADH to over three. At the high end are bacteria that have a respiratory chain like that found in mitochondria, such as *Paracoccus* (a bacterium actually related to mitochondria). It pumps 10 H+/NADH: 4 by the NADH dehydrogenase, 4 by the cytochrome b/c complex, and 2 by cytochrome oxidase (see Figure 7.13). This yields 3.3 ATP/NADH, as ATP synthase requires the entry of three protons to make one ATP.

Most microbes have less efficient electron transport systems for one or more reasons: NADH dehydrogenase may not be a proton pump, or it may pump fewer than four H+ per NADH. Cytochrome b/c complex may be absent, or it may pump fewer than four H+ per NADH, or the terminal oxidase may not act as a proton pump. For instance, *E. coli* makes two different NADH dehydrogenases; one pumps four protons per NADH, and the other does not pump any. It also makes two different quinol oxidases: one pumps four protons per NADH and the other one pumps two. Thus, each individual *E. coli* cell performs a mixture of up to four different simultaneous respirations, depending on the ratios of the dehydrogenases and oxidases in its membrane, which in turn depends on the O_2 concentration, with the less efficient complexes predominating at low O_2 and the more efficient ones

1. High and intermediate O₂

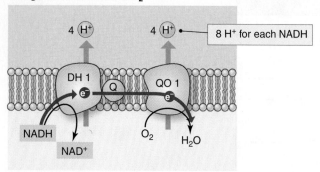

3. Intermediate O₂

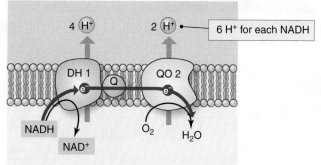

2. Intermediate O₂

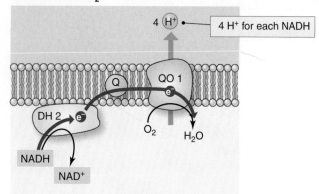

4. Low and intermediate O₂

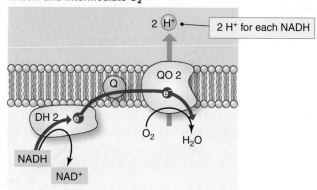

FIGURE 7.17 The four different electron transport systems in *E. coli*.

predominating at the high O_2. The stoichiometries range from two to eight $H^+/NADH$, equivalent to 0.67–2.5 ATP/NADH (Figure 7.17).

Eucaryotic microbes are generally fairly efficient because their respiration is conducted by mitochondria, which have the same efficient electron transport chain as *Paracoccus* (10 $H^+/NADH$). They only make 2.5 ATP/NADH, however, because it takes them four protons rather than three to make an ATP. This is because phosphate and ADP have to be transported into the mitochondrion in order for them to be formed into ATP, and the ATP then has to be exported (Figure 7.18). Although ATP and ADP have a simple exchange system that does not require any energy input, phosphate is transported into the mitochondrion in symport with a proton. Thus, one extra proton enters the mitochondrion for each ATP made.

Chemoautotrophic Respiration

Chemoautotrophs oxidize inorganic electron donors for energy, and as autotrophs, they need a reductant for CO_2 fixation. Thus, their inorganic electron donor has to do two things: provide electrons for energy-yielding electron transport and provide electrons for CO_2 reduction. Both processes are generally quite inefficient. The inefficiency of chemoautotrophic respiration stems largely from the fact that most of the compounds used as electron donors are not very good reducing agents (Table 7.3).

Although inefficient, this type of metabolism allows organisms to occupy a unique ecophysiological niche, where there is no competition from the hoards of heterotrophic organisms competing for dissolved organic compounds. Chemo-

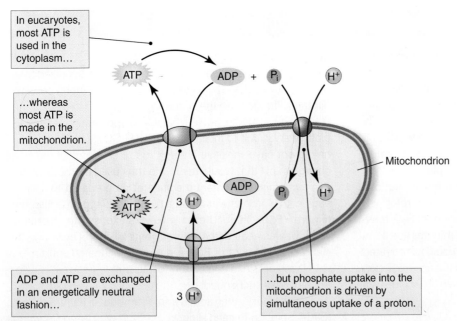

In eucaryotes, most ATP is used in the cytoplasm...

...whereas most ATP is made in the mitochondrion.

Mitochondrion

ADP and ATP are exchanged in an energetically neutral fashion...

...but phosphate uptake into the mitochondrion is driven by simultaneous uptake of a proton.

FIGURE 7.18 ATP synthesis in the mitochondrion.

autotrophic growth is a major specialization of procaryotes; both archaea and bacteria contain many groups that specialize in the oxidation of one or a few inorganic compounds.

7.25 Electrons from inorganic substrates normally enter the electron transport system at the level of quinone or cytochrome c

Because the reduction potentials of most chemoautotrophic electron donors are more positive than NADH, they cannot directly reduce NAD^+, and hence, they have to inject electrons into the electron transport chain at a lower redox level than NADH. In most cases, this is either to quinone or (in the case of oxidase positive organisms) to the periplasmic cytochrome c (Figure 7.19).

Even those electron donors, like hydrogen or carbon monoxide gases, that have standard reduction potentials more negative than NADH normally enter the electron

Table 7.3	Standard Reduction Potentials for Chemoautotrophic Electron Donors	
Electron Donor	**Oxidized Product**	**E_0^1 (mV)**
CO	CO_2	−540
H_2	$2H^+$	−410
NADH	NAD^+	−320
H_2S	S^0	−270
S^0	SO_4^{2-}	−200
NH_4^+	NO_2^-	+340
NO_2^-	NO_3^-	+430
Fe^{2+}	Fe^{3+}	+770
H_2O	O_2	+810

In 1889, the Russian microbiologist Sergius Winogradsky did an elegant series of very simple experiments that convinced him that some bacteria could grow at the expense of the oxidation of sulfur. He had observed that the aerobic filamentous organism *Beggiatoa* (Figure 7B.1) often contained numerous globules of sulfur when observed in natural habitats, in which the presence of hydrogen sulfide was conspicuous by its smell. When he removed a tuft of *Beggiatoa* cells, placed them in water on a microscope slide, and observed them, he noticed that the sulfur globules disappeared and were not reformed (Figure 7B.1b and c). If he added a small amount of H_2S, however, sulfur globules reappeared within 2 to 3 minutes and completely filled the cells within a few hours. This could be repeated many times, and H_2S presence always led to the accumulation of sulfur globules, which then disappeared when the sulfide was depleted. During these transformations, the *Beggiatoa* filaments grew longer. Clearly the sulfide was being oxidized to sulfur, which was then deposited into granules. Where, however, did the sulfur go once the sulfide was depleted?

Winogradsky suspected that the disappearance of the sulfur globules was due to their oxidation to H_2SO_4 (sulfuric acid), but he did not have sensitive enough methods to detect the small change in pH that would result. He thus used methods of inorganic chemistry to detect the sulfate. He prepared two slides, each with a drop of water and a big tuft of *Beggiatoa* filaments (Figure 7B.2). One of these slides was exposed to chloroform vapor or was heated; these treatments killed the cells, and this slide acted as a control. Both slides were incubated until the one with living cells had completely lost its sulfur globules. The control slide was unchanged.

Winogradsky then added a small amount of barium chloride ($BaCl_2$) solution to each slide. In the drop with living *Beggiatoa* cells, numerous distinctive crystals of barium sulfate ($BaSO_4$) formed, confirming that the sulfur had been converted to sulfate. In the control drop, there was barely detectable sulfate, confirming that the sulfate in the experimental drop had been produced by the living *Beggiatoa*. In several repetitions of the experiment, he made crude quantitative estimates of the amount of sulfate generated on successive days by comparing the number of crystals generated in the experimental drops to those generated from known concentrations of sulfate, generating a rough time course for the oxidation. Winogradsky did not graph his results, but we can (Figure 7B.3); the results are remarkably consistent with what we would now expect from what we know of microbial growth (see Chapter 9).

By these simple experiments, most of which involved tufts of *Beggiatoa* filaments on a microscope slide, Winogradsky combined careful microscopic observation with simple chemical experiments to show that an entirely novel mode of metabolism—chemoautotrophy—was used in this group of organisms. It would be many decades before much more would be learned about them.

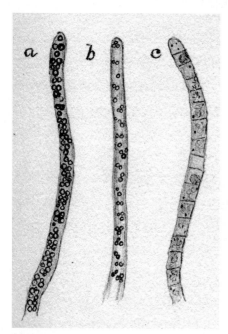

FIGURE 7B.1 *Beggiatoa.*

transport system at the quinone level. This is because the *actual* reduction potentials vary with concentration. At the 1 atm pressure of standard conditions, these gases are potent reductants; however, at the 10^{-3} atm or less that are commonly encountered in natural habitats, they are considerably less effective and hence cannot directly reduce NADH.

7.26 Oxidation of inorganic electron donors usually occurs in the periplasm

In the oxidation of sulfide, the substrate is oxidized in the periplasm. This is common for chemoautotrophic substrates; a majority of them are oxidized by enzymes in the periplasm that pass electrons to quinone or cytochrome c. This serves two purposes:

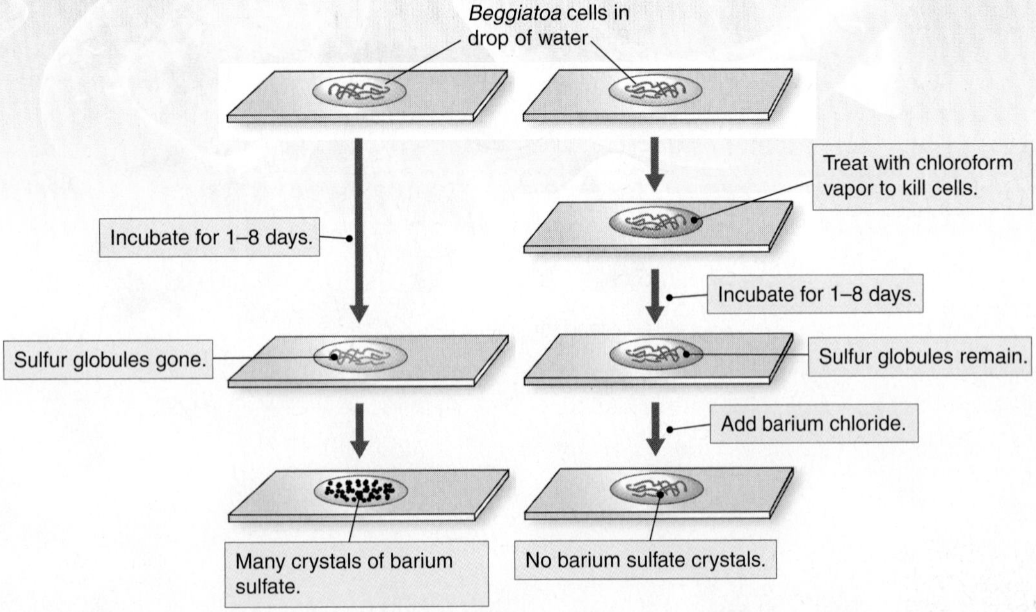
Beggiatoa cells in drop of water
Treat with chloroform vapor to kill cells.
Incubate for 1–8 days.
Incubate for 1–8 days.
Sulfur globules gone.
Sulfur globules remain.
Add barium chloride.
Many crystals of barium sulfate.
No barium sulfate crystals.

FIGURE 7B.2 Diagram of Winogradsky's experiment.

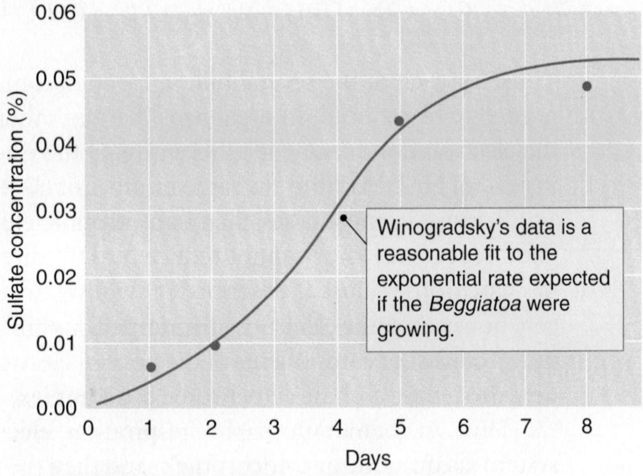
0.06
0.05
0.04
0.03
0.02
0.01
0.00
Sulfate concentration (%)
0
1
2
3
4
5
6
7
8
Days
Winogradsky's data is a reasonable fit to the exponential rate expected if the *Beggiatoa* were growing.

FIGURE 7B.3 Winogradsky's results.

first, there is no need to transport the substrate across the membrane, frequently an energy-demanding process, and second, it contributes to the Δp because protons are often produced during the oxidation reactions. Because protons will always be consumed on the inside of the membrane when oxygen is reduced to water, this is the equivalent of pumping 2 $H^+/2e^-$. This is a considerable contribution to the Δp.

7.27 Reverse electron transport is necessary for chemoautotrophs to generate reductant

Because chemoautotrophs use CO_2 as their carbon source, they need an electron donor for carbon dioxide reduction as well as for electron transport. It takes two

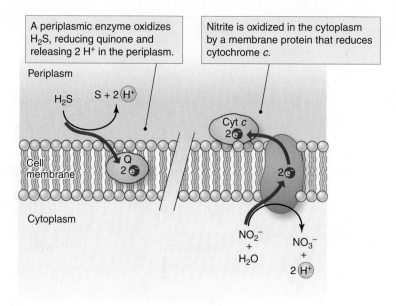

A periplasmic enzyme oxidizes H_2S, reducing quinone and releasing 2 H^+ in the periplasm.

Nitrite is oxidized in the cytoplasm by a membrane protein that reduces cytochrome c.

Periplasm

H_2S S + 2 $\text{(H}^+\text{)}$

Cyt c
2e

Cell membrane   Q
2e 2e

Cytoplasm

NO_2^-
+
H_2O NO_3^-
+
2 $\text{(H}^+\text{)}$

FIGURE 7.19 Oxidation of sulfide and nitrite.

pairs of electrons to reduce CO_2 to cell material. Most chemoautotrophs use the Calvin-Benson cycle for CO_2 fixation, in which the overall reaction is as follows:

$$CO_2 + 2NADPH + 2H^+ + 3ATP \rightarrow CH_2O + 2NAD^+ + 3ADP + 3P_i$$

Thus, for each of the several billion carbon atoms of the bacterial cell, two $NADP^+$ needed to be reduced. In organisms that use other pathways (see Sections 8.19–22), the electron donor may be different (e.g., the electron-carrying protein ferredoxin instead of NADPH), but the same number of electrons is needed.

Because in most cases chemoautotrophic electron donors are not particularly good reductants, they cannot reduce $NADP^+$ directly. Rather, they reduce quinone or cytochrome c, just as described previously. Instead of flowing "downhill" to oxygen, however, some electrons are pumped "uphill" to NAD^+. The components of this pump consist of components of the normal electron transport chain, such as NADH dehydrogenase and the cytochrome b/c complex, operating in reverse (Figure 7.20).

Thus, in chemoautotrophic respiration, electrons enter the electron transport system via quinone or cytochrome c, and then the flow splits: some of it flows downhill to oxygen, generating a Δp in the process. This Δp is used to generate ATP and also to pump the other stream of electrons uphill to generate reductant (Figure 7.21).

7.28 Transhydrogenase couples NADPH production to proton entry

Reverse electron transport produces NADH, not the NADPH needed for CO_2 fixation. In principle, it is a simple matter to produce NADPH from NADH, as both compounds have the same reduction potential. The reaction

$$NADH + NADP^+ \leftrightarrow NAD^+ + NADPH$$

thus has an equilibrium constant of 1; however, generally cells want to keep their NAD pools largely oxidized (mainly in the form of NAD^+ rather than NADH) in order to more efficiently do the oxidation reactions for which NAD^+ acts as electron

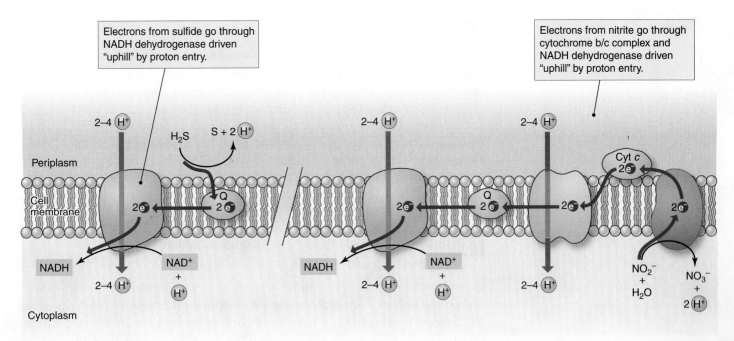

Electrons from sulfide go through NADH dehydrogenase driven "uphill" by proton entry.

Electrons from nitrite go through cytochrome b/c complex and NADH dehydrogenase driven "uphill" by proton entry.

FIGURE 7.20 NADH reduction by sulfide and nitrite.

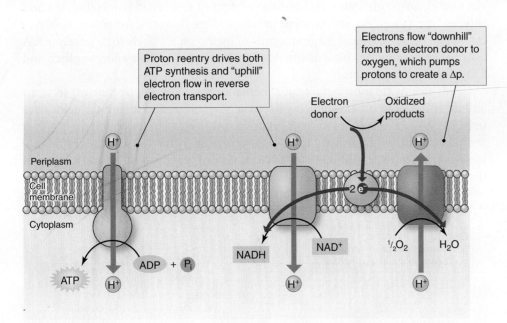

Proton reentry drives both ATP synthesis and "uphill" electron flow in reverse electron transport.

Electrons flow "downhill" from the electron donor to oxygen, which pumps protons to create a Δp.

FIGURE 7.21 Flow of electrons in chemoautotrophic respiration.

acceptor. On the other hand, they want to keep their NADP pools largely reduced (mainly NADPH rather than NADP$^+$) to more efficiently do the reduction reactions in biosynthesis for which NADPH is the usual electron donor. In order to shift the equilibrium of the reaction to the right, cells use the membrane enzyme **transhydrogenase** (Figure 7.22) that couples the reduction of NADP$^+$ by NADH to the entry of protons:

$$2H^+_{out} + NADH + NADP^+ \rightarrow NAD^+ + NADPH + 2H^+_{in}$$

The equilibrium of this reaction lies strongly to the right.

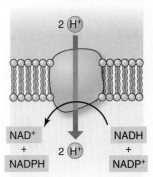

FIGURE 7.22 The transhydrogenase reaction.

Transhydrogenase couples NADPH production to proton entry

7.29 Chemoautotrophic respiration is very inefficient

Chemoautotrophic respiration is a very inefficient metabolism, largely for two related reasons: (1) because electrons enter the electron transport chain at the level of quinone or cytochrome c, few protons are pumped per pair of electrons, and (2) much of the Δp is expended on generating reductant rather than ATP. Thus, their growth rates tend to be very slow, and it takes a large amount of inorganic substrate to generate a small amount of new biomass. Despite its inefficiency, however, it has opened up a unique set of ecophysiological niches, with relatively little competition compared to that experienced by heterotrophs. Chemoautotrophs are, therefore, abundant in the world.

7.30 Anaerobic chemoautotrophic respirations commonly use H_2 as electron donor

Most of the well-studied chemoautotrophic respirations are aerobic, with oxygen as terminal electron acceptor, although a few of these aerobes are capable of denitrification. There are also some obligate anaerobes, however, that are at least facultative chemoautotrophs. In these organisms, hydrogen gas is the most commonly used electron donor; hydrogen is a common product of fermentative metabolism, and thus, in any anaerobic environment in which there is significant amounts of fermentable organic material, large amounts of hydrogen are likely to be generated. A number of procaryotes have adapted to use hydrogen as their energy source and CO_2 as their carbon source.

Two groups are most notable: the methanogenic archaea (see the next section) and the sulfur- and sulfate-reducing bacteria. The sulfur- and sulfate-reducing bacteria are obligate anaerobes that are found in nearly all anaerobic habitats, but they are most prominent in the sediments of the coastal shelves because of the abundance of sulfate in seawater. Although they are capable of heterotrophic growth, most of them can also grow autotrophically with H_2 as their electron donor and sulfate or sulfur as their electron acceptor (Figure 7.23). The details of their electron transport

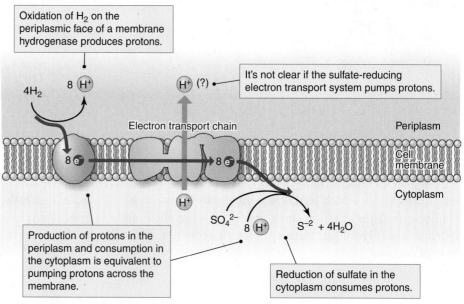

FIGURE 7.23 Chemoautotrophic respiration by sulfate reducers.

CHAPTER 7 MICROBIAL METABOLISM: FERMENTATION AND RESPIRATION

system have not been worked out, but they contain cytochromes and quinones and thus probably have a relatively ordinary system.

7.31 Methanogenesis pumps ions without a cytochrome-based electron transport system

Methanogenesis is the other major strictly anaerobic chemoautotrophic respiration. It is unique to the archaea and is the principal consumer of hydrogen in anaerobic soils, fresh water sediments, and the animal intestinal tract. In marine sediments, the methanogens cannot compete with the sulfate reducers, which have a more efficient hydrogenase.

In methanogenesis, CO_2 is reduced to methane, usually with H_2 as the electron donor:

$$4H_2 + CO_2 \rightarrow CH_4 + 2H_2O$$

Only the last few steps in the biochemical pathway of this reaction appear to be linked to the generation of a Δp. By this point, the CO_2 has been reduced to a methyl group, carried on the unique coenzyme, **coenzyme M.** The membrane enzyme **methyl-coenzyme M reductase** then catalyzes the transfer of the CoM group to the sulfhydryl group of another unique cofactor, **7-mercaptoheptanoylthreonine phosphate** (or HTP-SH), releasing methane, and the two coenzymes joined by a disulfide bond (Figure 7.24). Finally, the HTTP-CoM is reductively split to regenerate the two separate coenzymes.

Energy is conserved in this pathway at two points. First, the methyl-CoM reductase enzyme is a membrane protein that pumps sodium ions out of the cell simultaneously with the transfer of the CoM to HTTP to contribute to the electrical

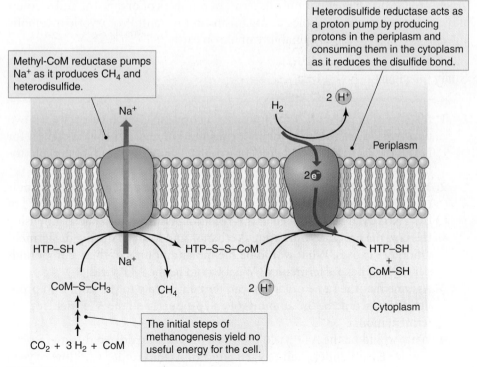

FIGURE 7.24 Biochemistry of methanogenesis.

potential across the cell membrane. Second, the enzyme that splits the CoM-HTTP, **heterodisulfide reductase,** is also a membrane protein, which pumps protons to contribute to the protonic potential. In some methanogens, electrons from H_2 pass through a short electron transport chain that includes a phenazine compound (that acts like a quinone) and a cytochrome. Thus there may be some additional proton pumping beyond that shown in Figure 7.24.

Thus, in methanogenesis, each CO_2 reduced to methane results in the pumping of only a couple of protons and a sodium ion, a quite inefficient form of metabolism, but one that capitalizes on two compounds found abundantly in anaerobic environments as fermentation products: CO_2 and H_2. Methanogens have thus been immensely successful in most anaerobic environments.

Summary

The chemotrophic procaryotes have a wide variety of types of ATP-generating metabolism—much more diverse than eucaryotes. Indeed, most of the diversity of procaryotes is biochemical, rather than the morphological diversity that characterizes eucaryotes. It is more what they do than what they look like that characterizes the many groups of procaryotes.

A number of modes of metabolism are unique to procaryotes, or nearly so. These include anaerobic respiration, chemoautotrophic respiration, and anoxygenic photosynthesis. Many specific pathways are also unique to procaryotes: many individual fermentation pathways, many pathways of aerobic oxidation of complex organic compounds, and all pathways of autotrophic CO_2 fixation other than the Calvin cycle (discussed in the next chapter).

Because of this, procaryotes are the most important organisms on the planet when it comes to the cycling of nutrients through the biosphere, as we discuss in Chapter 16. If all of the eucaryotes were eliminated from earth, the procaryotes would do just fine by themselves (as they did for nearly two billion years). If the procaryotes were eliminated, however, all life on earth would die out. This is, thus, a largely procaryotic planet not only in terms of numbers of organisms and amount of biomass, but also in the roles of organisms in nutrient. Procaryotic metabolic diversity is crucial to the sustainability of life on earth.

Study questions

1. A respiratory chemoheterotroph uses its organic growth substrate for two purposes: as an electron donor for respiration and to provide carbon for biosynthesis. What do chemoautotrophs, photoautotrophs, and photoheterotrophs use their growth substrates (organic or inorganic) for?
2. Many facultative anaerobes induce a novel enzyme under anaerobic conditions that converts pyruvate to acetyl-CoA by producing formate instead of CO_2. No NADH is produced in this reaction. Assume that a bacterium using this enzyme produces only acetate, ethanol, and formate during the fermentation of hexoses. What would be the molar amounts of these compounds per mole of hexose fermented? What would be the ATP yield?
3. Assume that the same organism can ferment lactate to the same end products. What would be the amounts of end products and the ATP yield for this fermentation?
4. What would be the ATP yield per mole of NADH oxidized during respiration by *E. coli* under high oxygen conditions and under very low oxygen conditions?

5. What would be the ATP yield per mole of NADH oxidized during respiration by *E. coli* under intermediate oxygen conditions, such that the high-oxygen dehydrogenase and oxidase complexes were present in the membrane at half the concentration of the low-oxygen ones?

6. What would be the theoretical maximum ATP yield per mole of NADH in *Paracoccus* growing under high-oxygen conditions? Compare this to the maximum theoretical yield of a respiratory eucaryotic cell?

7. The diagram below is a shorthand way of diagramming an electron transport system.

$$NADH \rightarrow NADH\ DH \rightarrow Q \rightarrow cyt\ b/c \rightarrow cyt\ c \rightarrow CO \rightarrow O_2$$

Make comparable diagrams for *E. coli* and *Paracoccus* growing under intermediate oxygen concentrations and for *Paracoccus* denitrifying.

8. How many moles of H_2S would a sulfur-oxidizing chemoautotroph have to oxidize to fix one mole of CO_2 in the Calvin cycle. Assume that the organism has an oxidase-positive electron transport system and that the proton-pumping stoichiometries of the protein complexes are NADH dehydrogenase pumps $4H^+$, the cytochrome complex pumps 4, and the oxidase complex pumps 2. Also assume that H_2S is oxidized only to elemental sulfur, a two-electron oxidation.

9. How many moles of nitrite would a nitrite-oxidizing chemoautotroph have to oxidize to fix one mole of CO_2 in the Calvin cycle. Assume the same electron transport system and stoichiometries as in the previous question.

8 Microbial Metabolism: Photosynthesis, Autotrophic Growth, and Nitrogen Fixation

I N THIS CHAPTER, WE continue our discussion of energy yielding metabolism by considering phototrophic metabolism, and we finish with mechanisms of CO_2 and N_2 fixation.

Photosynthesis

The term *photosynthesis* technically is a synonym for photoautotrophic growth (the "photo" refers to energy generation from light, and the "synthesis" refers to the synthesis of organic material from CO_2). Thus, it is a narrower term than phototrophic growth, which would include photoheterotrophic growth as well; however, it is commonly used in a broader sense, in which the "synthesis" refers to the synthesis of macromolecules from small molecules, either organic or CO_2. In this sense, it is synonymous with phototrophic growth; it is in this broad sense that we use the word here.

8.1 Photochemistry is mediated by membrane-embedded reaction centers

The primary light reaction in photosynthesis is mediated by reaction centers, embedded in a membrane. In almost all procaryotes, this is the cell membrane, or invaginations of it (Figure 8.1). The only exception is the cyanobacteria, which have intracellular membrane vesicles called **thylakoids,** in which the reaction centers are embedded.

Reaction centers have several prosthetic groups that can undergo reversible oxidation and reduction; thus, they can act as electron carriers. **Chlorophyll** (Figure 8.2)—a tetrapyrrole with a Mg atom in the center—is present in all reaction centers, as are iron–sulfur centers. Some reaction centers also contain **pheophytins** (chlorophyll molecules that lack the central magnesium atom) and tightly bound quinones (Figure 8.3).

8.2 Reaction centers are associated with pigment antennas

Because there are only one or two chlorophyll molecules in a reaction center, reaction centers themselves are not very efficient at absorbing light. Thus, each reaction center has associated with it a **pigment antenna:** a large group of pigment molecules that

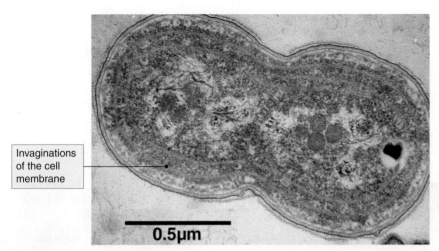

Invaginations
of the cell
membrane

0.5μm

FIGURE 8.1 Photosynthetic bacterium.

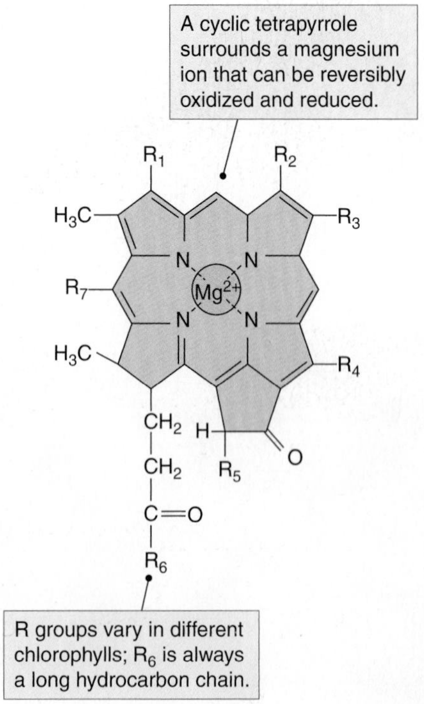

A cyclic tetrapyrrole
surrounds a magnesium
ion that can be reversibly
oxidized and reduced.

R groups vary in different
chlorophylls; R_6 is always
a long hydrocarbon chain.

FIGURE 8.2 General structure of a chlorophyll.

absorbs light. These pigment molecules are bound to proteins, and there are usually many pigment molecules per protein. Usually the pigment molecules are chlorophylls, but in the cyanobacteria, several linear tetrapyrroles called **phycobilins** are used (Figure 8.4). Usually the pigment–protein complexes are embedded in the membrane clustered around the reaction centers. When they absorb a photon of light, the energy is transmitted from pigment to pigment through the pigment antenna until it comes to the reaction center chlorophyll (Figure 8.5).

Two groups of bacteria have distinctive anatomical structures to house their pigment antennas: the cyanobacteria have **phycobilisomes** containing their phycobilin–protein complexes. The phycobilisomes are attached to the surface of the thylakoid

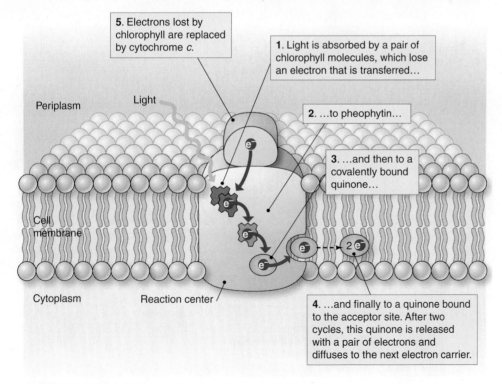

5. Electrons lost by chlorophyll are replaced by cytochrome *c*.

1. Light is absorbed by a pair of chlorophyll molecules, which lose an electron that is transferred…

2. …to pheophytin…

3. …and then to a covalently bound quinone…

4. …and finally to a quinone bound to the acceptor site. After two cycles, this quinone is released with a pair of electrons and diffuses to the next electron carrier.

Periplasm

Light

Cell membrane

Cytoplasm

Reaction center

FIGURE 8.3 A reaction center.

Reaction centers are associated with pigment antennas

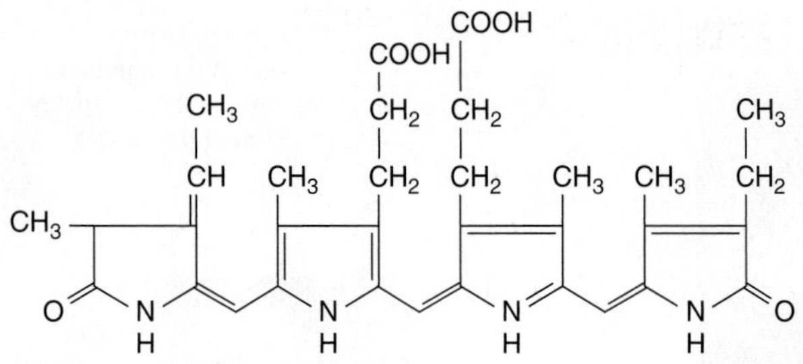

FIGURE 8.4 Phycobilin structure.

membranes on the cytoplasmic surface, in contact with the underlying reaction centers (Figure 8.6).

The green sulfur bacteria have structures termed **chlorosomes** attached to the inner surface to the cell membrane. The chlorosome has a single-layer lipid membrane around its outside, within which are stacks of the chlorophyll–protein complexes. Membrane proteins link the pigment antenna to the reaction centers in the cell membrane (Figure 8.7).

8.3 Photosynthesis is based on the cyclic transformations of chlorophyll through three different states

The reaction center chlorophyll undergoes a cyclical process of excitation, oxidation, and reduction, which is the basis of most photosynthesis (Figure 8.8). The ground-state chlorophyll is a very poor reducing agent, with a standard reduction potential in the range of +500 to +1,000 mV (reaction centers from different organisms differ). When the chlorophyll absorbs a photon of light, however, one of its electrons is promoted to an outer orbital, where it is less strongly attached to the chlorophyll, and hence can be more easily lost. Thus, the excited chlorophyll is an extremely powerful reducing agent—typically having a reducing potential of −800 to −1,200 mV.

In chlorophyll in solution, the electron instantly drops back to the ground state, re-emitting the energy it absorbed as a mixture of heat and light. The speed with which this happens (typically about a picosecond, 10^{-12} sec) is much faster than any chemical redox reaction. In a reaction center, however, the electron can move to nearby prosthetic group as fast as it can return to the chlorophyll, and thus, the

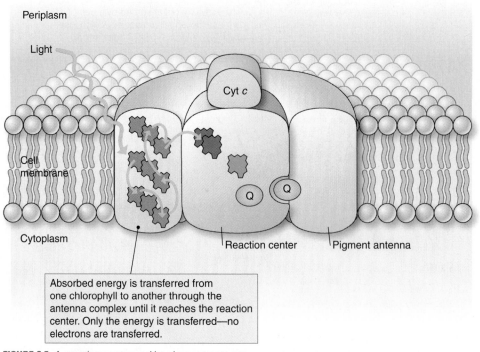

Absorbed energy is transferred from one chlorophyll to another through the antenna complex until it reaches the reaction center. Only the energy is transferred—no electrons are transferred.

FIGURE 8.5 A reaction center and its pigment antenna.

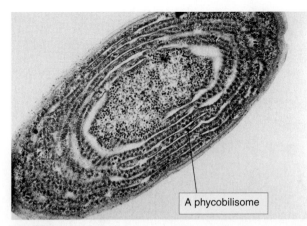

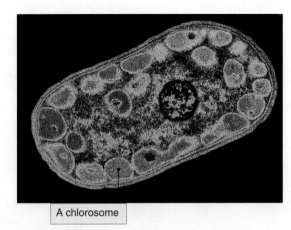

A phycobilisome

A chlorosome

FIGURE 8.6 Phycobilisomes.

FIGURE 8.7 Chlorosomes.

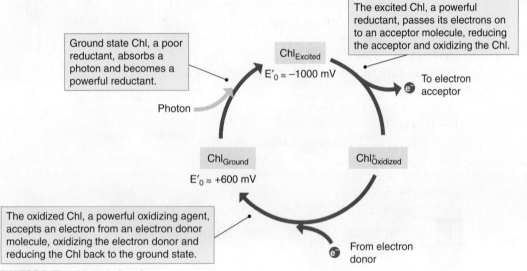

Ground state Chl, a poor reductant, absorbs a photon and becomes a powerful reductant.

$Chl_{Excited}$

$E'_0 \approx -1000$ mV

The excited Chl, a powerful reductant, passes its electrons on to an acceptor molecule, reducing the acceptor and oxidizing the Chl.

To electron acceptor

Photon

Chl_{Ground}

$E'_0 \approx +600$ mV

$Chl^+_{Oxidized}$

The oxidized Chl, a powerful oxidizing agent, accepts an electron from an electron donor molecule, oxidizing the electron donor and reducing the Chl back to the ground state.

From electron donor

FIGURE 8.8 The chlorophyll cycle.

structure of the reaction center allows the physical separation of the electron from the chlorophyll, thus oxidizing the chlorophyll.

The oxidized chlorophyll is a potent oxidizing agent and can be readily reduced by electrons from a variety of sources. In bacteria, the immediate donor is a cytochrome c, except for the cyanobacteria, in which it is plastocyanin, a copper-containing protein. This regenerates the ground state, and the reaction center is ready for another round.

8.4 Cyclic photophosphorylation generates energy in the light with no material input

Reaction centers can act as either electron donor or as electron acceptor, depending on whether they are in the excited or oxidized state. It is thus possible to construct a proton-pumping system in which the reaction centers act as both donor and acceptor to an electron transport chain (Figure 8.9). No material input is necessary, and as long as light is present, protons can be pumped and ATP made. Such a system exists in all phototrophs and is called **cyclic photophosphorylation.**

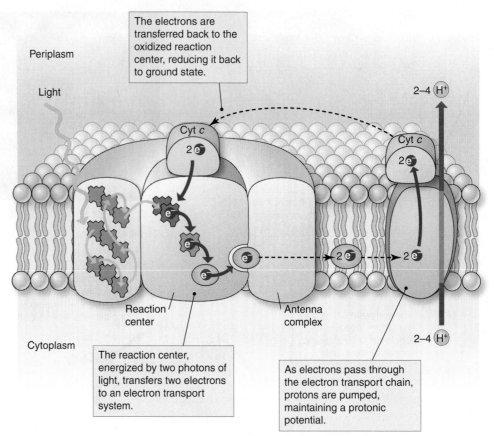

The electrons are transferred back to the oxidized reaction center, reducing it back to ground state.

Periplasm

Light

2–4 H^+

Cyt c

2 e^-

Cyt c

2 e^-

e^-

e^-

e^-

e^-

2 e^-

2 e^-

Reaction center

Antenna complex

Cytoplasm

2–4 H^+

The reaction center, energized by two photons of light, transfers two electrons to an electron transport system.

As electrons pass through the electron transport chain, protons are pumped, maintaining a protonic potential.

FIGURE 8.9 General scheme of cyclic photophosphorylation.

8.5 Electron donors and noncyclic photophosphorylation are needed for autotrophic growth only

Cyclic photophosphorylation is independent of electron donors, as the reaction center acts as both electron donor and acceptor. Thus, phototrophs need no exogenous compounds for continued production of a Δp and ATP; all that is needed is continuous illumination to drive cyclic photophosphorylation.

Autotrophic growth, however, imposes the need for an exogenous electron donor to reduce the CO_2 to organic material. Because the average redox level of cell material is CH_2O, this requires two pairs of electrons for each CO_2 molecule fixed:

$$CO_2 + 4H^+ + 4e^- \rightarrow CH_2O + H_2O$$

There is, thus, a requirement for substantial amounts of the electron donor to provide the large number of electrons necessary to reduce the large amounts of carbon needed for biosynthesis.

The reductants that participate in the pathways of CO_2 fixation are usually either NADPH or reduced ferredoxin. Thus, the challenge to phototrophs is to use their electron donor, often a weak reducing agent such as water or H_2S, to reduce $NADP^+$ or ferredoxin. Because this is a strongly endergonic reaction, an input of energy is necessary. The mechanism for putting energy into this reaction is to route

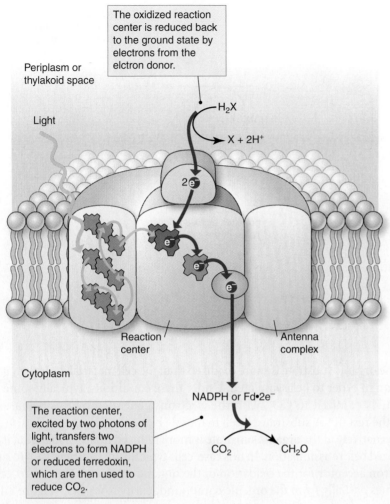

The oxidized reaction center is reduced back to the ground state by electrons from the elctron donor.

Periplasm or thylakoid space

Light

H_2X

$X + 2H^+$

$2e^-$

Reaction center

Antenna complex

Cytoplasm

NADPH or Fd•$2e^-$

The reaction center, excited by two photons of light, transfers two electrons to form NADPH or reduced ferredoxin, which are then used to reduce CO_2.

CO_2 CH_2O

FIGURE 8.10 General scheme of noncyclic photophosphorylation.

the electrons through the reaction center (Figure 8.10); thus, light provides some or all of the necessary energy.

8.6 Photoheterotrophic growth does not need an electron donor

Many photosynthetic microbes can grow photoheterophically when they get the opportunity. Most cyanobacteria and green sulfur bacteria are obligate photo-autotrophs, but many of the rest can use organic compounds as their source of carbon, driving the assimilation of the material with light energy. Because they are usually anaerobic phototrophs, they grow as photoheterotrophs only in illuminated anaerobic environments. The organic compounds available here are mainly fermentation end products. Cells growing this way use cyclic photophosphorylation to generate energy, and they assimilate organic compounds from the environment as their source of carbon. The organic material is rearranged to produce the necessary precursor metabolites and then all of the molecules of the cell. If the organic substrate is at the same oxidation level as cell material, which in nature is probably often the case, then no net reduction (or oxidation) is necessary, and thus, no electron donor (or acceptor) is needed—for instance, the assimilation of lactic acid (Figure 8.11).

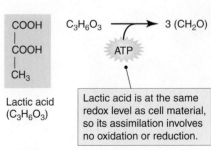

COOH
|
COOH
|
CH_3

Lactic acid $(C_3H_6O_3)$

$C_3H_6O_3 \longrightarrow 3 (CH_2O)$

ATP

Lactic acid is at the same redox level as cell material, so its assimilation involves no oxidation or reduction.

FIGURE 8.11 Assimilation of lactate.

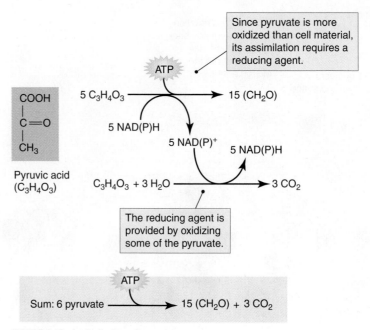

FIGURE 8.12 Assimilation of pyruvate.

If the organic substrate is more oxidized than the cell material, however, it needs reduction in order to be assimilated. Under these conditions, typically some of the substrate is oxidized to CO_2, with the electrons generated by this being used to reduce the rest of the substrate to cell material (Figure 8.12).

Alternatively, if the organic substrate is more reduced than cell material, it has to be oxidized before assimilation. In this case, cells typically fix sufficient CO_2 to provide an electron acceptor for the oxidation of the organic substrate. Thus, their cell carbon comes partially from the organic compound, partly from CO_2 (Figure 8.13).

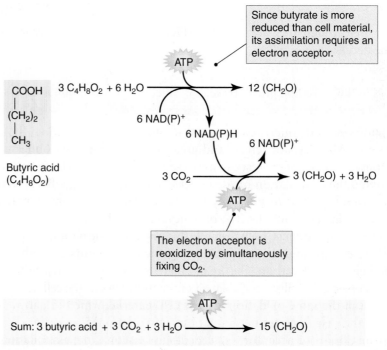

FIGURE 8.13 Assimilation of butyrate.

In all of these cases, the energy necessary for assimilation of the organic compound is provided by cyclic photophosphorylation. There is no need for noncyclic photophosphorylation.

8.7 There are two fundamentally different types of reaction center

Although all reaction centers contain chlorophyll, which goes through the same cycle of excitation, oxidation, and reduction, there are two biochemically distinct types. **Type I reaction centers** are **ferredoxin reductases,** for which the electron acceptor is the very electronegative iron–sulfur protein **ferredoxin.** With a very negative reduction potential (around −400 mV), this is a very potent reducing agent, and so much of the energy of the photon is conserved in the form of this powerful reductant.

Type II reaction centers, on the other hand, are **quinone reductases,** that reduce quinones to quinols. Because quinols are much weaker reducing agents (reduction potentials around 0 to −100 mV), much of the energy of the photon of light is dissipated within the reaction center. These reaction centers are thus less efficient than type I ones, a fact that has important biochemical consequences, as we will see.

8.8 There are three distinct types of photosynthesis

There are three principal types of photosynthesis, depending on the type of reaction center used: **type I photosynthesis** uses only type I reaction centers; **type II photosynthesis** uses only type II reaction centers; and **type I/II photosynthesis** uses both.

Both type I photosynthesis and type II photosynthesis are anoxygenic, meaning that they do not use water as electron donor and do not produce oxygen. Both are in fact usually anaerobic processes as well—not only do they not produce oxygen, but they cannot be performed under aerobic conditions. Type I/II photosynthesis is, on the other hand, oxygenic; it both produces oxygen and occurs almost always under aerobic conditions (even if conditions are initially anaerobic, oxygen production would quickly change that).

8.9 Photosynthesis is confined to the bacteria and their descendants the chloroplasts

There are six different groups of photosynthetic procaryotes (Figure 8.14). Type I photosynthesis is done by the green sulfur bacteria and the heliobacteria (a subgroup of the gram-positive bacteria); type II photosynthesis is done by the purple sulfur bacteria, the purple nonsulfur bacteria, and the green nonsulfur bacteria. Type I/II photosynthesis is done by the cyanobacteria (and by chloroplasts).

This distribution is difficult to explain. Photosystem I appears to be homologous in all of the organisms that possess it; thus, it presumably evolved only once, and all extant photosystems I descended from the same common ancestor. The same appears to be true of photosystem II. Thus, there are two major possibilities: (1) both photosystems evolved early, but were lost numerous times, or (2) the genes for photosynthesis were transferred among organisms by genetic exchange. Of course, the reality may have been a combination of the two.

We also cannot forget that the ability to do photosynthesis involves a large number of genes besides those for the reaction center polypeptides. It also requires genes

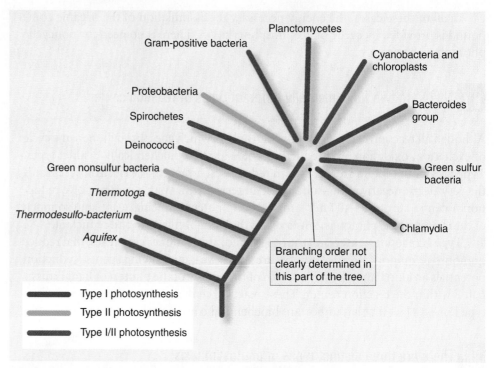

FIGURE 8.14 Distribution of photosynthesis within the bacterial tree.

for the antenna pigment polypeptides; for the cytochrome polypeptides; for the enzymes to synthesize chlorophylls, quinones, and hemes; for the enzymes to oxidize the electron donor; for the enzymes of the CO_2 fixation pathway; and for the regulatory polypeptides to control all of this. All photosynthetic eucaryotic organisms—algae and plants—are photosynthetic because of their possession of chloroplasts, which are clearly descendants of cyanobacteria.

Type I Photosynthesis

Type I photosynthesis is found in the green sulfur bacteria (but not in the green nonsulfur bacteria, which have type II photosynthesis). It is also found in the heliobacteria. It is characterized by cyclic photophosphorylation and noncyclic photophosphorylation (when growing autotrophically) using photosystem I only. The pigment antenna consists of chlorosomes in the green sulfur bacteria or membrane-embedded protein-chlorophyll complexes in the heliobacteria. It is anoxygenic and strictly anaerobic.

8.10 Cyclic photophosphorylation in type I photosynthesis uses photosystem I

The cyclic system in type I photosynthesis is quite simple (Figure 8.15). The reaction center (type I) reduces ferredoxin, which reduces in turn quinone and then the cytochrome b/c complex. The b/c complex then reduces the periplasmic cytochrome c, which diffuses through the periplasm to the oxidized reaction center and reduces it back to the ground state. The cytochrome b/c complex functions as a proton pump, as it does in respiration; thus, as long as cells are illuminated, continuous maintenance of a protonic potential is a simple matter.

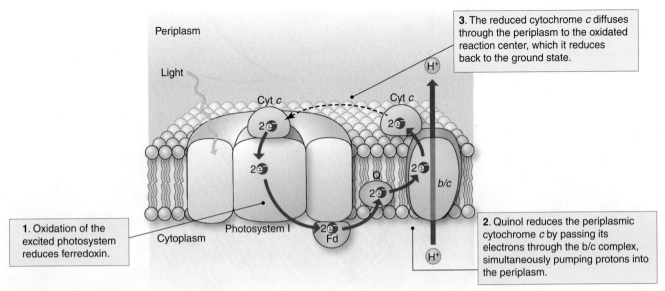

3. The reduced cytochrome *c* diffuses through the periplasm to the oxidized reaction center, which it reduces back to the ground state.

Light

Cyt *c*

Cyt *c*

H⁺

b/c

1. Oxidation of the excited photosystem reduces ferredoxin.

Photosystem I

Cytoplasm

Fd

H⁺

2. Quinol reduces the periplasmic cytochrome *c* by passing its electrons through the b/c complex, simultaneously pumping protons into the periplasm.

FIGURE 8.15 Cyclic photophosphorylation in type I photosynthesis.

8.11 Noncyclic photophosphorylation in type I photosynthesis uses photosystem I only

To generate the NADPH needed for autotrophic CO_2 fixation, green sulfur bacteria use H_2 or reduced sulfur compounds such as H_2S as electron donor (the heliobacteria do not grow autotrophically, and thus, they do not do this). Electrons from these compounds reduce the periplasmic cytochrome c, which then passes them on to the oxidized reaction center (Figure 8.16). From the excited reaction center, they are transferred to ferredoxin. Because ferredoxin has a reduction potential more negative than NADPH, it is a straightforward matter to pass the electrons from ferredoxin to $NADP^+$, a reaction catalyzed by the membrane **ferredoxin-$NADP^+$ reductase.**

Type II Photosynthesis

Type II photosynthesis is found in the purple bacteria and in the green nonsulfur bacteria. It is characterized by cyclic photophosphorylation and noncyclic photophosphorylation (when growing autotrophically) using photosystem II only. The pigment antenna consists of membrane-embedded protein–chlorophyll complexes in the purple bacteria or chlorosomes in the green nonsulfur bacteria. It is anoxygenic but can be used either aerobically or anaerobically. Most known type II phototrophs only grow phototrophically under anaerobic conditions, but the ability to grow photoheterotrophically under aerobic conditions appears widespread among marine bacteria.

8.12 Cyclic photophosphorylation in type II photosynthesis uses photosystem II

The cyclic system in type II photosynthesis is basically the same as in type I photosynthesis, except that their reaction center (type II) reduces quinone, instead of ferredoxin (Figure 8.17). From there, the system is the same as in type I

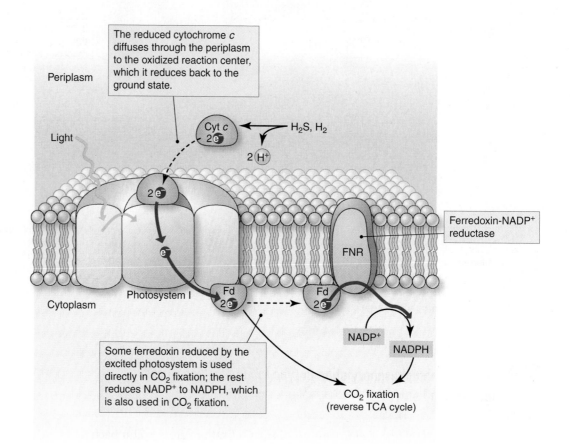

The reduced cytochrome c diffuses through the periplasm to the oxidized reaction center, which it reduces back to the ground state.

Periplasm

Light

Cyt c 2e⁻ ← H_2S, H_2

2 H⁺

2e⁻

Ferredoxin-NADP⁺ reductase

FNR

e⁻

Fd 2e⁻ - - - → Fd 2e⁻

Photosystem I

Cytoplasm

Some ferredoxin reduced by the excited photosystem is used directly in CO_2 fixation; the rest reduces NADP⁺ to NADPH, which is also used in CO_2 fixation.

NADP⁺

NADPH

CO_2 fixation (reverse TCA cycle)

FIGURE 8.16 Noncyclic photophosphorylation in type I photosynthesis.

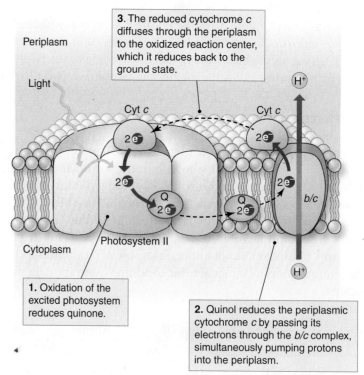

3. The reduced cytochrome c diffuses through the periplasm to the oxidized reaction center, which it reduces back to the ground state.

Periplasm

Light

H⁺

Cyt c   Cyt c

2e⁻   2e⁻

2e⁻   2e⁻   b/c

Q 2e⁻   Q 2e⁻

Photosystem II

Cytoplasm

H⁺

1. Oxidation of the excited photosystem reduces quinone.

2. Quinol reduces the periplasmic cytochrome c by passing its electrons through the b/c complex, simultaneously pumping protons into the periplasm.

FIGURE 8.17 Cyclic photophosphorylation in type II photosynthesis.

photosynthesis, and it too acts as an inexhaustible source of energy so long as cells are illuminated.

8.13 In type II photosynthesis, noncyclic photophosphorylation requires reverse electron transport

To generate the NADPH needed for autotrophic CO_2 fixation, purple bacteria or green nonsulfur bacteria use H_2 or reduced sulfur compounds such as H_2S as electron donor. Electrons from these compounds reduce the periplasmic cytochrome c, which then passes them on to the reaction center. From the excited reaction center they are transferred to quinone. This poses a problem for the cell: the resulting quinol is too weak a reductant to reduce $NADP^+$ (the quinol has a reduction potential between 0 and -100 mV, whereas that of NADPH is -320 mV). Thus, even though two photons of light have already been invested (one for each of the two electrons on the quinol), more energy has to be put into the system in order to move the electrons "uphill" from the weak reductant quinol to the strong reductant NADPH.

The mechanism for investing energy is to couple the redox reaction between quinol and $NADP^+$ to the entry of protons into the cell (Figure 8.18). The actual reaction reduces NAD^+ instead of $NADP^+$ and uses the NADH dehydrogenase that we have already seen in respiration and in the reverse electron transport of chemoautotrophs (see Chapter 7).

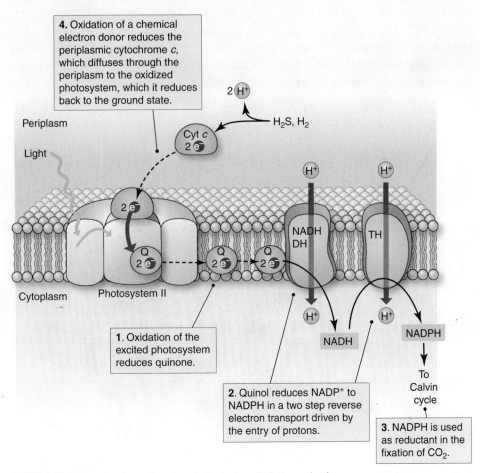

FIGURE 8.18 Noncyclic photophosphorylation in type II photosynthesis.

As in reverse electron transport in chemoautotrophs, transhydrogenase is involved in the further transfer of electrons from NADH to NADP$^+$ (see Section 7.28). Thus for each NADPH produced, purple bacterial photosynthesis has to invest the energy of at least four Δp-driven protons (at least two enter via NADH dehydrogenase and another two enter through transhydrogenase). This is equivalent to 1.3 ATPs (in addition to the two photons of light needed). Because type II photosynthesis requires reverse electron transport to generate reductant, it is less efficient than type I photosynthesis.

Type I/II Photosynthesis

Type I/II photosynthesis is distinctive in several ways: (1) it combines two photosystems into a single cell; (2) reaction centers and electron transport components are embedded in the thylakoid membrane rather than the cell membrane; and (3) it is oxygenic. This type of photosynthesis is confined to the cyanobacteria and their descendants the chloroplasts.

8.14 Cyclic photophosphorylation in cyanobacterial photosynthesis uses photosystem I only

Cyclic photophosphorylation in the cyanobacteria is essentially identical to that of type I photosynthesis (Figure 8.19). A few minor variations have no functional significance. These include a cytochrome b/f complex instead of cytochrome b/c (cytochrome f is a variant of cytochrome c). And the small water-soluble protein that accepts electrons from the b/f complex and returns them to the oxidized reaction center is a copper-containing protein **plastocyanin** instead of cytochrome c.

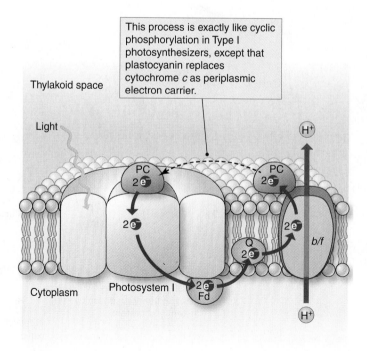

FIGURE 8.19 Cyclic photophosphorylation in cyanobacterial photosynthesis.

CHAPTER 8 MICROBIAL METABOLISM

8.15 Noncyclic photophosphorylation in cyanobacterial photosynthesis uses both photosystem I and II

The reducing end of noncyclic photophosphorylation in cyanobacteria and chloroplasts is essentially the same as in green sulfur bacteria. Photosystem I reduces ferredoxin, which can directly reduce the NADP$^+$ to produce the NADPH needed for CO_2 fixation. The oxidized reaction center is then reduced by electrons from plastocyanin.

In type I/II photosynthesis, however, the immediate source of electrons for plastocyanin is not the ultimate electron donor, but rather, electrons from photosystem II that have come through quinone and cytochrome b/f (Figure 8.20). Thus, the transfer of electrons from water to NADP requires two photons for each electron—one each to excite photosystems II and I.

The involvement of quinone and the cytochrome b/f complex also means that the cyanobacterial system of noncyclic photophosphorylation combines the proton-pumping activity of the electron transport chain of cyclic photophosphorylation, with the reductant generation of noncyclic. In these organisms, the ratio between noncyclic and cyclic photophosphorylation is greater than in the purple or green bacteria, as noncyclic makes more of a contribution to the energy metabolism. Even so, noncyclic photophosphorylation cannot come close to meeting all the energy needs of the cell, and thus, a substantial proportion of the light activity must still be cyclic photophosphorylation.

Although noncyclic photophosphorylation in type I and type II photosynthesis does not involve an electron transport chain, the process still contributes to the cell's energy needs because the electron donors are commonly oxidized in the periplasm, which generates protons (see Figures 8.16 and 8.18).

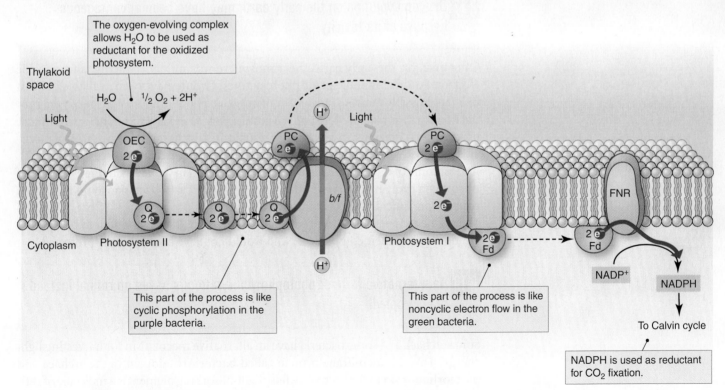

FIGURE 8.20 Noncyclic photophosphorylation in cyanobacteria.

8.16 The oxygen-evolving complex, combined with a high-potential photosystem II, allows cyanobacteria chloroplasts to use water as electron donor

One of the most spectacularly successful innovations in type I/II photosynthesis is the protein complex that acts as the electron donor to photosystem II. This is attached to photosystem II on the interior of the thylakoid membrane and has a unique arrangement of manganese atoms that constitute an electron-carrying prosthetic group. The complex oxidizes two water molecules to O_2 and four H^+, with the four electrons being held by the four Mn atoms and being transferred one at a time to the oxidized photosystem as it goes through four successive excitations and oxidations.

For water to serve as reductant to photosystem II, the reduction potential of the ground-state photosystem must be close to or more positive than that of water (+ 810 mV). Most photosystems II (in the purple and green nonsulfur bacteria) have reduction potentials considerably more negative than this (around + 500 mV is common). The cyanobacterial photosystem II is unusual in having a reduction potential of over + 900 mV, which makes it thermodynamically feasible for it to be reduced by water. These variations in reduction potential are a function of how the chlorophyll molecules that are excited are attached to the protein portion of the reaction center.

Thus, the combination of an unusually high-potential photosystem II, combined with the unique oxygen-evolving complex, makes it possible for cyanobacteria (and chloroplast-containing eucaryotes) to grow autotrophically in niches where other reduced electron donors such as sulfide or hydrogen gas are absent. This is a major advantage in today's oxidizing conditions, where such reduced compounds are uncommon outside of anaerobic environments.

8.17 Oxygen evolution on the early earth may have been advantageous because of its toxicity

Of course, on the early earth the atmosphere was mildly reducing, and suitable electron donors for anoxygenic photosynthesis would have been available in most habitats. Possibly the initial advantage of oxygenic photosynthesis was the oxygen production itself, as an early form of "chemical warfare." Oxygen is quite toxic because of its potent oxidizing capacity, and all aerobic organisms today have a series of mechanisms to detoxify O_2 and some of its even more toxic derivatives (see Section 9.21). At least some of these mechanisms must have evolved in parallel with the evolution of the oxygen evolving complex; otherwise, primitive cyanobacteria would have been committing suicide by their intracellular production of oxygen. Once protected against the oxygen they produced, however, they would have found it to be a potent weapon against their competitors, who would not have been similarly protected.

8.18 An alternative form of photophosphorylation is based on retinal instead of chlorophyll

Some archaea and some bacteria have an alternative mechanism for harvesting light energy. This is a membrane protein called **bacteriorhodopsin** in the archaea and **proteorhodopsin** in the bacteria, whose light-absorbing pigment is a molecule of **retinal** covalently bound to the protein. Bacteriorhodopsin pumps a proton across the

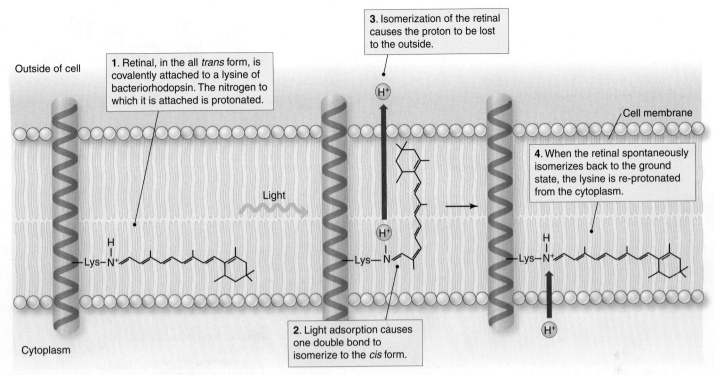

1. Retinal, in the all *trans* form, is covalently attached to a lysine of bacteriorhodopsin. The nitrogen to which it is attached is protonated.

3. Isomerization of the retinal causes the proton to be lost to the outside.

4. When the retinal spontaneously isomerizes back to the ground state, the lysine is re-protonated from the cytoplasm.

2. Light adsorption causes one double bond to isomerize to the *cis* form.

Outside of cell

Cell membrane

Light

Cytoplasm

FIGURE 8.21 Bacteriorhodopsin function.

cell membrane when it absorbs light (Figure 8.21). Thus, bacteriorhodopsin combines the functions of a reaction center and an electron transport chain, acting directly as a light-driven proton pump.

This protein is found in some of the halophilic archaea, in which it provides a means of energy metabolism when these obligate aerobes find themselves temporarily in anaerobic conditions (see Section 14.12). Recently, it has been shown that the gene for the proteorhodopsin is widely distributed in marine bacteria, suggesting that this may constitute a common mechanism of photophosphorylation in marine environments. Most likely, it is used as an energy source to drive photoheterotrophic growth, allowing efficient use of the sparse organic compounds in the surface layers of the open ocean.

Carbon Dioxide Fixation

In this section, we discuss the pathways of fixation of carbon dioxide. This is the distinguishing feature of all autotrophs—chemotrophic as well as phototrophic. A number of different pathways underlie the nutritional capacity for autotrophy, and it is probable that several new ones remain to be discovered.

8.19 The Calvin-Benson cycle is the most common pathway of CO_2 fixation

The Calvin-Benson cycle, or the ribulose–bisphosphate cycle (Figure 8.22), is the most widespread pathway for autotrophic CO_2 fixation. It is found in all bacterial aerobic chemoautotrophs, the purple bacteria, and the cyanobacteria. Because it is the cyanobacterial pathway, it is also the pathway of CO_2 fixation in chloroplast-containing eucaryotes.

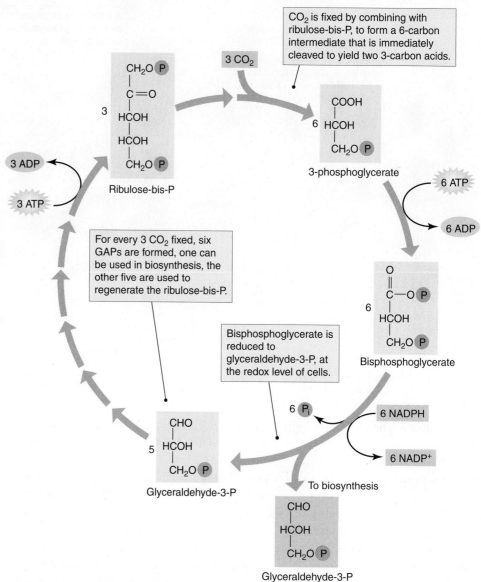

FIGURE 8.22 The Calvin-Benson cycle.

Within the figure:

CO$_2$ is fixed by combining with ribulose-bis-P, to form a 6-carbon intermediate that is immediately cleaved to yield two 3-carbon acids.

3 CO$_2$

3-phosphoglycerate

Ribulose-bis-P

3 ADP

3 ATP

6 ATP

6 ADP

For every 3 CO$_2$ fixed, six GAPs are formed, one can be used in biosynthesis, the other five are used to regenerate the ribulose-bis-P.

Bisphosphoglycerate

Bisphosphoglycerate is reduced to glyceraldehyde-3-P, at the redox level of cells.

6 P$_i$

6 NADPH

6 NADP$^+$

Glyceraldehyde-3-P

To biosynthesis

Glyceraldehyde-3-P

Most of the enzymes of the pathway are common enzymes of central metabolism, involved in the interconversion of various three-, four-, five-, six-, and seven-carbon sugars. Two enzymes are unique, however: the enzyme that phosphorylates ribulose-5-P (**phosphoribulokinase**) and the enzyme that adds CO$_2$ to ribulose-bis-phosphate, simultaneously splitting the six-carbon intermediate. This enzyme is commonly called **ribulose-bis-phosphate carboxylase/oxygenase,** or **RuBisCO** for short.

8.20 The reverse TCA cycle is used by a diverse group of autotrophic bacteria and archaea

The green sulfur bacteria, some sulfate-reducing bacteria, and a number of different archaea use the TCA cycle (citric acid cycle) for CO$_2$ fixation (Figure 8.23). To do this, the cycle is run in a direction opposite the way it functions in respiratory chemoheterotrophs. Instead of producing CO$_2$ and reductant (NADH)

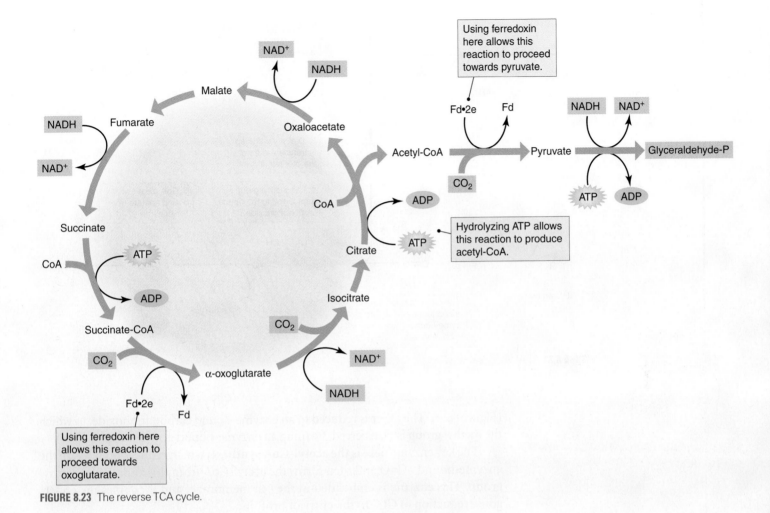

FIGURE 8.23 The reverse TCA cycle.

Boxes in figure:
- Using ferredoxin here allows this reaction to proceed towards pyruvate.
- Hydrolyzing ATP allows this reaction to produce acetyl-CoA.
- Using ferredoxin here allows this reaction to proceed towards oxoglutarate.

while consuming acetyl-CoA, it consumes CO_2 and reductant while producing acetyl-CoA.

Two reactions of the TCA cycle are normally irreversible: α-oxoglutarate dehydrogenase and citrate synthase. Cells that use the reverse TCA cycle thus have to have alternate enzymes that catalyze slightly different reactions to make these steps function in the opposite direction. This is done by using a more power-ful reductant than the NADH that is the product of the oxidative decarboxylations in the TCA cycle. Reduced ferredoxin is the usual reductant used in the reverse TCA cycle; ferredoxins have reduction potential that are typically more negative than −400 mV (compared with the −320 of NADH).

8.21 The Ljungdahl-Wood pathway is common in strictly anaerobic autotrophs

Some sulfate-reducing bacteria, the acetogenic bacteria, and a number of archaea, including the methanogens, use a novel, noncyclic pathway for CO_2 fix-ation. This pathway condenses two molecules of CO_2 into a molecule of acetyl-CoA. The pathway is called the **acetyl-CoA pathway,** or the **Ljungdahl-Wood pathway.**

In the Ljungdahl-Wood pathway, one of the molecules of CO_2 is bound to an enzyme cofactor of the *pterin* type and is reduced to a methyl group in successive steps

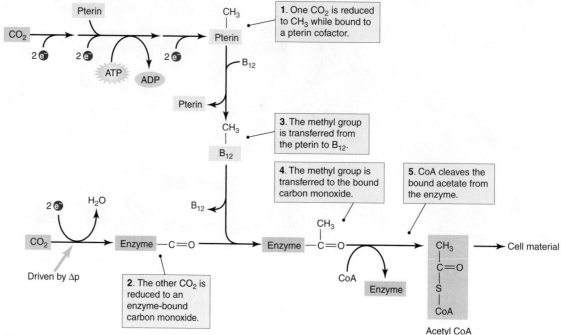

FIGURE 8.24 The Wood pathway.

1. One CO_2 is reduced to CH_3 while bound to a pterin cofactor.

3. The methyl group is transferred from the pterin to B_{12}.

4. The methyl group is transferred to the bound carbon monoxide.

5. CoA cleaves the bound acetate from the enzyme.

2. The other CO_2 is reduced to an enzyme-bound carbon monoxide.

Acetyl CoA

(Figure 8.24). The other is reduced to an enzyme-bound carbon monoxide, to which the methyl group is transferred, forming an enzyme-bound acetyl group.

The key enzyme here is the **acetyl-CoA synthase,** which reduces one CO_2 to the enzyme-bound $-C=O$ and then forms the acetyl-CoA from this and the B_{12}-methyl group. This enzyme is embedded in the cell membrane, and it couples the endergonic reduction of CO_2 to the entry of protons.

The electron donor for the reduction reactions varies in the different groups that use this pathway; however, the ultimate electron donor is almost always H_2. Thus, the overall reaction is as follows:

$$2\ CO_2 + 4\ H_2 \rightarrow acetate + 2H_2O$$

Interestingly, this is an exergonic reaction because H_2 is such a good reductant (the standard reduction potential at pH 7 is −410 mV). Thus, organisms using the Ljungdahl-Wood pathway for CO_2 fixation are in the fortunate position that the CO_2 fixation contributes to their energy conservation, rather than being a drain on it. The exact mechanism is not clear; however, the cleavage of the acetyl unit from the enzyme by CoA is thought to pump protons, and the production of 8 H^+ in the periplasm by H_2 oxidation probably also contributes.

It is clear that sufficient energy can be generated to more than pay back the investment of ATP and Δp required by the initial steps. Indeed, there are some bacteria that use this pathway for all of their energy needs.

8.22 The hydroxypropionic acid cycle is used by the green nonsulfur bacteria

A unique pathway is used by the green nonsulfur bacteria: the hydroxypropionic acid cycle (Figure 8.25). This pathway uses some of the TCA cycle enzymes, and

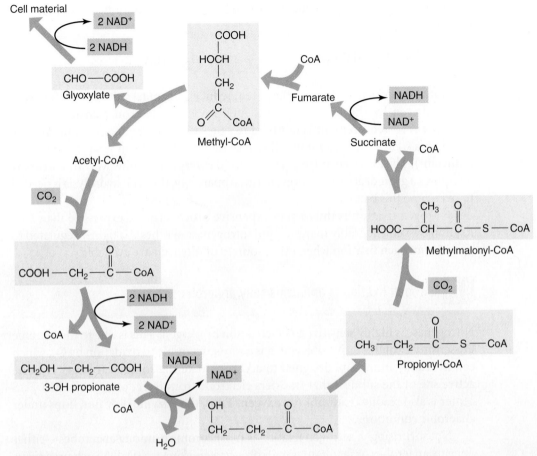

FIGURE 8.25 The hydroxypropionate pathway.

some novel ones, to condense two molecules of CO_2 into a molecule of glyoxylate. There are several biochemical routes that can then reduce the glyoxylate to compounds in central metabolism.

Nitrogen Fixation

One of the bioelements that most frequently limits the ability of microbes to grow in a particular location is nitrogen. Most microbes require nitrogen in the form of nitrate or ammonia; some require organic nitrogenous compounds. With nearly all of the microbes in a habitat competing for the available nitrogenous compounds, there is frequently very little available. Thus, in these nitrogen-poor habitats, it is a significant advantage to be able to use N_2 gas, which constitutes about 80% of the atmosphere and is thus the most abundant form of nitrogen in the biosphere. A limited number of procaryotes, widely distributed in many different bacterial and archaeal lineages, are capable of this.

8.23 Nitrogen fixation is a very expensive process

The enzyme that is responsible for nitrogen fixation is **nitrogenase.** It is a complex enzyme with multiple subunits and an unusual molybdenum–iron–sulfur center. It catalyzes the simultaneous reduction of one N_2 and $2\,H^+$ to ammonia and a molecule of hydrogen gas.

$$N_2 + 8H^+ + 8e^- + 16ATP \rightarrow 2NH_3 + H_2 + 16ADP + 16P_i$$

The immediate electron donor is the potent reducing agent ferredoxin, and the reaction is driven by the hydrolysis of 2 ATP for each electron transferred. This immense ATP requirement is not because the reaction is endergonic; in fact, the reaction is significantly exergonic (because reduced ferredoxin is such a powerful reductant). The nitrogen-nitrogen triple bond is very stable, however, and breaking it requires a great deal of activation energy. Apparently, the ATP hydrolysis is needed to overcome this barrier.

Nitrogen fixation is thus a very expensive process (more expensive than CO_2 fixation by a considerably margin), and nitrogenase synthesis is tightly regulated to prevent nitrogen fixation when other sources of nitrogen are available.

8.24 Nitrogen fixation is an intrinsically anaerobic process

Nitrogenase is highly sensitive to inactivation by oxygen. This is due to the extreme O_2-sensitivity of the active site, which is easily susceptible to oxidation by O_2, which can fit readily into a site designed to accommodate the somewhat larger N_2. The active site of the subunit that transfers electrons from ferredoxin to the Mo-Fe/S center is also readily oxidizable by oxygen. Thus, the enzyme only functions under anaerobic conditions.

Not surprisingly, nitrogen fixation is most common among anaerobes—either obligate anaerobes or facultative anaerobes growing anaerobically. Such organisms need no special protective mechanisms to allow nitrogen fixation to take place, except that in facultative anaerobes nitrogenase induction typically requires not only the absence of other nitrogen sources, but also the absence of oxygen as well.

There are a number of aerobes that can fix nitrogen, however, and their nitrogenases are as oxygen sensitive as those from strict anaerobes. These organisms must have mechanisms to prevent O_2 from gaining access to nitrogenase. A number of different mechanisms are known.

In most cases, aerobic respiration is a component of the protective mechanism. Some aerobes only fix nitrogen under microaerophilic conditions, where the rate of respiration (which, of course, occurs at the cell periphery) is sufficient to make the cytosol nearly anaerobic. Others, in the azotobacter group, make a less efficient respiratory electron transport chain when they are fixing nitrogen. This forces them to respire faster to maintain their Δp, and thus consume oxygen faster. The rapid oxygen consumption allows them to fix nitrogen under full atmospheric oxygen tensions.

Respiration is supplemented in many organisms by other protective mechanisms. Most nitrogen-fixing cyanobacteria make special cells in which nitrogen fixation takes place. Many bacteria make copious capsular layers that reduce the rate of oxygen diffusion into the cell by reducing convection currents at the cell surface, and some bacteria make special protective proteins that bind to nitrogenase and stabilize it; how these work on the molecular level is still unknown.

Summary

Photosynthesis is a widespread ability within the bacteria. Six different groups of procaryotes, in five different phylogenetic lineages, contain members capable of doing it. Outside of the bacteria, photosynthesis occurs only in eucaryotes that contain chloroplasts, which are themselves descendants of bacteria. Within the bacteria, there are several different types of photosynthesis, depending of the details of the reaction centers employed.

Bacterial autotrophs also have much more diversity in their pathways of CO_2 fixation. Eucaryotic autotrophs all use the Calvin-Benson cycle, which they acquired with the cyanobacterial symbionts that were the ancestors of their chloroplasts. In the bacteria, however, the Calvin-Benson cycle is only one of at least four different pathways of assimilating CO_2. Procaryotes (archaea as well as bacteria) are also the only organisms able to fix N_2 as a source of cellular nitrogen. Again we see that the procaryotes are more diverse in their metabolism than are the eucaryotes.

Study questions

1. Describe the properties of the three different states of chlorophyll.
2. Describe how reaction centers function.
3. Briefly summarize the similarities and differences among the three types of photosynthesis.
4. Construct a chart that shows the different groups of photosynthetic procaryotes, the types of photosynthesis they do, their antenna pigment structures, their electron donors, and their pathways of CO_2 fixation.
5. How many photons of light would it take the green sulfur bacteria to fix one molecule of CO_2? The green sulfur bacteria do Type I photosynthesis and use the reverse TCA cycle to fix CO_2. Assume that the cytochrome complex pumps two H^+ per pair of electrons that go through it.
6. How many photons of light would it take the cyanobacteria to fix one molecule of CO_2? The cyanobacteria do Type I/II photosynthesis and use the Calvin cycle to fix CO_2. Assume that the cytochrome complex pumps two H^+ per pair of electrons that go through it.
7. How many photons of light would it take the purple sulfur bacteria to fix one molecule of CO_2? The purple sulfur bacteria do Type II photosynthesis and use the Calvin cycle to fix CO_2. Assume that the cytochrome complex pumps two H^+ per pair of electrons that go through it. Further assume that NADH dehydrogenase and transhydrogenase both couple the movement of two H^+ across the membrane for each pair of electrons that are transferred through the complex.
8. What is bacteriorhodopsin/proteorhodopsin, and how does it contribute to procaryotic energy conservation?
9. Describe the reaction catalyzed by nitrogenase and any special features of the enzyme that affect its functioning and biological distribution.

Microbial Growth

L IKE ALL ORGANISMS, microbes grow and reproduce. Many of them, most notably many of the procaryotes, can do so at a phenomenal rate. This is a consequence of their high surface-to-volume ratio, their high-affinity permeases that concentrate nutrients, their extensive regulatory mechanisms that control the synthesis of protein and the activity of enzymes, and their cell cycle that allows overlapping rounds of DNA synthesis.

In this chapter, we discuss the growth and division of single cells, the growth of microbial cultures, and how growth is affected by various environmental factors.

Growth and Division of Individual Cells

As with eucaryotic cells, whose cell cycle you learned in introductory biology, individual procaryotic cells go through a cyclical series of processes that are collectively termed the cell cycle. The cell cycle includes a period of cell growth, followed by division; however, the procaryotic cell cycle differs profoundly from that of eucaryotes in ways that allow much more flexibility of growth rates.

9.1 The growth of individual cells is exponential

A procaryotic cell just produced by division is termed a **baby cell.** Its growth rate is slow but increases gradually through the cell cycle until shortly before division, at which time the growth rate slows to nearly zero as the **mother cell** prepares to divide into two baby cells (Figure 9.1).

This exponential growth is a simple consequence of the autocatalytic nature of the growth process. The process of growth involves the synthesis of new macromolecules from simple precursor molecules. This process requires catalysts: ribosomes and enzymes, yet these catalysts are themselves macromolecules. Thus, growth involves the synthesis not only of new structures but also of new catalysts for growth as well. Thus, the catalytic capacity of the cell increases steadily through the cell cycle, and hence, the absolute rate of growth increases steadily as well.

9.2 Most procaryotes divide by binary fission

The process of cell division in most procaryotic cells involves the inward growth of a **septum** in the approximate midline of the cell (Figure 9.2). The septum grows

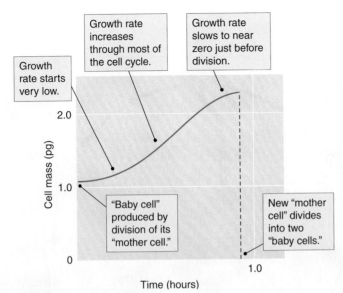

FIGURE 9.1 Cell growth during the cell cycle.

Growth rate starts very low.

Growth rate increases through most of the cell cycle.

Growth rate slows to near zero just before division.

"Baby cell" produced by division of its "mother cell."

New "mother cell" divides into two "baby cells."

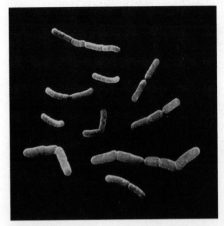

FIGURE 9.2 Dividing procaryotic cells.

inward until it is complete, and the two baby cells are entirely contained within their own cell membranes and separated by a wall. Normally the two cells then separate physically; however, sometimes they remain attached, forming a long chain of cells after repeated divisions.

The mechanism by which septation is controlled is still unclear, but we know that it involves a ring of the protein **FtsZ** that assembles just below the cell membrane around the equator of the cell when DNA replication finishes (Figure 9.3). This ring is anchored to the membrane by other special proteins. Apparently this pulls the cell membrane inward. The effect is to reorient peptidoglycan synthesis so that instead of adding new strands into the plane of the wall, new strands are added to extend the wall at right angles to the plane of the wall.

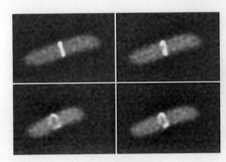

FIGURE 9.3 Initiation of septation.

9.3 Some cells divide by budding or multiple fission

Some procaryotic cells grow in a localized fashion, producing a **bud** that enlarges until it reaches approximately the size of the mother cell (Figure 9.4). Buds may be formed directly on the surface of the mother cell, or they may be formed at the tip of a hyphal tube (as in *Hyphomicrobium*).

A few procaryotes grow until the mother cell is many times the size of a baby cell and then undergo multiple rounds of division without any additional growth. This form of division is termed **multiple fission.** It is seen in some mycelial organisms, in which the mycelium fragments into unicellular rods after the growth period, and in a few organisms that grow into multinucleate spheres that then divide into numerous small cocci.

Most mycelial procaryotes produce specialized hyphae that undergo multiple fission to produce chains of spores (Figure 9.5). The rest of the mycelium does not fragment, however, and commonly dies during spore production.

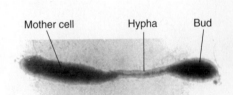

Mother cell   Hypha   Bud

FIGURE 9.4 Budding at the tip of a hypha.

9.4 The interdivision times of individual cells are quite variable

The period of time between the "birth" of a cell by division of its mother cell and its own division into two baby cells is called the **interdivision time.** This time is highly

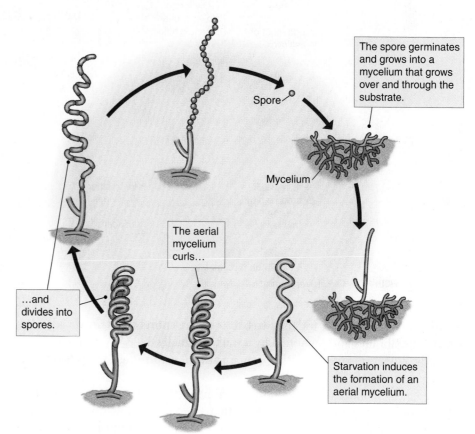

The spore germinates and grows into a mycelium that grows over and through the substrate.

Spore

Mycelium

The aerial mycelium curls…

…and divides into spores.

Starvation induces the formation of an aerial mycelium.

FIGURE 9.5 Streptomyces life cycle.

variable even for sister cells. This variability is easy to see when we start with a culture of cells, all of which are at the same point in their cell cycle. This is best achieved using a **baby machine** (see the MicroTopic on page 193), which generates a population of baby cells. If such a culture is started, it should show constant cell numbers for a time, as cells grow larger, and then the number of cells should double as all cells divide. If the doubling is very sudden, it indicates that all cells have very similar interdivision times; if the doubling is spread out over a period of time, it indicates that different cells are dividing at different times and thus have very different interdivision times. Experimentally, the latter is observed (Figure 9.6). Clearly interdivision times vary over at least a twofold range.

9.5 Procaryotic growth rates can be highly variable

When we talk about the growth rate of procaryotes under a specified set of conditions, we are speaking of the average for the entire population. Because we know that the interdivision times of individual cells vary over a wide range, the fact that a population will grow at a constant rate means that the *average* interdivision time remains constant, despite its great variability for individual cells. The most common measure of growth rate is the **generation time,** the time it takes for a population to double in number.

Procaryotic cells can multiply at widely different rates, depending on the conditions. Important conditions include the source of carbon and energy, the tem-

To study the bacterial cell cycle—for instance, to determine where in the cycle DNA synthesis occurs—one needs a population of cells all at the same stage. With such a population, cells can be exposed, for instance, to radioactive thymidine to measure the rate of DNA synthesis. A number of various different ways to synchronize cultures were tried, but these depended on manipulations that had the potential to disturb the cell cycle and create misleading results. Thus, the invention, by Charles Helmstetter and Stephen Cooper, of the simple and elegant "baby machine," was a significant advance. They began by filtering a culture of growing *E. coli* through a very fine-pore filter used normally to sterilize heat-sensitive media (see Section 3.19). It was known that bacteria not only cannot pass through the fine pores of such a filter but also actually stick to its surface (because of interactions between the LPS of the bacterial outer membrane and the surface of the filter). So Helmstetter and Cooper simply turned the filter over and trickled fresh medium through the filter

(Figure 9B.1). As the cells on the filter divided, they released baby cells into the medium flowing by. Thus, cells in the effluent from the filter would consist of baby cells only. Figure 9.6 shows the population growth of a sample taken from a baby machine; clearly the population is synchronized, although there is considerable heterogeneity in the interdivision times.

The baby machine produces baby cells that can be used to start synchronized cultures that can be used to study the cell cycle experimentally by, for instance, adding radioactive labels at various points in the cell cycle and determining the levels of incorporation into various macromolecules. Alternatively, an exponentially growing culture can be labeled first, then used to start a baby machine. The amount of radioactivity in cells released from the baby machine at different times then allows calculation of the rates of macromolecular synthesis in each segment of the cell cycle. It was this latter approach that was used to work out the details of overlapping cell cycles in bacteria, as described in Section 9.7.

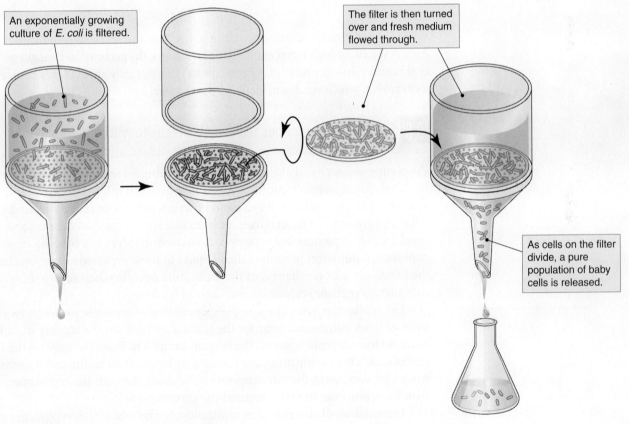

An exponentially growing culture of *E. coli* is filtered.

The filter is then turned over and fresh medium flowed through.

As cells on the filter divide, a pure population of baby cells is released.

FIGURE 9B.1 Diagram of the "baby machine."

perature, whether the medium is rich (with many biosynthetic precursor molecules available) or minimal, and whether there are toxic compounds present. Growth rates typically vary over about a 10-fold range. For instance, a population of *E. coli* can double in number as rapidly as 20 minutes in rich medium with glucose as carbon and energy source; in minimal medium with glycerol, it will take 1.5 hours, and

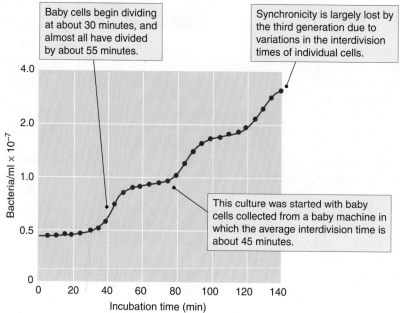

Baby cells begin dividing at about 30 minutes, and almost all have divided by about 55 minutes.

Synchronicity is largely lost by the third generation due to variations in the interdivision times of individual cells.

This culture was started with baby cells collected from a baby machine in which the average interdivision time is about 45 minutes.

FIGURE 9.6 Growth of a synchronized culture.

when the nutrient concentrations drop so low that the permeases in the membrane cannot maintain high intracellular concentrations, the generation time drops to several hours. Growth under these conditions is most easily studied in a device called a *chemostat,* which we discuss later in the chapter.

9.6 Bacterial chromosomes replicate bidirectionally from a single origin

Eucaryotic and archaeal chromosomes have multiple replication origins; however, bacterial chromosomes typically have a single origin. Replication begins at this spot at a regulated time in the cell cycle and proceeds in both directions around the circular chromosome. The **origin** is a distinctive region on the chromosome where special DNA sequences are repeated. **Initiation** involves the binding of multiple copies of an initiation protein (called **DnaA**) to these repeated sequences, bending the DNA into a loop. The rest of the replication proteins then add to this complex to form two **replisomes,** one on each side of the loop.

The replisomes are quite complex, consisting of multiple proteins: two molecules of DNA polymerase (one for the leading and one for the lagging strand), primase (to initiate replication on the lagging strand), helicase (to unwind the DNA), and others. These replisomes are thought to be attached to the cell membrane at particular sites; thus, the chromosome is threaded through the replisomes, rather than the replisomes traveling around the chromosome.

Termination of chromosome replication occurs when both replisomes reach a special region, the **terminus.** At this point, termination proteins bind and release the replisome proteins.

Although bacteria replicate their chromosomes bidirectionally from a single origin, most archaea appear to use multiple origins. Sequencing shows that archaeal replication proteins are more like those in eucaryotes than those in the bacteria. In bacteria, chromosome segregation occurs in parallel with replication. Shortly after initiation, the cytoskeleton begins to separate the newly replicated regions, so that chromosome segregation and chromosome replication finish nearly together. This

CHAPTER 9 **MICROBIAL GROWTH**

is quite different from the situation in eucaryotes, in which chromosome segregation by mitosis is distinct from replication.

9.7 In bacteria successive cell cycles can overlap

The amount of time it takes for the bacterial chromosome to be replicated is relatively constant regardless of growth rate; in *E. coli,* the time is about 40 minutes. This is termed the **C period** and is equivalent to the S period in the eucaryotic cell cycle. There is also an obligatory period of 20 minutes between the completion of chromosome replication and cell division, the **D period** (equivalent to the eucaryotic G_2 period). There is no obligatory G_1 period; however, when the environment constrains growth to a generation time longer than 60 minutes, there is a period equivalent to G_1. As the growth rate gets shorter, however, the gap gets shorter until it disappears entirely at a generation time of 60 minutes (Figure 9.7).

So far this is not particularly distinctive; many eucaryotic microbes (e.g., yeast) can grow so rapidly that there is no G_1; however, bacteria can grow with generation times significantly shorter than the sum of their C and D periods (S and G_2), a feat that no eucaryote is capable of. Because the C and D periods are essentially invariant, this means that the C period must be initiated before the cell is born, in the mother cell before it divides.

For example, let us consider *E. coli* growing at a generation time of 50 minutes. Because the D period takes 20 minutes, completion of DNA replication must occur 30 minutes after the cell was born. In order to do this, DNA replication had to be initiated 10 minutes before the cell was born; in other words, the C period began in the mother cell, overlapping the mother cell's cycle (Figure 9.8).

As the population grows increasingly quickly, the initiation time for DNA replication is pushed back further in the previous cell cycle. For instance, when the generation time is 40 minutes, DNA replication is initiated immediately when the previous round of replication is complete (Figure 9.9). Cells growing at this rate replicate DNA continuously; there is no gap separating rounds of replication, as there always is in eucaryotes.

At even shorter generation times, not only do the cell cycles overlap, but also rounds of DNA replication overlap. Cells synthesize DNA continuously throughout the cell cycle, but the rate varies depending on how many replication forks are active. At the maximum growth rate for *E. coli,* 20 minutes, the initiation of DNA

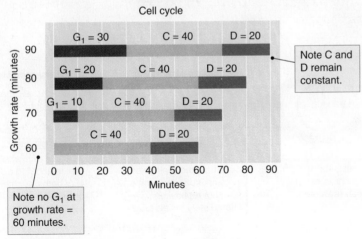

FIGURE 9.7 The cell cycle at growth rates above 60 minutes.

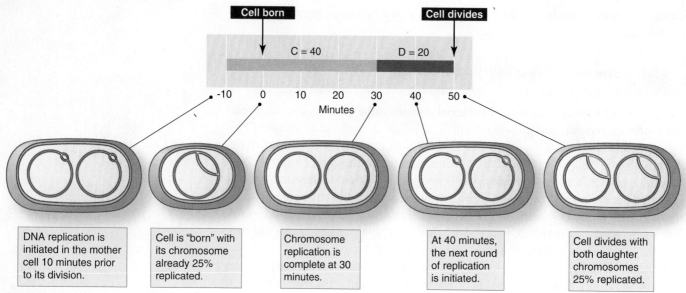

FIGURE 9.8 The cell cycle at 50 minutes generation time.

replication has been pushed back two full cell cycles, thus overlapping not only with the previous cell cycle but also with the one before as well (Figure 9.10).

Thus, the ability of bacteria to dispense with G_1 (which some eucaryotic microbes can also do) and to overlap their cell cycles (which no eucaryote can do) is a major reason for their rapid growth rates. In *E. coli,* a typical procaryote, the intrinsic events of the cell cycle take 60 minutes, yet the organism can multiply with a generation time as fast as 20 minutes.

It is also worth recognizing that the length of both the C and the D periods is variable in individual cells: 40 minutes and 20 minutes represent the averages. Because both of these vary, their sum is quite variable, thus explaining much of the variability of interdivision times that we discussed previously.

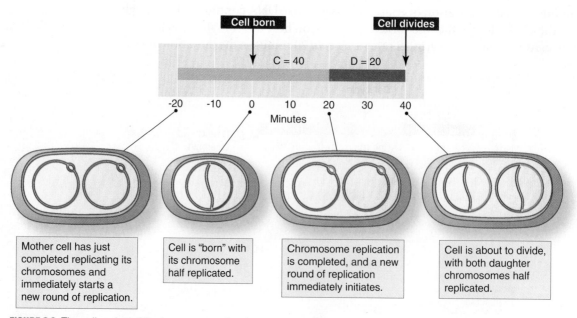

FIGURE 9.9 The cell cycle at 40 minutes generation time.

CHAPTER 9 MICROBIAL GROWTH

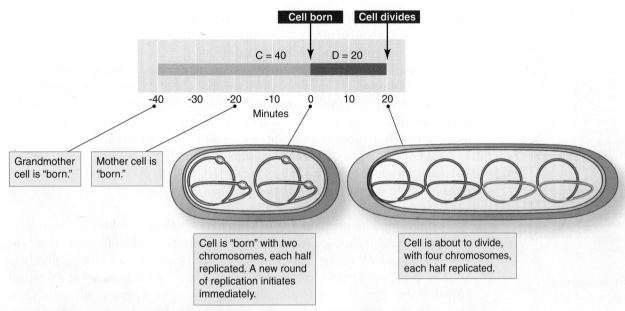

FIGURE 9.10 The cell cycle at 20 minutes generation time.

Microbial Population Growth

Microbiologists most frequently work with large populations—billions of cells rather than individuals. This is because microbes, particularly procaryotes, are so small that they present great technical challenges to work with individually. Thus, almost all microbiological experiments are done on populations.

When we speak of growth of microbial populations, we generally mean an increase in the number of cells per milliliter. This is different from the meaning of growth of a single cell, which refers only to an increase in size and mass and is distinct from multiplication.

9.8 Microbial population growth is normally measured by spectrophotometry

In order to be able to reproduce experiments, it is important for microbiologists to be able to measure accurately the size and growth kinetics of microbial populations. Most experiments on microbial physiology and biochemistry, for instance, are poorly reproducible unless the cells have been in a state of active multiplication for several generations.

Thus, a quick and accurate measure of microbial population size is necessary. Almost universally, spectrophotometry is the method of choice. A spectrophotometer is an instrument that shines light of a chosen wavelength through a specimen and determines the amount that gets through. The **optical density,** or **OD** (sometimes called *absorbance*), of a suspension is defined as follows:

$$OD = \log(I_o/I)$$

Where I_o is the amount of light entering the specimen, and I is the amount of light emerging from it. OD ranges from 0 (when $I_o = I$) to infinite (when $I = 0$).

The OD of a solution of a chemical compound is due to absorption of the light by the compound. Compounds that absorb visible light are generally colored, and

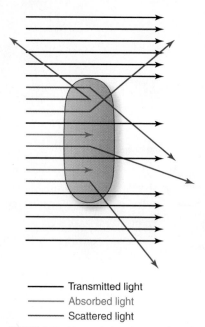

— Transmitted light
— Absorbed light
— Scattered light

FIGURE 9.11 Absorption and scattering of light by a bacterial cell.

light absorption is due to the interaction of the light quanta of particular wavelength with the electronic structure of the compound. Microbes, however, are particles in suspension, not compounds in solution; their interaction with light is considerably more complex. Some absorption always occurs, as microbes contain a vast number of chemical compounds, many of which absorb light of different wavelengths. The total concentration of these compounds is usually quite low, however, unless the organisms are colored (as with phototrophs, with their chlorophylls). Thus, absorption is generally a minor component of the OD of a microbial culture.

Most of the OD of a microbial culture is due to light scattering (Figure 9.11). This occurs when the photons of light are deflected by the microbial cells. The result is that the light still passes through the specimen, but it does not enter the photocell and thus is not detected.

Within limits, the amount of light scattered is strictly proportional to the amount of biomass present in suspension. Because cell number and biomass are also proportional to each other, the same method can also be used to measure cell number.

The principal limitation of spectrophotometric measurement of microbial growth is that it is not sensitive enough to measure small concentrations of cells. Spectrophotometry works only above approximately 10^7 cells per ml (where a suspension is faintly turbid).

9.9 Microbial population growth is exponential

Within a growing population, all cells are growing and dividing, with an average interdivision time equal to the population doubling time, or generation time. A growing population of microbes thus represents a mixture of cells at all stages in the cell cycle.

Because every cell divides into two cells, the instantaneous growth rate is proportional to the number of cells present:

$$dN/dt = kN$$

where k is the **growth rate constant.** Thus, when there are a thousand cells per milliliter, a period of time equal to the generation time will see the population increase by 1,000 cells; when there are a million cells per milliliter, the same amount of time will see an increase of one million cells.

The bacterial growth equation is usually used in the integrated form:

$$\int dN/N = k \int dt$$

$$\ln N - \ln N_0 = k\Delta t$$

$$\ln N = k\Delta t + \ln N_0$$

This is the equation for a straight line (Figure 9.12); if lnN is plotted against t, the result will be a straight line with slope k and an intercept on the y-axis of $\ln N_0$.

This is the most useful form of the equation, as most calculators today have natural log function. If working with logs to the base 10 is preferred, however, the equation can be simply converted:

$$\text{Log}_{10} N = k\Delta t/2.30 + \log_{10} N_0$$

In this case, a graph of $\log_{10} N$ versus t will have a slope of k/2.30.

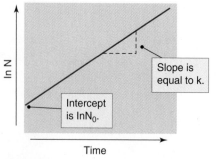

Slope is equal to k.

Intercept is $\ln N_0$.

ln N

Time

FIGURE 9.12 The growth rate equation graphed.

198

CHAPTER 9 MICROBIAL GROWTH

There is a simple relationship between the growth rate constant k and the generation time g. This can be demonstrated simply by setting $\Delta t = g$ and $N = 2N_0$ (i.e., the population has exactly doubled in an amount of time exactly equal to the generation time). We can then enter these values into the equation:

$$\ln 2N_0 = kg + \ln N_0$$

$$\ln(2N_0/N_0) = kg$$

$$\ln 2 = kg$$

$$kg = 0.693$$

Thus, if k is known, g can be easily calculated and vice versa.

Although the growth rate equation is usually written with N as the dependent variable, many other properties of the culture can be used, such as OD, μg/ml protein, or any other feature that increases proportionally to number.

9.10 Populations enter stationary phase when they run out of nutrients or accumulate toxic quantities of waste materials

Exponential growth cannot be continued for very long because the constantly increasing instantaneous rate of growth means that soon prodigious amounts of microbial cell mass would accumulate. Thus, an exponentially growing culture will soon exhaust one or more required nutrients and cease to grow. Alternatively, waste materials may accumulate to levels that inhibit growth, even if all nutrients are present in excess. The nongrowing state that results is termed **stationary phase.**

As cells approach stationary phase, they begin to reorganize their cellular composition. In particular, they stop DNA replication, but they may continue to undergo cell division if there are multiple nucleoids per cell (as is common in rapid exponential growth). The result is that in stationary phase cells are smaller and more of them have a single nucleoid. Other physiological changes occur to make the cell more resistant to environmental stresses, such as elevated temperature, desiccation, and ultraviolet light. The nature of these changes is not well understood.

9.11 Cells that have been in exponential growth for several generations are in balanced growth

When cells that have been in stationary phase for a while reenter the growth phase, there is a period of adjustment to the new conditions. This results in a **lag phase** before growth begins. The lag phase may be very short when the cells have not been in stationary phase long and when they entered stationary phase because of nutrient limitation; however, it can be several hours when cells have been in stationary phase for a long time or when they entered stationary phase because of the accumulation of toxic quantities of waste.

Cells in a culture in early exponential phase are still undergoing this adaptation and are changing their enzymatic composition and their size. Similarly, cells that are entering stationary phase are reorganizing their enzymatic composition, dividing without replicating DNA, and doing other things that change the cell composition and size. Physiological experiments performed on such cells are

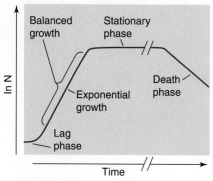

FIGURE 9.13 The standard growth curve.

difficult to reproduce, as it is difficult to catch rapidly changing populations in exactly the same state.

Between these two times, however, a culture growing exponentially remains quite constant in all of its important properties: average cell size, macromolecular composition, etc. The culture that has reached this point is said to be in **balanced growth,** and physiological experiments done on such cultures are much more reproducible (Figure 9.13).

9.12 Maintenance metabolism is necessary for cells in stationary phase

Even in stationary phase, cells must expend energy to maintain their integrity. There is continual leakage of materials across the cell membrane, necessitating continuous activity of permeases to maintain normal concentrations of solutes. Various kinds of radiation cause damage to DNA that necessitates continuous repair. Continued motility, advantageous because in nature starvation conditions may be local, requires a continual expenditure of energy. These and various other energy-requiring processes mean that even nongrowing cells must remain metabolically active at a low level (with the notable exception of the bacterial endospore, discussed in Chapter 14). The energy requirement of nongrowing cells is called **maintenance energy,** and it typically amounts to several percent of the energy requirement of an actively growing cell.

If cells have ceased growing because of nutrient limitation, they must use their own internal sources for maintenance energy. For this reason, actively growing cells often accumulate intracellular reserve materials, which are then used in stationary phase for maintenance energy generation (see Section 4.31); however, after these reserves run out, cells must start metabolizing other cellular constituents, such as proteins and rRNA. Obviously, this can only continue for a limited time until it does so much damage to the cell that it dies.

Microbial death is generally considered to be the loss of the ability to multiply. Thus, it is possible for a cell to die but to remain metabolically active and even motile for a while.

9.13 Microbial death is exponential

The stationary phase lasts for a variable time, depending mainly on the particular microbe but also on the particular conditions. In some cases, the stationary phase may last weeks; in other cases, the population begins to die immediately after it ceases growth, and there is no detectable stationary phase. In any case, after the cells begin to die, they do so with exponential kinetics, just like their growth but with a negative slope. Thus, initially, when the population is very large, many cells die in a given period of time. Later, when few cells are left, very few die in the same period of time. This implies that the actual event that causes cells to die is a random event whose probability remains constant and does not increase as the cells age. What this event might be remains mysterious.

Exponential death kinetics has practical consequences in achieving sterility. Because death is exponential, no period of treatment—be it autoclaving, ultraviolet irradiation, or any other—can be *guaranteed* to achieve sterility. There is only a greater or lesser *probability* of sterility (Figure 9.14). As a practical matter, sterilization conditions in laboratories and hospitals are determined by assuming a high level of contamination by a very hardy organism (bacterial endospores are usually chosen) and then choosing conditions that give very low odds of any survivors (e.g., 10^{-4} to 10^{-6}).

CHAPTER 9 MICROBIAL GROWTH

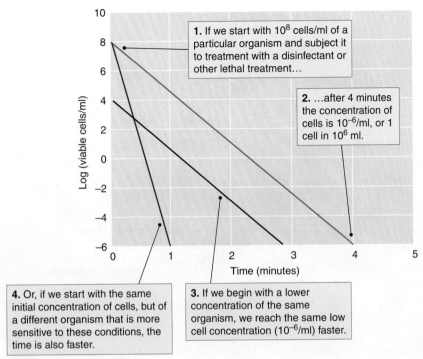

1. If we start with 10^8 cells/ml of a particular organism and subject it to treatment with a disinfectant or other lethal treatment...

2. ...after 4 minutes the concentration of cells is 10^{-6}/ml, or 1 cell in 10^6 ml.

4. Or, if we start with the same initial concentration of cells, but of a different organism that is more sensitive to these conditions, the time is also faster.

3. If we begin with a lower concentration of the same organism, we reach the same low cell concentration (10^{-6}/ml) faster.

FIGURE 9.14 Bacterial death rates.

9.14 A subpopulation of stationary phase cells can survive for months

When microbial populations are killed by the addition of disinfectants, by high temperature, or by other lethal conditions, cell death continues with exponential kinetics until the last cell in the population dies. When stationary phase cells begin to die spontaneously, however, a portion of them have undergone adaptations that prepare them for very long-term survival. Thus, most cells die in a relatively short period of time, leaving a portion that persists. For instance, in *E. coli* grown in rich medium, stationary phase contains about 5×10^9 cells and lasts about 3 days. The ensuing death phase sees the population size drop to about 10^7 cells, at which point the population size stabilizes and the remaining cells are viable for weeks or months.

9.15 Growth at low nutrient concentrations is studied in the chemostat

In a standard laboratory culture, often termed **batch culture,** microbial growth throughout the entire exponential phase occurs in the presence of excess nutrients. Only during the short transition into stationary phase is the growth rate limited by the nutrient supply, yet in nature, growth is often limited by low nutrient concentrations. It is thus of interest to study growth under these conditions; this is done in a device, the **chemostat,** that allows the continuous, nutrient-limited slow growth of a culture for hundreds or thousands or more generations (Figure 9.15).

The growth rate of a culture in a chemostat is set by setting the dilution rate. This is because the dilution rate D (in culture volumes per hour, usually) can be shown to be the same as k, the growth rate constant. Thus, to grow a culture more slowly, one simply slows down the rate at which new medium is added.

$$dN/dt = kN = DN$$

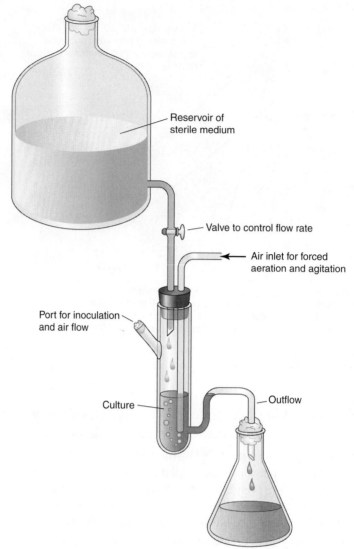

FIGURE 9.15 The chemostat.

This allows the study of physiology of cells grown at low growth rates, similar to those expected in the wild. It also allows very interesting evolutionary studies to be done, as the chemostat can be operated over thousands of generations.

9.16 Growth in nature is usually continuous like a chemostat, or episodic like successive batch cultures

In nature, there are probably two principal patterns of microbial growth. In some habitats, the microbes are attached to a solid substratum, and nutrients flow by (usually at a low concentration). This is the case, for instance, in most aquatic habitats, where water flow continually renews the supply of nutrients. In such habitats, the microbes multiply at a low rate determined by the nutrient concentration. As the microbial mass gets larger, some of it typically sloughs off and is washed away to die or perhaps to settle and attach to form a new colony elsewhere. This pattern of growth is very similar to that seen in a chemostat, where a constant population grows at a slow, nutrient-limited rate, with the excess cells continually washed out.

The other principal habitat is one in which the influx of nutrients is episodic, and cells alternate between a short burst of growth when nutrients are added and a stationary phase during the periods in between. Such a situation is common in soil habitats, and it is characteristic of most animal intestinal tracts as well (an exception is animals that eat nearly continuously, such as ruminants—see Section 17.3).

The Effect of Environmental Conditions on Growth

Of course, in nature, microbial growth kinetics are affected by the environment, both physical and biotic. The principal physical factors affecting microbial growth are temperature, pH, salinity, and oxygen concentration. The principal biotic factors are the presence of compounds excreted by other organisms, competition for nutrients, and predation. In this section, we briefly discuss the effects of the physical factors.

9.17 Each strain is characterized by a set of cardinal temperatures

Most procaryotes are capable of growth over about a 30°C range. Within this range they grow at variable rates: very slow near the **minimum,** increasing gradually to the **optimum,** and then falling off steeply to the **maximum** (Figure 9.16).

The reasons that microbes do not grow below a given temperature are not clear. It is thought that at low temperatures a variety of complex cellular processes go subtly wrong, and when too many of these processes are deranged, growth ceases. The principal processes affected are thought to be gene regulation, and the assembly of complex organelles such as ribosomes. Both of these are affected because the proteins involved do not have sufficient thermal energy to make needed conformational changes. In addition, membranes may lose too much fluidity and become leaky, or transport proteins may become dysfunctional.

The reasons that the growth rate of a particular strain increases with temperature are a function of the more rapid rate of all chemical reactions as temperature rises. Because increasing temperature means increasing speed of molecular movements, substrate molecules and enzymes collide with each other more frequently, and thus, the reaction rate increases. The rate of increase is exponential.

The growth rate does not increase indefinitely with temperature; at some point, the rate of increase slows down, stops, and then rapidly reverses direction and growth rate falls precipitously. This is thought to be due to increasing rates of denaturation of proteins, as the thermal energy begins to break the weak bonds that hold proteins in their proper tertiary and quaternary conformations. In a few cases, we know what protein is the one involved in setting the maximum. In *E. coli,* for instance, the maximum temperature in rich medium is 48°C, whereas in minimal medium it is 43°C. Experiments have shown that adding the amino acid methionine to minimal medium can mimic the effect of rich medium and that the first enzyme of the methionine biosynthetic pathway denatures rapidly above 40°C. Thus, in minimal medium, the first protein to denature and thus limit growth is this enzyme (homoserine succinyl transferase). In rich medium, some other protein or proteins are responsible.

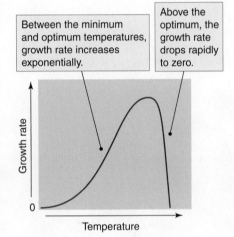

Between the minimum and optimum temperatures, growth rate increases exponentially.

Above the optimum, the growth rate drops rapidly to zero.

FIGURE 9.16 Microbial growth rate versus temperature.

9.18 Collectively, procaryotes grow over the entire range of temperature from −10°C to about 120°C

Although individual strains of microbe are usually limited to a growth range of about 30°C, collectively they can grow over an extraordinary range. Conventionally,

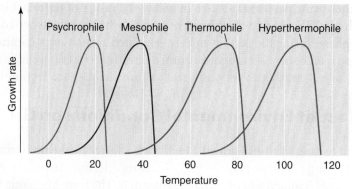

FIGURE 9.17 Comparison of the growth rates versus temperature of different categories of microbe.

they are divided into four different classes, depending on their optimum growth temperature (Figure 9.17).

Some of them are adapted for growth at cold temperatures, frequently having optimum rates at around 15°C or below, minimum temperatures for growth around or below 0°C, and maximum temperatures below 20°C. These are called **psychrophiles.** Others have optima in the range from 25°C to 40°C, with minima above 5°C and maxima below 45°C. These are called **mesophiles.**

Two categories of thermophiles are distinguished: **thermophiles** have optima from about 45°C to around 80°C, and **hyperthermophiles** have optima above 80°C. The highest growth rate optimum so far recorded is 113°C for the archaeon *Pyrolobus fumarii,* but this will undoubtedly be extended. Recently, another archaeon, an anaerobic organism that respires formate with iron as electron acceptor, has been recorded growing at 121°C (the temperature of an autoclave!), and it presumably has an optimum temperature of over 115°C.

9.19 Procaryotes grow over a range of pH values while maintaining constant internal pH

Each procaryotic microbe typically grows over a pH range of about two to four units, a range of 100- to 10,000-fold variation in H⁺ concentration. Within this range, they typically grow at a high rate over much of the range, growing significantly more slowly only at the extremes. Where on the pH scale the zone of maximal growth falls allows organisms to be divided into several rough categories (Figure 9.18): **neutro-**

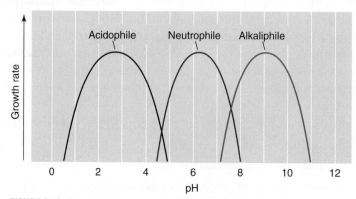

FIGURE 9.18 Growth rate versus pH of different categories of microbe.

philes grow best in the range of pH 6 to 8; **acidophiles** grow best at pH values less than about 4, and **alkaliphiles** grow best at pH values above 9.

Despite the rather wide range of pH values over which most microbes can grow, they typically maintain a fairly constant internal pH. For most, the cytosolic pH remains between 6 and 7 regardless of the environmental pH; however, alkaliphiles frequently have cytosolic pH values in the basic range (as high as 9).

Alkaliphiles have a particular problem in maintaining a Δp, as the external pH can be more alkaline than the cytosol. When the difference is minor, a Δp can still be maintained because proton pumping out of the cell results in the cytosol being negatively charged relative to the outside, even though there is no osmotic component to Δp under these conditions. When the pH difference is great (more than a pH unit), however, it becomes difficult or impossible to maintain a Δp. Thus, many alkaliphiles maintain a sodium gradient rather than a proton gradient, and they have permeases, ATP synthases, flagellar motors, etc., that are driven by Na^+ entry, rather than H^+ entry.

Although individual microbes typically grow over a range of only a few pH units, collectively the procaryotes grow over the entire range from less than 0 to more than 11—nearly a trillion-fold range of H^+ concentrations!

9.20 Osmoregulation in procaryotes involves the synthesis or uptake of solutes to keep the cytosol more concentrated than the environment

Collectively, the procaryotes grow over a wide range of osmolarities, ranging from essentially zero, to saturated NaCl. No single organism can grow over this entire range, however. Rather, each organism has a range of salinities that it tolerates; outside of this range it cannot grow. We divide microbes into three rough groups (Figure 9.19) **Nonhalophiles** do not have a salt requirement, grow well in dilute media, and usually cease growth when the osmolarity nears 1. **Halophiles** (commonly marine organisms) grow best in media with osmolarities around 2 and often have a specific requirement for Na^+ that prevents them from growing in media not specifically supplemented with this ion. **Extreme halophiles** typically do not grow in media with osmolarities below 1.5 and grow well in highly saline media (up to NaCl saturation). For comparison, sea water has an NaCl concentration of about 0.5 molar (total osmolarity is about 1.1); NaCl-saturated lakes have about 3.5 M NaCl and a total osmolarity of about 7. Extreme halophyles are generally restricted to aquatic systems with no outlet so that water is lost only through evaporation, concentrating the salts that are washed into the system.

Because most procaryotes have a rigid cell wall, growth in dilute media does not present problems for them. Without coping mechanisms, however, exposure to hypertonic medium would rapidly kill them due to membrane damage caused by shrinkage of the protoplast away from the wall. Furthermore, procaryotic cells rely on their internal osmotic pressure to grow; it is this pressure that makes the cell enlarge during the interdivision period, as new cell wall is being made. Without such pressure, the growth and division cycle is disrupted. Thus, to preserve membrane integrity and to grow and divide, nearly all procaryotes need to maintain their internal osmolarity higher than that of the external medium.

Normally, this is accomplished by adjusting the internal concentration of one or a few specific solutes—called **compatible solutes**—because high concentrations of them are compatible with continued macromolecular functioning. Compatible solutes may be synthesized when the external osmolarity goes up, or they may be taken up from the medium. K^+ is a

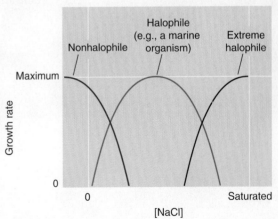

FIGURE 9.19 Growth rate versus salinity of different categories of microbe.

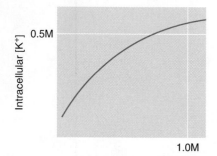

Intracellular [K⁺]

0.5M

1.0M

Osmolarity of medium

FIGURE 9.20 Cytoplasmic [K^+] versus medium osmolarity for *E. coli*.

common compatible solute taken up from the medium. *E. coli*, for example, concentrates K^+ to a level that is proportional to the external osmolarity (Figure 9.20).

Other common compatible solutes, whose internal concentrations are adjusted by regulating the rate of their synthesis, are the amino acids glutamate (especially in gram-negative bacteria) and proline (especially in gram-positive bacteria). A number of other amino acids, sugars, and derivatives of these are also used.

9.21 Aerobic growth requires several protections against the toxic effects of oxygen

We tend to think of oxygen as benign because we die within minutes in its absence; however, it is an extremely toxic gas because it reacts with various other compounds to form exceptionally powerful oxidizing agents, notably singlet oxygen ($^1O_2$) super-oxide (O_2^-) and hydrogen peroxide (H_2O_2). These problems are exacerbated in the light. Indeed, as we mentioned in Chapter 8, the toxic effects of O_2 might have been the principal evolutionary advantage to oxygenic photosynthesis on the early earth as a form of chemical warfare. Most microbiologists accept that when oxygenic photosynthesis became widespread there must have been an enormous wave of extinction, as many microbial species died out due to their inability to cope with the toxic new conditions. Increasingly, old forms of life were restricted to those few habitats from which oxygen was excluded, and new forms, adapted to an oxidizing environment, occupied the newly vacant habitats. Today, anaerobic habitats are principally sediments of lakes and oceans and the intestinal tracts of animals. These places are anaerobic not because oxygen cannot penetrate to them, but because facultative anaerobes that live there use it up as fast as it can diffuse in, thus preserving anaerobiosis and permitting the growth of strict anaerobes.

All aerobes must have a set of mechanisms to deal with the toxic forms of oxygen, and life for them requires the continual repair of oxidative damage. Phototrophs, and other organisms that live in brightly illuminated habitats, are especially vulnerable because singlet oxygen is produced by the reaction of O_2 with light-activated compounds. Singlet oxygen has its two unpaired electrons in different orbits and is an extremely powerful oxidizing agent. Protection against its effects is one of the major functions of **carotenoid pigments** (Figure 9.21). These are orange, yellow, or red pigments (Figure 9.22) that have an extensive system of double bonds so that when they react with singlet oxygen, the energy is distributed over the entire molecule and can be dissipated without damage.

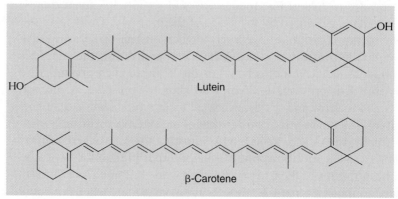

FIGURE 9.21 Carotenoid structure.

FIGURE 9.22 Salt evaporation pond containing dense populations of extremely halophilic archaea with abundant carotenoids.

Superoxide and hydrogen peroxide are detoxified by specialized enzymes present in virtually all aerobes. The most important are **superoxide dismutase,** which catalyzes the reaction

$$2O_2^- + 2H^+ \rightarrow H_2O_2 + O_2$$

and **catalase,** which catalyzes the reaction

$$2H_2O_2 \rightarrow H_2O + O_2$$

Some organisms have a different mechanism of detoxifying hydrogen peroxide, either instead of catalase or in addition to it. They use NADH to reduce it, catalyzed by the enzyme **peroxidase:**

$$NADH + H^+ + H_2O_2 \rightarrow 2H_2O + NAD^+$$

The combination of these defenses allows microbial life in air, and they must have evolved very early, probably initially in the ancestors of the cyanobacteria that first used oxygenic photosynthesis. Their evolution in other organisms allowed these others to accompany the cyanobacteria as they spread over the earth.

Summary

Procaryotic cells are generally adapted for rapid growth at the expense of dissolved nutrients. Their cell cycle allows DNA synthesis to begin in the mother cell or even in the grandmother cell, and, thus, under conditions that permit very rapid growth, a cell may be born with its chromosome already almost completely replicated. Such adaptations allow stupendous rates of multiplication when high concentrations of nutrients are available. Procaryotic cells also are adapted to long-term survival under

Aerobic growth requires several protections against the toxic effects of oxygen starvation conditions so that a portion of the population can persist when the majority of the population has died, increasing the chances that the organism will eventually encounter changed conditions that allow multiplication to resume. Of course, there are many exceptions to these generalizations; however, for many procaryotes, their spectacular evolutionary success is a result of their ability to exploit transient nutrient availability rapidly and to then persist until the next feast.

Taken as a whole, the procaryotes have revolutionized our understanding of the limits of life. Habitats with conditions that would seem to prohibit life totally turn out to be richly populated with microbes, and the more extreme are populated only with procaryotes. Procaryotes thrive in habitats with temperatures from −5°C to above 115°C, pH values from one normal acid to caustic concentrations of base, highly dilute to saturated brines, and so on. This range of adaptations has allowed procaryotes to colonize an extraordinary range of habitats on earth.

Study questions

1. Compare and contrast binary fission, budding, and multiple fission.
2. Explain how you can tell that the variation in interdivision times of cells in a culture is not genetically determined. In other words, a cell that divides in an interval significantly shorter than the average does not produce baby cells that themselves divide with the same short interdivision times.
3. When in the cell cycle would you find initiation and termination of DNA replication in a culture growing with a generation time of 35 minutes?
4. When in the cell cycle would you find initiation and termination of DNA replication in a culture growing with a generation time of 25 minutes?
5. For the culture in question 3, graph the rate of DNA synthesis (in arbitrary units) versus time in the cell cycle.
6. For the culture in question 4, graph the rate of DNA synthesis versus time in the cell cycle.
7. Assume you are growing a bacterial culture whose generation time you know to be 45 minutes under the conditions you are using. You plan to inoculate your culture at 9 AM with an already actively growing inoculum, and you wish to use the culture for an experiment at 1 PM, with a final cell density of 5×10^8/ml. What concentration of cells should you start with?
8. When you go to inoculate the culture described in question 7, you find you do not have an actively growing culture to use as inoculum. So you use a stationary phase culture from several days earlier, inoculating to the cell density you calculated for question 7. When you return at 1 PM, you find your culture is only at 2×10^8 cells/ml. You attribute this to a lag before the culture started growing. How long was the lag?
9. You start a culture with 5×10^7 cells/ml, and 3 hours later there are 4×10^8 cells/ml. What is the generation time?
10. You start a culture with 2×10^7 cells/ml, and 4 hours later there are 3×10^8 cells/ml. What is the generation time?

Procaryotic Genome Organization and Regulation

10

A S IN ALL CELLULAR organisms, microbes have a genetic system based on DNA as the information storage molecule, with transcription and translation systems that involve mRNA, tRNA, and ribosomes. Microbes also closely regulate the expression of most genes, often in response to their environment. In this chapter, we focus on the structure of the genome of procaryotic microbes and on the regulation of their gene expression.

Procaryotic Genome Structure and Transcription

Procaryotic genomes differ from eucaryotic genomes in their structure and in many details of their replication and transcription. We discussed chromosome replication in Chapter 9; here we review genome structure and some of the features of the transcriptional apparatus.

10.1 Most procaryotes have a single circular chromosome

The most common chromosomal organization in procaryotes—both bacteria and archaea—is a single, covalently closed, circular chromosome. This is in striking contrast to the eucaryotic genome, which consists of multiple linear chromosomes; however, there are many exceptions. For instance, the spirochete *Borrelia*, which causes Lyme disease, has a single linear chromosome. *Agrobacterium tumefaciens*, which causes tumors on plants, has one linear and one circular chromosome. Several bacteria, including *Vibrio cholera*, which causes the disease cholera, have two circular chromosomes, and *Burkholderia cepacia*, a plant pathogen, has three circular chromosomes.

10.2 The DNA in procaryotic chromosomes is supercoiled

Procaryotic DNA is **supercoiled**—as if the DNA has been enzymatically cut, one end twisted, and then the DNA resealed. This supercoiling is a consequence of the action of a class of enzymes called **DNA gyrases,** which maintain the proper amount of supercoiling. For reasons that are not entirely clear, supercoiling is essential to proper functioning of the nucleoid as a template for transcription. Probably one reason is that the direction of supercoiling is such that it tends to unwind the DNA helix; thus, supercoiled DNA is easier for RNA polymerase to unwind when transcription is initiated.

The chromosome is organized into about 50 discrete **domains.** Domains do not correspond to the loops of DNA in the nucleoid (see Section 4.25); rather, each domain corresponds to many of these very transitory loops. Domains are identified as regions of the genome whose supercoiling can be eliminated by nicking the DNA (breaking one of the strands), allowing it to rotate. If a supercoiled plasmid is nicked, the entire plasmid loses its supercoiling. If the bacterial chromosome is nicked, however, only about 2% of the chromosome loses its supercoiling. This implies that there are about 50 sites at which the chromosome is immobilized, preventing rotation. A domain is thus the stretch of DNA between two of these sites.

10.3 Most of the procaryotic genome is coding

Procaryotic chromosomes typically have relatively few introns, repeated sequences, transposons, or "junk" DNA. Thus, most of the genome encodes proteins or RNA molecules. There are exceptions, but none of them comes anywhere near the situation in mammals and higher plants, in which 90% or more of the genome does not encode a product.

Many procaryotes contain no introns; others contain a few, generally in the genes for ribosomal or transfer RNAs. Because procaryotes lack spliceosomes, all procaryotic introns are of the self-splicing type: a spontaneous intramolecular chemical reaction leads to the elimination of the intron.

Procaryotes also typically have relatively few nucleotides between genes. The median intergenic spacer distance varies from only three nucleotides (in *Pelagibacter ubique*) to several hundred (Figure 10.1). The median value for all procaryotes is about 60 to 70. Many plants and animals typically have tens or hundreds of thousands of nucleotides between genes.

10.4 Procaryotic chromosomes contain multiple transposons

The procaryotic chromosome, like that of eucaryotes, contains variable numbers of **transposons,** although there are generally fewer of them—a dozen or so in many bacterial genomes compared with hundreds of thousands in mammalian

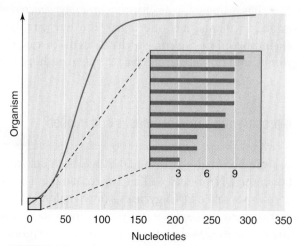

FIGURE 10.1 Intergenic distances in procaryotes.

CHAPTER 10 PROCARYOTIC GENOME ORGANIZATION AND REGULATION

genomes. Transposons are genetic sequences that can change their location on a DNA molecule. Their basic minimal structure consists of a segment of DNA containing the gene for the protein that catalyzes transposition, flanked by inverted repeat sequences (sequences that are the same or very similar, except that they have opposite orientations). When they transpose, the entire unit—the inverted repeats and everything in between—moves together.

There are two basic types of transposition: **replicative,** in which one copy of the transposon remains at its original site while a second copy is inserted elsewhere, and **nonreplicative,** in which the transposon is physically removed from its original site and reintegrated into the new site. Replicative transposition is considerably more common than nonreplicative because it is a more successful evolutionary strategy; other things being equal, a transposon that employs replicative transposition will leave more copies of itself in future generations than one that uses the nonreplicative mode.

Transposons normally transpose only rarely. This is undoubtedly due to the fact that frequent transposition would reduce the evolutionary fitness of its host cell by a large amount, as every transposition causes a mutation. If the mutation is in a noncoding sequence of DNA or in a rarely used gene, then the cell will remain viable. If the insertion is in an essential gene, however, the cell is killed. Thus, transposition frequency must be under very tight evolutionary selection: too rare and the transposon is unlikely to survive because it will not spread effectively in nature; too frequent and it will not survive because it will kill its host cells.

Transposons encode a **transposase** enzyme that is responsible for their transposition. Frequently, they also encode a repressor of transposase synthesis; this keeps the frequency of transposition down and prevents the cell that harbors the transposon from suffering too many mutations. If DNA sequences containing a transposon are transferred into a new cell by any of the several ways of gene transfer discussed in Chapter 12, however, the transposon will suddenly find itself in a cytoplasm free of repressor. This leads to a burst of transposase synthesis before repression is reestablished, and, thus, transposition frequency is temporarily elevated and the transposon quickly inserts several copies of itself into other DNA molecules in the cell.

Transposons that do not encode a repressor normally have a very weak promoter for their transposase; thus, even without repression, the frequency of transposition is very low. These transposons do not show a burst of transposition after genetic transfer.

The simplest transposons are called **insertion sequences,** as they were originally discovered as sequences that inserted randomly into genes, thereby causing mutations. Insertion sequences are typically around one or a few thousand base pairs in length and usually encode only a transposase or transposase and repressor. Insertion sequences are given names that consist of the abbreviation IS followed by a number. For instance, *E. coli* contains 5 to 10 copies of IS1, 4 to 8 copies of IS2, about 5 copies of IS3, 10 to 12 copies of IS5, and about 2 copies of IS10.

Other transposons are called **composite transposons,** and their names begin with the abbreviation Tn followed by a number. They combine transposon function with other genes, most commonly genes for antibiotic resistance. The ability of these transposons to move has been a significant factor in the spread of drug resistance. Most of the plasmid-encoded antibiotic resistance mentioned previously is because the plasmid contains a composite transposon. Of course, when such a plasmid is introduced into a new cell, the transposon and its drug resistance gene can hop to the chromosome or to any other plasmids that are present. In this way, transposon-encoded antibiotic resistance genes spread through the procaryotic world.

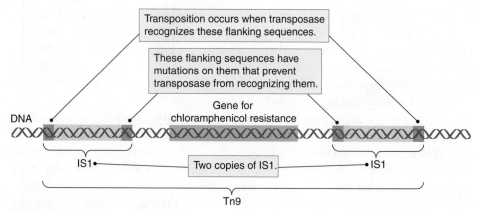

FIGURE 10.2 Composite transposon structure.

Some composite transposons are like large insertion sequences that contain antibiotic resistance genes in addition to transposase and repressor genes. Other composite transposons consist of two identical or nearly identical insertion sequences on either side of the antibiotic-resistance genes (Figure 10.2). These transpose as a unit, rather than the individual IS elements transposing individually.

The most dramatic example of transposon-mediated gene acquisition is *Enterococcus faecalis,* a normally benign inhabitant of the mammalian intestine that can on occasion cause serious infections. In *E. faecalis,* more than a quarter of the genome consists of transposons, IS elements, integrated phages, and integrated plasmids. Many of the transposons contain genes encoding antibiotic resistance or genes involved in disease causation. Clearly, in this organism, transposons have been a major force in genome evolution.

10.5 Many procaryotes have plasmids that encode dispensable functions

Although most procaryotes have a single chromosome, many have, in addition, a number of **plasmids.** Plasmids are, like the chromosome, usually circular molecules of supercoiled DNA. They differ from the chromosome in a number of ways, however: they are usually much smaller (typically 1/10th to 1/100th the size of the chromosome), they are present in some members of a species but not in others, and they encode no functions that are essential to the cell under all conditions of growth (they are dispensable). Like the chromosome, most plasmids replicate bidirectionally from a single origin; however, some plasmids of gram-positive bacteria replicate by a rolling circle mechanism, similar to the chromosome of many phages (see Section 6.13).

The kinds of functions typically encoded by plasmids fall into several distinct classes. Many encode enzymes conferring antibiotic resistance. These plasmids presumably have their origin in soil ecosystems where actinomycetes and fungi produce small amounts of antibiotics. Thus, soil bacteria may occasionally find antibiotic resistance selectively advantageous. With the discovery of the clinical utility of antibiotics and their widespread use in human and veterinary medicine beginning in the 1950s, many disease-causing bacteria acquired such plasmids, and they have become a major problem in medicine. The rapid spread of antibiotic-resistance plasmids among bacteria in hospitals and similar settings is evidence for significant amounts of interspecific and even intergeneric gene exchange among bacteria, by the mechanisms discussed in Chapter 12.

Another common category of plasmid encodes the ability to degrade exotic organic compounds, often ones with aromatic rings, as a source of carbon and energy. Strains possessing these plasmids can use such compounds, whereas strains lacking them cannot.

A third category of plasmid-borne genes consists of toxin genes. Pathogenic bacteria often cause specific disease symptoms as a result of the production of particular protein toxins (see Chapter 19). Often these toxins are encoded by genes on a plasmid. For example, *Bacillus anthracis,* the causative agent of the animal and human disease anthrax, has a large plasmid that encodes several different protein toxins. Strains of *B. anthracis* that lack the plasmid do not cause disease.

Plasmids generally encode at least some of their own replication proteins, and they replicate independently of chromosomal replication. They also encode their own segregation machinery, independent of the bacterial cytoskeleton. Many also encode a toxin/antitoxin system, which produces simultaneously a toxin that can kill the cell containing the plasmid and an antitoxin that prevents cell killing. This appears to be a system designed to kill any cell that fails to inherit the plasmid due to a defect in plasmid replication or segregation. The toxin and antitoxin molecules are made constitutively, and thus, they are present in the cytoplasm of all cells containing the plasmid. If an error in replication or segregation occurs, the plasmid-free daughter cell will still contain both toxin and antitoxin. Because the antitoxin is less stable than the toxin, after a while, it will be degraded, leaving behind the toxin, which kills the plasmid-free cell.

Many plasmids also encode the machinery for their own cell-to-cell transfer. Transfer is accompanied by replication, and, thus, the donor cell retains a copy while a copy is transferred to the recipient cell. These **transferrable plasmids** can, thus, spread in a population. Transfer is most common to other members of the same species, but frequently, transferrable plasmids can transfer to recipient cells in other genera or even different evolutionary lineages. As well as catalyzing their own transfer, transferable plasmids can occasionally effect the transfer of other plasmids, or even the chromosome (see Chapter 12). These plasmids are, thus, important in procaryotic evolution, as they constitute a mechanism for the transfer of genetic information across large evolutionary distances.

10.6 Procaryotic genes are often organized into operons

A **transcriptional unit** is a stretch of DNA that is transcribed into a molecule of RNA. In eucaryotes, these are commonly single genes. In procaryotes, however, multiple genes are commonly grouped together into a single transcriptional unit, under the control of a single promoter; such a transcriptional unit is called an **operon.** This grouping is feasible in procaryotes because procaryotic ribosomes bind to particular sequences (called **Shine-Dalgarno sequences**) on mRNA to initiate translation. Messages with multiple genes have a ribosome-binding site (Shine-Dalgarno sequence) at the beginning of each, thus allowing the synthesis of multiple proteins from a single mRNA. This is not feasible in eucaryotes because the ribosome binding site includes the 5′ cap on the mRNA; there is thus a single ribosome binding site per mRNA, and only a single polypeptide can be initiated per message.

Transcriptional units can be oriented in either direction on the circular chromosome; thus, for some operons, the "clockwise" strand is the one that is transcribed, whereas for other operons, the "counterclockwise" strand is transcribed; however, there is a tendency for transcriptional units to be oriented so that they are transcribed in the same direction as the chromosome is replicated (Figure 10.3). Because

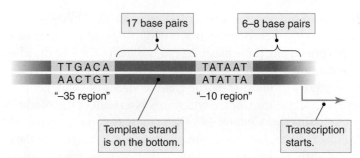

FIGURE 10.3 Consensus sequence for the *E. coli* standard sigma factor.

the chromosome is replicated bidirectionally from a single origin, most operons on one side of the chromosome are transcribed clockwise, whereas those on the other side are transcribed anticlockwise. This nonrandom orientation is thought to reduce the frequency with which transcribing RNA polymerases collide with a replisome. Such collisions are wasteful because they displace the RNA polymerase and result in an incomplete mRNA.

10.7 Promoter recognition by RNA polymerase is different in bacteria than in archaea and eucarya

Bacteria have a single RNA polymerase that has four core subunits: two α subunits and two very similar β subunits (β and β′). A fifth subunit (σ) associates with the enzyme before promoter binding but dissociates after transcription is initiated. This sigma subunit is responsible for recognizing promoters. RNA polymerase binds nonspecifically to DNA and then slides along it scanning for a promoter sequence. When one is detected, the polymerase stops, transcription is initiated, and the sigma subunit dissociates.

There are several different sigma subunits in bacteria, each with a different sequence specificity for promoter recognition. Bacterial cells regulate transcription of entire categories of gene by controlling whether particular sigma factors are made. For instance, in *Bacillus,* a number of genes for endospore formation are not expressed in growing cells because their promoters are not recognized by the sigma factor used by growing cells. As the cells enter stationary phase, however, another sigma factor is made, which recognizes some of these spore-specific promoters, and, thus, sporulation genes begin to be transcribed (see Chapter 11).

Promoter recognition in archaea and eucaryotes is fundamentally different. Their RNA polymerases have multiple subunits, none of which is functionally equivalent to sigma. Thus, they are not capable of recognizing promoters. Instead, specific proteins called **transcription factors** recognize and bind to sites adjacent to promoters. When the RNA polymerase slides along the DNA, it stops and initiates transcription when it encounters a bound transcription factor.

10.8 Promoters may be "weak" or "strong"

A large number of bacterial promoters have been sequenced, and although they are not the same, they are similar enough in sequence to support the notion that RNA polymerase with its bound sigma factor recognize a particular base sequence as the signal to bind and begin transcription. Because they are not all the same, we talk of a **consensus sequence** as the "ideal" sequence. Each base in the consensus sequence

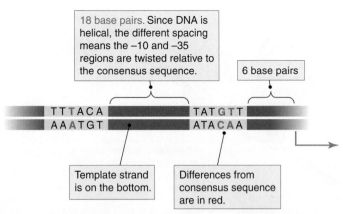

18 base pairs. Since DNA is helical, the different spacing means the −10 and −35 regions are twisted relative to the consensus sequence.

6 base pairs

TTTACA
AAATGT

TATGTT
ATACAA

Template strand is on the bottom.

Differences from consensus sequence are in red.

FIGURE 10.4 Sequence of a weak promoter (the *lac* promoter).

is the base that most commonly appears at that position (Figure 10.3); if all bases appear with roughly equal frequency, we assume that it does not matter which is at that position. For instance, in *E. coli*, the major sigma factor recognizes the consensus sequence shown in Figure 10.3.

The consensus sequence is one which the RNA polymerase will recognize easily and at which it will efficiently initiate transcription. Such a promoter is called a **strong promoter** because almost every time an RNA polymerase molecule encounters it, transcription ensues. As we said, however, many promoters have significant differences from the consensus sequence, and these **weak promoters** are recognized less effectively by RNA polymerase (Figure 10.4). The result is that frequently the polymerase fails to recognize the promoter, and no transcription results. These transcriptional units are transcribed less often, and thus, their products are present at lower concentration in the cell.

There are two principal strategies by which evolution has optimized promoter strength for each transcriptional unit. In most cases, the promoter strength is matched to the maximum amount of gene product the cell will need. Thus, proteins needed in large amounts will be transcribed from a strong promoter, whereas proteins needed in only small amounts will be transcribed from a weak promoter. If less than the full amount of gene product is needed, gene regulation can *reduce* the amount under appropriate circumstances.

The other option, which applies, for instance, to the *lac* operon whose promoter sequence is given in Figure 10.4, is to match genes whose products are needed at high amount to a *weak* promoter. In such a case, a transcription factor is needed, in addition to the RNA polymerase with its bound sigma factor, in order to achieve efficient recognition; thus, regulation in these cases serves to *increase* the amount when appropriate. We discuss this mechanism later in this chapter.

10.9 There are two types of bacterial transcription termination

When transcription of a transcriptional unit is complete, the RNA polymerase releases the RNA and dissociates from the DNA. Transcription termination is poorly understood in archaea and eucaryotes but has been well studied in bacteria. There are two basic mechanisms in bacteria by which the polymerase recognizes the proper point to terminate.

One type involves an RNA-binding protein called **Rho.** This protein binds to C-rich sequences in RNA that is not being translated (ribosomes protect RNA from rho binding) and then slides along the RNA in a 5′ to 3′ direction (Figure 10.5).

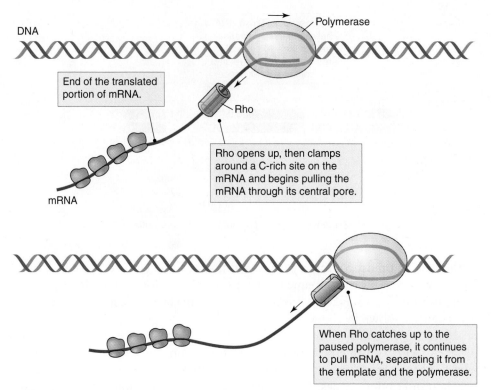

End of the translated portion of mRNA.

Polymerase

Rho

Rho opens up, then clamps around a C-rich site on the mRNA and begins pulling the mRNA through its central pore.

DNA

mRNA

When Rho catches up to the paused polymerase, it continues to pull mRNA, separating it from the template and the polymerase.

FIGURE 10.5 Rho-dependent transcription termination.

When it catches up to the RNA polymerase (normally after moving about 20 to 40 nucleotides), it pulls the transcript free of the polymerase and template, releasing it. This happens at sequences that cause the ribosome to pause briefly, thus allowing Rho to catch up. Rho-dependent termination thus can stop transcription when the polymerase has moved beyond the end of the last gene in an operon.

In Rho-independent termination, the RNA polymerase falls off the template when it is paused over a series of A-U base pairs (A in the template DNA and U in the mRNA). The hydrogen bonding of A-U base pairs has insufficient energy to keep the complex together for long, and the whole thing spontaneously dissociates. The pausing is caused by a stem-loop structure in the RNA immediately behind the polymerase (Figure 10.6). Thus, Rho-independent terminators have a recognizable

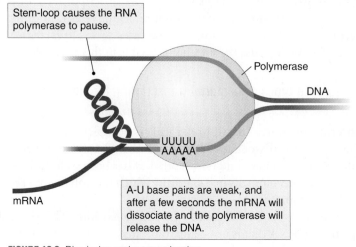

Stem-loop causes the RNA polymerase to pause.

Polymerase

DNA

UUUUU
AAAAA

A-U base pairs are weak, and after a few seconds the mRNA will dissociate and the polymerase will release the DNA.

mRNA

FIGURE 10.6 Rho-independent termination.

structure: a run of adenosines in the template strand, immediately preceded by two short sequences that are complementary to each other. Such terminators are also called **intrinsic terminators.**

10.10 Every transcriptional unit has a common set of signals

The discussion so far allows us to identify the signals that are common to all transcriptional units that encode proteins (Figure 10.7). Transcriptional units that are not translated—for instance, those for rRNA and for tRNA—lack the translational control signals, but have transcriptional control signals just like operons for mRNAs.

In addition to these control sequences, most transcriptional units also have one or more *regulatory sites* that allow the rate of transcription or the rate of translation to be controlled. These regulatory mechanisms fall into several broad categories:
- The rate of transcription initiation can be controlled.
- Transcription termination can be suppressed to transcribe additional sequences further away from the promoter.
- The rate of translation initiation can be controlled.

Each of these different mechanisms requires a different set of regulatory sequences in the DNA. Often, two or more controls can be imposed on a single transcription unit. In the remainder of this chapter, we examine regulatory systems that represent each of these different categories.

Regulation of Individual Transcriptional Units

Many genes in procaryotes are regulated to some extent, as most produce products that are needed in varying amounts at different times—for instance, in different environments or at different times in the cell cycle. It is convenient to consider the topic of gene regulation in two parts, corresponding to two different levels of control. Each regulated operon has its own specific control system that modulates expression in response to some environmental signal. Superimposed on these myriad specific control systems are a series of **global control** systems, in which many different operons can be shut down or turned on simultaneously in response to a particular environmental signal. Here we describe several of the mechanisms by which individual operons are regulated; global control mechanisms are described in subsequent sections.

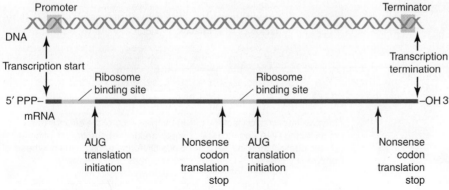

FIGURE 10.7 Control sequences in a polycistronic message.

10.11 Transcription initiation is controlled by allosteric proteins—repressors or transcription factors

Control of transcription initiation is a common mechanism of control of gene expression in procaryotes, both operon specific and global. This is commonly, but not always, mediated by allosteric proteins. These allosteric proteins have a binding site for a specific sequence of DNA and another binding site for a signal molecule that determines whether transcription is turned on or off. Allosteric proteins that bind to the DNA and turn *off* transcription are called **repressors,** and this mode of control is termed **negative control**. Allosteric proteins that bind to DNA and turn *on* transcription are called **transcription factors** (or sometimes **activators**), and this mode of control is termed **positive control** (Figure 10.8).

Repressors and transcription factors are in equilibrium between two different states: an active, DNA-binding form and an inactive form that cannot bind to the DNA. The equilibrium can be shifted by the presence or absence of a small **effector molecule** that binds to the protein and changes its conformation into the alternative form. Which form is inactive and which is active depends on the specific system. Generally, an operon is called **inducible** if the presence of a small molecule (called an **inducer**) turns the system *on* (Figure 10.9); it is called **repressible** if the presence of a small molecule (called a **corepressor**) turns the system *off* (Figure 10.10).

Inducible operons

Negative control—the *lac* operon

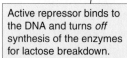

Active repressor binds to the DNA and turns *off* synthesis of the enzymes for lactose breakdown.

In the presence of the effector (a β-galactoside sugar), the repressor is inactivated and enzyme synthesis proceeds.

Positive control—the arabinose operon

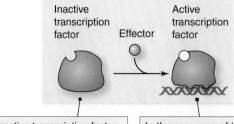

Inactive transcription factor cannot bind DNA and turn on the operon; thus, no enzyme synthesis.

In the presence of the effector (the sugar arabinose), the transcription factor binds DNA and turns *on* the synthesis of the enzymes of arabinose breakdown.

Negative control

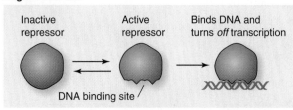

Inactive repressor → Active repressor → Binds DNA and turns *off* transcription

DNA binding site

Positive control

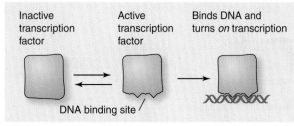

Inactive transcription factor → Active transcription factor → Binds DNA and turns *on* transcription

DNA binding site

FIGURE 10.8 Positive and negative control.

FIGURE 10.9 Inducible systems.

CHAPTER 10 PROCARYOTIC GENOME ORGANIZATION AND REGULATION

Repressible operons

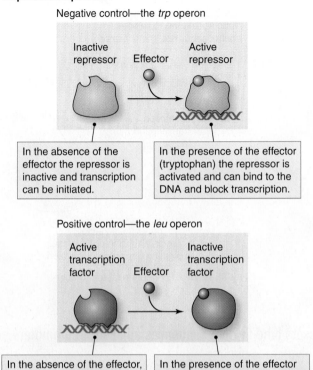

Negative control—the *trp* operon

Inactive repressor   Effector   Active repressor

In the absence of the effector the repressor is inactive and transcription can be initiated.

In the presence of the effector (tryptophan) the repressor is activated and can bind to the DNA and block transcription.

Positive control—the *leu* operon

Active transcription factor   Effector   Inactive transcription factor

In the absence of the effector, the transcription factor is active and can bind to DNA, stimulating transcription.

In the presence of the effector (leucine) the transcription factor is inactivated and transcription is repressed.

FIGURE 10.10 Repressible systems.

Generally, inducible operons are ones that encode enzymes to break down compounds as a source of carbon and energy. This makes sense, as the enzymes are not needed unless the compound is present. Thus, the presence of a particular carbon source can be expected to turn on enzyme synthesis.

Repressible systems tend to be ones for biosynthesis. Again, this makes sense, as cells need to make all their amino acids, purines and pyrimidines, vitamins, etc., *unless* they are present in the medium. Thus, in this case, the presence of a particular molecule can be expected to turn off enzyme synthesis.

10.12 Repressors and transcription factors bind to DNA sequences in the major groove

We tend to think about the specific differences in base pairs (A-T, T-A, G-C, and C-G) in terms of the arrangement of functional groups that engage in base pairing. Proteins that bind to specific DNA sequences, however, such as RNA polymerase, repressors, and transcription factors, cannot use these differences for recognition because these functional groups are buried in the middle of the double-helical DNA. What the proteins have to recognize, therefore, is the difference between the base pairs viewed from the side.

The two antiparallel strands of DNA are wound around each other in an asymmetrical way, leading to two grooves around the outside of the helix that differ in size. The larger **major groove** is the one that almost all DNA-binding proteins use to recognize base sequence.

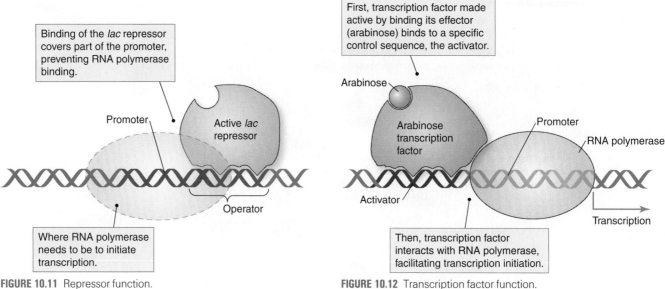

Binding of the *lac* repressor covers part of the promoter, preventing RNA polymerase binding.

Promoter

Active *lac* repressor

Operator

Where RNA polymerase needs to be to initiate transcription.

FIGURE 10.11 Repressor function.

First, transcription factor made active by binding its effector (arabinose) binds to a specific control sequence, the activator.

Arabinose

Arabinose transcription factor

Promoter

RNA polymerase

Activator

Transcription

Then, transcription factor interacts with RNA polymerase, facilitating transcription initiation.

FIGURE 10.12 Transcription factor function.

10.13 Repressors bind to DNA sequences adjacent to promoters and prevent RNA polymerase binding

The active form of a repressor binds to a particular DNA sequence, called an **operator,** that is located adjacent to the operon's promoter. Obviously, each different operon has a different operator sequence. Because the operator is adjacent to the promoter (sometimes even overlapping), when the bulky repressor is bound there, it interferes sterically with the binding of RNA polymerase (Figure 10.11). When the polymerase slides along the DNA, it hits the repressor before it occupies the promoter; thus, it dissociates from the DNA rather than initiating transcription.

10.14 Transcription factors bind to DNA sequences adjacent to weak promoters and allow RNA polymerase to recognize them

Transcription factors also bind adjacent to promoters, but they have an effect exactly opposite to that of repressors: rather than interfering with polymerase binding, they enhance it. This is because transcriptional units controlled this way have weak promoters that RNA polymerase rarely recognizes. In the absence of a bound transcription factor, the polymerase can slide right over the promoter without stopping. If a transcription factor is bound, however, it stops the polymerase and positions it correctly to allow it to initiate transcription (Figure 10.12).

10.15 Repressors and transcription factors often bind at two sites to form DNA loops

Many repressors and transcription factors are dimers or tetramers with two identical DNA-binding sites. Binding at one site facilitates binding at the other, and a

CHAPTER 10 PROCARYOTIC GENOME ORGANIZATION AND REGULATION

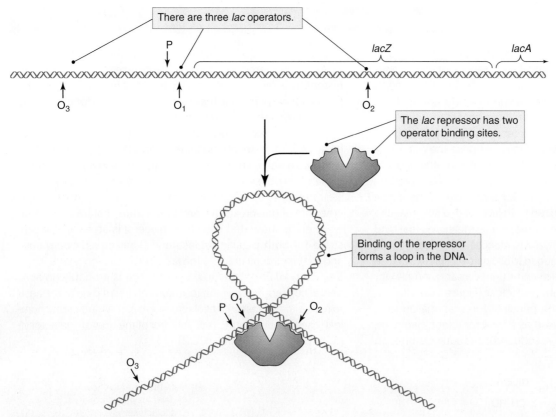

FIGURE 10.13 DNA looping during *lac* operon repression.

protein bound to DNA through both sites is much less likely to dissociate than one bound at only one site. Transcriptional units controlled by such proteins typically have two or more operator sites or activator sites. For instance, the *lac* operon (see MicroTopic on the next page) has three operators (Figure 10.13). The main one overlaps the promoter, and this is the preferred binding site. The other repressor binding site then binds to one of the other operators: in the middle of the *lacZ* gene or a couple of hundred base pairs upstream.

The function of these DNA loops is not clear. In most cases, it is suspected that the curvature either aids or inhibits the binding of other proteins, such as RNA polymerase or other regulatory proteins, by changing the geometry of the major groove.

DNA looping can also bring two different proteins bound at distant sites together. This is the basis for the action of transcription factors bound to sites known as **enhancers.** Enhancers are distant from the operon whose transcription they enhance, often by thousands of nucleotide pairs in eucaryotes, but generally a hundred or so in bacteria. Transcription factor binding at the enhancer activates transcription when the DNA loops around, bringing the bound transcription factors into contact with RNA polymerase (Figure 10.14).

10.16 Repressors and transcription factors can control multiple operons

Genes that encode enzymes of a single biochemical pathway are often clustered together into an operon—for instance, all eight genes for the histidine biosynthetic

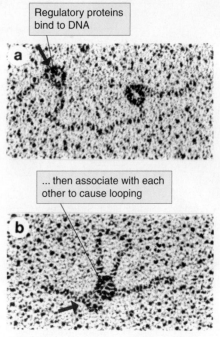

FIGURE 10.14 Electron micrograph of DNA looping.

The *lac* operon, which you learned about in introductory biology, draws its name from the fact that the possession of these genes allows *E. coli* to grow using the sugar lactose (galactosyl-glucose) as a source of carbon and energy. It was naturally assumed that the ability to use lactose was the selective advantage that led to the evolutionary acquisition and retention of this operon. However, there are some awkward facts that raise the possibility that this may not be the case. First, lactose is not a natural inducer of the operon, because it does not bind well to the repressor. Rather, a derivative of lactose called allolactose is the actual inducer. Allolactose is produced from lactose in a side reaction catalyzed by β-galactosidase—most lactose molecules that bind to β-galactosidase are hydrolyzed to glucose and galactose, but a small percentage are rearranged to allolactose. This allolactose then is the actual inducer. This seems unnecessarily complex for an operon designed to metabolize lactose (Figure 10B.1).

Second, lactose is found principally in mammalian milk, which, until recently, was used as a food for infant mammals, rarely by adults. *E. coli*, however, is found in the adult mammalian intestine, not in the infant intestine. This suggests that *E. coli* rarely encountered lactose over the course of its evolutionary history. This raises the possibility that β-galactosidase and the β-galactoside permease may have been selected for their ability to act on something other than lactose.

This puzzle was solved by Winfried Boos, who noted that galactosyl lipids are abundant in chloroplasts, and when these lipids are digested by lipases in the mammalian gut, galactosyl-glycerol is released. Thus, any mammal that consumes significant amounts of vegetable matter in its diet produces large amounts of galactosyl-glycerol in the intestinal tract. He then went on to show that galactosyl-glycerol is an excellent substrate for both β-galactosidase and β-galactoside permease and that it directly induces the *lac* operon to high levels. Thus, the ability of *E. coli* to grow with lactose is probably an accidental consequence of the fortuitous ability of β-galactosidase to rearrange lactose into allolactose, which has sufficient chemical resemblance to the "real" inducer of the operon, galactosyl-glycerol, to work (Figure 10B.2).

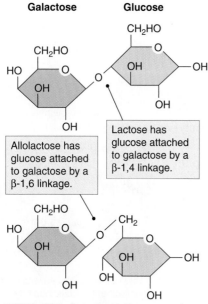

FIGURE 10B.1 Structures of lactose and allolactose.

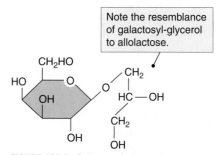

FIGURE 10B.2 Galactosyl-glycerol.

pathway form a single operon. This makes sense because all eight enzymes are needed, or not needed, simultaneously.

Sometimes, however, the genes of a biochemical pathway are scattered among many transcriptional units; for instance, the eight genes for enzymes of the arginine biosynthetic pathway are clustered into six *arg* transcriptional units, spread around the chromosome (Figure 10.15). In this case, each of these is controlled by the same repressor. Thus, the genes are all expressed or repressed in concert, just as if they were clustered into the same operon. A group of transcriptional units controlled by the same regulatory protein is called a **regulon.**

10.17 Some proteins can be both repressors and transcription factors

Generally, DNA-binding regulatory proteins act either to suppress or to enhance transcription initiation; however, some proteins can play both roles. Sometimes a protein is a repressor when not bound to its effector, but becomes a transcription factor in the presence of the effector. This is the case for the arabinose operon; the regulatory protein binds to an operator and functions as a repressor in the absence of the inducer arabinose (Figure 10.16). In the presence of the inducer, however, a

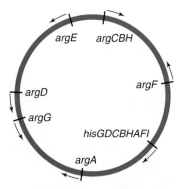

FIGURE 10.15 Location of the *E. coli* chromosome of the *arg* and *his* operons.

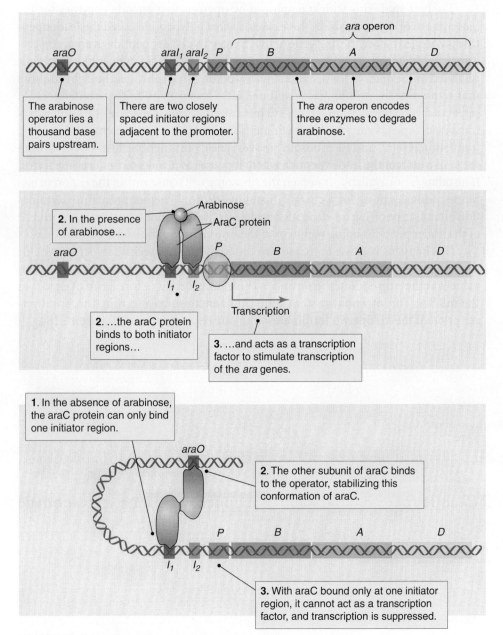

FIGURE 10.16 Regulation of the arabinose operon.

conformational change of the protein allows it to bind to a slightly different site and function as a transcription factor.

10.18 Many biosynthetic operons are controlled by attenuation, regulating early termination

Many operons encoding genes for the biosynthesis of amino acids and other essential compounds have a terminator shortly after the transcription start point and before the beginning of the first gene in the operon. Thus, transcription can be terminated almost immediately after it is initiated, and the genes in the operon are not expressed. Expression of the operon thus requires that this premature termination be suppressed. There are several known mechanisms to do this, collectively called **attenuation.**

The best studied is widely employed by operons that encode enzymes for the biosynthesis of amino acids. In these cases, a short polypeptide called the **leader peptide** is encoded before the first gene of the operon; this leader peptide overlaps the terminator. The leader peptide contains several of the amino acids whose biosynthesis is catalyzed by the operon; for instance, the *his* operon, which encodes the eight enzymes that synthesize the amino acid histidine, has a leader peptide that contains seven sequential histidines (Figure 10.17). Thus, whether this leader can be successfully translated depends on whether histidine is available. If there is plenty of histidine, translation is rapid and the ribosome moves quickly through the histidine codons to the stop codon of the leader peptide where it pauses as it waits to bind release factor. If histidine is not available, however, the ribosome will pause within the region of histidine codons, waiting for a charged histidinyl-tRNA. Whether premature termination of transcription occurs depends on whether the ribosome is paused in the middle of the leader peptide coding region or at the stop codon of the leader peptide.

This happens because there are two different possible secondary structures for the leader region of the mRNA. Two stem-loop structures can form when the ribosome is at the stop codon; the second of these occurs right before a run of uracils in the mRNA. This **attenuator** stem-loop thus functions as a terminator, and transcription of the *his* operon terminates prematurely at this point (Figure 10.18).

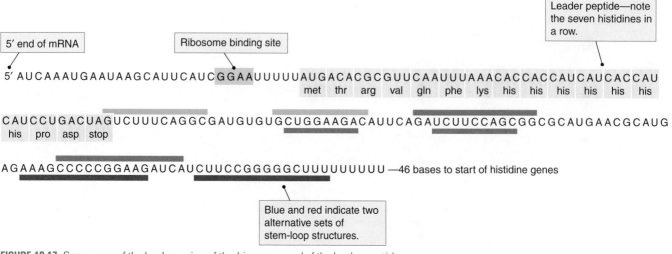

FIGURE 10.17 Sequences of the leader region of the *his* operon and of the leader peptide.

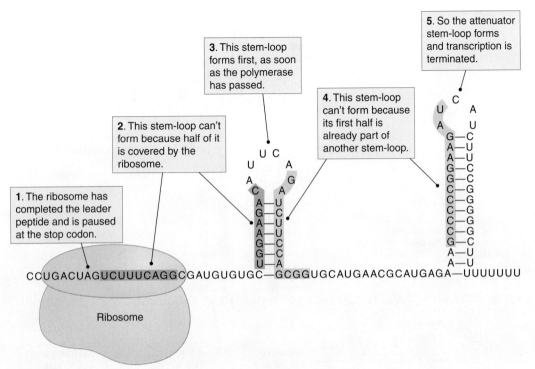

FIGURE 10.18 Secondary structure of the *his* operon mRNA with excess histidine.

The callout boxes in the figure read:

1. The ribosome has completed the leader peptide and is paused at the stop codon.

2. This stem-loop can't form because half of it is covered by the ribosome.

3. This stem-loop forms first, as soon as the polymerase has passed.

4. This stem-loop can't form because its first half is already part of another stem-loop.

5. So the attenuator stem-loop forms and transcription is terminated.

Ribosome

If the ribosome is paused in the middle of the *his* leader peptide, however, an alternate set of stem-loop structures forms (Figure 10.19). In this pattern there is no transcriptional terminator, and the RNA polymerase continues on to transcribe the operon.

Attenuation may be combined with control by repressors and operators (as in the tryptophan operon) but more commonly is the only major regulatory mechanism. Clearly it is not designed to turn operons fully on or fully off (something repressors do very well), but rather to fine tune the rate of transcription. With only 10 molecules of tRNA total per ribosome, transfer RNAs have to be charged (loaded with amino acids) as fast as they are used; even a slight decrease in the charged tRNA concentration for a single amino acid can dramatically slow protein synthesis for the entire cell. Attenuation measures the real-time demand for charged tRNA, and can fine tune the concentration of biosynthetic enzymes.

Attenuation requires coupling between transcription and translation; it is thus a regulatory mechanism restricted to procaryotes.

10.19 Antitermination is another mechanism regulating termination

Other instances of regulation of premature termination also occur. These involve special **antiterminator proteins** that bind to the DNA at special sequences before a terminator (Rho-dependent or intrinsic). These recognition sequences can be hundreds of bases upstream of the terminator, and thus, they appear to act not by preventing terminator formation or by blocking Rho, but rather by altering the RNA polymerase as it goes by. Binding of the antiterminator protein to its specific site on the DNA causes a conformational change in the protein, which can then bind to RNA polymerase when it passes, causing a conformational change that renders the polymerase resistant to terminators, probably by making it resistant to pausing.

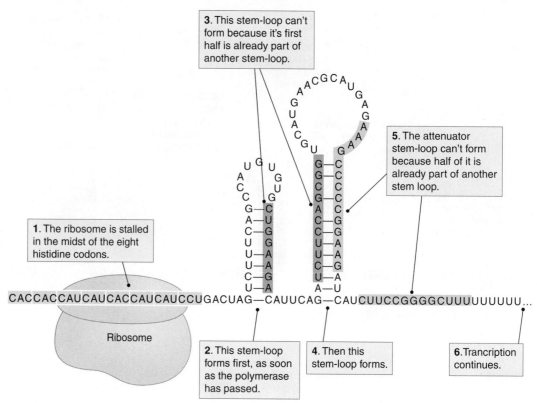

3. This stem-loop can't form because it's first half is already part of another stem-loop.

5. The attenuator stem-loop can't form because half of it is already part of another stem loop.

1. The ribosome is stalled in the midst of the eight histidine codons.

CACCACCAUCAUCACCAUCAUCCUGACUAG—CAUUCAG—CAUCUUCCGGGGCUUUUUUUUU...

Ribosome

2. This stem-loop forms first, as soon as the polymerase has passed.

4. Then this stem-loop forms.

6. Trancription continues.

FIGURE 10.19 Secondary structure of the *his* operon mRNA with limiting histidine.

An example is the synthesis of the Int protein of phage λ. This is the protein that integrates the λ prophage into the bacterial chromosome. It is important that this protein be made early in infection, but not immediately. The *int* gene is part of a transcriptional unit that includes an antitermination protein called N (Figure 10.20). Immediately after infection, the N protein is made, but transcription terminates before the *int* gene. Only after a several-minute delay does the N protein accumulate to a sufficient level to suppress termination and allow synthesis of the Int protein.

10.20 Procaryotic mRNA is rapidly degraded, and thus, continued gene expression requires continued transcription

The regulatory mechanisms that we have described so far control the transcription of genes. Thus, for example, the rate of transcription of the *lac* operon can increase more than a thousand-fold in a matter of seconds after exposure to inducer. Within a couple of minutes, translation is producing the lac enzymes at the new, fully induced level. Thus, transcriptional control is capable of very rapid increases of protein synthesis in response to environmental signals. However, an additional mechanism is necessary to ensure that when the signal is removed (e.g., the β-galactoside in the medium is exhausted) the response is also rapid. Even if transcription is immediately shut off, the cells have copious amounts of mRNA made during the period of induction, and if this mRNA remains continually available for translation, unnecessary protein synthesis will continue for many generations.

This does not happen in procaryotes because of the very rapid degradation of mRNA. A variety of enzymes degrade single-stranded RNA. These include a

CHAPTER 10 PROCARYOTIC GENOME ORGANIZATION AND REGULATION

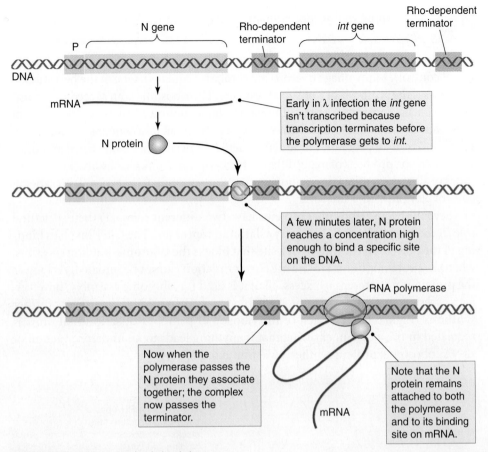

Early in λ infection the *int* gene isn't transcribed because transcription terminates before the polymerase gets to *int*.

A few minutes later, N protein reaches a concentration high enough to bind a specific site on the DNA.

Now when the polymerase passes the N protein they associate together; the complex now passes the terminator.

Note that the N protein remains attached to both the polymerase and to its binding site on mRNA.

FIGURE 10.20 Antitermination in lambda.

5′ exonuclease that removes nucleotides one at a time from the 5′ end of the molecule, a **3′ exonuclease,** and several different **endonucleases.** Endonucleases cut the mRNA internally to produce oligonucleotide fragments that are subsequently degraded by exonucleases to individual nucleotides (which are reused in biosynthesis). The combination of these enzymes results in rapid degradation of mRNA, typically such that half of the mRNA in the cell is degraded within a couple of minutes. Clearly, continued gene expression requires continued replacement of the mRNA and thus continued transcription.

Of course, the rate of degradation of specific RNA molecules varies over a wide range. Ribosomes inhibit cleavage by endonucleases by shielding about 20 to 30 nucleotides of the mRNA in their immediate vicinity. Thus, mRNAs that have very effective ribosome binding sites are likely to survive somewhat longer than those with weak sites, as they will, on average, have more ribosome bound to them at any time and thus be more effectively shielded from endonucleolytic cleavage.

Extensive secondary structure inhibits nucleolytic cleavage, and, thus, extensively base-paired RNA molecules, such as tRNAs, are highly resistant to cleavage. tRNAs also have a large proportion of their nucleotides chemically modified, which also reduces their susceptibility to nucleases. Thus, procaryotic cells are able to have two categories of single-stranded RNA: stable RNA, with extensive secondary structure and many modified nucleotides, and highly labile mRNA, with limited secondary structure and few or no modified nucleotides.

10.21 Translational repressors prevent translation of mRNA

A number of procaryotic operons are regulated at the translational level (translational control). Many phage operons are similarly regulated. One of the best studied examples is the synthesis of ribosomal proteins. The ribosome is an exceedingly complex organelle, consisting of two subunits. The large subunit consists of 2 molecules of rRNA and over 30 different protein molecules; the small one contains 1 molecule of rRNA and over 20 proteins. These components are needed in exactly equal amounts, yet they are the products of many different transcriptional units. Coordinating the synthesis of all of these different components is thus a substantial challenge. Translational control plays an important role.

Several specific ribosomal proteins have two different roles: (1) their structural role in the ribosome and (2) as a **translational repressor.** The latter involves binding of the protein to the mRNA at a site that blocks the ribosome-binding site; thus, when the protein is bound to the mRNA, translation cannot be initiated. As long as the protein is not present in excess, all of it is used for ribosome assembly; however, if an excess occurs for any reason, it will bind to its own mRNA and prevent further synthesis (Figure 10.21). Each of the ribosomal protein transcriptional units is regulated in this fashion, ensuring that if anything leads to an underproduction of rRNA, ribosomal protein synthesis will immediately stop.

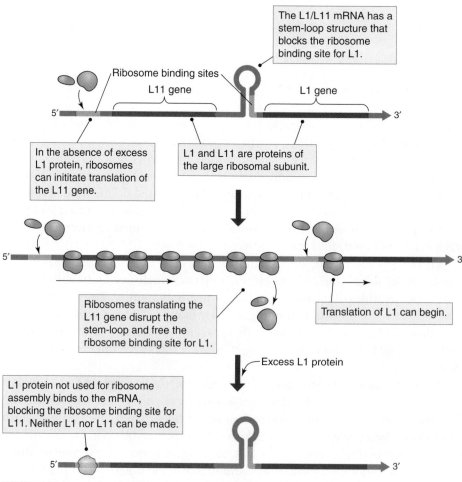

The L1/L11 mRNA has a stem-loop structure that blocks the ribosome binding site for L1.

Ribosome binding sites

L11 gene

L1 gene

In the absence of excess L1 protein, ribosomes can inititate translation of the L11 gene.

L1 and L11 are proteins of the large ribosomal subunit.

Ribosomes translating the L11 gene disrupt the stem-loop and free the ribosome binding site for L1.

Translation of L1 can begin.

Excess L1 protein

L1 protein not used for ribosome assembly binds to the mRNA, blocking the ribosome binding site for L11. Neither L1 nor L11 can be made.

FIGURE 10.21 Translational control of the L1/L11 operon.

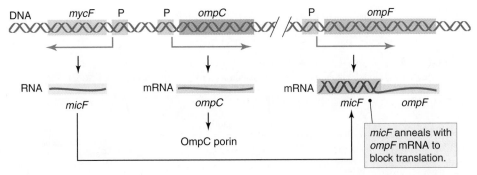

FIGURE 10.22 *ompF* regulation by *micF*.

10.22 Antisense RNA can prevent translation of mRNA

At least one case of bacterial translational regulation is known to involve **antisense RNA:** RNA that anneals to mRNA and blocks translation. This involves regulation of porin synthesis in *E. coli,* which has two different porin genes, each encoding an outer membrane protein with a slightly different pore size. In environments of low osmolarity, cells make the OmpF porin, with a pore diameter of 1.2 nm (pore area of 1.13 nm^2); in environments with high osmolarity, a different porin is made, OmpC, with a pore diameter of 1.1 nm (pore area of 0.95 nm^2).

The genes *ompC* and *ompF* are regulated by a number of different proteins in a very complicated and still poorly understood manner; however, one aspect of this regulation is the translational repression of *ompF* mRNA by the antisense RNA encoded by the *micF* gene (Figure 10.22). *micF* is adjacent to *ompC* and appears to be regulated in concert with it, and thus, when *ompC* is expressed, so too is *micF*. Thus, if there is any *ompF* mRNA in the cytosol, the *micF* RNA will hybridize with it, blocking the ribosome binding site and preventing any further translation.

10.23 Some operons are turned on and off by DNA rearrangements

A small proportion of specific gene regulation in procaryotes involves recombination events within the DNA to actually rearrange the genome. Most such rearrangements are involved with the differentiation of specialized cell types and are discussed in Chapter 11; however, several instances are known of gene rearrangement in the control of a gene not involved in cell differentiation.

The best studied of these is the system of control of which particular flagellin protein the bacterium *Salmonella* makes. The flagellar protein is a major surface structure on flagellated cells and is one of the targets of the vertebrate immune system (see Chapter 20). It is also the attachment site for some bacteriophage. *Salmonella,* a parasite of vertebrate intestinal tracts, thus has good reason to make different types of flagella, each with a different flagellin protein, a phenomenon known as **phase variation.** Typically, a culture of *Salmonella* contains a mixture of cells, some making flagella with the HagA flagellin protein and some with the HagB flagellin. If one starts a culture with a single cell, making only one kind of flagellum, the rate of change can be measured: approximately 1 of every 10^3 to 10^5 cells will spontaneously switch to the other type of flagellum.

The flagellin genes are separate on the *Salmonella* chromosome. One, *hagB,* is in a transcriptional unit by itself, controlled by a repressor HagR (Figure 10.23). The

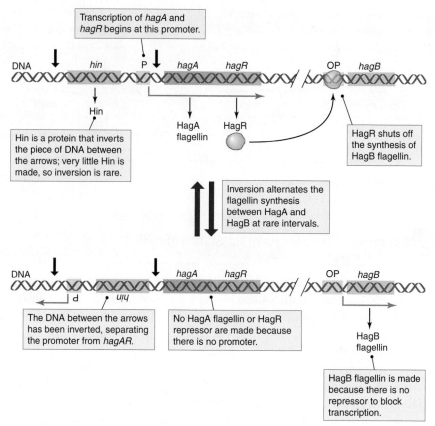

FIGURE 10.23 Mechanism of phase variation.

other, *hagA,* is in an operon along with *hagR,* the gene for the repressor of HagB. Thus, when the *hagAR* operon is turned on, *hagB* is turned off because its repressor is actively being made. Under these conditions, cells make flagella with the HagA flagellin.

A switch to the synthesis of HagB involves shutting off the *hagAR* operon by a recombinational event that separates the promoter from the operon; without its promoter, the operon cannot be transcribed. This recombinational event is an inversion, in which the segment of DNA on which the promoter lies flips over, separating the promoter from the operon it serves.

This inversion is catalyzed by an enzyme that is encoded by a gene (*hin*) within the invertible segment. Thus, this piece of DNA looks very much like a transposon, except that instead of moving to a different spot it flips over in place.

Global Control by Modulons

Specific control systems for turning on or off particular operons are an essential part of the efficient physiology of procaryotic cells. It enables them to economize on protein synthesis by not making proteins that are not essential under the particular conditions at the time; this is particularly important because proteins are especially expensive to make (each peptide bond costs four ATPs, in addition to the cost of synthesizing the amino acids); however, many genes make proteins that are part of a larger-scale response system to specific conditions, such as starvation, nitrogen limitation, and high temperature. Coordination of the many separate transcriptional units involved in each of these responses is the function of **global**

control systems, which affect the expression of many different operons simultaneously. A group of operons, each with its own specific control, that shares a common secondary control system that modulates the expression of them all, is called a **modulon.**

A variety of mechanisms underlie the different global control systems. Sometimes the many different transcriptional units of a global control system are all controlled by promoters that require a different sigma factor for their recognition. Synthesis or repression of that sigma factor thus turns them on or off as a block. Alternatively, they may all be repressed or activated by the same repressor or transcription factor.

10.24 There are many different global control systems

A number of different global modulons have been identified so far; many more probably await discovery by genomics. Some of the ones described to date are as follows:

- **Catabolite repression** turns off many different operons that encode enzymes of catabolism, such as the *lac* operon, when glucose is present.
- **Aerobic/anaerobic transition** involves large numbers of enzymes and electron carriers, grouped into several different modulons (e.g., one for aerobic respiration, one for anaerobic respiration, one for fermentation).
- **Nitrogen starvation** elicits a coordinated response of many different genes encoding enzymes of nitrogen uptake and metabolism.
- **Phosphate starvation** does the same for genes encoding enzymes of phosphate uptake and metabolism.
- **Heat shock**—temperatures near the maximum for growth—induces a modulon, including many different chaperones (proteins that help other proteins fold correctly) and proteases that degrade misfolded proteins.
- The **SOS response** is triggered by damaged DNA and involves the simultaneous induction of a number of different pathways of DNA repair.
- **Osmotic shock** induces a set of proteins including outer membrane porins, potassium permeases, compatible solute synthesis, etc.
- **Growth rate control** regulates the rates of synthesis of the protein synthesizing system (ribosomes, tRNAs, initiation and elongation factors, activating enzymes, etc.).
- The **stringent response** regulates the transition from high to low growth rate, and vice versa, and the initial response to starvation.
- **Cell division** appears to be controlled by at least one modulon that controls hundreds of genes involved with DNA synthesis and septum formation.
- **Stationary phase** is a period of starvation-enforced nongrowth, and cells adapt to it by inducing a large number of genes with unknown function important to survival.

These different modulons undoubtedly overlap—that is, some response proteins are probably part of two or more different modulons. They may also be coordinated by the sharing of regulatory mechanisms or by including the regulatory proteins of one modulon among the regulated proteins of another.

In the following sections we consider several representative modulons: catabolite repression because it is the best understood and because it is a good example of a modulon that operates via a common transcription factor; the heat-shock response, as an example of a modulon that works through an alternate sigma factor; growth rate control because of its relevance to Chapter 9 (growth) and because it provides an example of the common phenomenon of operons with more than one promoter; and the stringent response because of its relevance to Chapter 11.

10.25 Catabolite repression is mediated by the transcription factor CRP

Catabolite repression is the name given to the inhibitory effect that a preferred carbon and energy source has on the induction of enzymes for the degradation of other carbon and energy sources. For instance, in *E. coli*, the preferred carbon source is glucose. If a culture is grown in the presence of glucose and lactose, it will use the glucose first, and even though lactose is continuously present, the *lac* operon will be only slightly induced until after the glucose is all gone (Figure 10.24). This **diauxic growth** is indicative of catabolite repression.

Thus, there is more to inducing the *lac* operon than just removing the repressor from the operator. Remember that earlier (see Figure 10.4) we gave the *lac* promoter as an example of a weak promoter. Even with the repressor out of the way, RNA polymerase has a hard time initiating transcription at the *lac* promoter. In order to do so efficiently, the transcription factor **CRP**, for **cyclic AMP receptor protein** (also called **CAP** for **catabolite activator protein**), has to be bound at an adjacent activator site.

CRP is the modulator for the global control of carbon source usage. It is an allosteric protein, whose effector is the nucleotide **cyclic-3′, 5′-AMP.** This compound is made from ATP by the enzyme **adenyl cyclase** and is degraded to AMP by a **phosphodiesterase.** When it is present in high concentration, it binds to CRP, and the resulting active transcription factor can then bind to the activator sites adjacent to the promoters of many different operons that encode the enzymes necessary to break down other sugars, amino acids, etc. Thus, any of those operons whose specific inducer is also present will then turn on (Figure 10.25).

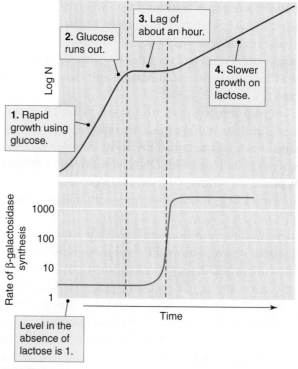

FIGURE 10.24 Diauxic growth.

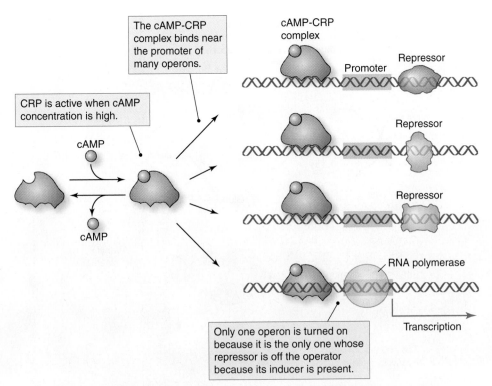

The cAMP-CRP complex binds near the promoter of many operons.

CRP is active when cAMP concentration is high.

cAMP

cAMP

cAMP-CRP complex

Promoter

Repressor

Repressor

Repressor

RNA polymerase

Transcription

Only one operon is turned on because it is the only one whose repressor is off the operator because its inducer is present.

FIGURE 10.25 CRP action on multiple operons.

10.26 Glucose permease indirectly controls the cAMP concentration

It is the activity of adenyl cyclase that controls the cellular level of cAMP, as the rate of degradation is relatively constant. Thus, the activity of adenyl cyclase must be inversely related to the availability of glucose: when glucose is abundant, adenyl cyclase is inactive, and the concentration of cAMP is very low. Thus, CRP is largely in the inactive form, and catabolic operons in the modulon cannot be turned on. When there is no glucose available, however, adenyl cyclase becomes active, the cAMP pools rise, CRP becomes active, and all of the operons in the modulon are ready to be turned on fully by their individual inducers.

The activation of adenyl cyclase is due to phosphorylation; in the nonphosphory-lated state it is inactive, but when phosphorylated, it becomes active. Phosphorylation is done by a protein kinase that is part of the glucose permease system. This is a phos-photransferase type of system (see Section 7.10), and the protein kinase that trans-fers a phosphate to the glucose permease is also capable of phosphorylating adenyl cyclase. When glucose is available and the glucose permease is working at its maxi-mum rate, most of the phosphorylation is directed towards glucose; it is only when glucose is not being actively transported that there is significant phosphorylation of adenyl cyclase (Figure 10.26).

10.27 The heat shock response is mediated by an alternate sigma factor

When bacterial cells encounter temperatures very near their maximum, they increase the rate of synthesis of a set of proteins involved in protein folding and protein

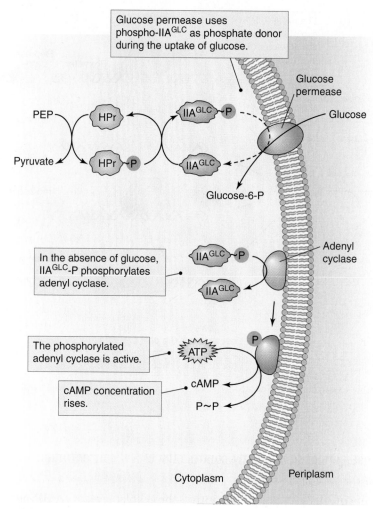

Glucose permease uses phospho-IIA^{GLC} as phosphate donor during the uptake of glucose.

Glucose permease

Glucose

PEP

HPr

IIA^{GLC} ~ P

Pyruvate

HPr ~ P

IIA^{GLC}

Glucose-6-P

In the absence of glucose, IIA^{GLC}-P phosphorylates adenyl cyclase.

IIA^{GLC} ~ P

Adenyl cyclase

IIA^{GLC}

The phosphorylated adenyl cyclase is active.

ATP

P

cAMP concentration rises.

cAMP

P~P

Cytoplasm

Periplasm

FIGURE 10.26 Activation of adenyl cyclase.

turnover. There is a constant turnover of proteins in the cell. Some misfold as they are made. Some denature spontaneously, so that in every generation, a couple of percent of the proteins of the cell are hydrolyzed back to amino acids by proteases in the cell. Alternatively, a denatured protein might be refolded into its correct confirmation, rather than degraded. This refolding is the activity of a set of **chaperone** proteins that unfold misfolded proteins and allow them to refold. The proteases and chaperones that mediate turnover and folding are part of the normal machinery of macromolecular synthesis of the cell and are needed at all times; however, higher amounts of them are needed at very high temperatures, where protein denaturing and misfolding is much more common.

The **heat shock modulon** includes the genes for a number of chaperones, proteases, and related proteins, whose synthetic rate is substantially increased by shift to high temperatures. The mechanism for modulating the rate of synthesis is via a special sigma factor needed for effective recognition of the promoters for these genes: σ^{32} (the superscript refers to the molecular weight in thousands of the sigma factor, to distinguish it from others, such as σ^{70}, the common *E. coli* sigma factor for most mRNA). σ^{32} is itself highly unstable, denaturing with a half-life of about a minute; thus, continued transcription of the heat-shock genes requires continued transcription of the σ^{32} gene.

CHAPTER 10 PROCARYOTIC GENOME ORGANIZATION AND REGULATION

10.28 Growth rate regulation is principally the regulation of the ribosome supply

Growth represents the combined activities of the several thousand proteins of the cellular **proteome**—the total cellular complement of proteins. These activities have to be quite closely balanced in order to not build up toxic levels of chemical compounds or derange the timing of cell cycle activities. Thus, it is no surprise that elaborate regulatory mechanisms are involved. Because proteins make up by far the most abundant molecule in the cell other than water—2,000 different proteins, 2 million individual molecules, comprising more than 50% of the cellular dry weight (see Section 1.11)—the regulation of ribosome synthesis is a prominent part of growth rate regulation. Ribosomes are the sites of protein synthesis, and thus, the number of ribosomes sets an upper limit on the rate at which cells can make protein; to grow faster, cells have to make more ribosomes, as well as all the other components of the protein-synthesizing system. Although the details are not known yet, it appears that growth rate regulation is the work of a modulon whose principal target is the protein-synthesizing system.

10.29 Rapidly growing cells are bigger and have more ribosomes than slowly growing ones

Rapidly growing cells are larger than slowly growing ones; they have a higher concentration of ribosomes, and their ribosomes translate mRNA more effectively. This has been studied extensively in *E. coli* by setting the growth rate at various levels in a chemostat (see Section 9.15). Table 10.1 shows that as the growth rate increases by a factor of 4, the cell mass increases by a factor of nearly 6, and the number of ribosomes per cell by more than 10.

Because rapidly growing cells are heavier and contain more ribosomes, they are larger than cells of the same strain growing more slowly. Part of the increase in size is accomplished by an increase in diameter (Figure 10.27). Normally, growing cells increase their length only; diameter remains constant. When procaryotic cells are increasing their growth rate, however, they increase in diameter until they reach the value appropriate to the new growth rate, after which diameter remains constant.

Clearly, the 10-fold increase of ribosome number when the cell mass has increased only sixfold results not just in an absolute increase in the number of ribosomes, but in a 40% increase in their concentration as well. This, in turn, appears to lead to more efficient translation initiation, as indicated by the closer spacing of

Table 10.1	Protein Synthesis and Growth Rate in *E. coli*				
G (min)	100	60	40	30	24
K (min⁻¹)	0.42	0.69	1.0	1.4	1.7
Dry wt/cell (pg)	0.15	0.26	0.42	0.64	0.87
Ribosomes per cell	6,700	13,400	25,000	45,000	71,000
Ribosome spacing (nucleotides)	119	127	97	78	61
Number of ribosomes translating each mRNA	27	33	50	71	94

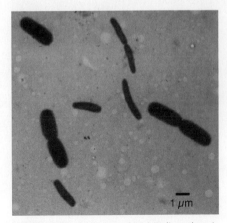

FIGURE 10.27 A mixture of rapidly (large) and slowly (small) growing *E. coli* cells.

ribosomes on mRNA. At slow growth rates, ribosomes are about 120 nucleotides apart, whereas at rapid growth rates they are 60 to 70 nucleotides apart. This effect presumably contributes substantially to the increase in the rate of protein synthesis. The closer spacing directly raises the rate of protein synthesis, as more ribosomes traverse the mRNA in a given period of time, and second, the closely spaced ribosomes extend the life of the mRNA somewhat by protecting it against endonucleases (see Section 10.20). The combined result is a threefold increase in the number of ribosomes that translate a message before it is degraded.

10.30 Slowly growing cells retain a reservoir of ribosomal subunits

A number of studies similar to the one shown in Table 10.1 have established that the number of ribosomes is proportional to the square of the growth rate. If the square root of the number of ribosomes per cell is graphed versus the growth rate constant (k), a straight line results (Figure 10.28). This line extrapolates through the origin, from which we can infer that all ribosomes are normally actively translating, and that the pool of inactive ribosomal subunits is small.

This relationship breaks down at slow growth rates, however. Thus, at slow growth rates, cells maintain a pool of inactive ribosomes, such that even cells that are barely growing have at least 2,000 ribosomes. This allows them to commence more rapid growth immediately when conditions improve.

10.31 Multiple rRNA operons are transcribed from two promoters

The synthesis of ribosomal proteins, as we have already seen (Section 10.21) is controlled by translational repression so that it is coordinated with the rate of rRNA transcription. Thus, controlling the number of ribosomes in a cell basically amounts to controlling the amount of rRNA. In order to supply sufficient rRNA to meet the needs of rapidly growing cells, almost all organisms have to have multiple copies of the genes, as even continuous transcription of a single operon from a very strong promoter would not produce enough rRNA to allow cells to grow at their maximal rate. *E. coli,* for instance, has seven copies of the rRNA operon. Each operon has the genes for the 23S, 16S, and 5S rRNAs, as well as one to three tRNA genes (Figure 10.29).

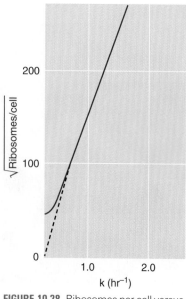

FIGURE 10.28 Ribosomes per cell versus growth rate.

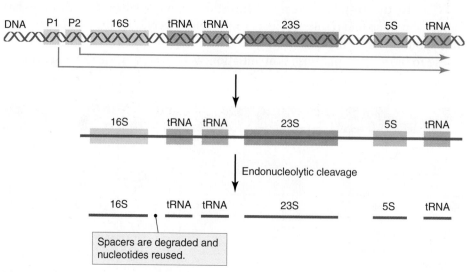

FIGURE 10.29 *rrn* operon transcript and products.

CHAPTER 10 PROCARYOTIC GENOME ORGANIZATION AND REGULATION

The entire thing is transcribed into a single RNA molecule, which is then cut by specific endonucleases to produce the three rRNAs in exactly equal amounts, along with several tRNA molecules.

Each of these **rrn operons** is transcribed from two promoters. P1 is a controlled by an unknown mechanism by the growth rate. At high growth rates, transcription initiation at this promoter is frequent, whereas at low growth rates, it is infrequent. Thus, rRNA is made in large amounts at high growth rates and in small amounts at slow growth rates.

The second promoter, P2, is a weak promoter that appears to be unregulated. At high growth rates, its contribution to *rrn* transcription is negligible, but at low growth rates, it is the principal source of rRNA. This promoter is responsible for the growth rate–independent synthesis of ribosomes by slowly growing cells.

10.32 The stringent response is mediated by ppGpp

When *E. coli* is shifted from a rich medium to a poor medium, there is a period of adaptation, during which cell size changes. Part of this adaptation period is governed by the **stringent response.** This response is triggered by two different situations: when there is a sudden decrease in the rate of protein synthesis due to the lack of charged tRNAs or starvation for a carbon and energy source.

The mediator of the stringent response is the nucleotide guanosine tetraphosphate (more accurately 5'-diphospho-3'-diphosphoguanosine), generally known as **ppGpp.** This is made by two pathways in the cell, one activated by ribosomes paused with an open A site (waiting for a charged tRNA) and the second activated by starvation (Figure 10.30).

The result of the increase in ppGpp concentration is to block transcription of rRNA and tRNAs, block initiation of DNA synthesis, and activate transcription of genes for enzymes of biosynthesis and catabolism. How ppGpp achieves its effects remains obscure. It is also unclear which of the many effects of starvation or protein synthesis arrest are directly due to the stringent response and which are indirect effects.

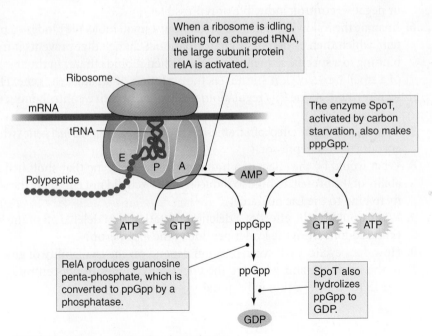

FIGURE 10.30 Pathways of ppGpp formation.

Summary

The procaryotic genome is quite streamlined compared with that of macroscopic eucaryotes. It contains very little in the way of repetitive sequences, introns, transposons, and junk DNA. Most of the genome thus encodes RNAs or proteins or is devoted to regulation of the coding sequences.

Although bacteria appear simple through the microscope, they are exceedingly complex at the molecular level, and nothing exemplifies this more than the elaborate mechanisms by which they regulate gene expression. Most major mechanisms of gene regulation were first discovered in bacteria, and there are very few that are not found there. Some mechanisms are unique to procaryotes—attenuation, for instance, which depends on the coupling between transcription and translation that is possible because of the lack of a nuclear envelope. Not only are control mechanisms inherently complex, but they are multilayered as well, with global control systems layered on top of local, operon-specific systems. The streamlined genome and the elaborate mechanisms to adjust gene expression to fit precisely the needs of the moment and the local environment suggest that efficiency at the molecular level has been one of the major driving forces of procaryotic evolution.

Study questions

1. What features distinguish the procaryotic genome?
2. What functions do plasmids encode, and how are they maintained in bacterial cells?
3. How are polycistronic messenger RNAs translated? Why are they not found in eucaryotic cells?
4. Describe the processes of transcription initiation and termination in bacteria.
5. Imagine the following system of regulation: a small molecule binds to a protein, which then undergoes a conformational change that allows it to bind to a specific sequence in DNA. When it binds there, it stimulates transcription of an adjacent gene. How would you describe this regulation: positive or negative control; inducible or repressible?
6. Imagine the following system of regulation: a small molecule binds to a protein, which then undergoes a conformational change that prevents it from binding to a specific sequence in DNA. When it binds there (in the absence of a small molecule), it stimulates transcription of an adjacent gene. How would you describe this regulation: positive or negative control; inducible or repressible?
7. What would be the effect of a frameshift mutation early in the *lacI* gene (which encodes the Lac repressor)?
8. What would be the effect of a mutation in the *lacI* gene that abolished the ability of the protein to bind to inducer, but which did not eliminate the ability to bind to the Lac operator?
9. What would be the effect of a deletion mutation that deleted all of the histidine codons from the leader peptide of the *his* operon?
10. How does cyclic AMP concentration depend on the availability of glucose in the medium, and what are the effects of changes in concentration of cAMP for transcription in bacteria?

Microbial Behavior and Development

11

W E GENERALLY THINK OF procaryotic microbes as unicellular organisms with very limited behaviors and no differentiation of different types of cells. Certainly, in comparison to plants and animals, this is true; nevertheless, an effective repertoire of behaviors is found in most microbes, and some have a limited degree of differentiation and specialization among cell types. The simplicity of these phenomena in procaryotes has allowed a rapidly increasing number of them to be understood in great detail at the molecular level, something with which we have had very limited success in more complex organisms.

In this chapter, we survey some of the behavioral patterns of procaryotes and then some of the best studied systems of procaryotic development. A common theme will be the underlying mechanisms of gene regulation that control these activities.

Tactic Behavior

Tactic responses are directed movements in response to a stimulus. Like all other organisms, procaryotes have a powerful interest in being able to flee unfavorable environments and seek out ones that allow rapid multiplication. Over the course of billions of years, those procaryotes that live in environments that are changeable in any important way have developed mechanisms to sense environmental change and then respond by moving toward or away from the stimulus. This repertoire of responses is the most obvious and best studied set of procaryotic behaviors.

11.1 Tactic responses can be positive or negative

A wide range of environmental stimuli can elicit a tactic response from one or another procaryote. The most common are oxygen concentration, light intensity, and the concentration of certain chemical compounds. Different procaryotes will typically respond to stimuli that are important to them in the context of their metabolism and their particular habitats. Responses can be **positive**—that is, the resulting movement is *toward* the stimulus—or **negative**—movement *away* from the stimulus. For instance, aerobes can be expected to move toward higher oxygen concentrations, whereas anaerobes would move away from it.

Tactic responses can also switch between positive and negative, depending on the intensity. For instance, most phototrophic procaryotes have a positive phototactic response; they move toward increasing light intensity. Above a given intensity, however, they move away to protect themselves against the very damaging effects of very high light intensity.

11.2 Tactic responses depend on reversals of direction or on biased random walks

Some organisms implement tactic responses by continuing to move forward if conditions are improving or steady and reverse direction when they sense a degradation of conditions. For instance, the anoxygenic phototroph *Thiospirillum* is phototactic, but this is the result not of a positive taxis toward increasing light intensity, but rather of a negative taxis toward decreasing light intensity. As long as it is in the light it swims continuously, with only rare reversals of direction. If it swims into the dark (or significantly dimmer light), however, it immediately reverses and returns to the light.

Other cells implement tactic responses with a **directed random walk.** A random walk is a process of movement in which direction changes are random; thus, the moving object tends to stay in roughly the same spot. A **biased random walk** is the same process, but with the added feature that movement in one direction tends to be prolonged. Thus, although each change of direction is random, there is overall progress (Figure 11.1).

Procaryotes that have tactic responses like this swim for a period of time, then stop and tumble for a few seconds, and then swim again. Because during the tumble they randomly change their orientation, the direction in which they resume swimming is random. Tactic responses are achieved by delaying tumbling when they are moving in a desirable direction and increasing the frequency tumbling when moving in an undesirable direction. Thus, the average **run** (the path between tumbles) tends to be longer when the cell is moving toward a desirable environment and shorter when it is moving away.

11.3 Tumbling is caused by a reversal of flagellar rotation

The periodic tumbles that occur during the motility of procaryotes that have tactic responses mediated by directed random walks are caused by a momentary reversal of the direction of rotation of their flagella. This is true of both polarly and peritrichously flagellated procaryotes; however, the reasons that a reversal of flagellar rotation causes tumbles are different.

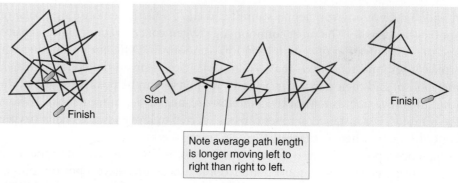

Note average path length is longer moving left to right than right to left.

FIGURE 11.1 Random walk and biased random walk.

CHAPTER 11 MICROBIAL BEHAVIOR AND DEVELOPMENT

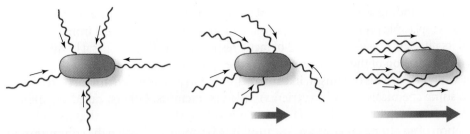

FIGURE 11.2 Commencement of movement in a peritrichously flagellated cell.

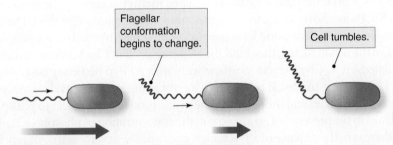

Flagellar conformation begins to change.

Cell tumbles.

FIGURE 11.3 Tumbling in a polarly flagellated cell.

In peritrichously flagellated procaryotes, swimming involves all flagella coming together in a bundle behind the cell and rotating at the same rate so that they do not interfere with each other. This is not an actively controlled process. When all flagella are rotating such that they push the cell, it goes nowhere until random Brownian motion or other hydrodynamic forces start it moving in one direction. This slight movement tends to sweep the flagella to the rear, and the push from the flagella then accelerates the movement, which in turn presses the flagella more firmly to the rear (Figure 11.2). Thus, after the flagella begin to push, in a matter of a few tenths of a second, the cell begins to move toward one end or the other.

When the flagella are rotating such that they pull rather than push, no such cooperation is possible. Even if all flagella pull in the same direction, the water flow tends to pull the bundle apart. Thus, when the flagella are pulling, they naturally tend to pull in contrary directions, and the cell cannot move in a sustained way; instead, it just sits and tumbles in place.

For polarly flagellated procaryotes, especially if there is a single flagellum, a different mechanism of tumbling has to operate, as reversing flagellar rotation should in principle simply reverse the direction of movement. In these procaryotes, tumbling occurs because the flagellum changes its conformation, beginning at the tip and progressing toward the basal body. During this transition, the flagellum is kinked and cannot exert much force on the cell. Thus, the cell comes to an abrupt stop, and Brownian motion then causes it to jiggle around until the flagellum reverses direction again and resumes its normal conformation (Figure 11.3).

11.4 Many tactic responses are mediated by two-component regulatory systems

One of the most common mechanisms mediating stimulus and response systems in procaryotes is the **two-component regulatory system.** These regulatory systems consist of two proteins. One of them—the **sensor**—is a membrane protein that undergoes a conformational change when it interacts with the environment in a specific way (e.g., by binding a particular chemical compound). The other—the **transmitter**—is a cytoplasmic protein that is responsible for transmitting the signal from the sensor to its ultimate target—in the case of tactic responses, the flagellar basal body.

Regulation of the tactic response has been most extensively studied for chemotaxis in *E. coli*. This organism, like many others, has a positive chemotactic response to some chemicals (mostly nutrients) called **attractants.** Others, called **repellents,** stimulate a negative reaction. Thus, when *E. coli* is swimming such that the concentration of an attractant is increasing, tumbling is suppressed; when the concentration is decreasing, tumbling frequency increases. The opposite is true of repellents.

The sensor proteins for chemotaxis are called **methyl-accepting chemotaxis proteins,** or **MCPs** for short (we see in the next section why they are called this). The MCPs are located in the membrane and have specific binding sites exposed on the periplasmic surface. On the inside surface they bind the response protein **CheA.** When bound to an unstimulated MCP, CheA reacts spontaneously with ATP to become phosphorylated (Figure 11.4). When the MCP is bound to an attractant on the outside, however, its conformation changes, and the CheA on the inside is no longer subject to phosphorylation. Thus, in the presence of an attractant, the concentration of phospho-CheA goes down, whereas in the absence of attractant, it goes up. The reverse is true of repellents.

Phospho-CheA acts, in turn, to phosphorylate a second cytoplasmic protein **CheY.** Phospho-CheY is the protein that acts to switch the flagellar motor into the clockwise rotation mode, which causes tumbling. Yet another protein, **CheZ,** is a phosphatase that splits the phosphate off of phospho-CheY. Thus, phospho-CheY acts only briefly because it is quickly hydrolyzed, and the flagellar motor resumes its normal counterclockwise rotation.

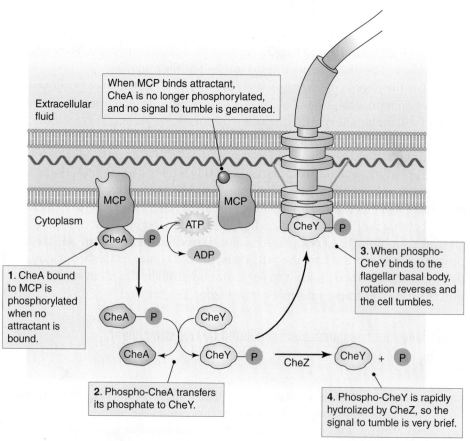

FIGURE 11.4 Chemotaxis system in *E. coli*.

The previous discussion explains how *E. coli* sense the presence or absence of attractants and repellents and how they affect the flagellar motor; however, this is very different from sensing a chemical *concentration gradient*. In order to do this, the cell needs to compare the concentration at two different points, rather than simply detect the presence of the compound. In the case of procaryotic cells, gradients are sensed by continuously comparing the current concentration to that of a few seconds ago; if it is higher, the cell is moving *up* the gradient; if it is lower, the cell is moving *down* the gradient. In other words, whatever the concentration was a few moments ago becomes the baseline, to which the chemotaxis system adapts. Any change in concentration is then sensed as an increase or decrease relative to that baseline. It is thus not the presence or absence of a chemical that is sensed, but its concentration relative to a few moments ago. This requires the cell to be able to adapt its system so that whatever concentration was sensed shortly before becomes the norm for the system. This process of **adaptation** is thus the critical feature of the system allowing a true sensory response.

Adaptation occurs as a result of the methylation of the MCPs (hence their name). Another Che protein, **CheR,** is a methylase that transfers methyl groups from a common methyl donor in all cells (S-adenosylmethionine) to four or five different glutamate residues on each MCP. CheR is present in very low amounts; thus, the rate of methylation is slow, but it is continuous. Methyl groups are removed from MCPs by a methylesterase enzyme, **CheB.** CheB activity is dependent on its being phosphorylated by CheA, just like the activity of CheY. Thus, when there is abundant phospho-CheA, both CheY and CheB are phosphorylated and active, whereas when CheA is largely unphosphorylated, both CheY and CheB are also unphosphorylated and inactive (Figure 11.5).

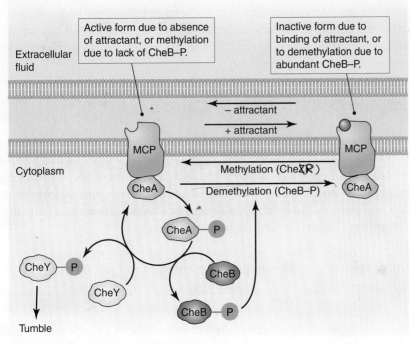

FIGURE 11.5 The two forms of MCP-CheA.

The effect of methylation on the activity of the MCP-CheA complex is exactly the reverse of the binding of attractants. That is, the more glutamates are methylated, the more CheA is phosphorylated, and the fewer the methyl groups, the less active is CheA phosphorylation. Thus, we can think of the MCP-CheA complex as existing in two principal states (with several intermediate ones): an active form in which CheA is rapidly phosphorylated and an inactive one in which it is not. The binding of attractant, or the removal of methyl groups, favors the inactive form and thus smooth swimming; the dissociation of attractant or methylation favors the active form and thus frequent tumbling.

Thus, this is a feedback system, in which an increase of attractant concentration leads to a decrease of phospho-CheA concentration, which in turn leads to *both* a decrease in tumbling (less phospho-CheY) *and* a decrease in the demethylation rate (less phospho-CheB). Then, over a period of tens of seconds, the gradual methylation of the MCPs restores the previous rate of tumbling. Thus, the cell is adapted to the new attractant concentration and can effectively sense a further increase or a decrease. In effect, what the methylation system does is to slowly bring the phospho-CheA concentration back into an intermediate range after it has substantially increased (absence of any attractants, or high concentration of repellents) or substantially decreased (high concentration of attractants, absence of any repellents).

11.6 The two-component chemotactic system integrates many signals through a single response protein

E. coli has several different MCPs, each of which can normally bind several different attractants and a few of which can also bind a repellent. Attractants include amino acids such as aspartate, glutamate, serine, alanine, and glycine; sugars such as galactose, glucose, and ribose; and a few other compounds. Repellents include heavy metals such as nickel and cobalt and the toxic aromatic compound phenol. Each of the different MCPs alternates between the active and inactive forms based on the presence or absence of attractant, repellent, and methylation level. Thus, each of several different MCPs contributes to the level of phospho-CheA, which in turn determines the tumbling frequency. Thus, the instantaneous level of phospho-CheA is determined by the combined effects of all the MCPs, reacting in an integrated fashion to the chemical environment.

Dimorphic Cell Division in *Caulobacter*

Caulobacter is a gram-negative bacterium found in dilute aqueous environments. Cell division produces two different types of cells because each of the poles of the dividing cells has different structures embedded in the cell envelope. Hence, we speak of a **dimorphic** ("two shapes") life cycle. This is a relatively simple form of differentiation and promises to be an ideal system in which to study how cells regulate the positioning of cellular structures in three dimensions.

11.7 Cell division in *Caulobacter* produces one stalked cell and one flagellated cell

One of the two types of *Caulobacter* cells is flagellated, with a single polar flagellum. Several pili are located adjacent to the flagellum, and chemotactic MCP proteins are

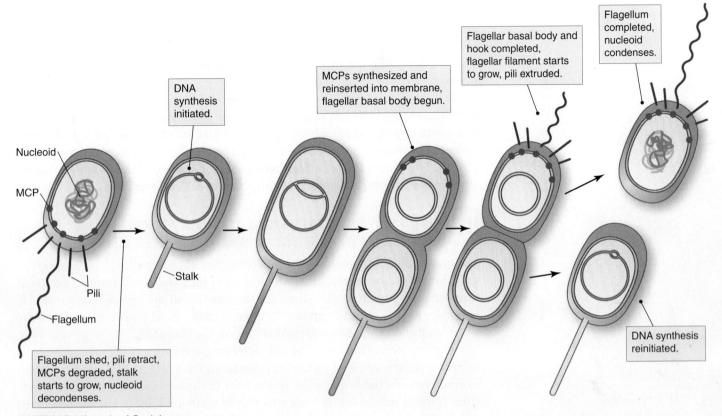

DNA synthesis initiated.

MCPs synthesized and reinserted into membrane, flagellar basal body begun.

Flagellar basal body and hook completed, flagellar filament starts to grow, pili extruded.

Flagellum completed, nucleoid condenses.

Nucleoid

MCP

Stalk

Pili

Flagellum

Flagellum shed, pili retract, MCPs degraded, stalk starts to grow, nucleoid decondenses.

DNA synthesis reinitiated.

FIGURE 11.6 Life cycle of *Caulobacter*.

embedded in the cell membrane at the flagellated pole. The chromosome is condensed into the compact form characteristic of stationary phase cells. This cell, termed a **swarmer cell,** swims for a defined period of time, approximately half an hour, responding chemotactically to nutrients, before it begins a process of development into a **stalked cell** (Figure 11.6).

The process of converting the swarmer cell into a stalked cell is signaled by the change of the nucleoid from the condensed form into the extended form and initiation of DNA synthesis. This is rapidly followed by loss of the flagellum (shed into the medium, and the basal body degraded), disassembly of the pili, degradation of the chemotactic proteins, and the synthesis of a stalk where the flagellum used to be. The stalk is an extension of the cell that contains cytoplasm and is surrounded by the cell membrane and wall (Figure 11.7); however, it contains no ribosomes or DNA. Its function is thought to be to increase the cell's surface/volume ratio to make nutrient uptake more efficient, an important consideration for cells that live in environments where nutrient concentrations are very low. It also serves as an attachment organelle, as its tip bears a **holdfast,** a patch of adhesive polysaccharide that serves to attach the stalked cell to a solid substratum, often another cell.

The stalked cell begins to grow as DNA synthesis is initiated; by the time chromosome replication is complete, the cell is nearly twice its original volume, and a septum is beginning to form. A flagellum and several pili develop at the end opposite the stalk, and chemotaxis MCP proteins are inserted into the cell membrane there as well. As division occurs, the nucleoid of the swarmer cell condenses, whereas the nucleoid of the stalked cell remains extended, and a new round of chromosome replication is initiated immediately after division.

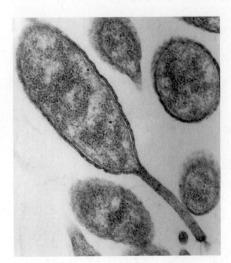

FIGURE 11.7 Thin section of *Caulobacter*.

Cell division in *Caulobacter* produces one stalked cell and one flagellated cell

Thus, division in *Caulobacter* produces two very different cells. One, the stalked cell or the "mother cell," has a stalk and is capable of immediately replicating its chromosome in preparation for the next division. The other, the swarmer cell, is flagellated and piliated and has chemotactic sensory proteins in its membrane. Its nucleoid is compact, and replication is delayed.

It is easy to imagine why such a dimorphic life cycle might serve *Caulobacter* well. Because it is normally found in very dilute environments, attachment to other cells could be a significant advantage. All cells leak small amounts of organic compounds, and an attached *Caulobacter* would be well placed to use these for its own growth. The production of a flagellated daughter cell allows the swarmer to swim away, avoiding competition between the mother cell and swarmer.

11.8 Flagellar synthesis is controlled by a regulatory cascade

The best studied component of the differentiation of cell types in *Caulobacter* is flagellar synthesis, the regulation of which is now pretty well understood. It occurs in three fairly discrete steps: (1) cell membrane constituents of the basal body are made and inserted into the membrane; (2) the rest of the basal body and the hook are assembled; and (3) flagellins are added to form the filament.

The first step, synthesis of the cell membrane components of the basal body, is subject to positive control by a transcription factor CtrA (Figure 11.8). This is present in swarmer cells, but is degraded by proteases in the cytosol as the swarmer differentiates into a stalked cell. It is not synthesized again until just before the process of cell division begins. Thus, the initial step in flagellar synthesis occurs late in the cell cycle, but before the two daughter cells are separated by a septum.

Another protein that is turned on by CtrA is an alternate sigma factor, σ^{54}. This sigma factor is required for the synthesis of the rest of the basal body proteins and the hook proteins. Thus, these components are synthesized after the cell membrane portion of the basal body.

Flagellin mRNA is also under the control of σ^{54} so that it is made simultaneously with the mRNA for most of the basal body and hook proteins; however, flagellin mRNA is under additional, poorly understood, translational control. This mRNA

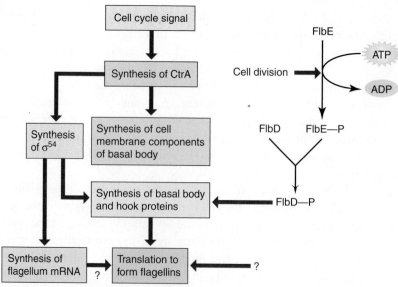

FIGURE 11.8 Regulatory cascade in flagellar synthesis.

CHAPTER 11 MICROBIAL BEHAVIOR AND DEVELOPMENT

is unusually stable, unlike most bacterial mRNA, and, thus, it accumulates, ultimately to be translated late in the cell cycle, after the basal body and hook have been completely assembled.

11.9 Transcription of flagellar genes is compartment specific

Most of the transcription of flagellar genes occurs immediately after the septum has formed between the stalked mother cell and the swarmer cell, but before they separate. This transcription is compartment specific; it occurs in one compartment and not in the other. This suggests that the two compartments differ in their macromolecular composition. One of the differences is beginning to be understood.

In addition to requiring σ^{54}, transcription of basal body, hook, and flagellin genes requires the transcription factor **FlbD.** The FlbD binding sites simultaneously *activate* basal body and hook protein transcription and *repress* transcription of the genes for the membrane components of the basal body (Figure 11.9). This is because FlbD binding sites overlap the binding sites for CtrA. Thus, when FlbD binds, CtrA cannot.

FlbD is the response protein of a two-component regulatory system. The sensor component is FlbE, some of which is specifically localized on one side of the ring of FtsZ that is responsible for localizing the division septum. Although the details are not yet fully worked out, it is thought that FlbE is activated to phosphorylate FlbD by the division process, and this phosphorylation is necessary for FlbD to act as a transcription factor (Figure 11.10).

Like all two-component systems, active response regulator protein does not last long. Thus, there is a brief burst of transcription of the basal body, hook, and flagellin genes, and then they shut down again as FlbD loses its phosphate and becomes inactive again. This is sufficient to provide enough basal body and hook proteins for the single flagellum that the cell makes, but there is a potential problem with flagellins.

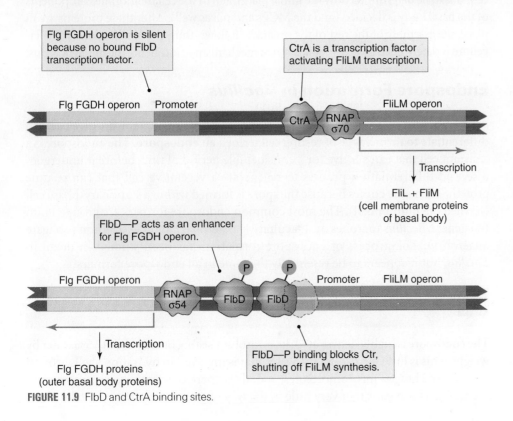

FIGURE 11.9 FlbD and CtrA binding sites.

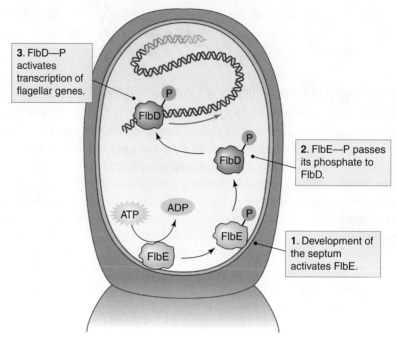

3. FlbD—P activates transcription of flagellar genes.

2. FlbE—P passes its phosphate to FlbD.

1. Development of the septum activates FlbE.

FlbD

FlbD

ATP

ADP

FlbE

FlbE

FIGURE 11.10 Action of FlbDE to activate transcription of flagellar genes.

The flagellins are needed in much larger amounts, and the assembly of the flagellar filament continues long after transcription of flagellin genes has been turned off. This is due to an exceptionally long half-life for flagellin mRNA in *Caulobacter* and by translational control of this mRNA.

This system explains part of the flagellum localization process, as it is only in the nascent swarmer cell that synthesis of most of the flagellum genes occurs; however, it remains to be determined how the initial placement of the cell membrane components of the basal body is decided (and the MCPs and pili as well). After these proteins are in place, the assembly of the rest of the organelle follows; thus, the localization of the flagellum is determined by an as yet unknown mechanism of localizing the initial proteins.

Endospore Formation in *Bacillus*

All members of a large, diverse, and very successful group of gram-positive bacteria differentiate to form a type of resting cell termed an **endospore.** The endospore is a resistant cell that can survive for a considerable period of time before it undergoes a second differentiation process to regenerate a vegetative cell that can resume growth. The name comes because the spore is formed *within* a stationary-phase cell (termed the **mother cell**). The most common endospore formers are grouped into the genera *Bacillus* (aerobes and facultative anaerobes) and *Clostridium* (obligate anaerobes). The process of endospore formation has been studied in detail in *Bacillus,* but it appears to be essentially the same in all endospore formers.

11.10 Endospores are highly durable and long lived

The endospore is highly dehydrated, its cytosol consisting of only about 15% water by weight. This is little more than the water that is involved in hydration shells around the various charged macromolecules, and, thus, there is essentially no free water in the spore. This means that very little in the way of chemical activity can take place, and the spore is termed **cryptobiotic,** meaning that there are no measurable activities that typically identify living organisms (no respiration, no biosynthesis, no DNA replication, no transcription or translation, no DNA repair, no motility, etc.).

Endospores are highly resistant to a wide variety of deleterious environmental conditions. They can often survive hours of boiling water, high doses of ultraviolet radiation or X-rays, highly acidic or basic conditions, long periods of extreme desiccation, and high concentrations of toxic chemicals. They are also often quite long lived. How long is not clear, but there are certainly well-documented examples of endospores being able to germinate after hundreds of years. Spores of *Thermoactinomyces* have germinated from 4,000-year-old sediments, and most startlingly, *Bacillus* endospores recovered from 20-million-year-old insects embedded in amber were able to germinate. Although there is a slight possibility that some of these reports were in error and the preparations were contaminated by modern spores, we can be confident that endospores are at least occasionally very long lived. It is unlikely that most spores are capable of such feats of longevity; however, several years of viability in soil and water are probably common.

11.11 Sporulation is a last-ditch response to starvation

The process of differentiation begins a couple of hours after cells enter stationary phase. In the phase-contrast light microscope, the first obvious sign is a dark **forespore** within the mother cell cytosol (Figure 11.11). After some time, the forespore becomes very bright because of its sudden loss of water. A few hours later the process is essentially complete, and the endospore is mature (Figure 11.12). The mother cell then ruptures and releases the spore into the environment.

The trigger for spore formation is nutrient deprivation—absence of carbon, nitrogen, or phosphorous. Although the endospore is resistant to a wide variety of environmental insults, such as high temperature and ultraviolet radiation, these conditions do not induce sporulation. Only nutrient deprivation is an effective trigger.

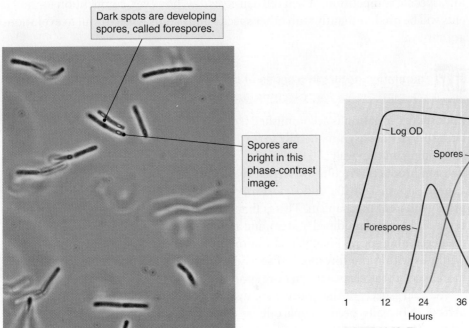

Dark spots are developing spores, called forespores.

Spores are bright in this phase-contrast image.

FIGURE 11.11 Phase-contrast micrograph of *Bacillus*.

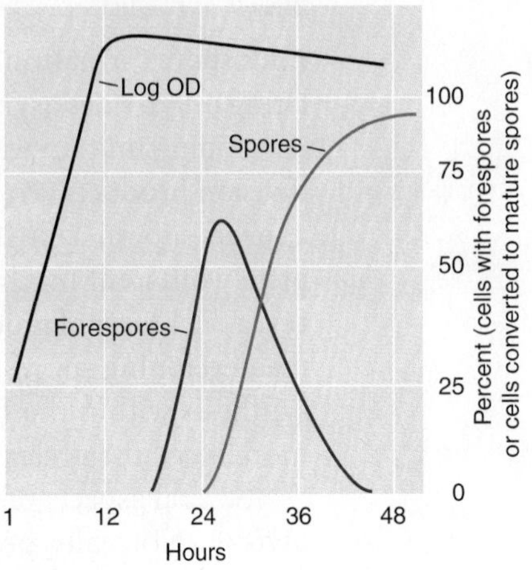

FIGURE 11.12 Time course of growth and spore formation.

Sporulation takes a considerable length of time, and a mature spore cannot germinate immediately, as we discuss in Section 11.16. Thus, any cell that sporulates is committing itself to a long period of time in suspended animation. It can thus be at a serious disadvantage relative to competitors that do not sporulate, but rather simply survive brief starvation periods in stationary phase, from which they can emerge in a matter of an hour or so as soon as conditions improve. For this reason, sporulation is not initiated immediately on entry into stationary phase, but is delayed for a number of hours. During this period, motile cells follow chemotactic signals to find more nutritive environments, and they induce a variety of modulons responding to nutrient deprivation. Only after these responses have had a chance to alleviate the starvation do the cells commit to sporulation. The mechanism by which sporulation is delayed is unclear, but it is partly related to cell density.

Quorum sensing refers to mechanisms that a number of procaryotes have to sense their cell density. A number of responses are only initiated if the cells are present at high density; sporulation is one of these. Thus, a few thousand *Bacillus* cells in a gram of nutrient-depleted soil will not sporulate, whereas a billion will. One of the components of starvation modulons is the gene for the synthesis of two oligopeptides that are excreted from the cell as they are made. These oligopeptides can be taken up by bacilli, and they are necessary for sporulation (we discuss how they act in Section 11.13). Thus, where cell density is low, the concentration of these oligopeptides never gets high enough to promote sporulation, and any bacilli remain in stationary phase. When cell density is high, however, the oligopeptide concentration can become high enough for sporulation to proceed. Of course, it takes some time for the external concentration to rise to the necessary level, and this explains part of the delay in sporulating.

Why cell density is an important part of the decision whether to sporulate is not immediately obvious; however, it is thought that it minimizes competition among closely related cells. When cell density is high, stationary phase cells are competing with each other as well as with other species for scarce nutrients. Under such conditions, it might be better for most of them to sporulate, thus suspending the intraspecific competition. When cell density is low, however, competition for nutrients will be predominantly with other species and is thus less harmful to evolutionary survival.

11.12 Sporulation occurs in a series of defined morphological stages

Endospore formation is accomplished by a series of discrete steps. First, the sporulating cell divides asymmetrically to form two daughter protoplasts within a single enveloping murein wall (Figure 11.13). Then the mother cell (the larger of the two division products) engulfs the forespore (the smaller one) in a unique process that appears superficially like endocytosis in eucaryotes but which probably has a completely different mechanism. This is then followed by the synthesis of a spore **cortex,** a thick layer of modified murein, and a **spore coat,** a tough layer of protein around the developing spore. Finally, maturation of the spore involves the uptake into or synthesis within the forespore of several of unique small molecules. Lysis of the mother cell then releases the mature spore.

Several aspects of this process are distinctive—first, of course, the asymmetric division. Normally, predivisional cells lay down a ring of the protein FtsZ around the middle of the cell, and this protein determines the site of the septum. In sporulating cells, the FtsZ ring is placed at both ends of the cell, in response to various regulatory

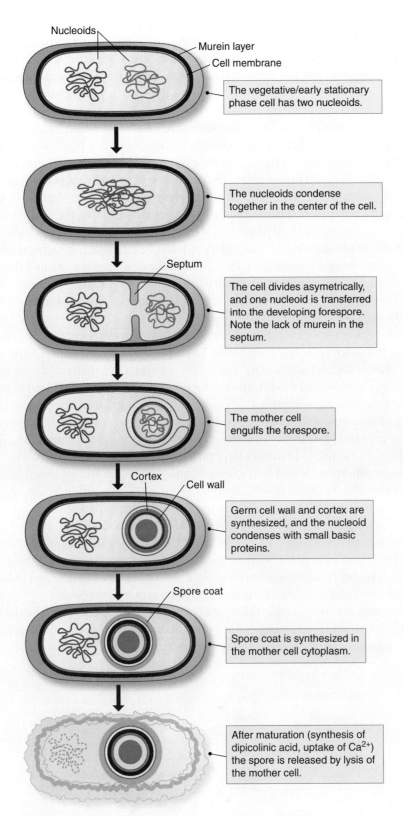

Nucleoids

Murein layer

Cell membrane

The vegetative/early stationary phase cell has two nucleoids.

The nucleoids condense together in the center of the cell.

Septum

The cell divides asymetrically, and one nucleoid is transferred into the developing forespore. Note the lack of murein in the septum.

The mother cell engulfs the forespore.

Cortex

Cell wall

Germ cell wall and cortex are synthesized, and the nucleoid condenses with small basic proteins.

Spore coat

Spore coat is synthesized in the mother cell cytoplasm.

After maturation (synthesis of dipicolinic acid, uptake of Ca^{2+}) the spore is released by lysis of the mother cell.

FIGURE 11.13 Stages of sporulation.

proteins of the sporulation process. Which of the two rings will be the site of actual septation appears random, and there are rare endospore formers that make two endospores per cell because both FtsZ rings are active.

The polar location of the septum is not the only unusual aspect of this process. Normally, a developing septum contains a thick layer of murein, and when the septum is complete, the two daughter cells are already separated by a thick wall. This would obviously prevent engulfment, and thus, the spore septum contains relatively little murein.

The engulfment process results in a forespore *within* the mother cell, surrounded by two concentric membranes with opposite orientation. The inner forespore membrane has a normal orientation. It was the cell membrane of the forespore before engulfment, and it will become the cell membrane of the new vegetative cell when the spore eventually germinates. The outer forespore membrane, derived from the mother cell's cell membrane, has an opposite orientation so that the outside surface of both membranes faces the space between them. This allows both forespore and mother cell to use their export machinery to transport materials into this space.

Between the two membranes there is a thin layer of murein, termed the **germ cell wall,** which will become the cell wall of the germinating spore. The germ cell wall is synthesized by the inner forespore membrane, using the small amount of murein from the division septum as the starting material.

This space is also the location of cortex synthesis, formed between the germ cell wall and the outer forespore membrane. The outer forespore membrane appears to be responsible for cortex synthesis. The cortex is formed of a material closely similar to murein, but chemical modifications of the muramic acid residues lead to greatly reduced cross-linking and thus a much more elastic, less rigid structure (Figure 11.14). Half of the muramic acid residues lack the acetyl group, which allows them to cyclize to form **muramic lactam,** which cannot carry a peptide side chain. Some of the remaining muramic acid residues have only an alanine for a side chain. The result of these modifications is that only a few percent of the muramic acid residues in the cortical peptidoglycan are cross-linked.

Simultaneously with cortex formation, large amounts of a set of unique, very small basic proteins (termed **small acid-soluble proteins [SASPs]**) are synthesized in the forespore cytoplasm. These bind to the DNA of the nucleoid and condense it into a tightly packed, donut-shaped structure. DNA complexed with these small proteins is highly resistant to ultraviolet light and other kinds of damage; thus, part of the spore's resistance to environmental insults is due to the SASPs.

The spore coat is formed in the mother cell, on the cytoplasmic side of the outer forespore membrane. It is composed of proteins with a high proportion of cysteine residues and is highly cross-linked with disulfide bonds. They also have a high

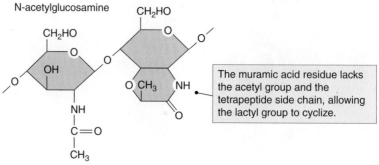

FIGURE 11.14 The principal repeating unit of cortex murein.

CHAPTER 11 MICROBIAL BEHAVIOR AND DEVELOPMENT

proportion of hydrophobic amino acids. They are highly resistant to proteases, making this a very tough layer, resistant to mechanical abrasion, caustic chemicals, and enzymes. Some species have a simple coat made of a single major protein; others have multilayered coats consisting of up to 20 different types of protein.

Maturation of the spore involves the synthesis of a large amount of dipicolinic acid and the uptake of an equimolar amount of calcium. It is thought that these two compounds form a complex within the cell (Figure 11.15).

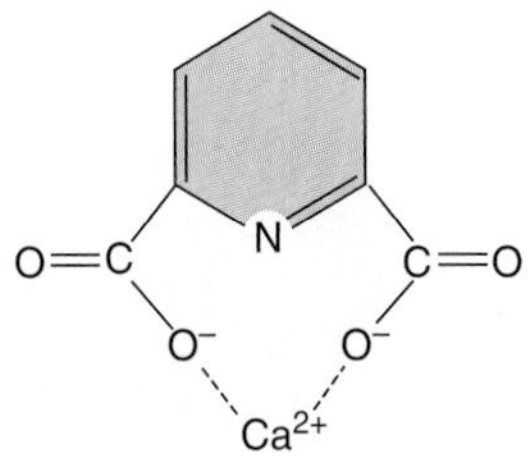

FIGURE 11.15 Ca-dipicolinate.

11.13 Initiation of sporulation involves a two-component regulatory system and a protein kinase cascade

The initiation of sporulation in *Bacillus* is controlled by a two-component system consisting of several sensor kinases and a response regulator, as in other two-component systems that we have discussed. In this case, however, the sensors do not directly phosphorylate the response regulator; rather, a series of intermediate protein kinases serve to transfer the phosphate (Figure 11.16).

There are two principal sensor proteins: **KinA** and **KinB**. KinA is a cytosolic protein, and what it responds to is still unknown. KinB is a membrane protein, and it is thought to respond to decreases in the protonic potential across the membrane, although that is not certain. In any event, these two sensor kinases become phosphorylated when they are stimulated by some signal related to nutrient deprivation, and they then pass their phosphate on to the protein **Spo0F**. Spo0F passes its phosphate on to Spo0B, which in turn passes the phosphate to the response regulator

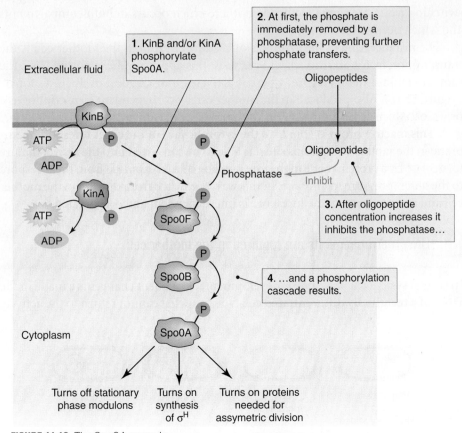

1. KinB and/or KinA phosphorylate Spo0A.

2. At first, the phosphate is immediately removed by a phosphatase, preventing further phosphate transfers.

3. After oligopeptide concentration increases it inhibits the phosphatase…

4. …and a phosphorylation cascade results.

Extracellular fluid

Oligopeptides

Oligopeptides

Phosphatase ← Inhibit

Cytoplasm

Turns off stationary phase modulons

Turns on synthesis of σ^H

Turns on proteins needed for assymetric division

FIGURE 11.16 The Spo0A cascade.

Spo0A. Spo0A is a transcription factor that is necessary for the synthesis of a series of proteins involved in early steps in sporulation, and it also acts as a repressor of stationary phase modulons.

Because the sensor kinases in this system appear to be stimulated by nutrient deprivation, this cascade is stimulated from the beginning of stationary phase. Sporulation, however, is not initiated immediately because several phosphatases hydrolyze the phosphate from Spo0F, thus preventing completion of the cascade. The extracellular oligopeptides that regulate the timing of the initiation of sporulation are specific inhibitors of these phosphatases. Thus, when the extracellular concentration of the oligopeptides gets high enough, they are taken up and block the phosphatases, thus allowing the cascade to be completed producing Spo0A phosphorylated into its active form.

11.14 Endospore development depends on the sequential action of a series of different sigma factors

Equally complex regulatory cascades occur at nearly every stage in sporulation. One of the results of these cascades is the synthesis of a series of different sigma factors in the mother cell and forespore. Much of the complexity of regulation of the different stages of sporulation has to do with regulating the synthesis of these sigma factors (Figure 11.17).

The complex pathway of endospore synthesis requires precisely timed synthetic activities in both forespore and mother cell. This implies that the two compartments continue to communicate even after their separation by the division septum. Some of these signals have been worked out, especially those involved in coordinating the switching from one sigma factor to another, which occurs in both compartments and which needs to be coordinated.

For instance, one of the products of σ^E transcription in the mother cell is the transcription factor SpoIIIA. After synthesis, this appears to be exported from the mother cell into the forespore, where it activates transcription of the gene for σ^G (Figure 11.18). Also produced in the mother cell by σ^E-dependent transcription is a proto-σ^K, which differs from active σ^K by having 20 extra amino acids at its N-terminal end. This inactive proto-σ^K binds to the cytosolic face of the outer forespore membrane in the mother cell. At this site it is able to be converted into the active, mature form of σ^K by a protein made in the forespore by σ^G and exported from the forespore to the outer forespore membrane. In this way, the switch from σ^E to σ^K in the mother cell and from σ^F to σ^G in the forespore is interdependent and coordinated.

11.15 DNA rearrangements are required in the mother cell

In several systems of procaryotic development, permanent changes are made in the DNA of a **terminally differentiated** cell type (one that cannot return to the actively

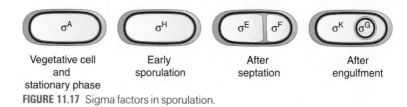

| Vegetative cell and stationary phase | Early sporulation | After septation | After engulfment |

FIGURE 11.17 Sigma factors in sporulation.

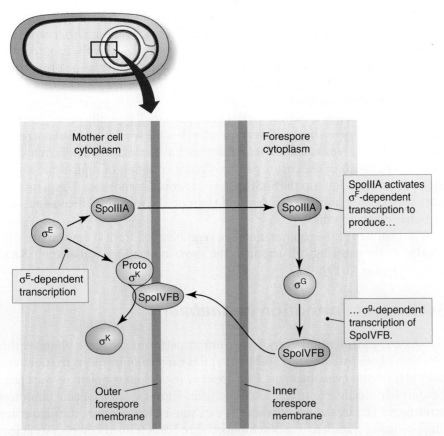

Mother cell cytoplasm

Forespore cytoplasm

SpoIIIA activates σ^F-dependent transcription to produce…

σ^E

SpoIIIA

SpoIIIA

σ^E-dependent transcription

Proto σ^K

SpoIVFB

σ^G

σ^K

SpoIVFB

… σ^g-dependent transcription of SpoIVFB.

Outer forespore membrane

Inner forespore membrane

FIGURE 11.18 Mechanism of sigma factor switching.

multiplying vegetative type of cell). Because these cells are terminal and will not multiply and leave progeny, permanent changes in their DNA are not inherited. Such a system is seen in endospore formation in some species, in which a specific sequence is deleted from the mother cell DNA. This deletion does not occur in the forespore DNA, and thus, it is not passed on to progeny.

The amino-terminal and carboxy-terminal portions of σ^K are encoded by DNA sequences separated by a 42-kb insert, which has to be removed before the gene can be transcribed. This insert has sequences reminiscent of a temperate phage, and its removal is catalyzed by a site-specific recombination enzyme encoded by the insert itself (as in the excision of many temperate phages). This recombinase, **SpoIVCA**, is transcribed just before the switch from σ^E to σ^K in the mother cell and clips out the insert from the σ^K gene before its transcription is activated.

11.16 Germination of endospores is a three-step process

Once formed, an endospore is unable to germinate immediately. The reasons for this are unclear, but its evolutionary advantage seems clear. Usually not all cells in a population sporulate; some remain in stationary phase. Thus, in a population of genetically identical bacilli under starvation conditions, some will have sporulated, and some will be in stationary phase. If the conditions improve shortly after sporulation is complete, the cells in stationary phase can reestablish exponentially growing populations. The spores will make the most effective contribution to long-term survival of the strain by remaining dormant, in case the improvement is only momentary.

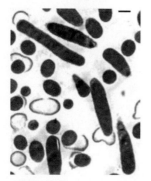

FIGURE 11.19 Outgrowth from a germinating endospore.

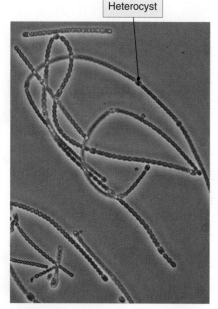

Heterocyst

FIGURE 11.20 Heterocystous cyanobacterial filaments.

Thus, there is an obligatory step of **activation,** in which the endospore is primed for germination. In the laboratory, the most effective activation is brief exposure to high temperature. In nature, however, simple aging is probably the most common activating stimulus; as a population of endospores ages, an increasing proportion of the spores will be able to germinate when conditions improve. The biochemical basis for activation is unknown.

The process of **germination** itself is generally rapid, typically taking less than a half hour. It is triggered by nutrient availability, but the sensing mechanism remains unknown. It involves a loss of refractility as the cytoplasm rapidly rehydrates, hydrolysis of much of the cortex peptidoglycan, excretion of dipicolinic acid and calcium, and breakdown of the SASPs as the nucleoid decondenses. The germinating spore swells in size, initially as a result of the influx of water and then as a result of the initiation of growth.

The final step in the regeneration of a vegetative cell from an endospore is **outgrowth,** as the growing cell ruptures the spore coat and a rod-shaped *Bacillus* emerges (Figure 11.19).

Heterocyst Formation in *Anabaena*

Anabaena is a filamentous cyanobacterium. It performs oxygenic photosynthesis with both photosystem (PS) I and PS II and is capable of nitrogen fixation, as are many filamentous cyanobacteria. This presents an obvious problem, as nitrogenase is exquisitely sensitive to inhibition by oxygen—how can an oxygenic phototroph fix nitrogen? The answer is that when they are starved for nitrogen, the filamentous, nitrogen-fixing cyanobacteria develop specialized cells termed **heterocysts** at intervals along the filament (Figure 11.20). These cells undergo a series of developmental changes that allow them to fix nitrogen and export the fixed nitrogen to adjacent vegetative cells.

Relatively few of the details of regulation of heterocyst formation are known. A number of genes that encode two-component regulatory proteins and transcription factors have been identified, but how they function and in what sequence remain unclear. It is likely that the details will be worked out rapidly now that the basic components are identified and the genome is sequenced.

11.17 Heterocysts lose photosystem II and the OEC and undergo significant morphological differentiation

The obvious incompatibility of oxygenic photosynthesis and nitrogen fixation is solved in large part by the heterocyst losing PS II, the oxygen-evolving complex (OEC), and most of the phycobiliproteins that serve as antenna pigments. These are degraded by proteases early in heterocyst differentiation, a process that presumably both serves to inactivate O_2 production and also to provide amino acids for new protein synthesis in the absence of a nitrogen source. Heterocysts are thus unable to perform noncyclic photophosphorylation and, consequently, generate no oxygen. Photosystem I remains, however, and cyclic photophosphorylation can continue to generate a protonic potential and ATP.

Another significant adaptation is the synthesis of several additional wall layers, thought to reduce the rate of gas diffusion. Because the atmosphere contains four times as much nitrogen as oxygen, a substantial reduction in the rate of gas diffusion could result in very little O_2 penetration, whereas enough N_2 still gets through to provide for nitrogen fixation.

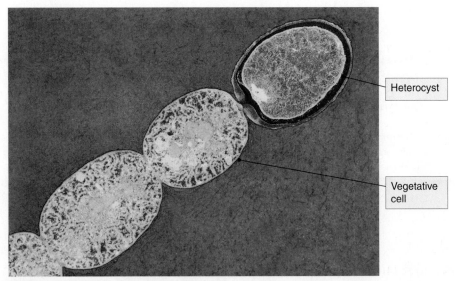

FIGURE 11.21 Thin section of a heterocyst.

The principal barrier to gas diffusion appears to be a thick layer of glycolipid (Figure 11.21). In addition, there is a polysaccharide layer external to the glycolipid layer, which is thought to function mainly to protect the glycolipids.

Significant alterations within the cytoplasm also occur. The thylakoids become fewer, and they are less organized than in vegetative cells. There are fewer ribosomes, and there is a plug of material at the end of the cell where it contacts adjacent vegetative cells. This plug is easily visible in the phase-contrast light microscope because it is highly refractile (see Figure 11.20). It is composed of **cyanophycin,** a polymer of arginine and aspartic acid, and functions as a nitrogen reserve material. Nitrogenase, the enzyme that reduces N_2, is induced to high levels in heterocysts but is absent in vegetative cells.

Thus, the development of heterocysts involves a series of changes: the loss of the components specific to noncyclic photophosphorylation flow, the synthesis of new wall layers, reorganization of the thylakoids, synthesis of cyanophycin, and induction of the nitrogenase genes.

These changes are irreversible; after a nascent heterocyst is visibly different from vegetative cells (called a **proheterocyst**), it is incapable of reverting back to the vegetative cell and resuming multiplication.

11.18 Nitrogen fixation requires material transport between heterocyst and adjacent vegetative cells

Nitrogen fixation requires both ATP and reductant. The heterocyst is capable of providing ATP through cyclic photophosphorylation, but the loss of photosystem II means that it is no longer capable of providing the necessary reductant. Thus, the process requires that reductant be provided by adjacent vegetative cells. It is provided largely as sucrose, produced by photosynthesis (Figure 11.22). The sucrose is transported into the heterocyst, where it is split into glucose and fructose and metabolized by the pentose phosphate pathway (an alternative biochemical pathway to glycolysis for converting sugars to pyruvate) and an incomplete tricarboxylic acid (TCA) cycle. Like many other obligate chemoautotrophs, cyanobacteria lack α-oxoglutarate dehydrogenase. Thus, the product of sucrose oxidation is α-oxoglutarate. The NADH generated in this process is used not only for nitrogen fixation, but for respiration as

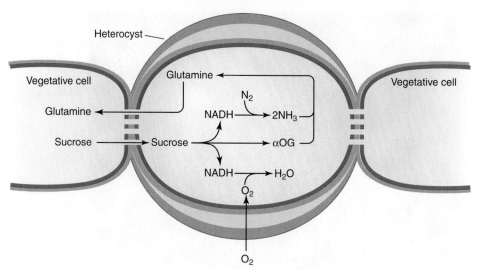

FIGURE 11.22 Material transport between heterocyst and vegetative cell.

well, thus consuming the residual amounts of oxygen that diffuse through the barrier wall layers. These are expensive processes; remember that nitrogen fixation requires immense amounts of ATP (Section 8.23), and the increased respiration rate that is necessary to keep the O_2 tension low in the heterocyst is also a significant drain on the filament's resources. It has been determined that approximately half of the primary productivity of the filament is transported into the heterocyst for these two functions.

Some of the carbon of sucrose is re-exported from the heterocyst back to the vegetative cells as glutamine by combining α-oxoglutarate with ammonia generated by nitrogen fixation. The glutamine then provides all of the nitrogen for growth and multiplication of all of the vegetative cells between heterocysts.

11.19 Heterocyst spacing is determined by an oligopeptide repressor of development

Heterocysts are spaced apart at a distance of approximately 10 cells (there is a lot of variability among strains in the spacing, but the average is usually approximately 10 or higher). This spacing appears to be established by the synthesis, early in heterocyst development, of an oligopeptide (17 amino acids in length) termed **PatS** that is produced by the developing heterocyst and diffuses along the filament. This oligopeptide inhibits heterocyst development, and, thus, all cells close enough to an existing heterocyst, or a developing one, to have a concentration of PatS above a certain level are inhibited from entering the development pathway.

Of course, as the vegetative cells grow and then divide, using the nitrogen exported from the heterocysts, the spacing increases by a factor of two, and thus, cells in the middle of the interval experience a significantly decreased concentration of PatS; one of them then begins the development process, and soon it begins secreting PatS, thus restoring the inhibitory concentration in its vicinity (Figure 11.23)

11.20 DNA rearrangements are required in the developing heterocyst

The many genes necessary to produce a functional nitrogenase are clustered into a couple of operons in all nitrogen-fixing bacteria, and *Anabaena* is no exception. In the filamentous cyanobacteria, however, there are several insertion mutations into

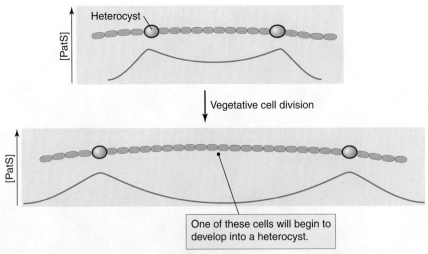

One of these cells will begin to develop into a heterocyst.

FIGURE 11.23 PatS concentration before and after vegetative cell division.

these genes (Figure 11.24): for example, an 11,000-base pair insert in the *nifD* gene (one of the polypeptides of nitrogenase) and a 55,000-base pair insert in the *fdxN* gene (a nitrogenase-specific ferredoxin). Each of these inserts contains a gene (*xis*) that encodes an excision enzyme that is specific to that particular insert. Late in heterocyst development these *xis* genes are induced, and they cut out the inserts just prior to the two operons being transcribed. Because heterocysts do not ever divide, these permanent changes in the DNA are not passed on to progeny.

Fruiting Body Formation and Sporulation in *Myxococcus*

Myxobacteria are a group of gram-negative gliding bacteria that have a unique life cycle that involves a distinctive set of behaviors and developmental processes. They typically live on solid substrata, where they glide around in search of nutrients. When they are starved, tens of thousands of individual vegetative cells come together to form a **fruiting body**—an aggregation of cells within which the developmental program leading to spore formation occurs (Figure 11.25). The best studied is *Myxococcus*, whose fruiting body is a simple mound or stalked sphere of cells.

Although the broad outlines of the behavior and developmental program are now known, the details of the regulation remain obscure; however, many of the molecular components have been identified (proteases, transcription factors, sigma factors, two-component regulatory systems, etc.), and progress can be expected to be rapid.

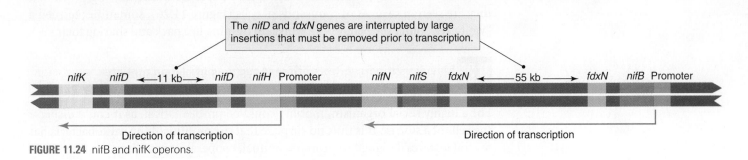

The *nifD* and *fdxN* genes are interrupted by large insertions that must be removed prior to transcription.

FIGURE 11.24 nifB and nifK operons.

DNA rearrangements are required in the developing heterocyst

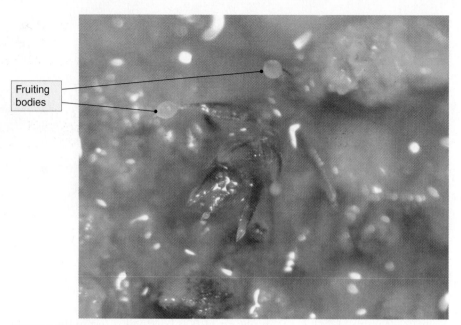

Fruiting bodies

FIGURE 11.25 Myxobacterial fruiting bodies.

11.21 Myxobacteria are highly social procaryotes

There are two distinct nutritional categories of myxobacteria. Most are predatory and prey on other microbes. These myxobacteria secrete a wide range of antibiotics capable of killing and lysing other microbes (often other procaryotes). They also excrete a number of proteases that break down the proteins of the lysed prey organisms, and the myxobacteria then use the resulting amino acids as their principal source of carbon and energy. This group is often termed the **proteolytic** myxobacteria. A few myxobacteria, termed **cellulolytic,** use cellulose, which they degrade into monosaccharides and disaccharides by cellulase enzymes that they excrete.

Both groups have evolved mechanisms to maintain dense populations. A single cell cannot raise the local concentration of antibiotics or extracellular enzymes very high or very fast. Even if it could, it would have to compete with all other organisms in the vicinity for the soluble products of hydrolysis—amino acids or sugars. Because these products will diffuse in random directions from the site of hydrolysis, a single cell will only be able to capture a minute fraction of them. A large group of cells, on the other hand, not only can produce enough extracellular enzymes and antibiotics to make a very high local concentration, but also by sheer numbers can compete effectively for much of the soluble product.

Thus, in nature, myxobacteria are almost never encountered singly—rather, they are seen in swarms of thousands of cells (Figure 11.26), sometimes termed a "wolf pack" (after the behavior of wolves of hunting in a pack and sharing their kill).

11.22 Myxobacteria have two different systems of gliding motility

For a highly social organism, motility could be problematical, as it could disperse the cells of a swarm. It is thus no surprise that the motility of the myxobacteria has several features designed to promote mutual, cooperative motility.

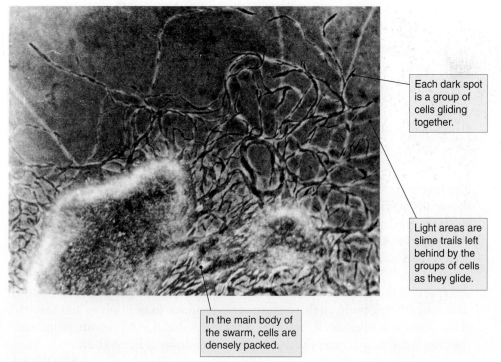

Each dark spot is a group of cells gliding together.

Light areas are slime trails left behind by the groups of cells as they glide.

In the main body of the swarm, cells are densely packed.

FIGURE 11.26 A myxobacterial swarm.

There are two largely independent systems of motility. Of the dozens of genes essential to gliding, only two are known to be common to both. One of the systems is termed **S motility,** for social motility; the other is **A motility,** for adventurous motility.

The social motility system allows groups of cells to move together while they are in direct contact with each other. Commonly, groups of several dozen to several thousand cells will move as a unit. Two extracellular structures are known to be essential to social motility: polar pili (Figure 11.27) and extracellular fibrils composed of protein and polysaccharide. The pili mediate the gliding (see Section 4.41), whereas the fibrils hold cells together.

The A motility system is powered by slime extrusion and allows cells to glide individually. Observation of the edge of a myxobacterial swarm shows cells continually gliding away from the mass of other cells into unpopulated territory; however, such adventurous cells tend to stop when they get 20 µm or so away from the swarm. They then either remain immotile until another cell glides up and contacts them (allowing them to resume motility), or they reverse direction and glide back to the swarm.

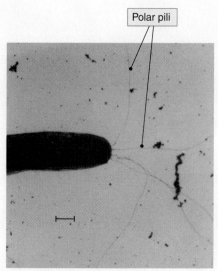

Polar pili

FIGURE 11.27 A myxobacterial cell.

11.23 Myxobacteria locate prey by sensing their physical presence

Homologues of all of the *che* genes of *E. coli* are found in myxobacteria, suggesting that they have a two-component system regulating chemotactic behavior; however, behavioral evidence for chemotaxis is available only for fatty acids, which act as attractants. Amino acids appear not to do so, despite their role as major nutrients. Under most conditions, myxobacterial cells reverse direction at a relatively constant rate of once every 6 to 8 minutes, independent of environmental conditions. It is generally thought that the chemotaxis system is principally involved in development and is probably not involved in nutrient searching by vegetative cells.

Although nutrient searching apparently does not involve chemotaxis, it does involve the ability of myxobacterial swarms to detect the physical presence of potential prey at a distance of up to 50 μm. That this is not chemotaxis was demonstrated by showing that the same response was obtained towards glass beads. The mechanism of this unique tactic response is unknown.

11.24 Aggregation is triggered by starvation and is mediated by a series of intercellular signals

The first stage in fruiting body formation is **aggregation,** the coming together of thousands of cells. It is triggered by nutrient deprivation, which leads to the synthesis and excretion of at least one special protease. What it hydrolyzes is unclear; probably other proteases previously released, which then provide nutrients for the swarm. This leads to a temporary increase in the local concentration of amino acids and peptides, which is used by the swarm as a mechanism of quorum sensing. If the increase is slight, it indicates that the cell density is low, and the swarm is too small to make a fruiting body. In this case, the cells continue to multiply as best they can. If the increase is substantial, however, it indicates that the cell density is high, and the cells of the swarm enter the developmental pathway (Figure 11.28).

We now understand at least part of the sequence of events that leads to this initial quorum-sensing signal in *Myxococcus*. In addition to using amino acids as a source of carbon, nitrogen, and energy, myxococci use them as direct precursors of protein. Several amino acids are essential because *Myxococcus* cannot synthesize them from other precursors. Thus, when these cells are starved for amino acids, there is an immediate shutdown of protein synthesis because of the lack of essential amino acids to charge tRNA. This activates the stringent response (see Section 10.32) and leads to an increase in the levels of ppGpp. PpGpp is required for synthesis of a two-component regulatory system, at least one transcription factor, and a sigma factor, all required for the synthesis of the protease responsible for quorum sensing.

S motility is the motility system used for aggregation, and chemotaxis of some kind is involved; aggregating cells reverse direction less frequently than feeding cells. As aggregation continues, other signals are excreted to which the aggregating cells respond; most of these are not characterized chemically and even the ones whose chemistry is understood act by unknown mechanisms. As a consequence of the signals, the cells begin to pile up in a mound, and these developing fruiting bodies are surrounded by immense rafts of cells gliding into them in a spiral fashion.

11.25 A subpopulation of cells is programmed to develop into myxospores—the rest lyse

Myxococcal swarms are composed of two genetically distinct variants: tan and yellow cells. Although the difference in color is not thought to be developmentally significant, it allows us to recognize these two different types. Cells can switch from tan to yellow at a fairly high frequency; the switch back is rare. Thus, myxobacterial swarms, which start out nearly all tan (for reasons that will become clear momentarily), gradually accumulate more and more yellow cells. The mechanism of this myxobacterial version of phase variation (see Section 10.23) is unknown, but probably involves a genetic rearrangement.

FIGURE 11.28 Scanning electron micrographs of stages of fruiting body formation.

At the time of fruiting body formation, it is common for the swarm to consist of about 70% to 90% yellow cells and the rest tan cells. As the fruiting body grows to its mature size, the yellow cells begin to lyse. Presumably this lysis is part of the developmental program, and their proteins are being used as nutrients to fuel the conversion of the tan cells into myxospores. Motility continues, with cells gliding continuously around the outside of the fruiting body. As tan cells begin to differentiate into myxospores, they lose their motility and are pushed into the center. Thus, the fruiting body has two distinct regions: an inner one of myxospores, surrounded by a surface layer of moving cells. When all of the outer layer cells have either lysed (yellow cells) or developed into myxospores (tan cells), the process is complete. At this stage, the fruiting body consists largely of myxospores, in a matrix of material left by the lysis of the rest of the cells.

At the base of the fruiting body there are usually a few thousand cells (both yellow and tan) that did not make it into the fruiting body. These are presumed to be cells that entered stationary phase physiology rather than the developmental pathway or that exited the developmental pathway before becoming committed. As with other sporulating procaryotes, the fact that not all cells commit to the developmental pathway, but rather some enter stationary phase, allows an alternative mode of survival that is probably important in evolution.

11.26 Fruiting body formation insures that spores remain together and can immediately reestablish swarms

The result of fructification in the myxobacteria is many thousands of myxospores held together in a mass. In some, such as the well-studied *Myxococcus*, the mass of spores is held together in a matrix of polysaccharide and other material derived from autolyzed vegetative cells. In others, such as *Chondromyces*, spores are enclosed within a **sporangiole,** a spore-containing sac surrounded by a fibrous membrane. In either case, the myxospores remain together until germination. Thus, at the time of germination, a swarm is instantly reconstituted, allowing the cooperative feeding that is characteristic of the myxobacteria.

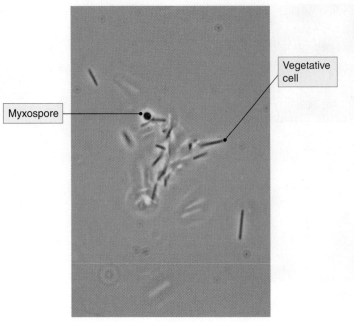

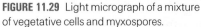

Myxospore

Vegetative cell

FIGURE 11.29 Light micrograph of a mixture of vegetative cells and myxospores.

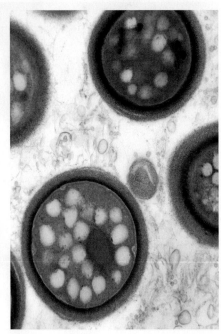

FIGURE 11.30 Thin section of a myxospore.

11.27 Myxospores are formed from entire vegetative cells

Unlike endospores, which are formed within a mother cell, myxospores are formed by a developmental process in which the entire vegetative cell converts to the spore. There is considerable variety among different myxobacteria in myxospore structure; only *Myxococcus* has been studied in any detail.

The myxococcal spore is round or oval, unlike the vegetative cell, which is a long rod (Figure 11.29). It is somewhat refractile due to modest dehydration and has nearly undetectable metabolic activity. It can remain dormant for months or years and is highly resistant to desiccation and to ultraviolet light, but is only modestly resistant to heat.

The molecular details of myxospore structure remain poorly understood. Presumably there is considerable reorganization of the murein layer during the conversion of the rod-shaped vegetative cell into the spherical spore. There is a thick coat of carbohydrate and a surface layer of protein (Figure 11.30).

Summary

Even unicellular procaryotes often show a considerable range of behaviors and often interact with their siblings through quorum sensing regulation of cell responses to their environment. Some go through elaborate processes of differentiation to be able to make spores or other kinds of specialized cells. The molecular regulation of these processes is now understood in broad outline.

Study questions

1. Describe the function and general strategy of procaryotic tactic responses.
2. What would be the effect of a mutation that abolished the functioning of one *E. coli* MCP?

3. What would be the effect of a mutation that abolished the functioning of CheA?
4. What is the likely advantage to *Caulobacter* of its dimorphic life cycle?
5. What is the evolutionary advantage of having sporulation in *Bacillus* subject to quorum sensing, and why do some stationary phase bacilli not sporulate even at high density?
6. What is the evolutionary advantage of the delayed activation of *Bacillus* endospores?
7. Summarize the major adaptations of the heterocyst to allow nitrogen fixation and its interactions with adjacent vegetative cells.
8. How is the spacing of heterocysts regulated?
9. What is the evolutionary advantage of the social systems that keep myxobacterial cells together in swarms?
10. Compare and contrast endospores and myxospores.

12

Microbial Genetics

NEARLY ALL FORMS OF life have mechanisms by which the genetic material of two individuals is recombined to give rise to new genotypes. This has been essential to evolutionary success because it is far more efficient than mutation in generating novel genotypes. The advantage of recombination over mutation is principally that it can generate novel combinations of alleles that have already proved functional, whereas mutation generates principally nonfunctional new alleles.

Most of this chapter focuses on procaryotic genetic recombination; at the end, a short section discusses some aspects of recombination in eucaryotic microbes. The reason that we spend less time on genetic processes in the eucaryotes is that in most cases the mechanisms of recombination are the same ones you already learned in beginning biology, involving meiosis and the fusion of haploid gametes.

12.1 Recombination is separate from multiplication in microbes

In most animals and plants, recombination and reproduction are part of the same process, and, thus, genetic recombination occurs with each generation. In the microbes, however, organismic multiplication is normally **asexual;** that is, it occurs without any genetic recombination and produces genetically identical offspring (except for any spontaneous mutations that might occur during DNA replication). Recombination occurs only occasionally, when the opportunity arises.

12.2 Genetic exchange in procaryotes is directional, incomplete, and polarized

In eucaryotes, fusion of two haploid cells leads to a zygote that contains equal genetic contributions from each parent: two complete sets of chromosomes. In procaryotes, however, a zygote with two complete procaryotic chromosomes from different parents is very rare. Almost always the zygote is incomplete, containing a complete chromosome from one parent and a fragment of the chromosome of the other. The zygote is thus only partially diploid, or **merodiploid** (Figure 12.1).

These incomplete zygotes, or **merozygotes,** then may (or may not) undergo a process of molecular recombination to exchange genes between the complete chromosome (called the **endogenote**) and the chromosomal fragment (called the **exogenote**). Such recombination generally requires that the two molecules be homologous—that they have very similar, although not necessarily identical,

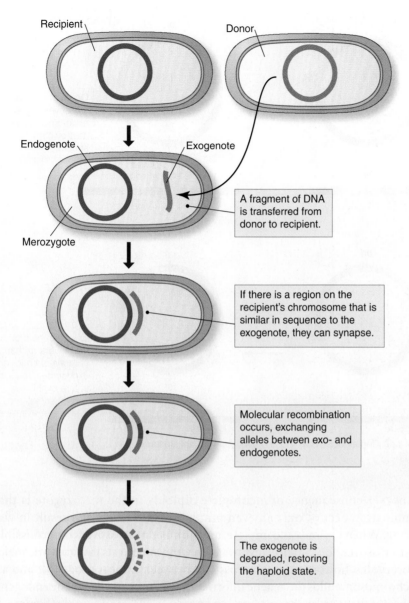

Recipient

Donor

Endogenote

Exogenote

A fragment of DNA is transferred from donor to recipient.

Merozygote

If there is a region on the recipient's chromosome that is similar in sequence to the exogenote, they can synapse.

Molecular recombination occurs, exchanging alleles between exo- and endogenotes.

The exogenote is degraded, restoring the haploid state.

FIGURE 12.1 Outline of genetic recombination in procaryotes.

DNA sequence. If they are not homologous, due to originating from parents of different species, then recombination is very unlikely.

Linear fragments of DNA are normally degraded rapidly so that there is essentially a race between the processes of degradation of the exogenote and recombination. In any event, the exogenote is soon hydrolyzed into individual nucleotides (which can be reused for biosynthesis), and the haploid state is regenerated.

Because the contributions of the two parents to the merozygote are asymmetric, we can think of the processes as directional: one parent, the **donor,** transfers a fragment of its chromosome to the other, the **recipient.** It is also polarized because the recombinant cell that results inherits most of its genes from the recipient and only a few from the donor. There are circumstances in which the genetic contributions of the two parents are more nearly equal, as we describe later; however, the norm is for the recipient to contribute the great majority of genes to the recombinant.

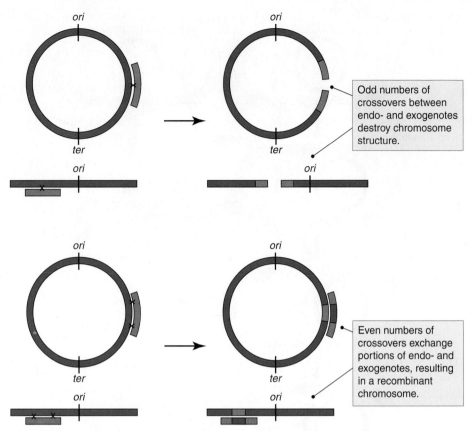

FIGURE 12.2 Consequences of odd and even numbers of crossovers.

Another consequence of incomplete diploidy in the merozygote is that if recombination occurs, only an even number of crossovers will result in viable offspring. When diploidy is complete, any number of crossovers is permissible—a single crossover, for instance, just swaps chromosome arms in eucaryotic meiosis. In procaryotes, however, a single crossover between a linear exogenote and a circular chromosome adds the genetic information of the exogenote to the endogenote, but it breaks the circular chromosome and results in a linear molecule (Figure 12.2). This is presumably lethal. Similarly, a single crossover between an exogenote and a linear chromosome breaks the chromosome into two unequal fragments. This too would probably be lethal, as only one of the fragments would have an origin of replication.

12.3 Genetic exchange often involves the transfer of a plasmid from donor to recipient

Another mechanism of genetic exchange is the transfer of a plasmid from one cell to another. Because plasmids are capable of independent replication, they do not need to recombine with the chromosome to become part of the recombinant genotype. Thus, they can be transferred among very distantly related cells, among which there is too little chromosomal homology to allow recombination with chromosomal genes. Plasmid transfer across very large phylogenetic distances has been documented, for example, between proteobacteria and gram-positive bacteria. Plasmid transfer between domains (Bacteria, Archaea, and Eucarya) is probably rare but has also been documented. Thus, plasmid transfer constitutes a mechanism by which genes can be exchanged across very great evolutionary distances.

Of course, plasmids normally only transfer those genes commonly found on plasmids (see Section 10.5). However, because chromosomal genes can occasionally be transferred to a plasmid, usually by being picked up by a transposon, it is possible for genes that are normally chromosomal to also be transferred to distantly related microbes. There is no doubt that this has been an important force in evolution.

12.4 There are three principal mechanisms by which procaryotes can exchange chromosomal genes

Three principal mechanisms are known by which procaryotes undergo genetic exchange. These are as follows:
- **Transformation,** in which DNA released by the donor cell is taken up by the recipient
- **Conjugation,** in which donor cells in direct contact with recipient cells transfer DNA directly into the recipient cytoplasm
- **Transduction,** in which donor DNA is packaged within a phage capsid, and then injected into the recipient cell

Interestingly, only transformation seems to be a mechanism that has been selected specifically for the purpose of genetic exchange of chromosomal material. Conjugation is principally a mechanism of plasmid transfer. Only occasionally does any chromosomal DNA get transferred in this process, and when it does, it seems to be by accident. Similarly, transduction is an accidental result of phage multiplication, certainly not an actively selected phenomenon. We discuss the details of these three processes later in this chapter.

Genetic Mapping

The various methods of genetic exchange in procaryotes are of interest for three reasons. One is their importance in evolution, as mentioned in the introduction to this chapter. A second reason, discussed later in this chapter, is that certain of them allow the construction of procaryotic strains that are diploid for some genes, an essential step in determining dominance relationships. Finally, a third reason for discussing these mechanisms is that they have been very useful for genetic **mapping:** the determination of the location and order of genes on the chromosome. Genetic mapping has, in turn, been critically important in understanding the regulation of gene expression. However, genetic techniques of gene mapping have been largely replaced, almost overnight, by sequencing and other physical methods of locating genes. Genetic mapping retains importance, however, as the foundation of most of what we know about genome organization and gene regulation in procaryotes. Working out the functioning of the *lac* operon, for instance, required the ability to map the genes involved (showing that the genes for β-galactosidase, permease, and transacetylase were clustered adjacent to the operator) and the ability to construct diploids for this region to test dominance relationships of repressor and operator mutants.

Genetic mapping remains necessary to determine the location of mutations. For instance, a mutation that renders a pathogen nonpathogenic (unable to cause disease) could be in any of a number of different known genes, or it could be in a gene not previously identified as involved in disease causation. Even if the pathogen's genome has been sequenced, mapping is still necessary to determine which gene is affected by the mutation. Even in well-studied organisms, about 20% of genes identified on the basis of sequencing have unknown function; to determine what they do, it is necessary to interfere with their function, and the process of isolating and mapping mutations remains an important approach to do this.

In this section, we discuss some of the basic elements of genetic mapping in procaryotes; additional information is given in the sections on each type of genetic exchange.

12.5 Mapping a gene requires that there be at least two known alleles of that gene

Gene mapping requires the ability to tell whether, after a genetic cross, the recipient cell has inherited the donor copy or the recipient copy of the gene. Thus, the two have to differ; in other words, they have to possess detectably different alleles for at least one gene. In haploid organisms such as procaryotes, recessive alleles are expressed, and mutant alleles are thus far less common than in diploid species. For instance, a mutation in a gene that encodes an enzyme for the biosynthesis of an amino acid will generally lead to the synthesis of an inactive enzyme. In a diploid organism, this will have little or no effect because the enzyme can still be made by translation of mRNA from the other, intact, copy of the gene. In a haploid organism, however, the effect will be to make the cell unable to make that amino acid, and, thus, in many cases such a mutant cell will be unable to compete in nature. The allelic diversity in haploid organisms is thus greatly restricted relative to diploid organisms.

For this reason, geneticists studying procaryotes have generally depended not on naturally occurring alleles, but on laboratory-produced mutant alleles. These mutant strains can survive in laboratory culture because the investigator can provide the required nutrients. For instance, a mutant that can no longer synthesize its own serine (an amino acid) is viable as long as the investigator includes serine in the growth medium.

This approach has allowed the identification of hundreds of genes that, when mutated, lead to a requirement for one or more **growth factors:** small molecules that must be provided in the medium in order for the mutant to grow. These growth factors include amino acids, purines and pyrimidines, and vitamins. Each of these is synthesized by a series of enzymes in **prototrophic** strains (strains that do not require growth factors) but is required by **auxotrophic** strains (strains that require one or more growth factors) as a result of a mutation.

Of course, not all mutations can be compensated by providing growth factors in the medium. For instance, enzymes involved in polymerizing monomers into macromolecules as their product (a particular kind of macromolecule) cannot be provided in the medium. To study genes for these enzymes, **conditional mutations** are used. Conditional mutations are those that have a mutant phenotype under certain conditions but wild-type phenotype under other conditions. For instance, **heat-sensitive** mutations are ones that lead to the synthesis of a product that is more sensitive to thermal denaturation than the wild-type product. Thus, the mutant strain can be grown at the **permissive** lower temperature (e.g., 30° C), but does not grow at the **restrictive** higher temperature (e.g., 37° C). The use of conditional mutants allows the isolation of mutations in essential genes whose loss cannot be compensated by additions to the growth medium.

12.6 Mapping usually involves three basic steps: crossing, selection, and scoring

Mapping genes by genetic crossing always requires as a first step that two parents possessing different alleles of at least two genes be crossed by whatever mechanism is to be used. Two genes are used because it is necessary to determine the amount of **co-inheritance** (simultaneous inheritance of the donor alleles of both genes) of two genes in order to determine whether they are close together or far apart.

The mechanisms of procaryotic genetic transfer normally are very inefficient, however, and typically only one in a thousand to one in several million recipient cells become recombinant. This means that some mechanism of **selecting** the few recombinant cells from among all the unaltered cells. This is done by imposing conditions after the cross such that only recombinant cells will grow, and neither the donor nor recipient cell will (Figure 12.3). This is done on solid medium, and, thus, the only colonies that appear are formed by recombinant cells.

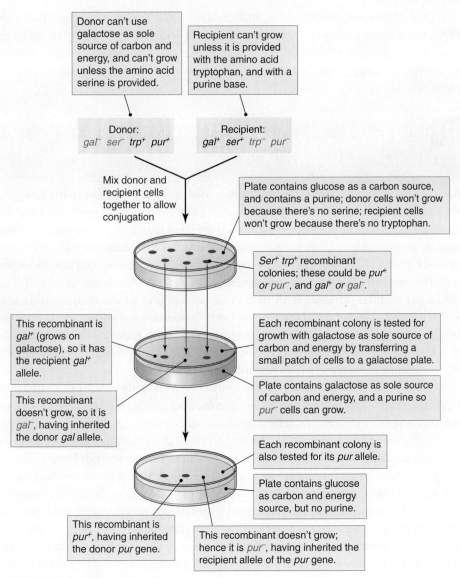

FIGURE 12.3 Steps in a conjugal cross.

When the cross is done by conjugation, selection of recombinants requires that one of the alleles of the recipient and one from the donor be selected to prevent either donor or recipient cells from forming colonies. In transformation or transduction, however, no live donor cells are present (only DNA or phage derived from the donor are used). Thus, in this case, it is only necessary to select one of the donor alleles, as neither free DNA nor free phage particles will form colonies.

After selection of recombinants, the extent of co-inheritance of the second donor allele is **scored** by determining how many of the recombinants carry the donor allele and how many carry the recipient allele.

Transformation

Transformation was the first genetic exchange mechanism to be discovered (by the English scientist Fredrick Griffith) in procaryotes. It was also bacterial transformation that provided the first solid evidence that the genetic material is DNA, in an elegant series of experiments by the American Oswald Avery, as a follow-up to Griffith's discovery.

Transformation mechanisms include features to maximize the uptake of DNA from closely related organisms, either via sequence specificity or through a quorum sensing mechanism. This suggests that transformation is a true system of genetic exchange, selected for its ability to encourage recombination among members of a species.

12.7 Gram-positive transformation is nonspecific and DNA enters the recipient cell in single-stranded form

Many gram-positive bacteria can be transformed, and they all seem to use basically the same mechanism. A few gram-negative cells also seem to use this system. Double-stranded DNA is bound nonspecifically (i.e., the DNA can be from any source) to the outside of the wall by a DNA-binding protein complex that spans the wall and membrane (Figure 12.4). The DNA is then cleaved at the site of binding, and uptake begins with the new end. One strand is hydrolyzed as the other is threaded into the cell. The single strand then synapses with its homologous region of the chromosome and undergoes recombination. If the DNA is from an unrelated organism, no homology will be present, and the fragment will simply be degraded. Indeed, it is quite possible that this form of transformation evolved originally for a nutritional function, to take up nucleotides from the environment in the form of DNA.

Gram-positive transformation generally results in fairly small pieces of DNA being taken up—averaging about 20 to 30 genes. Because this is less than 1% of a typical procaryotic chromosome, genes have to be very close together to be co-inherited.

In this type of transformation, the recombinant that results is a **heteroduplex**—having a region of double-stranded DNA in which the two strands have different origins (Figure 12.5). If there are genetic differences between the donor and recipient in this region, there will be places where the DNA strands are not complementary. These **mismatches** are repaired by special enzymes that scan the DNA and repair damage. The enzyme systems preferentially replace the base in the new strand, thus restoring the genotype of the recipient. If the region is replicated before it is repaired, however, one of the two daughter chromosomes will be recombinant, and the other will be identical to the recipient.

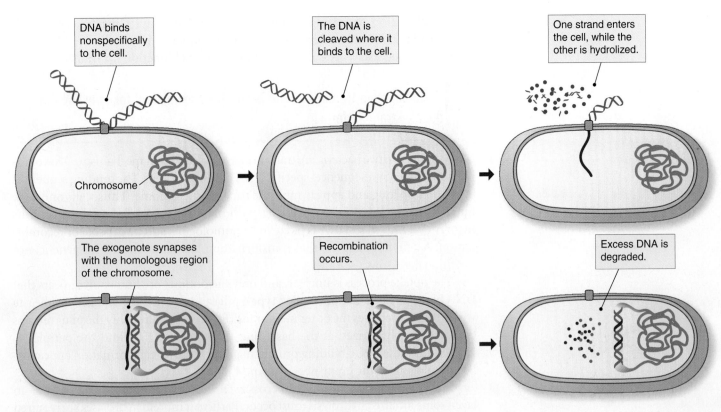

FIGURE 12.4 Gram-positive transformation.

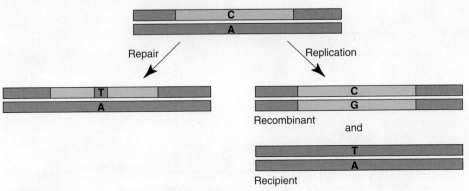

FIGURE 12.5 Resolution of a heteroduplex region.

12.8 Competence to be transformed is subject to a quorum sensing mechanism

Some gram-positive bacteria are constitutively competent to be transformed—that is, they are always receptive. Most, however, are only competent when they are present at medium to high density because of a quorum sensing mechanism. A small excreted protein termed **competence factor** is continuously excreted by all cells. At low cell density, the competence factor cannot accumulate to high concentration; however, at high cell density, the competence factor reaches a high enough concentration in the medium that it can bind to the sensor protein of a two-component regulatory system in the cell membrane. This binding activates a transcription factor that stimulates the transcription of about a dozen proteins needed for DNA uptake. This quorum sensing system ensures that the complex machinery of DNA uptake will not be made unless the density of potential donors in the neighborhood is high. This suggests a genetic rather than nutritional function for transformation.

12.9 Gram-negative transformation is species specific and takes in DNA in double-stranded form

Many gram-negative bacteria are transformable by a different mechanism. DNA binding is mediated by a sequence-specific DNA binding protein. The binding sequences are 8 to 11 bp long and appear multiple times in the genome. Thus, cells only take up DNA from closely related strains, with the same recognition sequence. This insures that imported DNA is likely to be homologous and be capable of recombination. Again, this suggests that transformation has been selected for genetic, not nutritional, reasons.

The uptake process is unclear, and only the outlines are known. It appears that DNA binds to particular pili (called **type 4 pili**) and enters the periplasm via the site where the pilus crosses the outer membrane (Figure 12.6). Probably the pilus retracts by being depolymerized at the base, thus drawing the DNA into the periplasm. Interestingly, the DNA-binding protein in gram-positive transformation appears to be homologous to the gram-negative type 4 pilin.

It appears that for the DNA to cross even the very thin gram-negative murein layer, some localized hydrolysis must occur. Particular murein hydrolases are required

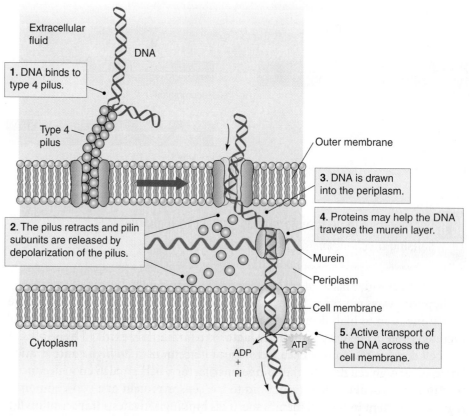

FIGURE 12.6 Gram-negative transformation.

CHAPTER 12 MICROBIAL GENETICS

for transformation, and it is thought that they act at this point to open a gap in the murein for the DNA to penetrate.

Finally, DNA crosses the cell membrane through a specific, energy-requiring, uptake protein. It remains double-stranded during passage, and thus, the recombination process is a normal one of recombination between two double-stranded DNA molecules.

12.10 Donor cells may actively release DNA

In the laboratory, transformation is studied with purified DNA obtained by lysing donor cells. It was assumed for a long time that in nature, lysed cells are also the source of DNA for transformation. It is unlikely, however, that there will be much lysis at the times in the growth curve where competence occurs, and the ones that do lyse may be the least fit and the least desirable as donors. Thus, it is no surprise that cells are now thought to release DNA actively.

If donor cells do actively release DNA, this process is probably also under the same quorum sensing control as competence, to ensure that release of DNA from donor cells, and recipient cell competence, occur at the same time.

Conjugation

Conjugation is a mechanism of plasmid transfer from one cell to another. Its importance in the spread of plasmids, such as antibiotic-resistance plasmids, is clear. It also incidentally facilitates occasional chromosomal gene exchange among closely related procaryotes, and, thus, it is thought to be a force in bacterial evolution as well.

12.11 Transmissible plasmids encode special pili that bind donor cells to recipient cells

The first step in conjugation is for donor and recipient cells to recognize and bind to each other. This recognition and binding is mediated by special pili, synthesized only by cells that harbor a transmissible plasmid (because the genes for the pili are on the plasmids). These **conjugal pili** (sometimes called **sex pili**) have adhesins at their tip that bind to a specific structure on the outer surface of potential recipient cells (Figure 12.7). Normally, such cells will be relatives of the donor, as they must share the same surface structure (because the plasmid had to be able to get into the donor in the first place). However, because these surface recognition structures are generally simple oligosaccharide or peptide sequences, they are occasionally found on very distantly related cells; thus, conjugation may sometimes transfer plasmids among very distantly related cells. There are even documented cases of conjugal DNA transfer from bacteria to plants and to animals.

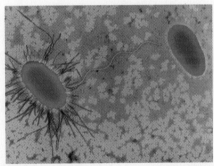

FIGURE 12.7 Donor and recipient cells linked by the F-pilus.

Donor cells do not bind to other donors carrying the same transmissibility plasmid. This phenomenon, termed **surface exclusion,** is due to another plasmid-encoded protein that embeds in the outer membrane and masks the surface receptor for the sex pilus. This prevents populations of donor cells from engaging in lots of unnecessary and energetically expensive mating with each other.

The sex pilus retracts after it binds to a recipient cell, probably by depolymerizing at the base, thus drawing the two cells into close contact with each other to form a **mating pair.** There appears to be some rearrangement of the structure of the outer membranes of the conjugating cells where they contact each other, but the nature of these changes is not known. The effect, however, is to open up a channel by which DNA can traverse the four membranes and two murein layers that separate donor and recipient cytoplasm.

12.12 DNA transfer to the recipient and plasmid replication by a rolling circle mechanism occur simultaneously

After the establishment of the mating pair, a special round of plasmid DNA replication is initiated. Unlike vegetative replication of the plasmid, which is generally bidirectional like the chromosome, transfer replication is by a rolling circle mechanism (see Section 6.13). The plasmid encodes the necessary proteins, and it has a special origin (*oriT*) for initiation of transfer replication.

As rolling circle replication proceeds around the plasmid, leading strand synthesis displaces the 5′ strand into the recipient, where lagging strand synthesis takes place (Figure 12.8). When replication is complete and *oriT* comes around again, another nick is made, cutting off the single-stranded copy of the plasmid and releasing it into the recipient. The conjugation bridge is disassembled and the mating pair separates. The result is that both cells now have a copy of the plasmid.

12.13 Transfer genes are normally repressed

Most transferable plasmids normally repress the 30 to 40 genes necessary for plasmid transfer, so that it is only an occasional cell in a population that is by chance derepressed and capable of transferring DNA. This probably reflects the energetic costs of making sex pili and other transfer proteins, and the improbability of encountering a recipient cell. The repression systems are complex and involve both repressors and transcription factors.

How these plasmids spread in nature is unclear. Even well-studied transmissible plasmids encode numerous proteins whose function we do not yet understand; it is likely that some of these are regulatory proteins that lift the repression of fertility under certain circumstances—high population density, for instance. Thus, under conditions such that mating is likely to be possible, the fertility system is probably derepressed and the plasmid spreads rapidly from cell to cell.

12.14 F is derepressed for transfer because it has an insertion mutation in a regulatory gene

The best studied transmissible plasmid is named **F** for "fertility" and is found in several genera of enteric bacteria—the group that includes *E. coli*, which appears to be the principal host for F. F is derepressed for transfer, and thus, it is capable of transfer at any time. We now know why F is derepressed: it has an insertion mutation in the gene *finO*, an essential component of the regulation of many transmissible plasmids. The system works as follows. TraJ is a transcription factor essential to the transcription of the 32 gene *tra* (for "transfer") operon; however, normally TraJ synthesis is translationally blocked by an antisense RNA that hybridizes with its mRNA and blocks its ribosome binding site (Figure 12.9). The antisense RNA is stabilized by binding to the protein FinO; in the absence of FinO antisense RNA half-life is about 2 minutes, but in its presence the half-life is over 40 minutes.

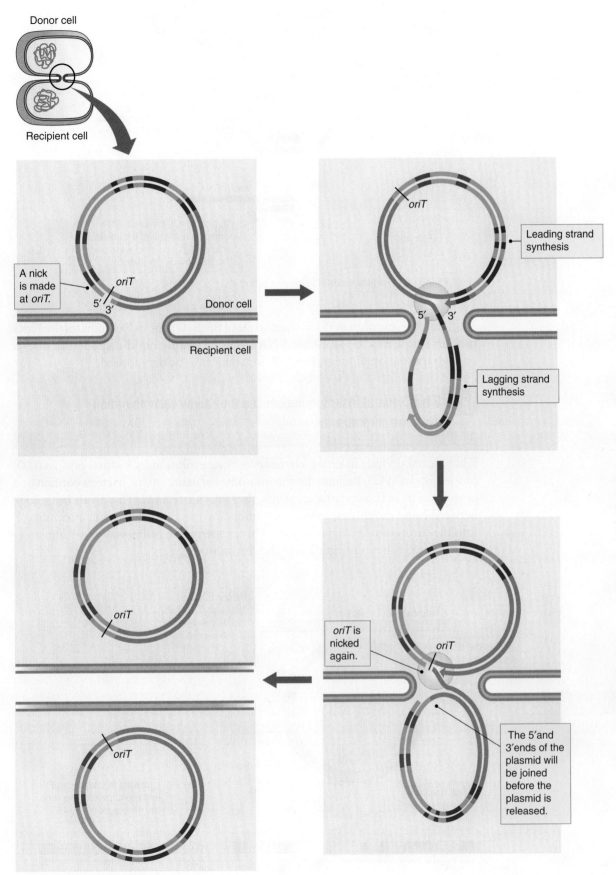

Donor cell

Recipient cell

A nick is made at *oriT*.

oriT

5' 3'

Donor cell

Recipient cell

Leading strand synthesis

oriT

5' 3'

Lagging strand synthesis

oriT is nicked again.

oriT

The 5' and 3' ends of the plasmid will be joined before the plasmid is released.

oriT

oriT

oriT

FIGURE 12.8 Plasmid transfer.

F is derepressed for transfer because it has an insertion mutation in a regulatory gene

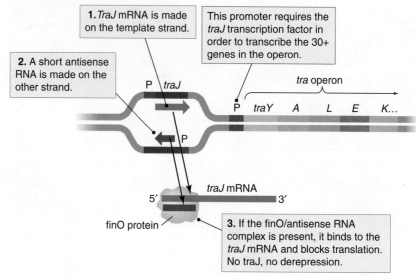

1. *TraJ* mRNA is made on the template strand.

This promoter requires the *traJ* transcription factor in order to transcribe the 30+ genes in the operon.

2. A short antisense RNA is made on the other strand.

tra operon

P \ *traJ*

P *traY*   *A*   *L*   *E*   *K...*

P

traJ mRNA

5' ——— 3'

finO protein

3. If the finO/antisense RNA complex is present, it binds to the *traJ* mRNA and blocks translation. No traJ, no derepression.

FIGURE 12.9 Fertility repression in F-like plasmids.

Thus, the insertion of IS3 in the middle of the *finO* gene in F eliminates the FinO protein and leaves the antisense RNA unprotected; its rapid degradation allows enough translation of the *traJ* mRNA for transfer to be derepressed.

12.15 F has several insertion sequences that allow recombination with the chromosome

F has picked up four insertion elements over its evolutionary history: one Tn1000, one IS2, and two IS3. Because the chromosomes of many enteric bacteria contain one or more of these transposable elements, they provide regions of base-pair homology between F and the chromosome. Recombination can thus occur here, integrating F into the chromosome (Figure 12.10). This property of F is unusual; most transmissible plasmids do not integrate into the chromosome.

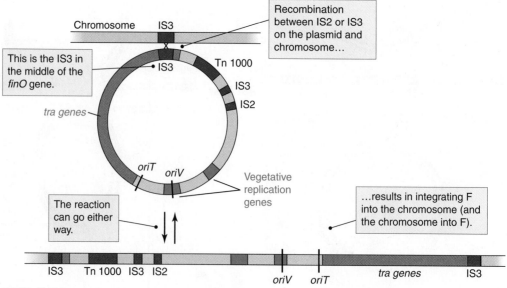

Chromosome   IS3

Recombination between IS2 or IS3 on the plasmid and chromosome...

This is the IS3 in the middle of the *finO* gene.

IS3

Tn 1000

IS3

IS2

tra genes

oriT *oriV*

Vegetative replication genes

The reaction can go either way.

...results in integrating F into the chromosome (and the chromosome into F).

IS3   Tn 1000   IS3   IS2   *oriV*   *oriT*   *tra* genes   IS3

FIGURE 12.10 Integration of F into the chromosome.

CHAPTER 12 **MICROBIAL GENETICS**

A cell with an integrated F in its chromosome is termed an **Hfr** (for high frequency of recombination—a term we explain shortly). In an Hfr, the vegetative *oriV* does not function, and F is replicated passively as part of the chromosome.

Thus, there are three principle F fertility types: **F⁻**, in which the cell lacks the F plasmid; **F⁺**, in which the F plasmid replicates autonomously; and **Hfr,** in which the F plasmid is integrated into the chromosome.

Because the insertion event that generates an Hfr is reversible, every culture of an F⁺ or an Hfr strain is mixed. In an Hfr culture, there will always be some cells in which recombination has excised the plasmid to restore the F⁺ state; and in an F⁺ culture, there will always be a minority of Hfr cells of various types depending on which IS2 or IS3 their F is integrated into.

12.16 Hfr strains donate chromosomal genes to recipients at high frequency

Although the functioning of *oriV* in Hfrs is repressed, the transfer genes continue to be expressed, and thus, normal initiation of mating occurs when an Hfr encounters an F⁻. As usual, the 5′ end of F passes first across the conjugation bridge, but the sequences that follow are bacterial, beginning with the genes adjacent to the insertion site and continuing in a linear fashion (Figure 12.11). Usually the mating pair spontaneously breaks apart before the entire bacterial chromosome can be transferred; recombinant cells therefore remain F⁻, as half of F, including all of the *tra* genes, never make it into the recipient.

12.17 Hfrs can be used to map genes on the bacterial chromosome

Because each Hfr strain transfers the chromosome from a specific point and in a specific direction, they can be used to map genes on the enterobacterial chromosome. There are several ways to do this; the most accurate is by **time-of-entry**. In this

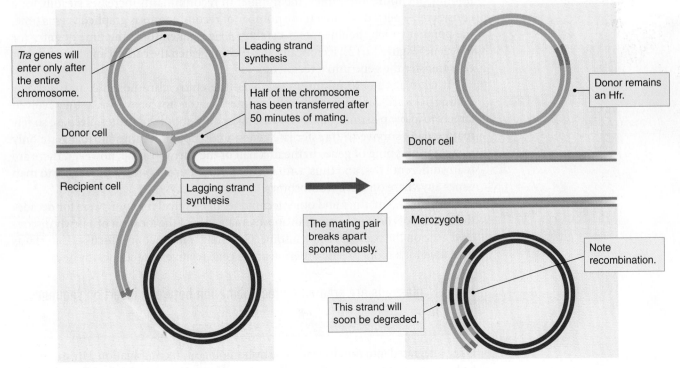

FIGURE 12.11 Hfr mating.

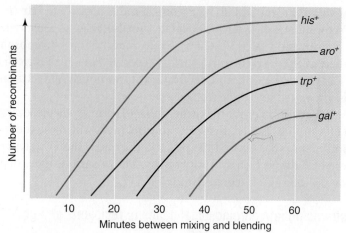

Donor: Hfr KL98 *serA⁻*

Recipient: *hisD⁻, aroD⁻, trpC, galK⁻*

Selection: *ser⁺* and either *his⁺, aro⁺, trp⁺,* or *gal⁺*

FIGURE 12.12 Interrupted mating.

method, the donor and recipient cultures are mixed, and then at various times after mixing, a sample is taken and agitated vigorously in a blender or homogenizer. This breaks the mating pairs apart and prevents any further gene transfer. The mixture is then plated on media that select recombinants of specific types. If the amount of time allowed for mating was too short for the selected gene to enter the recipient, no recombinants will be obtained; however, if the time was adequate for all donors to transfer the gene to the recipient, many recombinants will be obtained.

Because not all mating pairs are established at the same instant after the donor and recipient cultures are mixed and because there are small variations in the rate of DNA transfer, the appearance of recombinants for a given gene is not an all-or-nothing phenomenon; rather, the number of recombinants increases steadily for a considerable period of time. If the number of recombinants is graphed versus time, the point at which the line intersects the x-axis is taken to be the time of entry for that gene (Figure 12.12); technically, it is the time when the earliest and fastest donor cells transfer the gene into a recipient.

It takes about 100 minutes to transfer the entire chromosome, and very few mating pairs are stable for this length of time. They do not have to be, as they evolved largely to allow plasmid transfer, which takes only about 2 minutes. Because so few mating pairs survive to transfer late genes, interrupted matings normally are only useful for mapping of genes in the first half of the chromosome; however, there are many different Hfrs, and thus, with two or three different ones, it is possible to map genes anywhere on the chromosome.

Interrupted matings and other techniques of conjugal mapping were for decades the method of choice of getting an approximate idea of the location of a newly discovered gene on the bacterial chromosome. Generally transduction (discussed later) was then used for a more precise ordering of the gene relative to other nearby ones.

12.18 F′ plasmids are created by recombination between insertion sequences in an Hfr

Once integrated into the chromosome by homologous recombination, Hfr strains can be fairly stable. Of course, they periodically revert to F⁺ strains by an exact reversal of the insertion event that originally created them, and, thus, every culture of an Hfr contains F+ cells as well. Because there are many insertion sequences in the chromosome and several in F, however, there are many different possibilities for homologous recombination; only recombination between two particular IS elements (the same two that recombined during integration of F) will generate an F+. All of the other possibilities will either generate nonviable recombinants, or an **F′** ("F-prime"). An F′ strain is one in which the F replicates autonomously, but in which it contains some bacterial genes as well. If the recombination involves an IS element within F, then some sequences of F get left in the chromosome, in exchange for the bacterial sequences (Figure 12.13). If the recombination is between sequences on either side of F, however, then the F′ contains all F sequences plus some bacterial ones. In both cases, the chromosome is now missing some genes, but because those genes are carried on the plasmid and can still be normally expressed, the cell remains viable and can multiply, and because all essential replication and transfer sequences are retained on F, the F′ remains a donor.

When an F′ mates, it donates the F′ plasmid at high frequency. Thus, unlike both F+ and Hfr strains, F′s donate some bacterial genes at very high frequency but donate most bacterial genes at the low frequency characteristic of F+ strains. (F+ strains occasionally donate chromosomal genes because of the rare Hfr cells mixed in with the majority of F+ cells.) Furthermore, the recipient cells are converted into donors by conjugation with F′ cells because the F′ plasmid is still much smaller than the chromosome and can be transferred completely before many of the mating pairs break apart.

12.19 F′ plasmids can be used to construct stable partial diploids

Genes that are on the plasmid in an F′ are normally present in two copies in zygotes formed by an F′ × F− mating: the recipient alleles on the chromosome, plus the donor alleles on the plasmid. These merodiploid recombinants are known as **secondary F′** strains, to contrast them to the haploid **primary F′** strains where they first arose.

This has been very useful in testing dominance relationships of genes—particularly useful for deducing the role of regulatory genes. Different F′ plasmids carry different regions of the *E. coli* chromosome, and every region of the chromosome is represented on one or more F′. Thus, it is feasible to make any region of the *E. coli* chromosome diploid.

Transduction

Transduction is a mechanism of genetic exchange that depends on the accidental packaging of host cell DNA by a phage during its multiplication. This is obviously deleterious to the survival of the phage, as it decreases the number of phage particles that contain phage DNA and that can perpetuate the phage through evolutionary time. Thus, it is unlikely that this feature of phage multiplication has been selected; rather, it would appear to be an unavoidable consequence of the molecular mechanisms involved in phage replication. Given all of this, it is no surprise that transduction is rare—typically only a few percent of the phage particles contain host DNA. Nevertheless, this is sufficient to have a significant impact on the evolution of procaryotes.

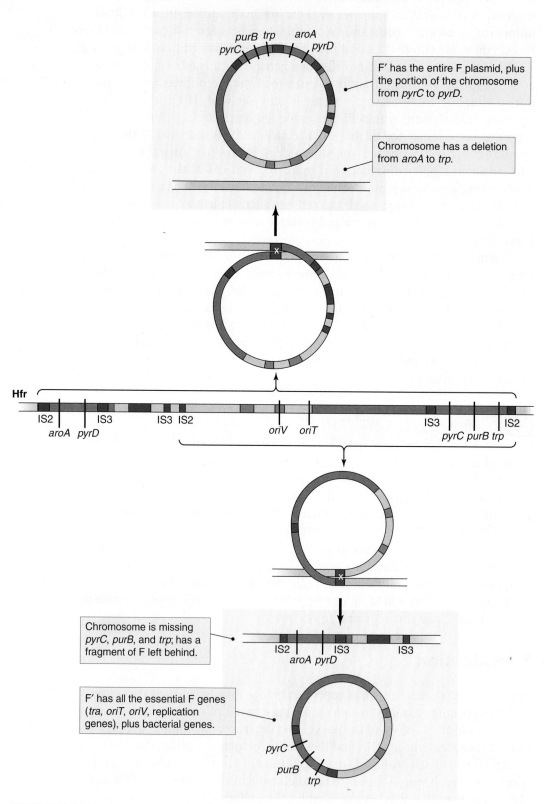

F′ has the entire F plasmid, plus the portion of the chromosome from *pyrC* to *pyrD*.

Chromosome has a deletion from *aroA* to *trp*.

Hfr

IS2 *aroA* *pyrD* IS3 IS3 IS2 *oriV* *oriT* IS3 *pyrC* *purB* *trp* IS2

Chromosome is missing *pyrC*, *purB*, and *trp*; has a fragment of F left behind.

IS2 *aroA* *pyrD* IS3 IS3

F′ has all the essential F genes (*tra*, *oriT*, *oriV*, replication genes), plus bacterial genes.

pyrC

purB

trp

FIGURE 12.13 Mechanism of generation of F′s.

Table 12.1 Comparison of Specialized and Generalized Transduction

Generalized Transduction	Specialized Transduction
Any host genes can be transduced	Only a few specific host genes can be transduced (ones adjacent to the integration site of the prophage).
Transducing particles contain *only* host DNA.	Transducing particles contain host DNA *and* phage DNA, covalently joined.
A single cell can produce a mixture of transducing particles *and* phage particles.	A single cell produces *either* transducing particles or phage particles.
Transducing particles are formed as a result of a mistake in DNA packaging into phage capsids.	Transducing particles are formed as a result of a mistake in the excision of a prophage.
Recombinants are haploid.	Recombinants are diploid for the specific genes transduced.

12.20 There are two principal forms of transduction, with different mechanisms

There are two principal forms of transduction: **generalized transduction,** in which any host gene can be transduced at low frequency, and **specialized transduction,** in which only a few specific genes are transduced (Table 12.1). **Transducing particles** (a term that refers to a phage capsid containing some host cell DNA) also differ in their DNA content: in generalized transduction, transducing particles contain only host cell DNA; in specialized transduction, they contain host *and* phage DNA, covalently joined. These fundamental differences between the two forms of transduction result from different mechanisms by which transducing particles are generated, as we discuss in the following sections.

12.21 Generalized transduction results from mistakes during packaging

Phages that use a headfull mechanism of packaging their DNA into the capsid (see Section 6.19) are often capable of generalized transduction. These phages produce an enzyme that cuts the concatemer of phage DNA (sometimes at a specific sequence, sometimes randomly), following which successive headfulls of DNA are encapsidated. Transducing particles are formed when the enzyme mistakenly cuts the host chromosome. This then leads to the packaging of sequential headfulls of host DNA, instead of phage DNA. This can happen simultaneously with the packaging of phage DNA by the normal process, so that the infected cell produces a mixture of normal phage and transducing particles.

   Thus, when the host cell lyses and releases hundreds or thousands of virions, a few may contain host DNA instead of phage DNA (Figure 12.14). If such a transducing particle encounters a new host cell, it can bind to the cell and inject its DNA. The DNA can then undergo recombination with the endogenote to produce a recombinant.

12.22 Generalized transduction is used for precise mapping of nearby genes

Conjugal mapping using Hfr strains gives a rough idea of the location of a gene, but there is considerable error associated with time-of-entry experiments (and other ways of using conjugation to map genes). Thus, when several genes are nearby, transductional crosses are used to determine order more precisely.

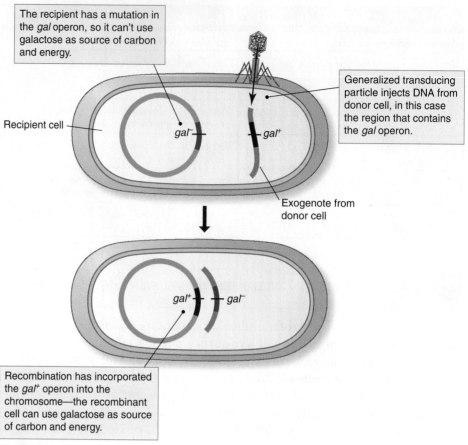

FIGURE 12.14 Generalized transduction.

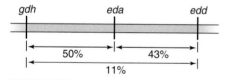

FIGURE 12.15 Two point crosses.

Sometimes the **cotransduction frequencies** (co-inheritance frequencies determined by transduction) of each pair-wise combination of genes give a clear indication of order. For instance, the cotransduction frequencies for the three genes shown in Figure 12.15 are consistent with only one gene order.

In other cases, the differences in cotransduction frequencies may not be sufficient to be sure of the gene order. In this case, **three point crosses** are used. Three point crosses are crosses in which three genes are involved, each with different donor and recipient alleles. In such crosses, there are four different possible combinations of the unselected alleles. If the selected allele is between the other two, all four combinations can be formed by two crossover events, one on each side of the selected allele (Figure 12.16). The frequencies of the four different combinations may be different, but they are not expected to differ dramatically. If the selected allele is one of the flanking genes, however, one of the combinations of unselected alleles will require four crossover events. This category is expected to be significantly less frequent than the others.

12.23 Specialized transduction results from a mistaken excision of an integrated prophage

Many phages are capable of lysogenizing their hosts by inserting the viral chromosome into the host chromosome, where it is passively replicated whenever the host chromosome is replicated. When the regulatory systems of the phage detect physiological stress to the host cell, indicating possible impending death, the prophage is excised

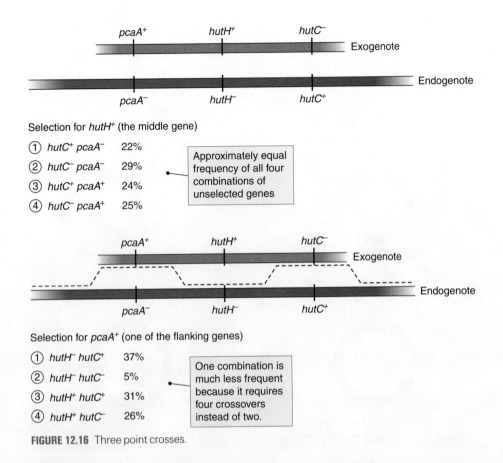

Selection for *hutH*⁺ (the middle gene)

1. *hutC⁺ pcaA⁻*   22%
2. *hutC⁻ pcaA⁻*   29%
3. *hutC⁺ pcaA⁺*   24%
4. *hutC⁻ pcaA⁺*   25%

Approximately equal frequency of all four combinations of unselected genes

Selection for *pcaA*⁺ (one of the flanking genes)

1. *hutH⁻ hutC⁺*   37%
2. *hutH⁻ hutC⁻*   5%
3. *hutH⁺ hutC⁺*   31%
4. *hutH⁺ hutC⁻*   26%

One combination is much less frequent because it requires four crossovers instead of two.

FIGURE 12.16 Three point crosses.

again, enters vegetative replication, and the infected host produces a burst of progeny virions (see Chapter 6).

Typically, the excision of the prophage is an exact reversal of the recombination event that inserted it (Figure 12.17); however, occasionally the excision is aberrant, and a few host cell genes are excised in place of some phage genes. These aberrant excisions are very rare (typically 1 in a million), and they occur at nearly random sites with very little sequence homology. The molecular mechanism of this illegitimate recombination is not understood.

Although some phage genes are missing from the replicating phage DNA, they are still present in the cell, and replication and packaging can occur. The result is the release of a burst of virions all of which contain a mixture of both phage and host DNA, covalently joined.

12.24 Specialized transduction produces partial diploids

When a specialized transducing particle injects its DNA into a recipient cell, the DNA can circularize and integrate normally (Figure 12.18). The result is a lysogen; however, normally, the integrated prophage is defective because it lacks the viral sequences that were left behind in the donor. Occasionally, the prophage is still capable of being induced and producing progeny virions, if the missing sequences are small and not essential.

In addition to the integrated prophage, the recombinant has duplicate copies of those host sequences that were incorporated into the phage genome by the illegitimate recombination in the donor. If these sequences carry different alleles, the phenotype of the recombinant will be that associated with the dominant allele.

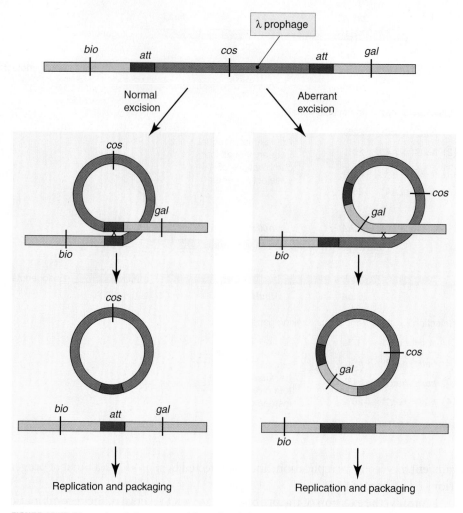

FIGURE 12.17 Normal and aberrant excision of lambda.

Genomic Mapping

For 40 years, microbial geneticists used conjugation, transduction, and transformation to map genes on procaryotic chromosomes. These techniques allowed the construction of chromosomal maps showing the location of more than a hundred genes in some species, most notably *E. coli* and its close relative *Salmonella typhimurium*. The last decade or so, however, has seen an explosion of biophysical methods of exploring chromosome structure, some of which can be used for very precise mapping. These physical methods largely supplanted genetic mapping techniques for procaryotic genes. More recently, sequencing methods became rapid enough to allow sequencing of entire genomes, and these genomic sequences have now made traditional methods of mapping obsolete.

12.25 Genomic sequencing allows the identification of genes on the chromosome

By now, several hundred procaryotes have had their genome completely sequenced. After the sequence is in hand, it is scanned by computer programs that identify **open reading frames** (commonly called **orfs**), each orf being a potential gene. These are identified by scanning for the standard transcriptional and translational signals.

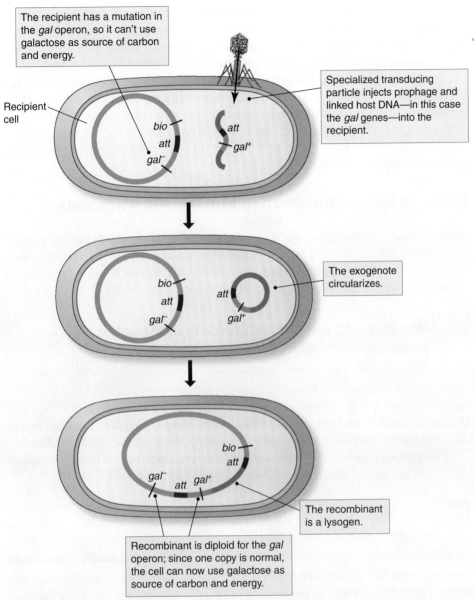

The recipient has a mutation in the *gal* operon, so it can't use galactose as source of carbon and energy.

Specialized transducing particle injects prophage and linked host DNA—in this case the *gal* genes—into the recipient.

Recipient cell

bio
att
gal⁻

att
gal⁺

The exogenote circularizes.

bio
att
gal⁻

att
gal⁺

bio
att

gal⁻
att
gal⁺

The recombinant is a lysogen.

Recombinant is diploid for the *gal* operon; since one copy is normal, the cell can now use galactose as source of carbon and energy.

FIGURE 12.18 Integration of lambda transducing particle.

Each orf is then compared by computer matching programs with a database of sequences of genes of known function. Because a large number of genes in all organisms are homologous to those in other organisms, this normally allows the identification of most of the basic genes for biosynthesis, ribosomal components, DNA replication proteins, structural proteins such as porins, many regulatory proteins, sensory and tactic systems, etc.

Genes that do not give an unambiguous match to genes of known function will often show part of their sequence that is similar to other partial sequences. Protein regions such as binding sites for ATP, GTP, NAD(P)H, and DNA can often be identified clearly, giving hints to the function of the unknown gene.

Even after the tentative identification of genes on the basis of full or partial homology, it is not uncommon for 10% to 30% of genes identified on the basis of sequence to have unknown function. Identifying their function can be quite difficult. There are techniques that allow mutations to be generated in specific genes, but often

Genomic sequencing allows the identification of genes on the chromosome there appears to be no phenotypic effect of mutationally knocking out a gene of unknown function. Scientists also use RNAi (see Section 18.11) to inactivate mRNA transcribed from such genes, but again, there is often no effect. Sometimes this is due to not testing the right conditions. For instance, knocking out a gene that is involved in long-term survival in stationary phase might not have any effect if the investigator is simply looking for an effect on growth rate under normal laboratory conditions. In other cases, the products of multiple genes may interact, and the lack of any one has only a subtle effect that is easily overlooked. Given these difficulties, it is likely that it will be some time before we come close to a complete functional understanding of any genome.

Genetic Recombination in Eucaryotic Microbes

Among eucaryotic microbes, the molecular mechanisms of genetic recombination are generally similar in broad outline to that in plants and animals, involving meiosis and the fusion of two haploid cells to form a diploid zygote. Because you studied these processes in general biology, we do not discuss them here.

12.26 Most eucaryotic microbes multiply mainly by asexual means; genetic recombination is rare

Most, probably almost all, eucaryotic microbes have systems of occasional genetic recombination. These systems, however, are not coupled to multiplication the way they are in animals and plants. Like procaryotes, most eucaryotic microbes reproduce themselves asexually, and genetic recombination is only an occasional event.

The asexual reproduction of eucaryotic microbes may be of haploid or diploid cells. In many cases, both are capable of multiplication, as we see with yeast.

12.27 Most unicellular eucaryotic microbes have morphologically indistinguishable mating types

In some microbes, genetic recombination involves the fusion of morphologically distinct gametes, just like in the animals. Typically, a small, motile "male" gamete seeks out and fuses with a larger, immotile "female" gamete. This is especially common in the multicellular representatives of the fungi and algae.

Most unicellular eucaryotic microbes, however, do not produce morphologically distinct gametes; rather, their mating involves the fusion of morphologically identical cells. Although the cells that fuse are indistinguishable morphologically, they differ genetically, typically at a single locus that determines their **mating type.** Typically, the mating type locus has two alleles, determining two different mating types—in yeast, these are called **MATa** and **MATα.** Mating occurs only between cells of different mating types.

The MATa/α loci encode alternative transcription factors that activate alternative sets of genes. One set of genes encodes proteins specific to a cells, and the other encodes α-specific proteins. By this simple mechanism, the many different genes necessary to determine a specific mating type are turned on and off by a single master switch.

Each mating type secretes a different short polypeptide pheromone, or signal molecule, termed **a-factor** or **α-factor.** Each mating type also has expressed on its membrane surface a receptor for the pheromone of the opposite mating type. Thus, each mating type signals its presence to the other, and each is able to sense the other.

When a yeast cell binds pheromones from the opposite mating type, the membrane receptor changes conformation and initiates a chain of molecular events that arrests the cell cycle in G_1 and prepares the cell to fuse. When cells of opposite mating types that are activated in this fashion collide with each other (yeast are immotile, and, thus, they depend on random collisions to come into contact), they adhere and undergo a change of shape related to the breakdown of cell wall material that is necessary for fusion. Finally, the cells fuse, and then their two nuclei fuse to form a diploid zygote.

The subsequent fate of the zygote depends on the environmental conditions. If nutrients are plentiful, it begins vegetative replication and multiplies as a diploid so long as nutrients are plentiful. If nutrients are scarce, the zygote immediately goes through meiosis to produce four haploid spores. This presumably helps survival, as it is more likely that one of the four small haploid cells will survive starvation than that the single large diploid cell will.

12.28 Yeast cells switch their mating type frequently

Yeast cells, and probably many other unicellular microbes with single-gene mating type determination, can switch from one mating type to the other. This switching is frequent in yeast—every division or two a cell switches mating types. This insures that a population founded by a single cell (which would, of course, without switching all be the same mating type) will rather consist of a mixture of a and **α** types and can engage in frequent genetic recombination.

Most yeast divide by a process of **budding,** in which a localized outgrowth on the surface of the **mother cell** enlarges until it is the size of a small cell. Mitosis occurs, and one of the daughter nuclei migrates into the bud. The bud then separates from the mother cell, becoming an independent **daughter cell.** Mating type switching occurs only in mother cells in G_1 immediately after division (Figure 12.19). Thus, after each division, the daughter cell remains the same mating type as its mother was, and the mother cell switches. Of course, as the daughter cell grows, buds, and divides, it becomes a mother cell itself and begins the process of switching back and forth after each division.

Thus, any haploid yeast culture consists of approximately equal numbers of a and α cells. The culture will continue multiplying asexually until it becomes dense enough for a-factor and α-factor to reach high concentrations and for cells to collide frequently. At this point mating begins, and the culture converts to the diploid form.

12.29 Mating type switching is a form of gene conversion

Mating type switching occurs by a process of **gene conversion**—in which the sequence of one region of the chromosome is converted to that of another. One of the yeast chromosomes has the genes for both a and α; these genes are transcriptionally inactive. On the same chromosome is the MAT locus, which contains a copy of either the a or the α sequence—whichever allele is located here determines the mating type. Mating type conversion begins when a specific endonuclease binds to MAT and makes a double-stranded cut at one end of the a or α sequence (Figure 12.20). The mating type gene is then degraded and replaced by copying the information from the silent copy of the opposite mating type. This is sometimes called the **cassette model** of gene regulation, and the copy of the mating type gene that is copied into the MAT locus is called a cassette.

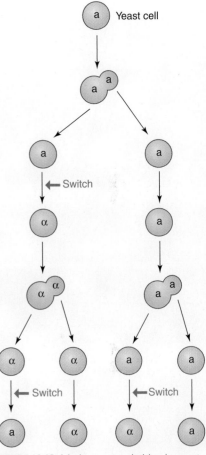

FIGURE 12.19 Mating type switching in yeast.

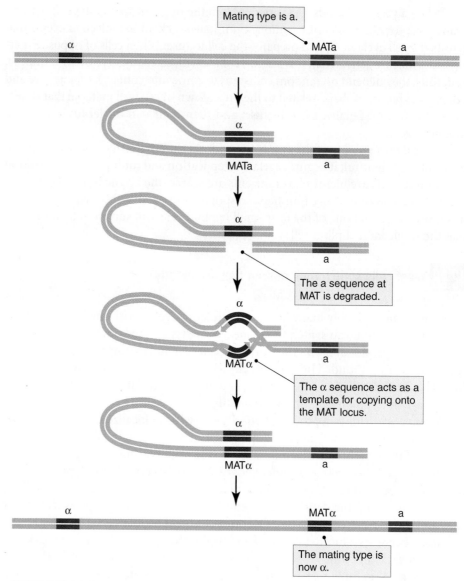

FIGURE 12.20 Mating type conversion in yeast.

The following labels appear in the figure:

Mating type is a.

The a sequence at MAT is degraded.

The α sequence acts as a template for copying onto the MAT locus.

The mating type is now α.

Only mother cells switch mating types because only mother cells have the specific endonuclease that initiates the process. This is because daughter cells lack a transcription factor that is essential for the synthesis of the endonuclease, because they make a repressor that represses synthesis of the transcription factor. Only daughter cells make this repressor because during budding, all of the mRNA that encodes the repressor is actively transferred into the bud; thus, at the time of division, the daughter cell is making repressor, but the mother cell is not.

Summary

Microbial cells reproduce asexually, and, thus, genetic recombination is separated from reproduction; however, most or all microbes do occasionally exchange genetic information. Among the protists, cell fusion and meiosis are involved, as they are in macro-organisms in which recombination and reproduction are combined. The major difference is that the cells that fuse are not normally specialized gametes, but vegetative cells that have responded to hormonal signals to behave as gametes. Mating in protists is rare or occasional and is not required for organismic multiplication.

In procaryotes, there are no mechanisms comparable to meiosis or gamete fusion, and when genetic exchange does occur, it nearly always results in the transfer of a fragment of the chromosome from a donor cell to a recipient cell to form a partial diploid. Recombination between the two molecules can produce genetic recombinants, and the haploid state is regenerated by degradation of the DNA fragment. Of the three known mechanisms of genetic exchange of chromosomal material, only transformation seems to have been selected as a mechanism of generating recombinants. Conjugation is fundamentally a mechanism of plasmid transfer, not chromosomal gene exchange, and transduction is simply the consequence of occasional, but inevitable, accidents during phage multiplication.

Despite the low frequency of genetic exchange among procaryotes, genomic sequences make it clear that it has been a major force in their evolution. Every procaryotic genomic sequence to date shows numerous regions that appear to have originated from a different source than the rest of the genome, by genetic exchange and recombination. Understanding that genetic exchange among procaryotes is frequent is not only an academically interesting insight, but has practical implications. For instance, it should be taken into consideration when considering the release of genetically engineered bacteria into the environment.

Study questions

1. Compare and contrast genetic recombination and reproduction in procaryotes, eucaryotic microbes, and eucaryotic multicellular organisms.

2. Why are even numbers of crossovers necessary in procaryotic genetic recombination but not in eucaryotes?

3. What are the three mechanisms of genetic exchange in procaryotes, and which of them appear to have been selected for their capability to catalyze such exchange?

4. Compare and contrast transformation of the gram-negative and gram-positive types.

5. What is the evolutionary reason that transferrable plasmids typically have their transfer genes repressed most of the time? Why is F transfer derepressed?

6. Some phages can perform both specialized and generalized transduction. Describe how that might be possible.

7. Imagine three genes, X, Y, and Z. You do a series of transductional crosses among mutant strains that are X⁻, Y⁻, or Z⁻ and find the following results: X-Y are cotransducible about 5% of the time, X-Z about 45% of the time, and Y-Z about 57% of the time. What is the gene order?

8. Another set of genes, A, B, and C, are all cotransducible with each other at high frequencies, and you are not confident that you can determine the gene order from your two-point crosses. So you cross an A⁻ B⁻ C⁺ recipient with a donor that is A⁺ B⁺ and C⁻. When you select for A⁺, you obtain transductants that inherit the donor alleles for B and C 29% of the time, donor allele for B and recipient allele for C 20% of the time, recipient allele for B and donor allele for C 32% of the time, and recipient alleles for B and C 19% of the time. What is the gene order?

9. You repeat the above crosses, but this time select for B⁺. What results do you expect?

10. To be certain, you would also like to repeat the cross selecting for C⁺. What would you have to do in order to do this cross?

13

Microbial Systematics

A S WE SAW IN Chapter 1, procaryotic microbes have been evolving continuously for 3.8 billion years or more. Eucaryotic microbes, although not present in the fossil record until much later, are thought from chemical evidence and molecular sequence evidence to be nearly as ancient. In contrast, multicellular organisms are much more recent; fungi and algae, with their poorly differentiated tissues, are perhaps 1.5 to 2.0 billion years old, and plants and animals half that. Thus, the amount of evolutionary diversification in the microbes surpasses that in the multicellular organisms by a substantial amount. In this chapter, we discuss the techniques of determining the evolutionary histories of the microbes.

13.1 Classification aims to group-related organisms

The basis of our current system of classifying and naming organisms was laid in the mid 18th century by the Swedish naturalist Karl Linné, who published his work in Latin under the name Carolus Linnaeus. Linnaeus's system had three important features: (1) It was **hierarchical;** that is, it established a hierarchy of categories, each more general and inclusive than the one below it. Thus, very similar organisms were grouped together into the same species. Similar species were grouped together into the same genus. Similar genera were grouped together into the same family, etc. (2) Organisms were assigned to categories based on a small number of **diagnostic characters;** for instance, Linnaeus defined the major groups of animals on the basis of the number of chambers in their hearts, whether they were warm or cold blooded and whether they bore live young or laid eggs. By judicious choice of these diagnostic characters, he found that the categories seemed to correlate with many other characters. (3) Individual organisms were given **binomial** names—composed of the genus and species names. These three features were revolutionary at the time and still characterize organismic classification.

The publication of *The Origin of Species* by Charles Darwin in 1859 provided an explanation for why groups of organisms share large numbers of features—because they are *related* by virtue of evolutionary descent from a common ancestor. They *resemble* each other for the same reason that members of a family resemble each other. Thus, the goal of biological classification became to group related organisms together, and the successively smaller categories grouped together organisms with increasingly close relationship.

13.2 Convergence can confuse the determination of relationships

Unfortunately, not all resemblances are due to common descent; sometimes unrelated organisms converge on the same properties by chance or in response to similar evolutionary pressures. For instance, birds and bats both have wings, but this is not because they are descended from the same winged ancestor. Birds are descended from an unwinged reptile and bats from an unwinged mammal. Such convergences are the bane of **taxonomists** (people who classify things). In the case of birds and bats, it is not much of an issue; birds and bats differ sufficiently in other characteristics that there is no confusion about their true relationships. In the microbial world, however, it is quite another matter. Microbial morphologies, especially for procaryotes, are much simpler and are much more likely to be the result of convergence. Physiological traits, for instance whether sucrose is used as a carbon and energy source or whether an organism uses nitrate respiration, are also highly subject to convergence. Thus, unlike the classification of plants and animals, with their complex morphologies and reproductive strategies, the morphologically simple, asexually reproducing microbes defied evolutionary classification for more than a century.

13.3 Molecular sequences avoid convergence and allow evolutionary classification of microbes

Over long periods of time, every population of organism gradually accumulates mutations in its DNA. Some of these have phenotypic effects and are either eliminated or spread through the population by natural selection. Others, called **neutral mutations,** have no effect on fitness. Each one of these neutral mutations will eventually either be lost entirely or become universal within the population by pure chance. Thus, over time, each population of organism accumulates mutations that differentiate it from other populations—some neutral, some selected.

Selection is, of course, the process that drives both the phenotypic divergence of populations and their convergence; however, neutral mutations should in principal never result in convergence, and it is generally thought that most mutations are neutral. This is because many mutations in protein coding genes do not change the amino acid encoded because of the redundancy of the code; others change the amino acid, but the change does not affect the protein's functioning. Others occur in stretches of DNA that do not encode a product. Thus, the sequence of DNA of related organisms should be a pretty good measure of relationship, unobscured by convergence.

The theoretical importance of molecular sequences as measures of relationship was first recognized in the 1950s by Linus Pauling and Emile Zukerkandle, who coined the phrase **molecular chronometer** to describe them. At that time, it was impossible to sequence nucleic acids, and, thus, they relied on protein sequences. Mutations that change the sequences of amino acids in a protein are less likely to be neutral than the average mutation. Even so, many changes in amino acid sequence appear to be selectively neutral, and thus, protein sequences are generally good indicators of relationship, free of convergence. When methods of sequencing gene-sized pieces of DNA became available, however, they supplanted protein sequences as the method of choice for determining relationships.

13.4 The small subunit ribosomal RNA gene is the most commonly used sequence for phylogeny reconstruction

A good molecular chronometer has several properties. First, it has to be universal—that is, every organism must possess the sequence. Otherwise, some organisms would be left out of the resulting evolutionary tree, or **phylogeny.** Second, it has to change slowly enough that there is measurable relationship between the most distantly related organism. Third, it must change rapidly enough that it differs even in closely related organisms. Fourth, it must be large enough to show enough differences to be statistically significant.

These requirements limit the genes that can usefully be used for phylogeny reconstruction. Most genes encoding proteins are fairly small, and many of them change quite rapidly, but the genes that encode the large rRNA molecules are perfect. The small subunit rRNA is about 1,500 nucleotides in length, and the large subunit rRNA is about 3,000 nucleotides (Table 13.1). Both are large enough to be reliable, yet short enough to be easily sequenced. For historical reasons, the most often used is the small subunit rRNA (when sequencing techniques were primitive, a 1,500 nucleotide molecule was much easier to sequence than a 3,000 nucleotide one.

Because the rRNA is part of a very complicated structure, forming a complex with many different proteins, it has changed quite slowly over evolutionary time. Even in the most distantly related organisms, approximately 60% of the positions in the sequence have the same nucleotide, far above the 25% that would be expected from chance alone. Thus, this sequence is very good for determining distant relationships.

Despite the overall conservative rate of change, some regions change more rapidly, allowing closer relationships to also be measured. For very close relationships (e.g., within the same species), however, ribosomal RNA sequences are usually too similar (even identical), and other sequences or other methods have to be used to measure relationships. We discuss one of these (DNA–DNA hybridization) later in this chapter.

13.5 Extensive secondary structure in rRNA helps with sequence alignment

All ribosomal RNAs have extensive secondary structure—internal regions of self-complementarity that allow the molecule to form double-stranded regions, or stem-loop structures (Figure 13.1). This secondary structure appears to be the most strongly conserved feature of the molecule; even when the sequences are significantly different, the secondary structure is substantially the same. Presumably

Table 13.1   Ribosome Composition (Exact Numbers Can Vary Significantly Among Different Organisms)

	Procaryotes	Eucaryotic Cytoplasm*
Small subunit	1 rRNA: 1,500 nucleotides 20 proteins	1 rRNA: 2,000 nucleotides 35 proteins
Large subunit	2 rRNAs: 120 and 3,000 nucleotides 35 proteins	3 rRNAs: 120, 160, and 5,000 nucleotides 45 proteins

*Most eucaryotes contain mitochondria, and many also contain chloroplasts. These organelles contain ribosomes that are highly variable in their composition, but usually resemble procaryotic ribosomes more than eucaryotic cytoplasmic ribosomes.

CHAPTER 13 MICROBIAL SYSTEMATICS

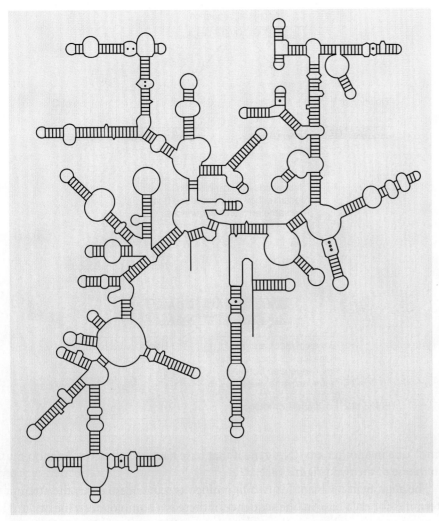

FIGURE 13.1 Small subunit rRNA secondary structure.

the various stem-loops interact with the various ribosomal proteins and thus must be conserved. This allows a much more accurate **alignment** of sequences from different organisms.

Alignment is the first step in a sequence comparison. In this activity, the two sequences are lined up next to each other to match each base in one with a base in the other. Because the total lengths can be different, due to deletion or insertion mutations, this is not an easy matter. At each position, one has to ask whether there is a corresponding base in the other sequence or whether a gap should be put there. Obviously, each gap slides the sequence alignment of all subsequent bases. Also, because a gap can be placed at any position in either sequence, the number of possible alignments is nearly infinite for a sequence the size of an rRNA. This presents a very difficult problem of determining the most accurate alignment.

Another way of viewing the alignment problem is that it is attempting to establish **positional homology** between two sequences. Positional homology refers to the concept that a base in a sequence is homologous with a base in another sequence if both are descended from the same base in the last common ancestor of the two sequences. Thus, not only can the sequences as a whole be considered homologous (the ssu RNA gene of all organisms are descended from an ancestral ssu RNA gene and hence are homologous), but each base in an ssu RNA gene will have a

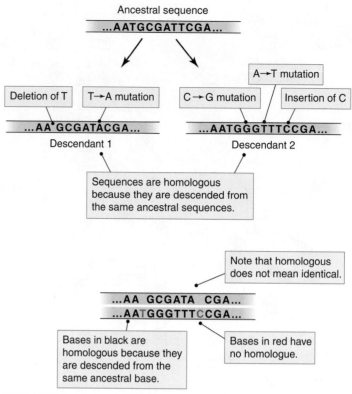

FIGURE 13.2 Positional homology.

homologue in another ssu RNA gene if that base has not suffered a deletion mutation in one of the two (Figure 13.2).

The alignment problem for two homologous sequences is thus the attempt to determine, for each base in one sequence, if there is a homologue in the other. This can be quite difficult because there are only four different bases in a DNA sequence and thus there is 25% chance that the bases will match even if the alignment is incorrect. This is where the secondary structure is helpful. Because we know the secondary structure of a newly determined rRNA sequence is probably the same as for previously determined sequences, we can find the internally complementary regions of the new sequence and match those regions with the internally complementary regions of the sequence we are trying to match with (Figure 13.3).

13.6 After homologous sequences are aligned, phylogenetic trees are reconstructed by computers

After many sequences have been aligned, the most likely phylogeny has to be inferred from them. The basic principle of all computer methods is to place the most similar sequences closest together on the tree and the most dissimilar sequences far apart; however, it is much easier to say that than to actually write a computer program that does it. Random fluctuations in mutation rates and different mutation rates in different organisms can make computer reconstruction very difficult. Currently, there is no single computer program generally agreed to be better than others, and different scientists use different programs. Thus, conclusions sometimes differ. The most careful scientists routinely use several different methods and compare the results, and they use statistical methods to evaluate how well the results

CHAPTER 13 MICROBIAL SYSTEMATICS

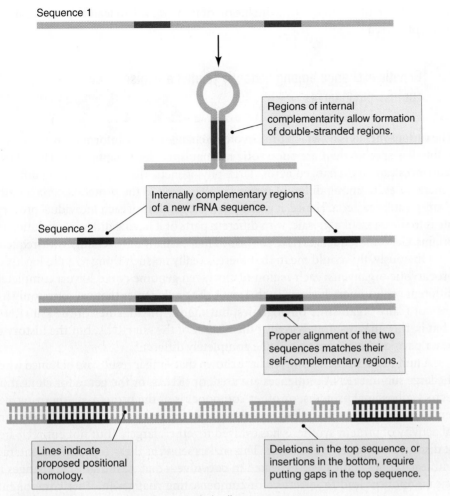

Sequence 1

Regions of internal complementarity allow formation of double-stranded regions.

Internally complementary regions of a new rRNA sequence.

Sequence 2

Proper alignment of the two sequences matches their self-complementary regions.

Lines indicate proposed positional homology.

Deletions in the top sequence, or insertions in the bottom, require putting gaps in the top sequence.

FIGURE 13.3 Using secondary structure to help align sequences.

match the data. With appropriate caution, phylogeny reconstruction can be quite reproducible.

There are three different classes of phylogeny reconstruction programs, each type generally having several different versions. **Distance methods** calculate an **evolutionary distance** between each pair of sequences (basically the number of differences between the two) and then try to construct a tree in which all the distances add up right. **Parsimony methods** take all of the possible trees and calculate for each how many mutations would be required to explain the evolution of the existing sequences. The best tree is considered to be the one requiring the fewest mutations. **Likelihood methods** start with a model that proposes certain frequencies for each of the different types of mutation (A to C, A to T, A to G, etc.; deletions; insertions). They then take all possible trees and evaluate the statistical fit of each tree to the model.

Often, shortcuts are required. For instance, in parsimony and likelihood models, the goal is to evaluate each possible tree; however, the number of possible trees becomes so large that when there are more than about 10 sequences to be placed on a tree, even supercomputers are unable to handle the computational demands. Thus, often only a subset of the possible trees is evaluated.

Hence for both theoretical and practical reasons, phylogenetic trees should be viewed as models, not established fact. Each phylogenetic tree is a hypothesis that fits the existing data best, given the method by which it has been constructed. The

After homologous sequences are aligned, phylogenetic trees are reconstructed by computers use of a different method, or the inclusion of more data, can lead to a different tree being preferred.

13.7 Genetic exchange among distantly related organisms can confuse phylogenies

The various methods by which procaryotes exchange genetic information are considerably less specific than are eucaryotic mechanisms (see Chapter 12). Thus, DNA sequences can be exchanged across large phylogenetic distances—among different genera, or even among different major lineages, such as the proteobacteria and the gram-positive bacteria. Hence, it is likely that the genome of each individual procaryote is to some extent a mosaic, with different parts of it having different evolutionary origins. Genomic sequences have confirmed that such distant exchanges are frequent.

Obviously, this could mean that there is really no such thing as a phylogeny of procaryotic organisms; each region of any given genome could have a completely different evolutionary history. In this view, the phylogenies that we commonly use are not really organismic phylogenies, but merely phylogenies of the ssu rRNA. That is, they may accurately reflect the history of the ssu rRNA, but the history of other parts of the genome would be completely different.

A number of studies, however, have shown that similar results are obtained when the large subunit rRNA sequences are used, or tRNAs, or the genes for elongation factors, ribosomal proteins, or other components of the protein-synthesizing system. Thus, it appears that a large portion of the genome devoted to the fundamental activity of protein synthesis has evolved together, largely (but not entirely) free of disruption by genetic exchange. This makes sense, as these components interact with each other and can be expected to **coevolve**—that is, mutational changes in one component lead to selection for compensating changes in other components with which the first interacts. When such coevolution is extensive, as it might well be with such a complex interactive system as the protein synthetic machinery, genetic exchange commonly reduces fitness, and thus, it is unlikely that the components have different evolutionary origins.

Thus, it seems likely that the procaryotic genome consists of a large fraction of co-evolved genes involved in protein synthesis (possibly including genes for DNA replication and transcription as well), combined with sequences from a variety of other sources that have been accumulated over the eons by genetic exchange. Thus, the phylogeny we show here probably represents the evolutionary history of the most stable, and largest, part of the genome.

Eventually, complete genome sequences can be expected to give accurate, composite phylogenies, showing the different evolutionary histories of each of the different regions of the chromosome. Currently, however, the software needed to do such an analysis is too primitive, but **bioinformatics** (the science of managing and interpreting large biological databases) is one of the most rapidly growing aspects of genomics, and we can expect to begin basing phylogenies on complete genomes fairly soon.

It is encouraging that a very crude beginning of genomic phylogenies shows the same results as ssu rRNA phylogenies. This approach simply analyzed the numbers of homologous genes in each pair of more than a dozen different procaryotes (and one eucaryote) and used this index of shared genes as the basis for constructing a tree. The results were almost identical to results obtained by traditional methods with ssu rRNA sequences. Thus, it may be that gene transfer has not been as much of an influence as some have thought. Final resolution of this question will, however, have to await the development of adequate bioinformatic tools.

13.8 ssu rRNA phylogenies show three major lines of descent, or "domains"

The difficulties we have outlined here require us to be cautious in accepting too uncritically any phylogeny; however, the voluminous evidence from molecular sequences indicates that it is most likely that life on earth consists of three major lineages, or **domains,** which have been named the Bacteria, the Archaea, and the Eucarya (Figure 13.4). The Bacteria consist of organisms with procaryotic cell structure and membranes based on fattyacyl glycerol diesters. The Archaea consist of organisms with procaryotic cell structure and membranes based on phytanyl glycerol diethers. The Eucarya consist of organisms with eucaryotic cell structure.

Each line segment on this tree is proportional to the amount of sequence change that occurred. This is not the same as the amount of time that elapsed, as the rate of mutation may have varied. In particular, we think that very early rates of mutation were probably considerably higher than they became later. Thus, the very early divergences probably happened considerably more quickly than would appear from the tree.

This is born out by dating of certain spots on the tree. This can be done for morphologically distinctive fossils by dating the rocks in which they occur. In other cases, chemical analysis of fossil-containing rocks shows chemical compounds that are diagnostic of particular organisms. This is possible because within a rock matrix organic compounds may be extraordinarily stable, particularly complex polycyclic compounds, like steroids, hopanoids, and tetrapyrroles. Thus, the identification in fossil-containing rock of compounds known today to be distinctive of a particular group of organisms should give us a rough date for the latest possible appearance of that group, or one of its extinct relatives. Several such dates are shown on the tree.

Perhaps the most anomalous feature of this tree is the age of the eucaryotic lineage. Eucaryotic fossils first appear less than 1.5 billion years ago, yet the sequence evidence and the chemical evidence indicate that their lineage diverged over 2.7 billion years ago. There are several possibilities to explain this discrepancy. First, and simplest, it may be that we just have not looked in the right places for eucaryotic fossils. The search for microfossils has not been very systematic or thorough yet, and it is likely that as more old rocks are examined, the age for morphologically identifiable eucaryotic fossils will be pushed back considerably. It is also possible that the ancestors of the modern eucaryotes had a cell structure that was not recognizably different from the procaryotes, at least in fossil form.

13.9 Each of the three domains contains multiple major lineages, or "kingdoms"

The bacteria and the eucarya each contain more than a dozen major sublineages, and the Archaea are currently recognized to contain three (undoubtedly more will be discovered). There has been a continuing problem of what to call these. This is sometimes called the "kingdom problem" because several of these lineages have traditionally been given the rank of kingdom. The kingdom was established by Linnaeus himself as the highest taxonomic rank, and Linnaeus recognized three kingdoms in the natural world: animals, plants, and minerals. After Linnaeus, the mineral kingdom was dropped, and for about a hundred years, two kingdoms

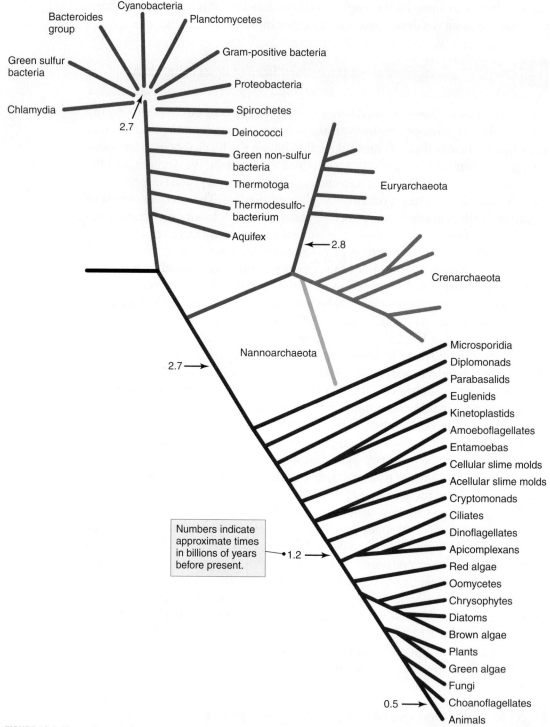

FIGURE 13.4 The universal phylogenetic tree.

were recognized. With increasing knowledge of the microbial world, however, it became apparent that dividing the microbes between the plants and animals was unsatisfactory, and in the late 1800s, Ernst Haeckel proposed a third kingdom, the Protista, for the microbes. This satisfactorily dealt with the problem for a few decades, but eventually, it too became insufficient, and two additional kingdoms were added, splitting out the Fungi and the Monera (procaryotes) from the Protista.

This five-kingdom scheme was widely accepted until the 1990s, when rRNA phylogenies, pioneered by Carl Woese, for the first time gave an objective basis on which to understand microbial phylogenies. Woese and his collaborators proposed two major revisions in the classical taxonomic scheme: the addition of the domain category above the kingdom as the highest taxonomic rank and the use of the term kingdom for each of the major sublineages within domains. These reforms have been widely accepted by microbiologists (although many prefer to use the term *phylum* rather than *kingdom* for the various procaryotic lineages) but are still controversial among zoologists and botanists. In this book, as in all other contemporary microbiology texts, we are following the Woese scheme.

13.10 Close relationships are measured by DNA/DNA hybridization

Some regions of the rRNA genes accumulate mutations rapidly enough that they can be used to measure relationships between closely related organisms—closely related genera, for instance; however, relationships closer than this are often too close for the method to reliably quantitate. Whether two organisms should be considered members of the same species or of different but closely related species is often not determinable by rRNA sequences. For these closely related organisms, **DNA/DNA hybridization** (or **DNA reannealing**) is the method of choice.

This method takes advantage of the double-stranded nature of DNA. In a nutshell, DNA from two different organisms is mixed and heated to separate the two strands and then cooled and allowed to reform the double helices (Figure 13.5). If there is substantial sequence similarity between the two different DNAs, then heterologous pairs will be formed in addition to homologous ones.

Values for DNA/DNA hybridization experiments are generally given as percentages. The value observed when the unlabeled DNA comes from the same organism is defined as 100%, and values for hybridization when the unlabeled DNA is from different organisms are expected to fall below this. Thus, this method gives a quantitative estimate of the genome-wide similarity of base sequence.

Although very simple in principle, these experiments are quite tricky in practice. The unlabeled DNA has to be present in great excess, and the radioactive DNA has to be present in quite small amounts. This ensures that the single strands of radioactive DNA will not reanneal with other radioactive strands (because their concentration is too low for them to find their complements within the time of the experiment). The high concentration of unlabeled DNA allows the labeled fragments to find unlabeled complementary strands to hybridize with, if any are present.

The choice of incubation temperature is critical. Typically it is 10 to 30 degrees C below the melting temperature (the temperature at which the strands separate). The high end of this range represents quite stringent conditions. At these temperatures, only duplexes with a nearly perfect match (i.e., almost perfectly complementary) can remain double stranded. At the other end of the range, the conditions are more permissive, and duplexes with a considerable number of mismatched base pairs can be stable.

Even at permissive temperatures, however, nearly 90% of the base pairs need to be matched to allow measurable duplex formation. Two strains in the same species can be expected to give a high level of hybridization (typically over 70%). Two strains that are in the same genus but not the same species can be expected to give a lower level, but still significant, of hybridization (typically 30% to 60%). Strains in different genera usually fail to show any hybridization at all.

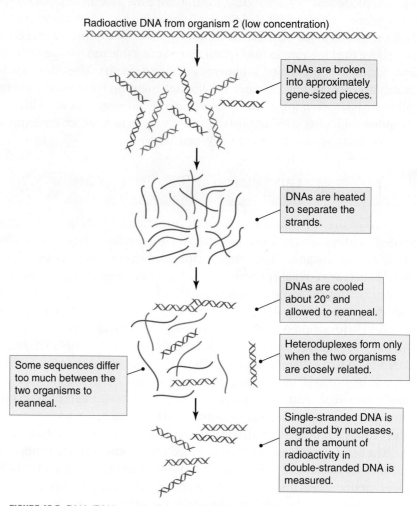

DNA from organism 1 (high concentration)

Radioactive DNA from organism 2 (low concentration)

DNAs are broken into approximately gene-sized pieces.

DNAs are heated to separate the strands.

DNAs are cooled about 20° and allowed to reanneal.

Heteroduplexes form only when the two organisms are closely related.

Some sequences differ too much between the two organisms to reanneal.

Single-stranded DNA is degraded by nucleases, and the amount of radioactivity in double-stranded DNA is measured.

FIGURE 13.5 DNA/DNA reannealing experiment.

13.11 The concept of a procaryotic species is fundamentally different from the traditional species concept

Traditionally, the concept of a species has been based on the capacity to interbreed. That is, organisms that can interbreed with each other and produce fertile offspring are considered to be members of the same species. Organisms that cannot interbreed, or whose interbreeding produces sterile offspring (such as donkeys and horses), are considered to be different species. This is a neat and simple definition and has the advantage of not only defining species but also of specifying a simple test for membership in the same species. Like everything in biology, however, the reality is much more complicated. Biologists have for a long time recognized that there are species that are capable of interbreeding to produce fertile offspring but that normally do not—sometimes because they live in different geographical areas or because they have different breeding times or because their habitats are different even though within a common range. In some cases, interbreeding is common where the ranges overlap. No one wants to lump such species together because in all other respects they seem perfectly distinct. Thus, biologists have always been pragmatic about the concept of a species; the philosophical species concept represents their ideal, but they are willing to put up with numerous exceptions where appropriate.

Even this imperfect scheme does not work for procaryotes, however, for the simple reason that their multiplication is entirely asexual. What mechanisms of genetic exchange they have are remarkably nonspecific, and "interbreeding" between different genera, and even kingdoms, occurs. Thus, the traditional species concept cannot be used.

As a practical matter, procaryotic species are usually defined as a group of similar related strains that differ more from other strains than they do from each other. In other words, the boundaries of species are defined by discontinuities in the degree of relatedness or in the degree of similarity. This has the practical consequence that some procaryotic species are quite homogeneous, with all strains being highly similar and markedly different from all others, whereas other species are very large and heterogeneous.

Another approach has been to try to define species and genera on the basis of the quantitative measure of DNA/DNA hybridization as described previously. This approach works well in some groups of organisms, where the levels of DNA/DNA hybridization correlate closely with phenotypic discontinuities and where hybridization values tend to fall into defined clusters: well above 70%, 30% to 50%, and zero. In other groups, however, there are too many organisms on the borderlines for this approach to be satisfying.

Currently, it is fair to say that there is no single concept of a procaryotic species, either in theory or in practice. The only thing that is clear is that a procaryotic species is very different from an animal or plant species. For this reason, comparisons of the numbers of species are misleading. Currently, there are about 250,000 described species of plant, over a million of animals, 150,000 of eucaryotic microbes, and only 5,000 of procaryotes (Bacteria and Archaea combined). This makes it appear that procaryotes have a very limited range of biodiversity; however, because each procaryotic species can contain organisms that are genetically as different as the members of different *families* of plant or animal, there is considerably more evolutionary and genetic diversity among procaryotes than among the macroscopic organisms on earth. The small number of procaryotic species compared with eucaryotes is an artifact of the differing ways of defining species.

13.12 Most procaryotes have never been cultured

Microbiologists have traditionally isolated new procaryotic species into pure culture, studied their phenotype, described them, and named them. However, this process, on which our understanding of procaryotic diversity rests, works only for those microbes whose nutritional and physical needs can be satisfied in the laboratory. Many lines of evidence suggest that many more microbes exist than have ever been cultivated. Many of these "uncultivatable" microbes are probably cultivatable with sufficient effort to identify their needs. Others may have requirements that cannot be provided to pure cultures.

The most dramatic demonstration of the hidden diversity of procaryotes has been provided by genomics. The effort to sequence the very large human genome led to the development of automated, high-speed sequencing technology and very sophisticated bioinformatic tools that allowed the assembly of the sequences of each of the 23 chromosomes of the human haploid genome from the many overlapping sequences of individual cloned regions. These tools have continued to be improved and refined, and they are now applied not just to pure cultures, but also to the entire assemblage of organisms in specific habitats. There is nothing different in principle or practice between assembling sequence information into the many different chromosomes of a single organism and assembling sequence information into the many different chromosomes of a mixture of organisms, an approach termed **metagenomics.** There are many pitfalls and difficulties in this approach, and to date, it has been most successful in habitats in which the species diversity has been low. Advances are rapid, however, and metagenomics will soon give us complete genomic sequences for many uncultivatable microbes. In turn, after those genomic sequences are in hand, an inventory of the genes present can often give us a good understanding of the physiological activities of the organism and its role in the ecosystem.

Summary

A natural, or phylogenetic, classification became the goal of taxonomy as soon as *The Origin of Species* was published, and it became clear that resemblances among organisms were evidence of relationships. However, procaryotes, with their simple morphologies, resisted evolutionary classification until the development of molecular tools that could measure the similarity of molecular sequences—proteins or nucleic acids. Since that time, molecular taxonomy has provided us with a comprehensive model for procaryotic evolution and relationships, as well as helping to sort out many taxonomic uncertainties regarding plants and animals. As metagenomic techniques increasingly reveal the full range of procaryotic diversity, we will undoubtedly be forced to significantly revise our understanding, but it seems clear that we are converging on a reliable systematic understanding of the vast diversity of the procaryotes.

Study questions

1. Why is convergence not a problem when using nucleic acid sequences?
2. What are the properties of a good molecular chronometer? Why are the rRNA genes good candidates?
3. How do the notions of homology of genes and homology of individual base pairs relate to each other?
4. How does the concept of a procaryotic species differ from that of a plant or animal species?

Procaryotic Microbes

I N THIS CHAPTER, we briefly survey the various groups of procaryotic microbes: the Bacteria and the Archaea. This is a huge and very diverse range of living organisms. We necessarily treat each group quite superficially.

Bacterial Diversity

The Bacteria are a large and evolutionarily diverse group. The earliest recognizable microfossils are of apparent cyanobacteria, and they are nearly three billion years old. Because the cyanobacterial lineage is not one of the earlier branching ones, the Bacteria must have been quite diverse already then (Figure 14.1). This immense time over which the group has been diversifying has led most of the constituent lineages to be physiologically, morphologically, and genetically diverse. Other lineages, on the other hand, show little variation, despite their antiquity, for reasons that we do not understand. One possibility, of course, is that these lineages are actually quite diverse, but that most representatives have yet to be cultivated. In such cases, we can expect metagenomics to reveal the true diversity.

Although most microbiologists consider each of the major lineages of the Bacteria to deserve kingdom rank, formal kingdom names have not been proposed. Thus, each of these lineages has an informal name, generally referring to one of the genera within the lineage or to one of the properties of the group. In this section, we briefly describe the properties of the major lineages.

14.1 Hyperthermophilic bacterial kingdoms

The most deeply branching lineages of the Bacteria contain a single genus each, although it is likely that others will be discovered. All three genera are hyperthermophilic, with optimum temperatures of growth in the 70°C to 85°C range. They have been isolated from marine and freshwater hydrothermal sites.

Aquifex is a rod-shaped hyperthermophilic aerobic chemoautotroph. It oxidizes elemental sulfur or H_2 with either oxygen or nitrate as terminal electron acceptor. Although aerobic, it is very sensitive to oxygen concentration and prefers microaerobic conditions. It is in fact quite unusual for hyperthermophiles to be aerobic at all, probably largely due to the very low solubility of gases in very hot water. The reverse TCA cycle is used for CO_2 fixation.

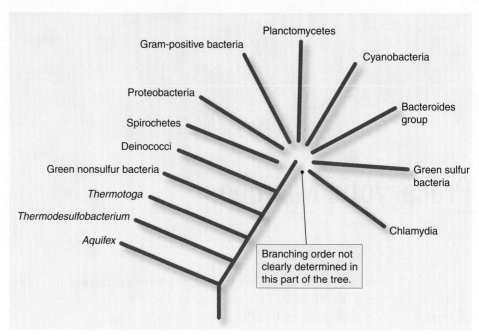

FIGURE 14.1 Phylogenetic tree of the Bacteria.

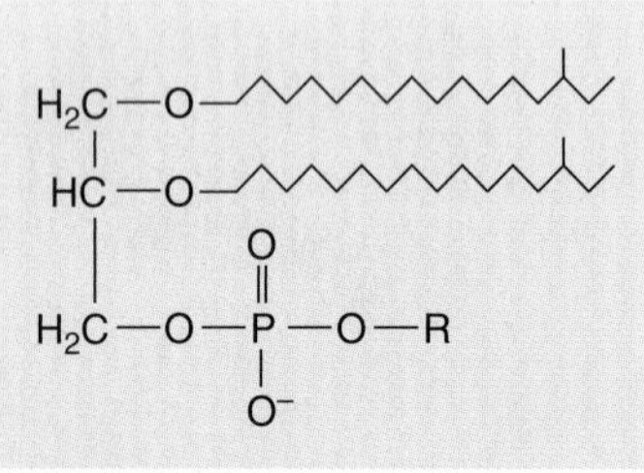

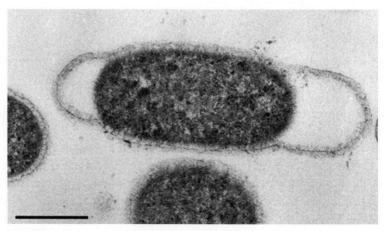

FIGURE 14.2 Thermodesulfo-bacterium lipids.

Thermodesulfobacterium is a rod-shaped anaerobic chemoheterotroph that uses sulfate as terminal electron acceptor. Compounds such as lactate, pyruvate, or ethanol are used as electron donors, but as is the case in many other sulfate reducers, they are only oxidized to acetate.

Thermodesulfobacterium has very unusual lipids in which two fatty acyl alcohols are attached to glycerol via ether linkages (Figure 14.2). Their membrane lipids, thus, have features of archaeal lipids (the ether linkages) and of bacterial lipids (the fatty acyl side chains).

Thermotoga is a fermentative chemoheterotroph that ferments polysaccharides and sugars to lactate, acetate, CO_2, and H_2. It has an unusual wall with a unique loose external layer, or **toga** (Figure 14.3).

FIGURE 14.3 Thin section of *Thermotoga*. (Bar = 0.5μm.)

CHAPTER 14 PROCARYOTIC MICROBES

14.2 Green nonsulfur bacteria

The green nonsulfur bacteria (Figure 14.4) consist mainly of filamentous, anaerobic, anoxygenic phototrophs. They are motile by gliding and have a single type of photosystem (PS II). Their accessory pigments are housed in membrane-bound sacs (termed chlorosomes) pressed against the inside of the cell membrane, which houses the reaction centers (see Figure 14.19). They are thermophilic (but generally not hyperthermophilic) and are usually found as components of dense mats of microbial biomass in hot springs, where they use their gliding motility to glide along the cells of other microbes in the mat to find the optimal conditions of light, anaerobiosis, and nutrients.

The green nonsulfur bacteria grow best as photoheterotrophs, but most can also grow as photoautotrophs, using H_2 or reduced sulfur compounds as electron donor, and the hydroxypropionate cycle to fix CO_2.

The group contains at least one nonphototroph, the chemoheterotroph *Thermomicrobium*. The most interesting feature of this member of the group is its membrane lipids, which consist principally of fatty diols, not linked to glycerol (Figure 14.5).

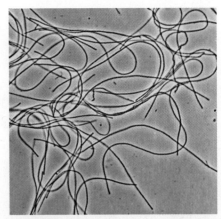

FIGURE 14.4 Light micrograph of *Chloroflexus*.

14.3 Deinococci

The deinococci contain several genera, the best studied of which is *Deinococcus*, known for its extreme radiation resistance, which results from extremely active DNA repair systems. Doses of ionizing radiation in the range of 3 million rad are tolerated—humans are killed by a 5 rad dose. *Deinococcus* is also resistant to ultraviolet (UV) radiation—more so even than bacterial endospores—and, thus, they are common as air contaminants (because airborne microbial cells are exposed to significant UV).

The deinococcal cell wall includes a gram-positive–like thick peptidoglycan layer, overlain by a gram-negative–like outer membrane (Figure 14.6). The peptidoglycan layer lacks teichoic acids, however, and the outer membrane lacks lipopolysaccharide; thus, the homologies of these layers to the classic bacterial gram-negative and gram-positive walls are not clear. The deinococci stain gram positive.

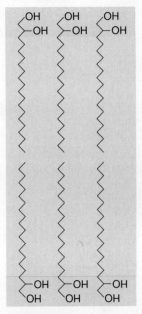

FIGURE 14.5 *Thermomicrobium* membrane.

14.4 Proteobacteria

The proteobacteria are the largest of the bacterial lineages and, as their name suggests, the most variable. Some of them were discussed in detail in Chapter 11—the myxobacteria and *Caulobacter*. Here we describe several other major subgroups.

The **enteric bacteria** are rod-shaped, peritrichously flagellated, facultatively anaerobic chemoheterotrophs (Figure 14.7). Most of them inhabit the mammalian intestinal tract, and their consumption of oxygen is principally responsible for maintaining anaerobic conditions in the intestinal tract. The group includes *Escherichia coli*, the best studied organism in the world, as well as a number of pathogenic organisms, such as *Salmonella* and *Shigella*, which cause intestinal disease, and *Yersinia pestis*, which causes plague. Closely related to this group is the **vibrio group**, which are physiologically similar to the enteric bacteria but which are generally free-living marine or estuarine organisms. They are generally curved rods, motile by polar or peritrichous flagella. One species, *Vibrio cholera*, causes the serious intestinal disease cholera.

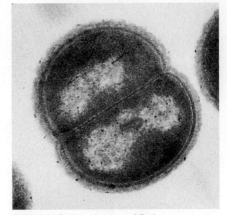

FIGURE 14.6 Thin section of *Deinococcus*.

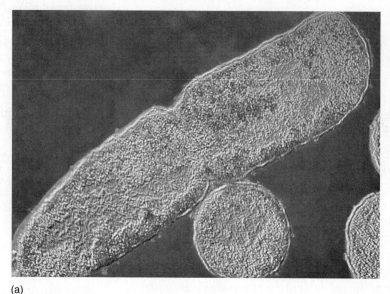
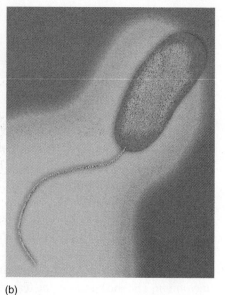

(a)   (b)

FIGURE 14.7 (a) *E. coli* and (b) *Vibrio.*

The **pseudomonads** are strictly respiratory, polarly flagellated rods (Figure 14.8). Most are strictly aerobic, but a few can also denitrify and thus grow anaerobically with nitrate as electron acceptor. They are probably responsible for much of the denitrification globally, and, thus, play an important role in the global ecology. There are about half a dozen genera that compose the pseudomonads, and each of the genera is more closely related to nonpseudomonads than to the other pseudomonads. The group is thus not a natural one. The pseudomonads are generally able to oxidize a wide range of organic compounds as sole source of carbon and energy, and, thus, play a major role in the soil ecosystem as agents of turnover of organic matter. Several plant pathogens are included, but only a few pseudomonads are capable of causing human or animal disease. One of these, *P. aeruginosa,* is an opportunistic pathogen, capable of causing disease in debilitated patients only. Nevertheless, it is a serious problem in hospitals, causing particularly nasty skin infections in burn patients. It is also a major cause of life-threatening respiratory infections in children with cystic fibrosis.

Most of the aerobic chemoautotrophs also are proteobacteria. The **nitrifying bacteria** (Figure 14.9) are mostly obligate chemoautotrophs that oxidize ammonia

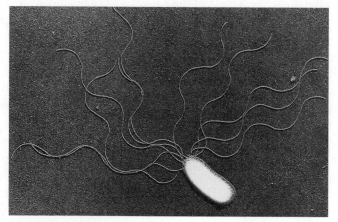

FIGURE 14.8 *Pseudomonas.*

or nitrate as their energy source. There are two subgroups: the ammonia oxidizers oxidize ammonia to nitrite, and the nitrate oxidizers oxidize nitrite to nitrate. They are, therefore, always found together; the ammonia oxidizers depend on the nitrite oxidizers to remove the toxic nitrite that they produce, and the nitrite oxidizers depend on the ammonia oxidizers for the nitrite that is their energy source. These bacteria are found in nearly all aerobic environments in which there is decomposition of organic matter, as the decomposition of proteins and nucleic acids releases abundant ammonia.

The **aerobic sulfur bacteria** are also proteobacteria. They are morphologically diverse, including gliding filamentous organisms and unicellular flagellated ones (Figure 14.10). They, too, are found wherever organic matter is decomposing aerobically, oxidizing the small amounts of H_2S released. They can also be found in great numbers at the oxic/anoxic interface of nutrient-rich sediments of estuaries and the continental shelf, using the abundant H_2S that is produced by the sulfate-reducing bacteria. When sulfide is abundant, they typically oxidize it only to elemental sulfur, which is deposited as globules in the periplasm of the filamentous forms, or extracellularly in the case of the unicellular ones. Some of the sulfur bacteria can also oxidize iron. Many are facultative autotrophs, growing as chemoheterotrophs in the presence of organic material.

The **sulfate-reducing bacteria** are ubiquitous in anaerobic environments and are particularly prominent in marine sediments. They are strictly anaerobic chemoheterotrophs or chemoautotrophs. As chemoheterotrophs, they oxidize fermentation end products such as lactate, ethanol, and acetate using sulfate (which is abundant in sea water) as their terminal electron acceptor. Many of them can also grow chemoautotrophically with H_2 as their electron donor, fixing CO_2 by the Wood pathway. Most of them are unicellular, and they are often flagellated (Figure 14.11).

Another group of anaerobic proteobacteria are the **purple bacteria,** anoxygenic phototrophs (Figure 14.12). The **purple sulfur bacteria** are mostly obligate photoautotrophs and obligate anaerobes. They grow in stagnant anoxic waters, where sulfate reduction in the sediments produces sulfide for them to use as electron donor. Like the aerobic sulfur bacteria, they store elemental sulfur in periplasmic globules when sulfide concentrations are high.

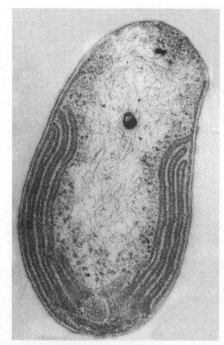

FIGURE 14.9 A nitrifying bacterium.

FIGURE 14.10 *Beggiatoa*, a filamentous sulfur oxidizer.

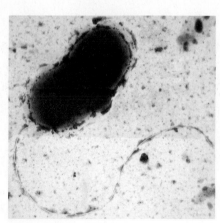

FIGURE 14.11 *Desulfovibrio*.

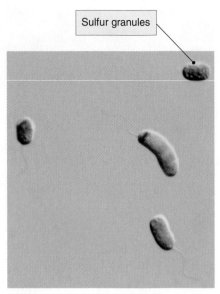

FIGURE 14.12 LM of *Chromatium.*

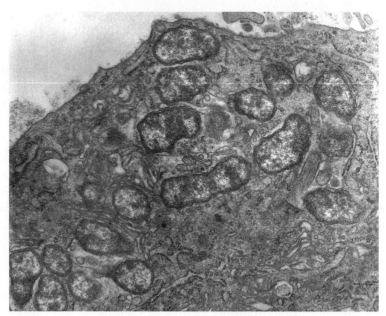

FIGURE 14.13 EM of *Rickettsia* within a mammalian cell.

The **purple nonsulfur bacteria** can also grow photoautotrophically using sulfide as electron donor (despite their name), but they tend to be inhibited by high sulfide concentrations and, hence, are rarely found in the same habitats as the purple sulfur bacteria. They do not make periplasmic sulfur globules. They are a widespread group of soil and water bacteria, probably because of their nutritional versatility. They grow very well as photoheterotrophs in illuminated anaerobic environments so long as the sulfide concentration is not too high, using fermentation end products such as lactate and acetate as their carbon source. They also grow well as aerobic chemoheterotrophs.

The **rickettsias** are a group of obligately intracellular parasites (Figure 14.13) that are generally passed from one vertebrate host to another by ticks, lice, or fleas. Many of them cause serious diseases, such as the tick-borne spotted fevers (*Rickettsia*) and Q-fever (*Coxiella*) and the louse-borne typhus (*Rickettsia*) and trench fever (*Rochalimaea*). *Coxiella* can also be transmitted through the air because it produces spores that are quite resistant to desiccation and UV light. The development of chemical insecticides after World War I (which saw major epidemics of typhus and trench fever among troops and civilians) brought most rickettsial disease under control, by allowing the vectors that transmit them to be targeted.

14.5 Gram-positive bacteria

The gram-positive bacteria have been recognized as a distinct group of bacteria for more than a century, and as we now know, they have a distinctive cell wall structure. They are diverse in both physiological and morphological terms. In addition to a heterogeneous group of fermentative, respiratory, or facultative chemoheterotrophic rods and cocci, many of them are medically or industrially important. They include three morphologically distinct groups: the **endospore formers,** the **actinomycetes,** and the **mycoplasmas.**

The **endospore formers** are defined by their ability to form a dormant spore (Figure 14.14) highly resistant to environmental hazards such as desiccation and UV

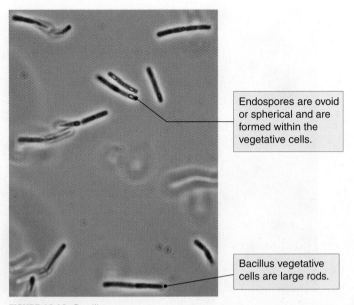

Endospores are ovoid or spherical and are formed within the vegetative cells.

Bacillus vegetative cells are large rods.

FIGURE 14.14 *Bacillus.*

irradiation (see Chapter 11). Most are rods and include aerobes or facultative anaerobes (genus *Bacillus*) and obligate anaerobes (genus *Clostridium*). Both *Bacillus* and *Clostridium* are enormously heterogeneous, and each will undoubtedly in the future be split into a number of genera. Both are principally soil organisms, but both contain species that are capable of causing disease in humans or animals.

Closely related to the endospore formers are the **mycoplasmas,** a group of bacteria that lack a murein-containing wall. They have traditionally been considered to be "naked" protoplasts, but that is clearly not the case; their cell membranes contain abundant glycolipids, and the polysaccharide portions form a dense layer on the external surface of the cell membrane. Hydrogen bonding among these strands is thought to strengthen the membrane, preventing osmotic lysis in the same way a wall does. Many of them are filamentous or helical (Figure 14.15), testifying to the strength of their glycolipid layer (because a sphere is the shape that turgor pressure would force the cell into if it were only barely contained). The mycoplasmas include respiratory and fermentative organisms; some are free living, but most are parasitic on plants or animals.

The **actinomycetes** are the only truly mycelial procaryotes. They are mainly soil organisms whose mycelia proliferate through organic debris. Under starvation conditions, this **substrate mycelium** begins to lyse, while an **aerial mycelium** is formed (see Figure 21.6). The aerial mycelium grows away from the substrate rather than within it and when mature produces a chain of conidia (spores). These conidia are then dispersed by wind or water; when one encounters nutrients, it germinates, and a new mycelium grows from it. The conidia are not completely dormant like endospores, and, thus, they are not as long lived; however, they are quite resistant to desiccation and UV light.

The actinomycetes produce many complex organic compounds during their sporulation phase. The distinctive smell of rich soil is, in fact, the odor of some of these compounds; sterile soil has no odor of its own. Most of these compounds are antibiotics, and they are thought to prevent other bacteria from scavenging the nutrients being released from the lysing substrate mycelium, which the aerial mycelium uses for its growth and sporulation. Many of the antibiotics used in medicine are produced by actinomycetes (see Chapter 21).

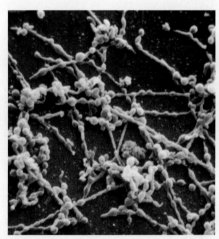

FIGURE 14.15 *Mycoplasma.*

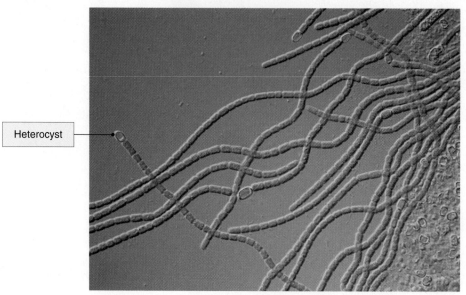

Heterocyst

FIGURE 14.16 Cyanobacteria.

14.6 Cyanobacteria

The cyanobacteria are a large and ancient group, morphologically diverse but physiologically uniform. They are unicellular or filamentous, and many are motile by gliding (Figure 14.16). They are obligate photoautotrophs and are the only procaryotes to perform oxygenic photosynthesis. Many filamentous cyanobacteria also fix nitrogen, sequestering nitrogenase into specialized cells called heterocysts, in which photosystem II synthesis is repressed, thus protecting nitrogenase from inactivation by oxygen (see Chapter 11).

The chloroplast is known to be derived from an endosymbiotic cyanobacterium, probably 1.5 to 2.0 billion years ago. It is clear that different groups of eucaryotic phototrophs are descended from different ancestral symbioses. The chloroplast of the red algae is the only one that shares the pigment composition and structure of contemporary cyanobacteria. This includes the use of linear tetrapyrroles (called phycobilins) as their major light-harvesting pigments and the organization of these phycobilins into phycobilisomes on the cytosolic surface of the thylakoid membranes.

There is one group of cyanobacteria, termed **prochlorophytes,** that has the same pigment composition as the green algae and plants—chlorophyll b in the thylakoid membrane as the major light-harvesting pigment. The molecular phylogenies suggest, however, that this group is not the one from which the green algal chloroplast is derived, despite their name.

14.7 Spirochetes

The spirochetes are the only group of bacteria traditionally defined purely on morphological grounds that is actually a coherent phylogenetic lineage. These are helical organisms that possess **endoflagella**—flagella that are confined within the periplasm (Figure 14.17). Spirochetes are the only bacteria known to have endoflagella. They originate at both poles of the cell and spiral around the cell outside the murein but inside the outer membrane, overlapping at the midpoint. The cell body and the endoflagella thus form a double helix, each winding around the other.

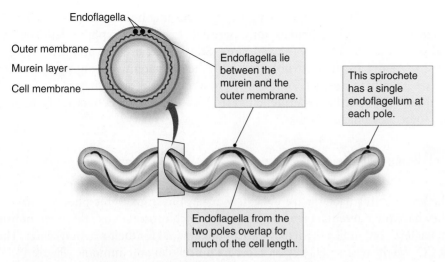

Endoflagella

Outer membrane

Murein layer

Cell membrane

Endoflagella lie between the murein and the outer membrane.

This spirochete has a single endoflagellum at each pole.

Endoflagella from the two poles overlap for much of the cell length.

FIGURE 14.17 Diagram of a spirochete.

When a spirochete prepares to divide, two new sets of endoflagella are formed at the midpoint of the cell, growing away from each other. The cell then divides between the new endoflagellar bundles, resulting in daughter cells with one set of endoflagella that are longer than the cell body and one set that are shorter and may not even reach the middle. Continued growth of both the cell and the flagella quickly restores the normal situation.

How endoflagella function remains a mystery. Several models have been proposed, but none is fully satisfactory. It is clear, however, that spirochete motility is a function of the endoflagella; mutants that lack them or that have abnormal flagellins are immotile. Spirochete motility is unusual in many respects. For instance, swimming is much more effective in highly viscous media. This may be an adaptation to live on the surfaces of animal mucous membranes, as many spirochetes are parasitic. It may also serve free-living spirochetes that live in highly viscous silty sediments.

Spirochetes also are capable of a gliding-like motility on solid surfaces, sometimes called "creeping," and they frequently lash around in solution without any overall directional movement. Because almost all of them are very thin, they look in the microscope like highly animated, fine, kinky threads. Once seen, they are instantly recognizable (Figure 14.18).

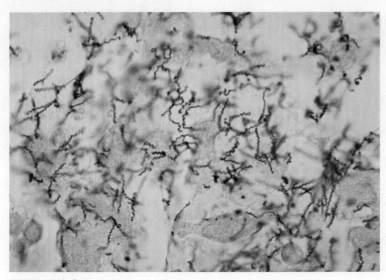

FIGURE 14.18 Spirochetes.

A number of spirochetes are pathogenic, causing human diseases such as the venereal disease syphilis (and its nonvenereal form endemic syphilis), leptospirosis, and skin diseases such as yaws and pinta. Spirochetes are also abundant as commensal organisms, living within the body but apparently doing no harm. The human mouth, for instance, is inhabited by a number of spirochetes.

Many spirochetes are fermentative and are anaerobes. Others are facultative, and some are obligate aerobes.

14.8 Green sulfur bacteria

The green sulfur bacteria are obligately anaerobic, anoxygenic photoautotrophs. They have their antenna pigments contained in chlorosomes like the green nonsulfur bacteria, but unlike them they have photosystem I for their reaction center. They fix CO_2 via the reverse TCA cycle. Most are unicellular and immotile (Figure 14.19), although there are a few that glide, and one filamentous form is known. They use H_2S as their electron donor, and they are very tolerant of high sulfide concentrations. Elemental sulfur, when it is accumulated, is excreted rather than accumulated in the periplasm. They are also very efficient at harvesting light and thus can grow at lower light intensities than other phototrophs. For these reasons, they are often found in anaerobic aquatic systems below the purple sulfur bacteria, where the light is dimmer and the sulfide concentrations higher. Many have gas vacuoles, although obviously regulated in number to position them deep in the water column.

14.9 Bacteroides group

This is a heterogeneous group that contains several distinctive subgroups, of which we will only mention two. *Bacteroides* is a genus of obligately anaerobic, rod-shaped bacteria, fermenting sugars, amino acids, or organic acids and producing succinate, propionate, and acetate (Figure 14.20). They are the most abundant organism in the mammalian intestinal tract and can number more than 10^{10} per gram of human feces (in contrast, *E. coli* is typically below 10^8 per gram). They are also found in the mouth and in anaerobic sediments.

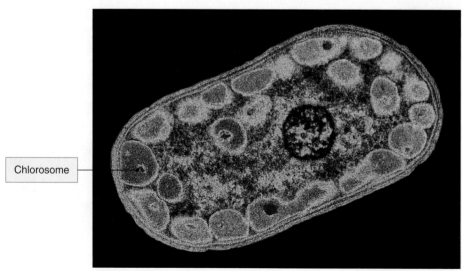

Chlorosome

FIGURE 14.19 *Chlorobium.*

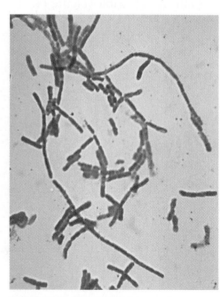

FIGURE 14.20 *Bacteroides.*

The **cytophagas** include several genera of unicellular gliding chemoheterotrophs, mostly strictly aerobic. They are generally found in soil, and they degrade complex macromolecules, most notably cellulose, chitin, agar, or proteins. Most are specialized for the use of only one of these classes of macromolecule. Several species of *Cytophaga* are pathogenic for fish and have a serious impact on aquaculture.

14.10 Planctomyces group

The planctomyces group is a heterogeneous group of poorly studied aquatic and marine procaryotes. Most are unicellular budding organisms, but at least one is filamentous. They are the only Bacterial lineage to lack murein entirely, having walls of protein. Some make protein stalks for attachment and produce daughter cells that are flagellated for a period before they grow a stalk. Their life cycle thus resembles that of *Caulobacter* (see Chapter 11), although this is presumably convergence, as the mode of cell division and the nature of the stalk are different. They are typically facultative anaerobic chemoheterotrophs.

Many of the planctomycetes have internal membranes that appear to divide the cell into compartments. Sometimes the compartment includes the nucleoid so that the cells appear to have a nucleus; in other cells the compartments do not include the nucleoid. The function of these membranes is completely unknown, as is their biosynthesis (Figure 14.21).

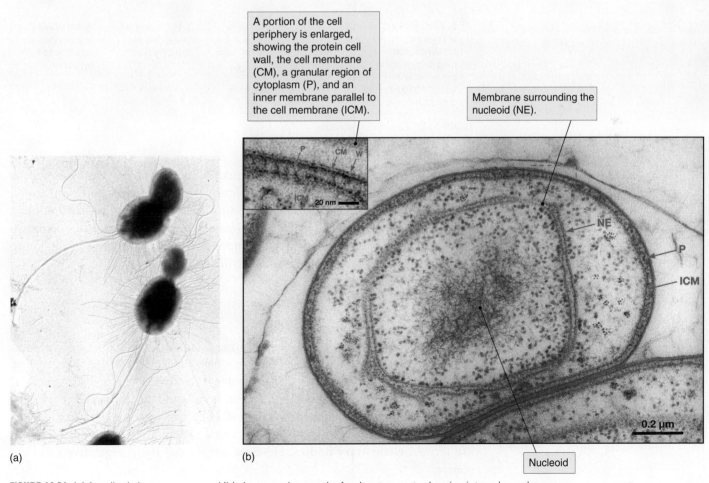

A portion of the cell periphery is enlarged, showing the protein cell wall, the cell membrane (CM), a granular region of cytoplasm (P), and an inner membrane parallel to the cell membrane (ICM).

Membrane surrounding the nucleoid (NE).

Nucleoid

(a)

(b)

FIGURE 14.21 (a) A stalked planctomycete and (b) electron micrograph of a planctomycete showing internal membranes.

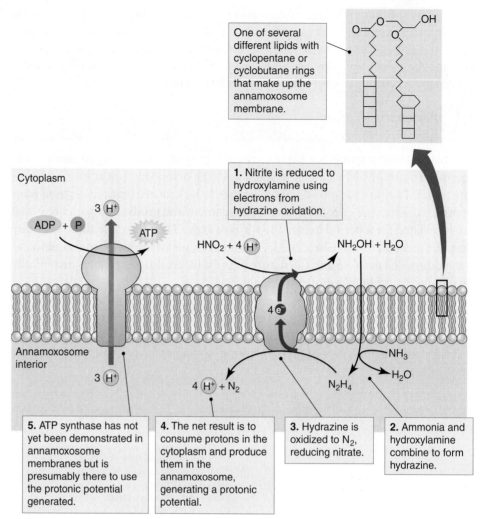

One of several different lipids with cyclopentane or cyclobutane rings that make up the annamoxosome membrane.

Cytoplasm

1. Nitrite is reduced to hydroxylamine using electrons from hydrazine oxidation.

$3\ H^+$

$ADP + P$

ATP

$HNO_2 + 4\ H^+$

$NH_2OH + H_2O$

$4\ e^-$

Annamoxosome interior

$3\ H^+$

NH_3

H_2O

$4\ H^+ + N_2$

N_2H_4

5. ATP synthase has not yet been demonstrated in annamoxosome membranes but is presumably there to use the protonic potential generated.

4. The net result is to consume protons in the cytoplasm and produce them in the annamoxosome, generating a protonic potential.

3. Hydrazine is oxidized to N_2, reducing nitrate.

2. Ammonia and hydroxylamine combine to form hydrazine.

FIGURE 14.22 The annamoxosome.

Another membrane-bound compartment found only in planctomycetes is the **annamoxosome,** organelles that anaerobically oxidize ammonia and nitrite to N_2 as a source of energy (Figure 14.22). These organelles have highly unusual lipids that contain linked cyclobutane rings, which appear to form a membrane that is unusually impermeable to small, nonpolar compounds. This is thought to function to contain the highly reactive intermediate hydrazine.

14.11 Chlamydia

The chlamydiae are a group of obligate intracellular parasites belonging to a single genus, *Chlamydia.* They have a life cycle in which a spore-like resting cell, involved in transmission between host cells, alternates with the actively replicating vegetative form (Figure 14.23). Unfortunately, the spores and vegetative cells have been given arcane names that date from a time when the biology of the group was badly misunderstood: the vegetative cells are termed **reticulate bodies,** and the spores are termed **elementary bodies.** Here we simply call them vegetative cells and spores.

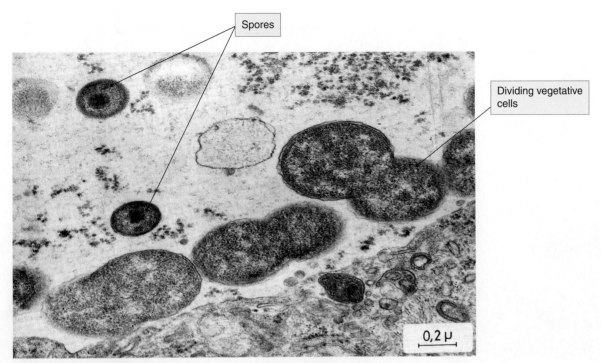

FIGURE 14.23 Chlamydiae.

Electron microscopy shows a typical gram-negative cell envelope, although muramic acid is lacking. It thus appears that they have an atypical peptidoglycan, not the murein commonly found in other gram-negative bacteria.

Metabolically, the chlamydiae are very limited. Their mode of ATP generation is not understood. They have an ATP/ADP exchange system, and it is generally thought that they depend on their host cells for phosphorylation of ADP; however, this remains controversial. They also have very limited biosynthetic capabilities and depend on their host cells for the synthesis of most monomers and vitamins.

The life cycle begins with the phagocytosis of the small, dense spore form, after which the spore develops into the vegetative form and begins dividing by binary fission. Presumably, this involves the production of proteins that make the phagosome membrane leaky so that host cytosolic constituents leak into the phagosome where they can be taken up by the chlamydial cells. Several rounds of division lead to several dozen daughter cells that then convert to the spore form. Spores are released by lysis of the host cell.

Chlamydiae are spread among host organism either by aerosol, as in the case of the respiratory disease psittacosis, or by direct contact as in the venereal disease lymphogranuloma venereum, or by flies, as in the case of the eye disease trachoma.

The Archaea

The Archaea are, relative to the Bacteria, remarkably lacking in evolutionary diversity (Figure 14.24). Only three lineages are known. Despite this lack of diversity, the Archaea are very successful. Some of them are widespread—for instance, nearly 30% of the microbes in the open ocean are archaea. Methanogens are ubiquitous in anaerobic environments, and some specialized habitats, such as very high temperature ones, may be predominantly populated with Archaea. As with the sparsely represented Bacterial lineages, metagenomics can be expected to reveal additional diversity.

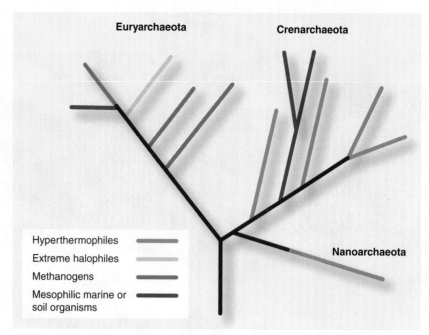

FIGURE 14.24 Phylogenetic tree of the Archaea.

Unlike the Bacteria, the sublineages of Archaea have been given formal kingdom names. At the time that the molecular phylogeny of procaryotes was being worked out by Woese and his collaborators and the concept of the Domain was introduced, the archaeal domain contained only two lineages, none of which had common names or alternative scientific names. This made the task of providing formal taxonomic description and Latin names more manageable.

Currently, three kingdoms of Archaea are recognized, with a fourth (the **Korarchaeota**) proposed on the basis of rRNA sequences detected in hot springs.

14.12 Euryarchaeotes

The kingdom Euryarchaeota includes two principal types of archaea: extreme halophiles and methanogens. The two physiological types do not appear to constitute two discrete sublineages; rather, they are intermixed with each other. It thus appears that the methanogens gave rise to halophilic descendants several times.

The **methanogens** are strict anaerobes that respire hydrogen gas with CO_2 as electron acceptor, producing methane (CH_4). Some also use acetate, producing a mixture of CO_2 and methane. They are common in most anaerobic habitats except marine ones, where sulfate reducers out-compete them for hydrogen. The methanogens include rods, cocci, and filaments and have a diversity of wall types, including all of the principal archaeal walls: pseudomurein, protein, and polysaccharide (Figure 14.25).

The methanogens are important in the global ecosystem for two related reasons. First, they are the terminal members of fresh-water anaerobic food chains, oxidizing acetate to gaseous end products that can then diffuse back into the aerobic zone (in marine environments, the sulfate reducers play this role). They thus prevent increasing amounts of organic material from being sequestered in anaerobic habitats, unable to be further metabolized.

Second, the methane they produce is a very powerful greenhouse gas. Most of it is oxidized to CO_2 in the anaerobic sediments where it is produced. No anaerobic methane oxidizers have been isolated in pure culture, but metagenomic analysis

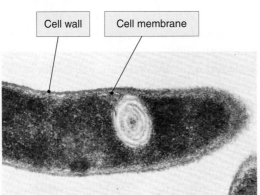
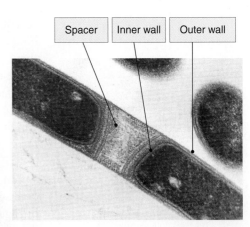

FIGURE 14.25 Methanogens.

suggests that a group of archaea derived from the methanogens uses a reversal of the methanogenic pathway to oxidize CH_4 to CO_2, doing so in symbiosis with sulfate reducers. Methane that escapes anaerobic oxidation is mostly oxidized at the interface between the aerobic and anaerobic zones by the **methanotrophic bacteria** (a subgroup of proteobacteria), but some escapes into the atmosphere where it contributes to global warming. This is an increasing problem, as human activities are increasing the numbers of animals that produce significant amounts of methane as an intestinal gas (principally ruminants such as cattle and sheep, whose numbers have increased along with human populations, and termites, whose distribution in Africa has expanded dramatically as humans cut down forests).

The **halophilic** archaea are a group of unicellular archaea that require high salt concentrations to grow—a minimum of 1.5 M (many require at least 2.5 M), with best growth near saturation (5.5 M). At these very high environmental osmolarities, plasmolysis would be lethal without some compensating mechanism. These bacteria concentrate K^+ to concentrations equal to the external osmolarity, thus maintaining cellular turgor pressure. Most halophilic archaea have proteinaceous cell walls, although some have walls of complex polysaccharides (Figure 14.26).

The halophilic archaea are aerobic chemoheterotrophs that respire organic compounds (most commonly amino acids) with oxygen as their terminal electron

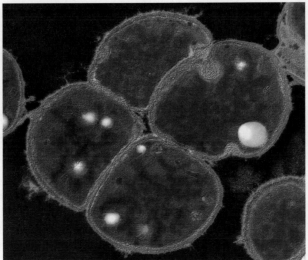

(a)   (b)

FIGURE 14.26 (a) *Halobacterium* and (b) *Halococcus*.

acceptor; however, saturated brines, their natural habitats, have low oxygen solubility, and they are often stagnant, making the rate of oxygen permeation low. Thus, these organisms often find themselves in microaerobic or anaerobic conditions. Some of them are motile and can swim to maintain themselves near the air-brine interface. Others have gas vacuoles that do the same thing.

Many of the halophiles make bacteriorhodopsin under anaerobic conditions, which acts as a light-driven proton pump. This allows them to maintain their protonic potential and hence their ability to continue to swim and seek out aerobic conditions.

In addition to the methanogens and halophiles, the euryarchaeotes include a diverse collection of hyperthermophiles, all of which appear to be early diverging lineages from the main line of methanogen evolution. Physiologically, they are similar to the hyperthermophilic crenarchaeotes. One group of marine archaea is also a member of this lineage.

14.13 Crenarchaeotes

The kingdom Crenarchaeota contains mainly hyperthermophiles, with optimum growth temperatures above 80°C and maximum growth temperatures near or above 100°C. The most hyperthermophilic organisms known belong to this group, with at least one having a maximal growth temperature above 120°C. Many experts expect that even more extreme hyperthermophiles will be discovered, able to grow at temperatures as high as 125°C. Of course, these organisms, like all living creatures, require liquid water; thus, the most extreme hyperthermophiles are generally deep sea organisms, found in geothermal areas at great depths where pressure keeps water liquid at temperatures of several hundred degrees centigrade (Figure 14.27). The less extreme hyperthermophiles, which grow at temperatures between 80°C and 100°C, are found in geothermal areas in shallow water and on land.

(a)

(b)

(c)

FIGURE 14.27 Chrenarchaeote habitats. (a) Deep sea hydrothermal vents. (b) Hot springs. (c) Solfatara—soils permeated with geothermal gases.

The hyperthermophilic crenarchaeotes are quite varied metabolically. The most common pattern is the oxidation of organic compounds, using anaerobic respiration with elemental sulfur or Fe^{3+} as terminal electron acceptor. Some use sulfate or nitrate as electron acceptor, and a few are capable of facultative aerobic respiration. Less common is fermentation. Many are facultative chemoautotrophs, using H_2 or elemental sulfur as electron donor, and the reverse TCA cycle to fix CO_2.

One sublineage, the **marine crenarchaeotes,** consists of psychrophilic, rather than hyperthermophilic, organisms. They are extremely common in the ocean, even in the frigid waters of the Arctic and Antarctic Oceans. Their physiology is not yet studied.

14.14 Nanoarchaeotes

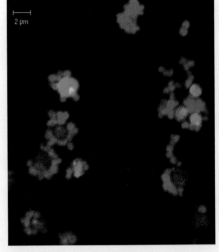

FIGURE 14.28 *Nanoarchaeum/Ignicoccus.*

The kingdom Nanoarchaeota contains a single organism that grows in association with another archaeon, the hyperthermophilic crenarchaeote *Ignicoccus* (Figure 14.28). It is unclear what the nature of the association is—whether *Nanoarchaeum* is a parasite or whether this is a symbiosis in which both partners benefit. All attempts to culture *Nanoarchaeum* separately have failed.

Nanoarchaeum has very small cells—a diameter of 0.4 µm, for a cell volume of slightly over 0.1 µm³. It also has the smallest genome of any known cellular organism, with 4.9×10^5 base pairs (see Figure 4.25). The genome lacks genes for most biosynthetic enzymes, as well as genes for catabolic enzymes. Presumably, necessary nutrients are provided by *Ignicoccus.*

Summary

This brief survey of the major groups of procaryotes describes the principal features of most of the well-studied and described groups; however, many more groups remain to be characterized, and many more branches on the procaryotic tree will undoubtedly be added in the coming years. The recent development of metagenomic tools, coupled with increasingly powerful bioinformatic methods, can be expected to lead to a rapidly expanding understanding of procaryotic diversity, how the various groups are related, and how they interact in their habitats. The last few decades have been an exciting period in microbial systematics; the next couple of decades promise to be even more exciting.

Study questions

1. To which lineage do the oldest identifiable microbial fossils belong? What does this imply about the diversification of the procaryotes?
2. From the phylogenetic history of life on earth, what can be said about the likely nature of the common ancestor of all life?
3. What can be inferred from the phylogenetic tree about the origin and evolution of photosynthesis?
4. Which groups of procaryotes contain human pathogens?
5. Which groups of procaryotes contain autotrophic representatives, and what pathways of CO_2 fixation does each use?
6. Which groups of bacteria lack murein?
7. Why are methanogens ecologically important?

15

Eucaryotic Microbes

THE EUCARYA consists of more than twenty major sublineages, three of them widely recognized as kingdoms (plants, animals, and fungi), whereas the others have been traditionally lumped together in a single kingdom. Some of the Eucarya are macroscopic: the plants and animals consist of mostly macroscopic, multicellular organisms with extensive tissue differentiation. The rest of the eucaryotes are in lineages that are at least in part microscopic. The term protist is often used to refer to these organisms.

There are three main subcategories of protist: **Fungi** consist of microscopic chemoheterotrophs, mostly multicellular, mycelial organisms without extensive tissue differentiation; however, many of them make macroscopic fruiting bodies called mushrooms (many scientists do not consider fungi part of the protists, but rather a separate lineage of eucaryotic microbe). **Algae** are photosynthetic organisms, most lineages consisting of microscopic forms, but several (notably the green, brown, and red algae) including both microscopic as well as macroscopic members. The macroscopic forms are multicellular, although generally with little tissue differentiation. **Protozoa** include many distinct lineages of chemoheterotrophic unicellular or colonial organisms.

The fungi consist of two lineages (although several protozoal lineages are traditionally included in the fungi also). The algae and the protozoa both consist of multiple sublineages; in some lineages, protozoal and algal types are intermixed. In the sections that follow, we describe some of the major lineages, focusing on their microscopic members. We ignore the green algae, as they are part of the plant lineage, and the brown algae, as they are mainly macroscopic.

Currently, there is no universally accepted phylogeny of the protists. There is a great deal of complexity of cell structure in the various subgroups, and many scientists prefer to rely on structural similarities as indications of relationships. Others prefer molecular sequences; however, different sequences may suggest different relationships, as can the same sequences compared with different computer programs. Sequences and morphologies often do not agree. Most scientists try to use both morphology and sequence information in a hybrid system, relying on whichever seems to make the most sense in a given situation. The phylogeny of most of the many dozen of eucaryotic lineages is obscure, and we only discuss those that are reasonably clear (Figure 15.1).

Metabolism among the protists is nowhere near as variable as among the procaryotes. Aerobic respiration and oxygenic photosynthesis are the dominant forms, with a minority of protists being fermentative. Aerobic respiration is performed by

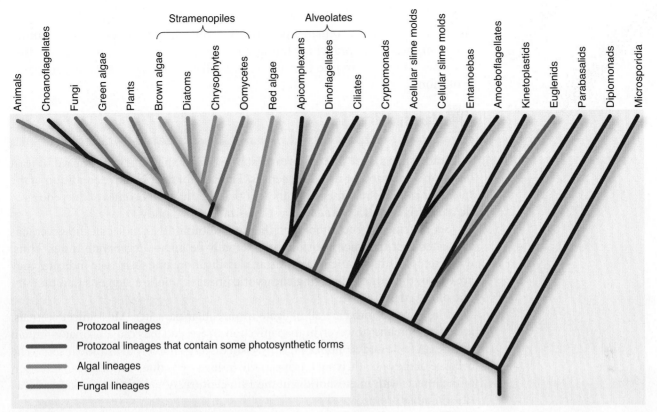

FIGURE 15.1 Detailed phylogenetic tree of the Eucarya.

Labels (left to right): Animals, Choanoflagellates, Fungi, Green algae, Plants, Brown algae, Diatoms, Chrysophytes, Oomycetes, Red algae, Apicomplexans, Dinoflagellates, Ciliates, Cryptomonads, Acellular slime molds, Cellular slime molds, Entamoebas, Amoeboflagellates, Kinetoplastids, Euglenids, Parabasalids, Diplomonads, Microsporidia

Stramenopiles (bracket over Brown algae, Diatoms, Chrysophytes, Oomycetes)
Alveolates (bracket over Apicomplexans, Dinoflagellates, Ciliates)

Legend:
— Protozoal lineages
— Protozoal lineages that contain some photosynthetic forms
— Algal lineages
— Fungal lineages

mitochondria, which have electron transport chains as discussed in Section 7.19. Unlike the procaryotes, there is little variation in electron transport components. Photosynthesis is Type I/II, as in the cyanobacteria, with the principal variation being in the nature of accessory light-harvesting pigments.

The chemotrophic members of the protists may take up dissolved nutrients, as in the fungi, or they may take in particulate matter via phagocytosis, as in most of the protozoa. The latter are predators, making their living by capturing and digesting other microbes—either procaryotic or eucaryotic. Indeed, some may ingest prey as large as themselves (Figure 15.2).

Many of the eucaryotic microbes have extraordinarily complex cells, with many specialized structures not seen in plant and animal cells. Most structures are not well understood at the molecular level, and, thus, there are years of detailed study yet to be done on these organisms, probably the least well understood of all the major categories of life.

Anaerobic Protozoa

There are a number of anaerobic protozoa. In some cases, entire groups are anaerobic, such as the ones described here; others are rare exceptions in a generally aerobic group, such as the ciliates or amoebas. The anaerobic, early-branching lineages lack mitochondria and hydrogenosomes. Thus, many scientists have speculated that they are descendants of the very earliest protoeucaryotes, before any of the endosymbiotic events that established the mitochondrial, hydrogenosomal, or plastid organelles. Sequencing, however, has shown that some of these organisms possess genes that are thought to be mitochondrial in origin. Thus, it

Predator

Prey

FIGURE 15.2 Ciliate ingesting another ciliate.

is possible that these early branching lineages once had mitochondria, but lost them. Nevertheless, it remains possible that the bacterial genes in the chromosomes of these early-branching lineages have a different origin and that the deep branches of the eucaryotic tree are in fact relics of an early stage in eucaryotic evolution.

15.1 Microsporidia

The microsporidia are one of the earliest branches of the eucaryotic lineage, at least based on small subunit rRNA sequences. The case for the primitive condition of the microsporidia is strengthened by the fact that they have ribosomes of the procaryotic type (see Table 13.1) and lack a recognizable Golgi body.

Sequencing of genes other than the small subunit RNA molecule, however, has suggested that the microsporidia may actually be highly degenerate fungi. Thus, they exemplify the problem of protistan classification: one set of data indicates they are one of the earliest diverging eucaryotic lineages; another suggests that they are one of the latest.

The microsporidia are all obligate intracellular parasites. Most infect animals, including humans; however, human infection is rare, and they cause no major human diseases. A few cause economically important diseases, notably of fish and insects. The collapse of the once-thriving European silk industry was due to widespread infection of the insects with microsporidia in the 19th century. A few microsporidia are parasitic on other protozoa.

The metabolism of microsporidia is fermentative, but they are oxygen tolerant, a necessity for infecting the oxygenated tissues of animals.

The microsporidia may have a single nucleus, or they may have two per cell. They have a life cycle in which a replicating vegetative form alternates with an environmentally resistant spore form. The spore is surrounded by a thick wall, including a layer of chitin. When it germinates, a long tube (typically about 0.1 μm in diameter and dozens of micrometers long) that is coiled up within the spore is everted at high speed. The tip of the tube penetrates through the host cell membrane, and the spore protoplast is extruded through the tube directly into the host cell cytoplasm. The vegetative cells divide by binary or multiple fission and appear to lack a wall of any kind. When the host cell cytoplasm is almost completely used up, they sporulate (Figure 15.3).

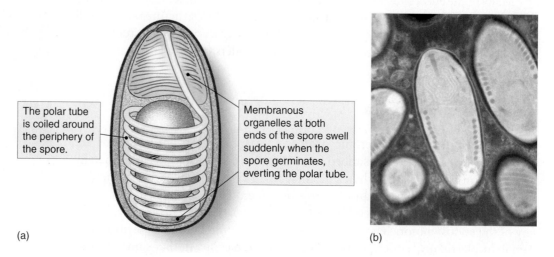

The polar tube is coiled around the periphery of the spore.

Membranous organelles at both ends of the spore swell suddenly when the spore germinates, everting the polar tube.

(a)

(b)

FIGURE 15.3 Diagram (a) and thin section (b) of a microsporidial spore.

CHAPTER 15 EUCARYOTIC MICROBES

15.2 Diplomonads

The diplomonads are another early-branching lineage that lacks mitochondria, hydrogenosomes, and a Golgi body. They are aerotolerant anaerobes, and most are inhabitants of the animal gut—mammals, rodents, birds, reptiles, and invertebrates. Some are pathogenic; one, *Giardia*, is a widespread human pathogen transmitted by the fecal–oral route. It is typically transmitted person to person in urban areas, especially among children in daycare centers (because of the poor hygiene of young children), and it infects hikers and backpackers through water contaminated by feces of domestic or wild animals.

The diplomonads have an extraordinarily complex cell structure. They are roughly the shape of a flattened pear, with the broad end the front. Typically, they have two nuclei, each associated with four flagella originating from basal bodies adjacent to the nucleus (Figure 15.4). Some of the flagella project forward, others backward, running through the cytoplasm before exiting the cell. A complex cytoskeleton of microtubules and microfilaments anchors the system.

Giardia is unique among diplomonads in having a sucking disk on its ventral surface that it uses to attach to the wall of the intestinal tract, thus avoiding being carried out by peristalsis (Figure 15.5). The disk is supported by a spiral of microtubules just beneath the cell membrane, and it is bordered by a ring of proteins, including actin and myosin.

Some diplomonads are predatory, ingesting particulate material such as bacteria through a **cytostome,** or cell mouth—a site on the ventral surface where phagocytosis takes place. Others consume dissolved nutrients and lack a cytostome. Organic compounds, whether taken up from the medium or from bacteria digested within phagosomes, are fermented to ethanol and acetate.

Diplomonads lack a cell wall or defined pellicle, but they make a thick-walled cyst that is the primary vehicle for transmission from one host to another. The cyst contains a tetranucleate cell that immediately divides into two when the cyst germinates. Division of vegetative cells is by binary fission.

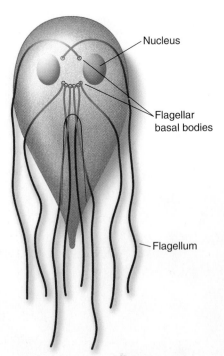

FIGURE 15.4 Diagram of a diplomonad.

Nucleus

Flagellar basal bodies

Flagellum

15.3 Parabasalids

The parabasalids have a prominent and well-developed Golgi body with up to 30 cisternae, sometimes called a parabasal body, adjacent to their anterior flagellar basal

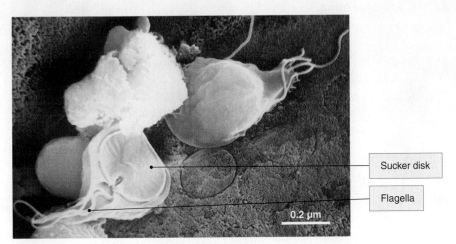

Sucker disk

Flagella

0.2 µm

FIGURE 15.5 Scanning electron micrograph of *Giardia*.

bodies and nucleus. Also associated with the nucleus is an **axostyle** that runs the length of the cell. The axostyle consists of a bundle of thousands of microtubules cross-linked to each other by motor proteins. Like flagella, it is capable of a wave-like motion because of the sliding of the microtubules against each other. This motion makes the cells undulate.

Parabasalids are another group of anaerobic, amitochondriate, early-branching protists. Unlike the diplomonads and microsporidia, however, they have hydrogenosomes. They ferment, with acetate, H_2, and either succinate or lactate as products. There are two major subgroups: the trichomonads and the hypermastigotes.

The trichomonads typically have four flagella originating from the anterior end of the cell. A fifth flagellum curves to the rear and supports a membranous flap that runs the length of the cell. As the flagellum beats, this flap moves in a sinuous pattern that has earned it the name **undulating membrane** (Figure 15.6).

Trichomonads inhabit the intestinal tract of a wide range of animals, from insects to humans. They are also found in the mammalian mouth and genital tract. Several percent of human beings worldwide are infected by *Trichomonas vaginalis;* most of these infections are asymptomatic, but occasionally the infection causes serious venereal disease (especially in women). Even though infection is not normally associated with disease, the sheer numbers of infected people make this one of the most common venereal diseases.

The hypermastigotes are typically inhabitants of the intestinal tract of wood-eating insects such as termites. They can be very large and have dozens to hundreds of flagella. Many have, in addition to the flagella at their anterior end, numerous spirochetes attached to special brackets on their surface (Figure 15.7). It is thought

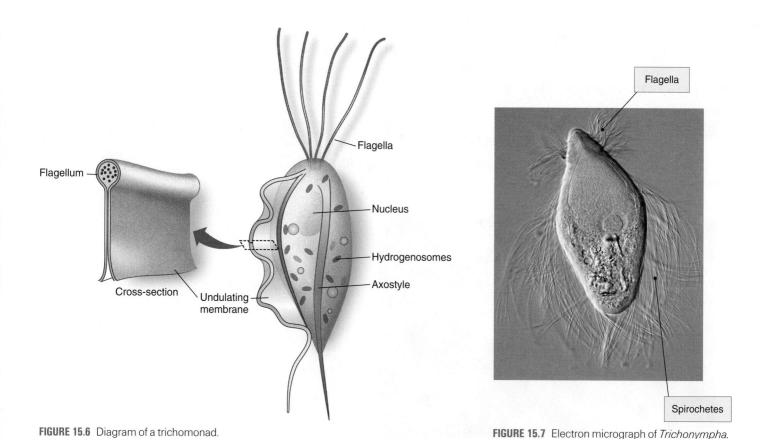

FIGURE 15.6 Diagram of a trichomonad.

FIGURE 15.7 Electron micrograph of *Trichonympha.*

CHAPTER 15 EUCARYOTIC MICROBES

that swimming movements by the spirochetes propel the cell, with its own flagella being used mainly to steer. They also have many endosymbiotic bacteria; these bacteria are responsible for synthesizing the cellulase enzyme with which the hypermastigotes digest phagocytosed cellulose.

Aerobic Protozoa

Most protozoa are aerobic; a few are facultative but most are obligate aerobes. They include all of the protozoal groups with photosynthetic representatives (which produce oxygen via oxygenic photosynthesis) and most of the chemotrophic protozoan groups.

15.4 Euglenids

The euglenids are elongated cells, with mitochondria and typical Golgi bodies. They usually have two flagella, although frequently one of the flagella is so short as to be invisible in the light microscope. Flagella originate at the anterior end, usually in a pocket. The flagella are normally thickened by a rod composed of protein lying adjacent to the flagellar axoneme (Figure 15.8).

Euglenid cells are flexible, but they have a **pellicle,** a system of proteinaceous plates just below the cytoplasmic membrane (Figure 15.9). These plates wind around the cell in a spiral fashion and overlap adjacent plates. The system is further strengthened by microtubules that run parallel to the plates. In many

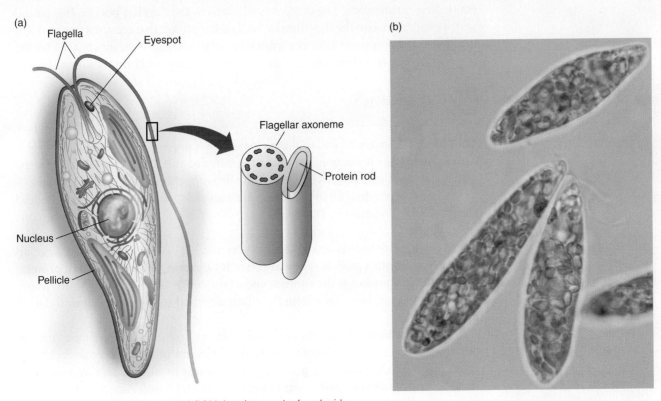

FIGURE 15.8 (a) Diagram of a euglenid. (b) Light micrograph of euglenids.

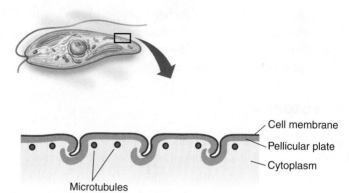

FIGURE 15.9 Diagram of the euglenid pellicle.

euglenids, the plates can move relative to each other, giving cells an earthworm kind of motility, independent of their flagella. The function of the pellicle is obscure; because most euglenids have a contractile vacuole, it is presumably not protection against osmotic lysis.

Most euglenids are aerobic chemoheterotrophs. They are predators, obtaining their food by phagocytosing bacteria or other protists. Phagocytosis occurs in the anterior sac where the flagella originate; food particles are swept there by the flagella. Other euglenids are phototrophic because of their possession of plastids. The plastids are contained within a third membrane, suggesting that they result from an endosymbiosis with a eucaryotic alga (see Section 5.26).

Most euglenids, even the chemotrophic ones, have a prominent light-sensing apparatus. The most visible component is the **eyespot**—a collection of vacuoles containing carotenoids. The eyespot is adjacent to the flagellar pocket. It is paired with a swollen area on the flagellum in the flagellar pocket that contains flavins. How the light-sensing apparatus works is not known; presumably, the eyespot shades the flagellar flavins, allowing determination of the direction of illumination.

15.5 Kinetoplastids

The kinetoplastids are related to the euglenids. They are aerobic chemoheterotrophs, with a single large mitochondrion near the anterior end of the cell, adjacent to the nucleus and the flagellar basal bodies. Within this mitochondrion is the **kinetoplast**— a mass of mitochondrial DNA consisting of thousands of copies of the mitochondrial chromosome. The mass of DNA is so large that it is visible in the light microscope (Figure 15.10).

Many kinetoplastids are otherwise very similar to euglenids. They have an elongated cell with a pellicle. They have two flagella, thickened by a rod of protein, originating in a pocket at the anterior end of the cell; one is often very short. Many of them, however, have their long flagellum attached to the cell as an undulating membrane.

The **trypanosomes** are kinetoplastids with an undulating membrane that are parasitic on a variety of mammals, including humans, in tropical areas worldwide (Figure 15.11). Generally, the genus *Trypanosoma* is most common in Africa and Asia, and *Leishmania* is most common in Central and South America. The principle human diseases are sleeping sickness, caused by *Trypanosoma*, and leishmaniasis. Sleeping sickness is characterized by fever, lassitude, coma, and death. Leishmaniasis

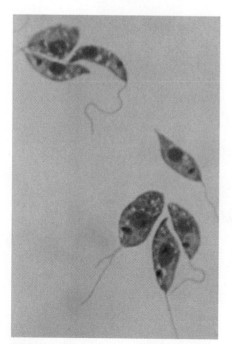

FIGURE 15.10 Light micrograph of *Leishmania*.

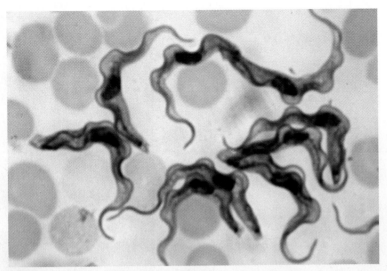

FIGURE 15.11 *Trypanosoma* in blood.

is usually a skin disease, characterized by persistent ulcers, although some forms are systemic and can be fatal. Kinetoplastids have complex life cycles, with several stages occurring in the blood-sucking arthropods that transmit the trypanosomes from one mammal to another.

15.6 Amoebas and amoeboflagellates

Several lineages of protists have amoeboid cells: the entamoebas, the amoeboflagellates, the slime molds, and several distinct lineages of amoebas. They are characterized by flexible cells and the use of **pseudopodia** for both motility and food capture. Pseudopodia are dynamic extensions of the cell that are not organized by a microtubular cytoskeleton. Pseudopodia can be coarse and thick or very fine and filamentous. The amoebas are divided into two main groups: the **naked amoebas** and the **testate amoebas.** The testate amoebas have a **test,** or shell, that is lacking in the naked amoebas (Figure 15.12).

The amoebas are predatory, consuming bacteria or other protists by phagocytosis, usually at the advancing tip of a pseudopod. In the case of amoebas with filiform pseudopods, prey is attached to the pseudopod, which then retracts, bringing the prey to the body of the amoeba where it is phagocytosed. Most are aerobic, but there are some anaerobic amoebas that inhabit the animal digestive tract (often called the **entamoebas**), or anaerobic sediments. Some of these can cause intestinal disease. Most amoebas and amoeboflagellates form cysts that allow them to survive periods of desiccation or, in the case of the entamoebas, exposure to air.

One group of marine testate amoebas, the **foraminifera,** forms elaborate, multichambered tests. They are composed of an organic matrix that is frequently calcified. These are very durable, and there are thus many thousands of fossil foraminifera known. Usually, the shells have one major opening like a mollusk shell and many fine pores over the whole surface. Pseudopodia extend both from the major opening and through each of the tiny pores (Figure 15.13).

Amoeboflagellates are characterized by a dimorphic life cycle, in which an amoeboid form alternates with a flagellated form, with two or four flagella. Multi-

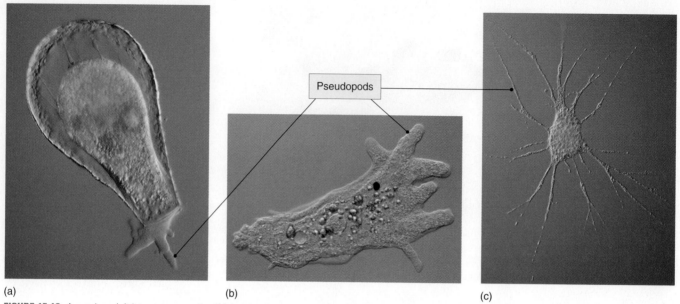

(a)   (b)   (c)

FIGURE 15.12 Amoebas. (a) A testate amoeba. (b) A naked amoeba. (c) An amoeba with filiform pseudopods.

plication is in the amoeboid form. Sudden changes in environmental conditions, such as temperature or osmolarity, induce the formation of flagella, which is followed by a period of swimming motility before the flagella are shed and the cell returns to the amoeboid form.

One of the amoeboflagellates, *Naegleria*, can cause human disease. It is an inhabitant of warm water and can penetrate the nasal epithelia of swimmers and then

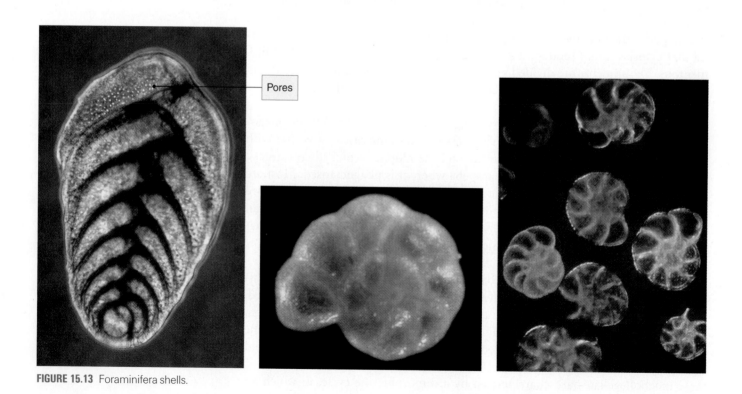

FIGURE 15.13 Foraminifera shells.

migrate along olfactory nerves to the brain. Multiplication in the brain can cause death in a few days.

15.7 Actinopoda

The actinopoda are a highly variable group of marine protists that superficially resemble the foraminifera. Their phylogenetic position is unclear, and they probably do not constitute a single lineage. They have elaborate skeletons composed of silica that lie just below the cell surface. Many filiform pseudopods, called **actinopoda,** extend from the cell surface, each stiffened by a microtubular cytoskeleton. The bundles of microtubules originate at the center of the cell and pass through perforations in the skeleton. These microtubular bundles usually have a precise arrangement that differs in different subgroups (Figure 15.14).

Prey is captured by extrusomes on the actinopodia, which then retract by sliding and disassembly of the microtubular bundle, bringing the prey to the cell surface for phagocytosis.

15.8 Slime molds

Two separate groups are included in the slime molds—the **cellular slime molds,** so called because most of their life cycle is spent as unicellular amoeboid cells, and the **acellular slime molds,** which spend most of their life as a multinucleate mass of cytoplasm, or a **plasmodium.**

The cellular slime molds are chemoheterotrophic amoebas that multiply asexually as they forage for prey in leaf litter in temperate forests. When they have exhausted the resources of a particular site and enter starvation conditions, chemotactic signals (including cAMP) cause the amoebas to come together to form a multicellular aggregate, or **slug** (sometimes called a **pseudoplasmodium**). The slug moves around for a while and then settles down and many of the cells differentiate

(a)

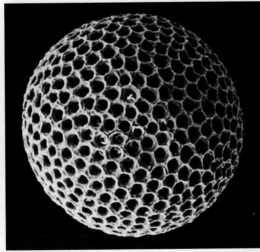

(b)

FIGURE 15.14 Living actinopod (a) and actinopod shells (b).

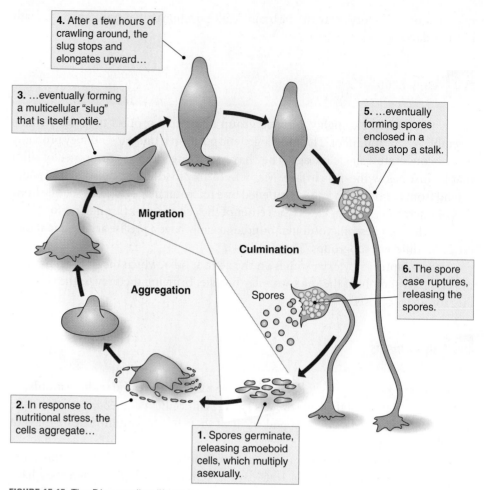

4. After a few hours of crawling around, the slug stops and elongates upward…

3. …eventually forming a multicellular "slug" that is itself motile.

5. …eventually forming spores enclosed in a case atop a stalk.

Migration

Culmination

6. The spore case ruptures, releasing the spores.

Aggregation

Spores

2. In response to nutritional stress, the cells aggregate…

1. Spores germinate, releasing amoeboid cells, which multiply asexually.

FIGURE 15.15 The *Dictyostelium* life cycle.

into spores. In some genera, the spore mass is born on a stalk; in others, it is directly on the substratum. The spores are then dispersed by wind or rain, and the cycle begins over again for those spores lucky enough to land in favorable conditions (Figure 15.15).

The **acellular slime molds** consist of a multinucleate cytoplasm, called a **plasmodium**—essentially a huge amoeba with thousands of diploid nuclei. They crawl slowly through the leaf litter of temperate forests, like the cellular slime molds, but as a flat, fan-shaped mass of protoplasm rather than a swarm of amoebas (Figure 15.16). They can reach huge sizes—as much as a meter or more across—but they are more commonly a few centimeters to a few tens of centimeters across. Numerous large vein-like channels run from the back end to the advancing edge of the fan. Cytoplasm is vigorously pumped back and forth in these channels by alternate contractions in the front and back ends of the plasmodium. The phenomenon is called **shuttle streaming** and is mediated by actin microfilaments and myosin motor proteins. Forward flow is more vigorous than reverse flow, and hence, the entire organism gradually flows forward as the cytoplasm shuttles back and forth.

This diploid plasmodium stage alternates with a unicellular haploid stage. Under starvation conditions, stalked sporangia are produced, which by a process of meiosis and cytokinesis produce haploid, uninucleate spores, either amoebas or flagellated cells, that multiply vegetatively. Fusion of haploid cells regenerates a diploid

FIGURE 15.16 *Fusarium.*

CHAPTER 15 EUCARYOTIC MICROBES

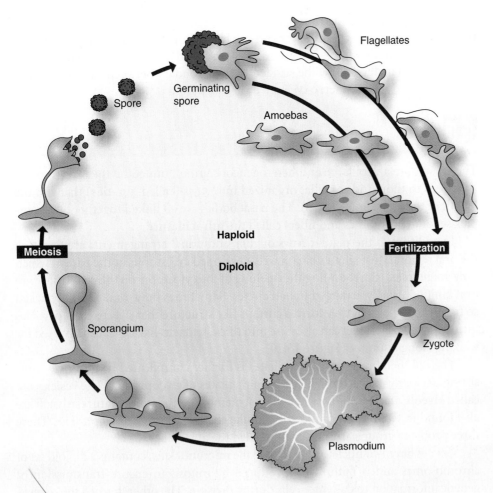

Flagellates

Germinating
spore

Spore

Amoebas

Haploid

Meiosis

Diploid

Fertilization

Sporangium

Zygote

Plasmodium

FIGURE 15.17 The *Fusarium* life cycle.

zygote, which then grows into the plasmodium by repeated mitotic nuclear divisions without cytokinesis (Figure 15.17).

15.9 Cryptomonads

The cryptomonads are a protozoal group that is largely unicellular, although some are colonial, with colonies consisting of individual cells held together in a mass of gelatinous slime. The cells are flattened ovals, with two flagella emerging from a flagellar pocket, like the euglenids. Also like the euglenids, cryptomonads often have an eyespot adjacent to the flagellar pocket. Just beneath the cell membrane is a pellicle consisting of hexagonal plates of organic material (Figure 15.18).

Most cryptomonads are photosynthetic, as a result of the presence of plastids that are clearly the remains of an endosymbiosis with a eucaryotic alga. The relict alga contains a functional nucleus, termed a **nucleomorph** to distinguish it from the host nucleus, as well as a plastid, and the entire thing lies inside the endoplasmic reticulum. Sequencing of rRNA genes from the nucleomorph places the ancestral endosymbiont within the red algae, and the pigment composition of the plastid is consistent with this. Like the red algae and the cyanobacteria, the principal photosynthetic pigments are phycobilins.

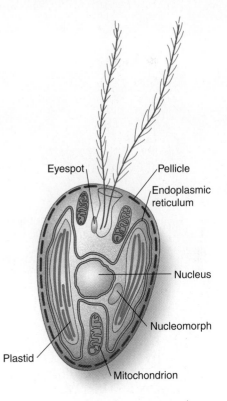

Eyespot

Pellicle

Endoplasmic
reticulum

Nucleus

Nucleomorph

Plastid

Mitochondrion

FIGURE 15.18 Diagram of a cryptomonad.

Cryptomonads

The cryptomonads are found in both freshwater and marine habitats. They are particularly common in cold water and can be the dominant primary producers in Arctic and Antarctic waters. They are also often found in a narrow band quite deep, as much as 20 meters below the surface, at the boundary between the cold, anoxic deep water and the warmer, oxygenated surface water.

15.10 Ciliates

The ciliates are a very large and heterogeneous group, united by the possession of hundreds to thousands of cilia, organized into rows (called **kineties**) that run the length of the cell (Figure 15.19). The basal bodies are all linked together by a complex set of microtubules and fibers called the **infraciliature.**

In addition to the regular rows of cilia, additional arrangements are found in the oral region. Most ciliates have a well-developed cytostome, at the base of which prey are ingested. Surrounding the opening of the cytostome on the cell surface are oral kinities, usually arranged in half circles. Sometimes these oral cilia are packed so tightly together that they form a curtain-like structure that sweeps food particles into the cytostome. Cilia can also fuse into thick, tentacle-like **cirri,** usually used for walking along a surface (Figure 15.20).

The cell membrane region of ciliates is further complicated by a pellicle, consisting of closely packed vesicles containing plates of glycoprotein. The vesicles are called **alveoli,** and they lie between the cell membrane and infraciliature. Alveoli are also found in two other groups: the dinoflagellates and the apicomplexans. These three protistan groups are collectively called the **alveolates.**

Ciliates have two nuclei. One, termed the **micronucleus,** contains a diploid set of chromosomes and is faithfully reproduced by mitosis to ensure transmission of genetic information to daughter cells during division. The other, termed the **macronucleus,** has multiple copies of chromosomes and is the site of most transcription. The macronucleus does not divide mitotically, but it is recreated in each generation by division of the micronucleus, followed by differentiation of one of the nuclei into a macronucleus.

Sex in ciliates involves meiosis of the micronucleus in response to mating factors produced by each of the many mating types. After meiosis, fusion of the cell

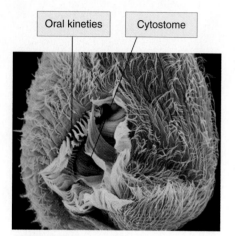

Oral kineties   Cytostome

FIGURE 15.19 A ciliate.

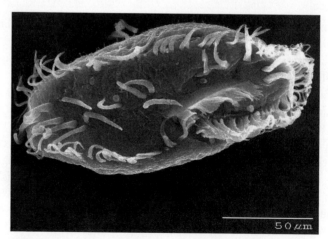

50 μm

FIGURE 15.20 A ciliate with cirri.

membranes of the mating ciliates forms a conjugation bridge, and the two partners each transfer one of their micronuclei. The conjugation bridge then closes, and fusion of the transferred micronucleus with one of the resident micronuclei restores the diploid vegetative state.

Ciliates are predatory, consuming bacteria or other protists. Prey is often captured by harpooning with extrusomes; oral cilia then sweep it into the cytostome.

Ciliates are common in aquatic systems (freshwater and marine) and are also found in moist soils worldwide. Most are motile, using their cilia to swim; some use cirri to walk over a solid surface, and some are sessile, attached to the surface with a stalk. Sessile ciliates use their cilia to create water currents that sweep prey organisms into capture range.

15.11 Dinoflagellates

The dinoflagellates are flagellated alveolates. Their pellicle includes large cellulose plates within the alveolar sacs in a characteristic arrangement. The cell is encircled by a groove around the middle, within which one of the two flagella lies. The other flagellum trails behind (Figure 15.21). It appears that forward motion is primarily due to the flagellum that encircles the cell. It forms a tightly coiled helix in the groove, and its inner edge is anchored to the groove. Thus, when it beats, the outer edge of this ribbon exerts force both to the rear and to the side. Thus, the cell is propelled forward, rotating as it goes (Figure 15.22). The trailing flagellum appears to be mainly for steering. Nonmotile, dormant cysts may be formed under starvation conditions.

Dinoflagellates include both heterotrophic and photosynthetic species. Phototrophs have plastids with three concentric membranes, suggesting that they result from an endosymbiosis with a eucaryotic alga. They commonly have an eyespot. In the oceans, dinoflagellates are probably second only to diatoms in their contribution to primary productivity.

Several phototrophic species of dinoflagellate produce **saxitoxin,** a complex organic compound that can accumulate in filter-feeding organisms, such as shellfish, that ingest large numbers of dinoflagellates (Figure 15.23). Human consumption of the shellfish leads to nerve damage, called **paralytic shellfish poisoning.** This is a seasonal problem in many areas of the world, including the west coast of North America, when warmer summer waters and high summer light intensities lead to blooms of toxic dinoflagellates. These blooms can be so substantial as to color the water, leading to the term "red tide" (dinoflagellates are often colored red because of large amounts of carotenoids in their plastids). Other toxins may also be produced, some of which are important agents of fish kills during dinoflagellate blooms.

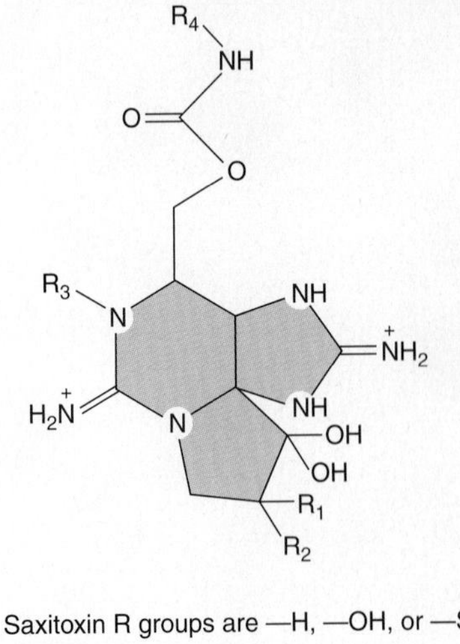

Some dinoflagellates, both phototrophs and chemotrophs, are bioluminescent. They are the principal causes of marine phosphorescence—a greenish, yellowish, or bluish glow to the water when it is disturbed, as by breaking waves, or the movement of fish or boats through it. Phosphorescence often appears to be due to tiny glowing particles; some of these are bits of organic detritus, on which heterotrophic dinoflagellates are feeding, but others may be the dinoflagellates themselves, some of which can be quite large. The heterotrophic, bioluminescent *Noctiluca* is about 2 mm in diameter.

FIGURE 15.21 Dinoflagellates.

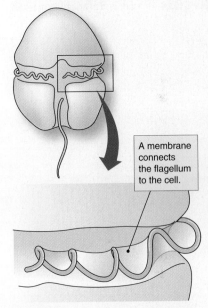

A membrane connects the flagellum to the cell.

FIGURE 15.22 Dinoflagellate flagellar arrangement.

Saxitoxin R groups are —H, —OH, or —SO$_3^-$
FIGURE 15.23 Saxitoxin.

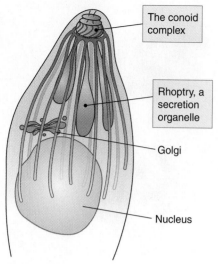

FIGURE 15.24 Anterior end of an apicomplexan.

Labels: The conoid complex; Rhoptry, a secretion organelle; Golgi; Nucleus

15.12 Apicomplexans

The third group of alveolates is the apicomplexans, named after the complex apical organ by which these intracellular parasites attach to and penetrate into host cells. The apical organ, or **conoid complex,** consists of a cone of microtubules and a series of rings, from which a system of longitudinal microtubules radiate. It is generally associated with sac-like excretion organelles called **rhoptries,** which pass through the conoid complex and empty at the tip. The Golgi apparatus is associated with the rhoptries, presumably supplying them with material to excrete (Figure 15.24).

The alveoli in apicomplexans do not contain pellicular plates as in the ciliates and dinoflagellates. Thus, they simply consist of a membranous sac that surrounds the cell immediately beneath the cell membrane.

The apicomplexans have complex life cycles, often completed in different host organisms. Malaria, for instance, is caused by an apicomplexan (*Plasmodium*) that multiplies vegetatively in humans or other animals, in either liver or red blood cells, but which requires ingestion by a mosquito for the haploid cells to develop into gametes, fuse to form diploid zygotes, and then undergo meiosis to regenerate the haploid vegetative cells. The final stage involves migration of *Plasmodium* cells to the mosquito's salivary gland, whence they are injected back into their vertebrate host (Figure 15.25).

15.13 Choanoflagellates

The choanoflagellates, named for the collar that surrounds their flagellum, are small marine or freshwater organisms. They may be unicellular or colonial. Many of them make a silicaceous shell. They are important predators of bacteria in marine ecosystems.

The collar is formed of a large number of fine, tubular, cytoplasmic projections, or **microvilli.** They encircle the flagellum, whose beating causes a current of water to flow through the microvilli. Bacteria suspended in the water get trapped by the

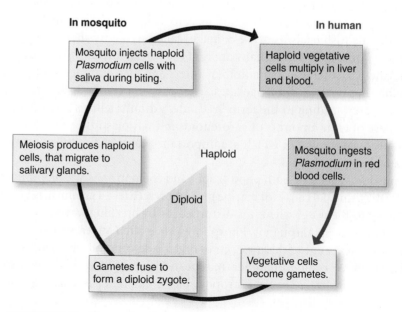

In mosquito   In human

Mosquito injects haploid *Plasmodium* cells with saliva during biting.

Haploid vegetative cells multiply in liver and blood.

Meiosis produces haploid cells, that migrate to salivary glands.

Haploid

Diploid

Mosquito ingests *Plasmodium* in red blood cells.

Gametes fuse to form a diploid zygote.

Vegetative cells become gametes.

FIGURE 15.25 The *Plasmodium* life cycle.

microvilli and then moved to the base of the collar where they are phagocytosed (Figure 15.26).

Although the choanoflagellates are a small group of inconspicuous organisms, they are important as the sister group of the animals. Indeed, sponges (the most primitive group of animals) filter feed using cells that are indistinguishable from choanoflagellates.

The Algae

Several lineages of protozoa that we have discussed contain algal representatives—the euglenids and dinoflagellates, for example. These are generally viewed as primarily nonphotosynthetic, with the photosynthetic representatives being more recently derived. In the algal lineages, the same situations are seen—nonphotosynthetic representatives may be intermixed with photosynthetic ones. In this case, however, the photosynthetic ability is seen as the ancestral state of the group, and nonphotosynthetic representatives are more recent derivatives.

Many algal groups contain macroscopic representatives as well as microscopic ones. Some of the macroscopic forms can reach huge sizes; giant kelp, for instance, can grow to more than 60 meters in length. We only discuss the microscopic forms here.

15.14 Red algae

The red algae may be unicellular or multicellular, and many of the multicellular forms are macroscopic (Figure 15.27). They are characterized by having plastids whose principal accessory pigments are phycobiliproteins as in the cyanobacteria. These pigments are red or pink, thus giving the name to the group. Red algae lack flagella completely, unlike most other algae, which have a flagellated stage in their life cycle. They have a cell wall containing cellulose, like most other algal groups, but their walls are more flexible than most. The cellulose fibers are embedded in a matrix of polymers of glucose and galactose derivatives. These polysaccharides are commercially quite valuable: agar is one of them (used in food and as a base for microbiological media), and carrageenin is another (also used in foods).

The nuclear biology of red algae is very complex. Cells may be uninucleate or multinucleate, and the individual nuclei may undergo repeated rounds of chromosome replication without mitosis, leading to polyploid nuclei. Meiosis and sexual reproduction systems are also very complicated. Asexual reproduction by the formation of spores is common but not universal; many red algae reproduce only sexually.

Cell division in the red algae is by constriction, mediated by the contraction of a ring of actin and myosin. In most cases, it is incomplete, leaving daughter cells connected by a small pore. The pore becomes filled with a plug of protein, called a **pit-plug.** The pit-plug may eventually be covered by a membrane, or by other layers (Figure 15.28).

Many red algae are parasites on other red algae. The parasitic red algae are usually unpigmented, although they have vestigial plastids. The parasitic process is unique and complex. Typically, when a parasite spore lands on the surface of a host alga, it germinates and forms a localized zone of fusion of the cell membranes of the host and parasite. The parasite nucleus is then transferred into the host cell, where it replicates faster than that of the host. The parasite nucleus either stimulates the

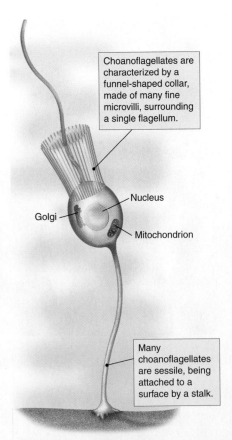

Choanoflagellates are characterized by a funnel-shaped collar, made of many fine microvilli, surrounding a single flagellum.

Nucleus

Golgi

Mitochondrion

Many choanoflagellates are sessile, being attached to a surface by a stalk.

FIGURE 15.26 A choanoflagellate.

(a)   (b)

FIGURE 15.27 Red algae (a) *Ceramium* sp. and (b) *Antithamnion plumula.*

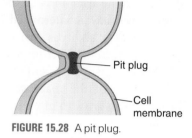

Pit plug

Cell
membrane

FIGURE 15.28 A pit plug.

host cell to multiply, or it induces the dissolution of the pit-plugs connecting host cells, allowing the rapidly multiplying parasite nuclei to spread through the host. Eventually, spore formation produces numerous progeny spores, each containing a parasite nucleus (Figure 15.29).

15.15 Stramenopiles

The next several groups—chrysophytes, diatoms, and oomycetes—are (along with the brown algae, which we are not considering here) all members of a single major eucaryotic lineage, called the stramenopiles. In addition to small subunit rRNA sequences, a number of features of their cell biology indicate that these groups are related. Most notably, they have a distinctive flagellar arrangement in which one of a pair of flagella is decorated with fine stiff tubular hairs of glycoprotein, and the other is shorter and naked (Figure 15.30). The term *stramenopile* refers to these hairs. Another commonly used term is **heterokont,** which simply means "different flagella." As we discussed in Chapter 13, most botanists and zoologists have not accepted the notion that each distinct lineage of microorganism deserves kingdom status; however, the stramenopiles are widely accepted as a kingdom.

There are a few protozoal lineages among the stramenopiles and one that was formerly placed in the fungi (the oomycetes). Most of them, however, are algal, with chlorophylls a and c, and a distinctive brown pigment fucoxanthin. Thus, the algal lineages are usually brown or golden brown; collectively, they are called the **ochrophytes** (ochre algae). One of the lineages, the brown algae, consists mainly of macroscopic forms; most other ochrophytes are microscopic. There is a great diversity of them, and we consider only a few of the groups here.

CHAPTER 15 EUCARYOTIC MICROBES

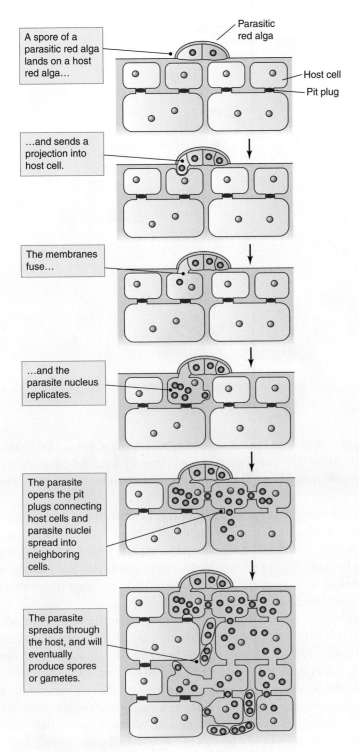

A spore of a parasitic red alga lands on a host red alga...

Parasitic red alga

Host cell

Pit plug

...and sends a projection into host cell.

The membranes fuse...

...and the parasite nucleus replicates.

The parasite opens the pit plugs connecting host cells and parasite nuclei spread into neighboring cells.

The parasite spreads through the host, and will eventually produce spores or gametes.

FIGURE 15.29 Red algal parasitism.

15.16 Chrysophytes

The chrysophytes are unicellular or colonial flagellated algae (Figure 15.31). Most have the flagellar arrangement typical of stramenopiles. An eyespot is normally present. A few are nonmotile or amoeboid, but even these produce flagellated

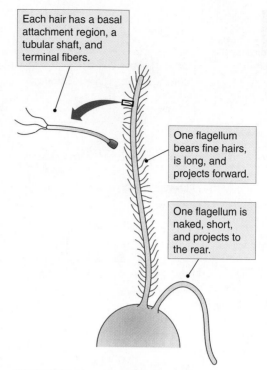

Each hair has a basal attachment region, a tubular shaft, and terminal fibers.

One flagellum bears fine hairs, is long, and projects forward.

One flagellum is naked, short, and projects to the rear.

FIGURE 15.30 Heterokont flagellar arrangement.

spores. Their plastids give the cells a golden brown color, hence their name (*chrysos* is Greek for gold). The chrysophytes are primarily aquatic, inhabiting predominantly nutrient-poor lakes. They are especially abundant in cold, slightly acidic waters.

Most chrysophytes are facultative chemoheterotrophs, ingesting particulate food by phagocytosis at the base of the flagella, or sometimes taking up dissolved organic material. A number of them seem to be photoheterotrophic—that is, they prey on bacteria or other protists as their carbon source, but they use cyclic photophosphorylation as their major energy source. They are often the principal predators in nutrient-poor lakes.

Chrysophytes produce distinctive resting cysts with a wall composed of silica (Figure 15.32). The cysts can be produced either as a result of sexual process (gamete fusion produces a zygote that encysts), or they can be a response to environmental factors. Cells form cysts by losing their flagella and becoming spherical. The silica wall is formed by the fusion of Golgi vesicles just beneath the cell membrane, to form a large vesicle completely surrounding the cell except for the location of the pore. Silica is deposited in this vesicle to form the wall. When complete, the cyst is enclosed within a silica wall, except for a pore, which is plugged with polysaccharide. When the cyst germinates, the polysaccharide plug is degraded, and one or more flagellated vegetative cells emerge through the pore.

15.17 Diatoms

The diatoms are one of the most successful algal groups. They are abundant in almost all marine and aquatic systems and are one of the most important primary producers in most of them. They are very diverse; over 10,000 species

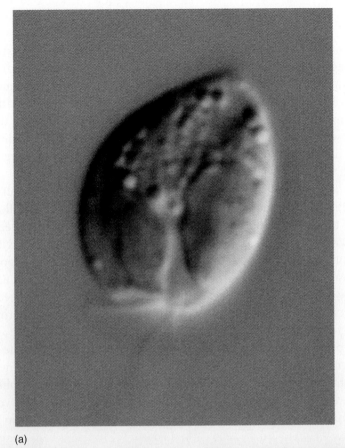

(a)

(b)

FIGURE 15.31 A unicellular (a) and colonial (b) chrysophyte.

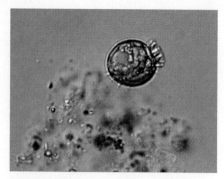
FIGURE 15.32 A stomatocyst.

have been described, and some scientists estimate that there are over a million species.

Diatoms are unicellular, or they may form chains of cells. Vegetative cells do not have flagella, but during sexual reproduction, they form sperm that have a single flagellum that is decorated with the typical tubular hairs of stramenopiles. Cell shape is highly variable, but there are two principal categories: **Centric** diatoms are radially symmetrical, with disc-shaped or cylindrical cells. **Pennate** diatoms are elongated, with bilaterally symmetrical cells (Figure 15.33). Centric diatoms have many discoid plastids, whereas pennate diatoms typically have two large elongated plastids.

The most striking feature of diatoms is their **frustule**—a cell wall composed of silica. Silica is resistant to enzymatic digestion, and, thus, it may allow them to pass safely through the digestive tracts of aquatic and marine invertebrates. The frustule is composed of two overlapping halves, or **valves,** like a box with separate top and bottom. The valves are perforated with thousands of tiny pores, which may be arranged in rows, and they can be strengthened with ridges as well. The pores allow transport of nutrients and wastes, and there are specialized pores for mucous secretion—living diatoms typically have a layer of slime coating the outside of the frustule.

Many pennate diatoms (but not centric ones) are rapidly motile by gliding on solid surfaces. Motile diatoms have a slit in one of the valves, termed a **raphe.** It is thought that polysaccharide filaments secreted through the raphe attach to the surface at one end and to the cell membrane at the other end move along the raphe by microtubules and motor proteins, thus moving the diatom relative to the surface.

Because of their rigid wall, diatoms cannot grow larger except by getting thicker. When a diatom has approximately doubled its thickness, it divides by mitosis and cytokinesis, and a new valve is formed for each of the daughter cells. Because the

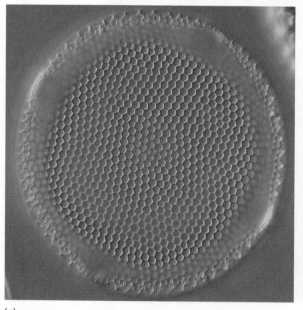
(a)

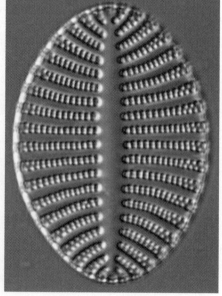
(b)

FIGURE 15.33 Centric (a) and pennate (b) diatoms.

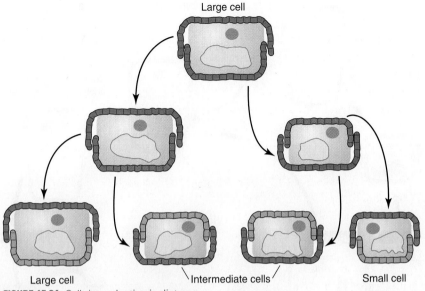

FIGURE 15.34 Cell size reduction in diatoms.

Large cell

Large cell Intermediate cells Small cell

new valve is always the inner one, the two daughter cells are different sizes (Figure 15.34). With each successive division, the average size of diatoms in a population gets progressively smaller. Eventually, some of the cells get so small—approximately one third of the maximum size—that they cannot undergo further division, and they die.

The principal mechanism of regenerating large-cell populations is sexual reproduction. Varying environmental cues—such as nitrogen limitation, increasing temperature, or increasing light levels—stimulate diatoms to undergo meiosis. In response to these cues, all diatoms of a particular species will simultaneously produce gametes, form zygotes, and then spores. When the spores germinate, they form large cells again (Figure 15.35).

15.18 Oomycetes

The oomycetes, or "egg fungi," were recognized in the mid 19th century as probably related to the algae, but were subsequently placed in the fungi until rRNA sequences showed that the early intuition was correct. We now recognize them as members of an algal lineage that lost chloroplasts and became chemoheterotrophic.

There are some unicellular oomycetes, but most are characterized by a mycelial habit of growth, in which a multinucleate cytoplasm is enclosed within a tubular wall consisting of polysaccharide, usually cellulose. Growth involves synthesis of new wall at the tips of the hyphae.

They are a diverse group, consisting of two major types: The aquatic oomycetes, often called the water molds, grow attached to decaying organic material in fresh water, and occasionally in salt water. A few water molds are parasitic on aquatic plants or animals, but most are saprophytic. In contrast, most terrestrial oomycetes are parasitic on plants and cause some serious plant diseases. Late blight of potato and downy mildew of grapes, for instance, both cause major economic damage. The former caused such a devastating outbreak in the potato crop in Ireland in the 19th

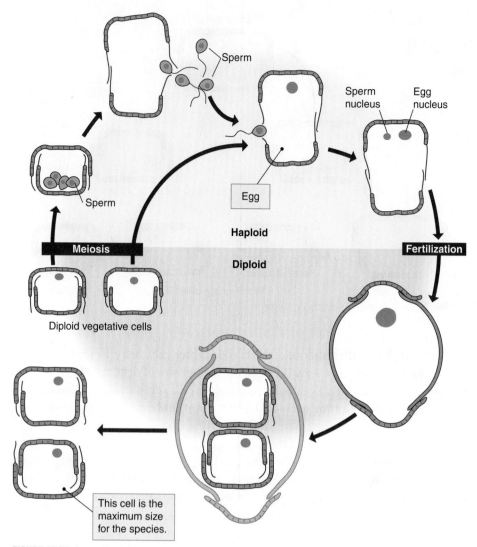

Sperm

Sperm

Egg

Sperm nucleus

Egg nucleus

Haploid

Meiosis

Diploid

Fertilization

Diploid vegetative cells

This cell is the maximum size for the species.

FIGURE 15.35 Sexual reproduction in diatoms.

century that mass starvation ensued, as well as mass emigration of the Irish to America.

The hyphae of water molds invade the decaying material on which they grow, releasing hydrolytic enzymes that break down macromolecules. The soluble nutrients that result are taken up by the water mold. The parasitic forms often have more specialized uptake organs, called **haustoria.** The haustorium is a short branch of the mycelium that invades a host cell by penetrating the host wall, but not the cell membrane. How it works is unclear, but it is thought to take nutrients from the cell, presumably by secreting proteins that make the host cell membrane leaky.

Reproduction in the oomycetes is typically both asexual and sexual, either sequentially or simultaneously. Asexual reproduction involves the formation of sporangia, usually on lateral branches of the mycelium, separated by a septum from the rest of the mycelium. Multiple rounds of cytokinesis produce uninucleate **zoospores,** each with the typical heterokont flagellation. The zoospores are released from the sporangium, swim for a while, and then loose their flagella and form cysts. When the cysts encounter an appropriate bit of detritus, or a host organism, they germinate and form a new mycelium (Figure 15.36).

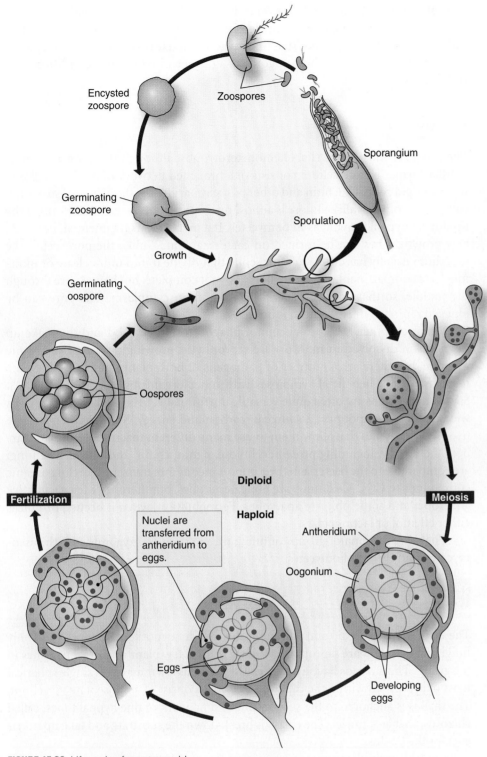

FIGURE 15.36 Life cycle of a water mold.

Sexual reproduction involves the formation of two specialized structures, separated from the mycelium by a septum: an **oogonium** and an **antheridium.** The hypha bearing the antheridium grows toward the oogonium, guided by hormones released by the oogonium. When they make contact, both undergo meiosis to produce haploid nuclei, and the oogonium normally undergoes cytokinesis to produce a variable number of haploid eggs. The antheridial wall then fuses with the wall of the oogonium, and one haploid nucleus is then transferred to each egg. Nuclear fusion regenerates the diploid state. The zygotes then form cysts termed **oospores,** which are released from the oogonium and germinate when conditions are suitable.

Fungi

The fungi are mostly mycelial chemoheterotrophs, although there are a few unicellular forms. The mycelium consists of a branched network of hyphae enclosed within a rigid cell wall. Chitin and other polysaccharides are the principal structural component of the wall; cellulose is almost never present. New cell wall forms at the tip during hyphal growth, as in oomycetes. If a mycelium is fragmented, however, new growth zones can be formed on each fragment. Unlike the oomycetes, the mycelium usually has cross-walls, or septa, that divide it into uninucleate or multinucleate compartments. Septa are normally not complete, but have a pore through the middle, so that adjacent cells have a common cytoplasm. The pore can be plugged rapidly if the mycelium is broken.

Because of their wall, fungi cannot phagocytose their food and thus take up nutrients in dissolved form. Most are saprophytic, growing on decaying organic material, but there are a number of pathogens of both plants and animals. Some form important beneficial symbioses with other organisms (see Section 17.11). Saprophytic forms are often quite versatile nutritionally, being able to metabolize a wide range of compounds as source of carbon and energy. Many of them secrete hydrolytic enzymes that allow them to use many different macromolecules for carbon and energy, including protein, cellulose, lignin, chitin, and others. The fungi are comparable to the bacteria in their importance as mineralizers, recycling organic compounds to CO_2. Most fungi are obligate aerobes, but there are a few facultative anaerobes and some obligate anaerobes. Fermentation, when it occurs, produces either ethanol or lactic acid.

There are four major groups of fungi: the chytrids, the zygomycetes, the ascomycetes, and the basidiomycetes.

15.19 Chytrids

The chytrids are aquatic and soil fungi that produce motile zoospores—the only fungi that do. Most are saprophytic, but there are a few plant or animal parasites in the group. The chytrids have a body (called a **thallus**) that may be a single spherical mass of cytoplasm within the cell wall, or it may consist of a nonseptate mycelium. The thallus is anchored to the substratum by a network of fine hyphal tubes, called **rhizoids,** that lack nuclei. The rhizoids break down the substrate and take up nutrients for the thallus.

Chytrid multiplication can be asexual or sexual. In asexual reproduction of chytrids with a simple spherical thallus, the haploid, multinucleate thallus undergoes multiple rounds of cytokinesis to produce zoospores and then ruptures to release them (Figure 15.37). In chytrids with a mycelial thallus, special sporangia are formed at hyphal tips or on short side branches.

Sexual reproduction is also variable. In simple chytrids, it occurs when the rhizoids of one mating type contact those of another and fuse together. The cy-

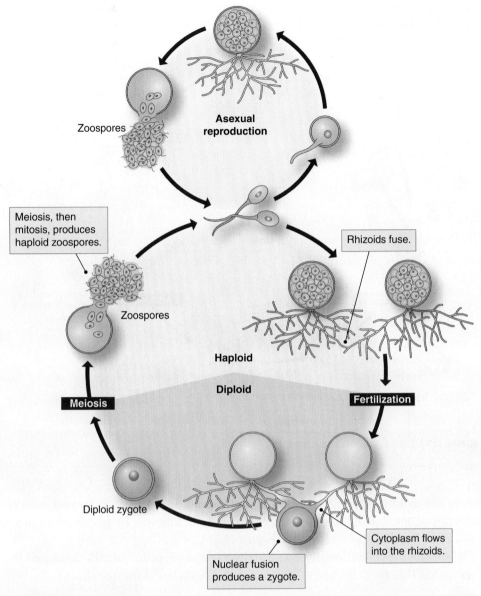

Zoospores

Asexual reproduction

Meiosis, then mitosis, produces haploid zoospores.

Zoospores

Rhizoids fuse.

Haploid

Diploid

Meiosis

Fertilization

Diploid zygote

Nuclear fusion produces a zygote.

Cytoplasm flows into the rhizoids.

FIGURE 15.37 Life cycle of a chytrid.

toplasm of the thalli then flows into the rhizoids and merges in a swelling at the site of original contact. Nuclear fusion generates diploid nuclei, within a thick-walled cyst. Meiosis and germination then release haploid zoospores. Mycelial chytrids have specialized cells that produce motile gametes, whose fusion forms a zygote.

15.20 Zygomycetes

The zygomycetes, or molds, are largely soil fungi, some saprophytic and some parasitic. They are characterized by the formation of a thick-walled **zygospore** as the product of sexual reproduction. Sexual reproduction usually does not involve gametes; rather, as in the simple chytrids, it is initiated by the fusion of mycelia of

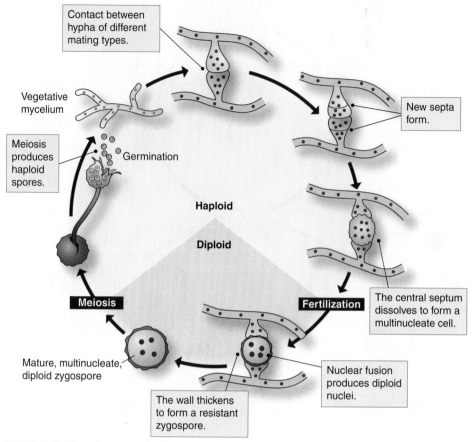

Contact between hypha of different mating types.

New septa form.

Vegetative mycelium

Meiosis produces haploid spores.

Germination

Haploid

Diploid

The central septum dissolves to form a multinucleate cell.

Meiosis

Fertilization

Mature, multinucleate, diploid zygospore

Nuclear fusion produces diploid nuclei.

The wall thickens to form a resistant zygospore.

FIGURE 15.38 Sexual reproduction in a mold.

opposite mating types. After fusion, a single diploid zygote cell is formed, which develops a thick wall. Meiosis inside the zygospore, followed by germination, produces the haploid mycelium again (Figure 15.38).

Asexual reproduction involves the production of spores, usually produced in sporangia that are at the tips of long aerial side branches of the mycelium.

15.21 Ascomycetes

The two most visible groups of fungi are the ascomycetes and the basidiomycetes. Many of them produce macroscopic fruiting bodies, colloquially called mushrooms, during sexual reproduction. These fruiting bodies produce haploid spores, but the process of spore formation differs and is the basis for the distinction between the two groups. In the ascomycetes, spores are formed inside a special cell called an **ascus,** Greek for sac.

Some of the ascomycetes are unicellular and are termed **yeasts.** They are commonly found in habitats high in sugar, such as plant nectars, sap oozing from plant wounds, and fruit surfaces. They are facultative anaerobes, switching between aerobic respiration and alcoholic fermentation depending on the availability of oxygen.

Yeasts are normally haploid, but can mate by cell fusion to form a diploid zygote (see Chapter 12). Sometimes the zygote multiplies vegetatively, and thus, there is a significant diploid phase to the life cycle. More often, however, the zygote immediately undergoes meiosis to form four haploid nuclei, each of which becomes enclosed within a spore wall to form ascospores. Thus, the entire yeast zygote becomes the ascus, rather than a special structure being formed as in the mycelial ascomycetes. Yeast reproduction is mostly asexual by the simple process of either budding or binary fission (Figure 15.39).

In the mycelial ascomycetes, vegetative growth is haploid, as in the yeasts. The mycelia are septate, and usually the individual cells are uninucleate. Asexual reproduction is by the fragmentation of the mycelium into short pieces, or by the formation of **conidia.** Conidia are spores produced at the tips of hyphae. Commonly, there are specialized hyphae, typically projecting away from the substrate into the air, and conidia are produced at the tips of these aerial hyphae (Figure 15.40).

Sexual reproduction in ascomycetes normally involves the formation of specialized cells in regions where mycelia from different mating types come together. One of the specialized cells then transfers a nucleus to the other to form a cell that has two different haploid nuclei in it, a condition termed **dicaryotic.** The dicaryotic cell then multiplies extensively to form an extensive branched dicaryotic mycelium. Eventually, the mycelial tips begin the process of ascospore formation (Figure 15.41).

The formation of the individual ascospores is a complex and poorly understood process. Each of the ascospore nuclei becomes surrounded by a double membrane, probably derived from the cell membrane. The spore wall then forms between these two membranes (Figure 15.42).

Some of the ascomycetes form macroscopically visible fruiting bodies, or mushrooms, made of mycelium matted together. Sometimes the asci are on the surface, and sometimes they are within pits that communicate to the surface through a pore. Regardless of the detailed arrangement, they have many square centimeters of ascus-bearing surface and produce millions of ascospores.

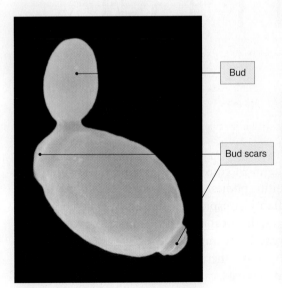

Bud

Bud scars

FIGURE 15.39 A yeast cell.

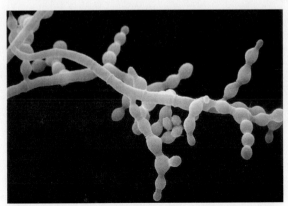

FIGURE 15.40 Chains of conidia.

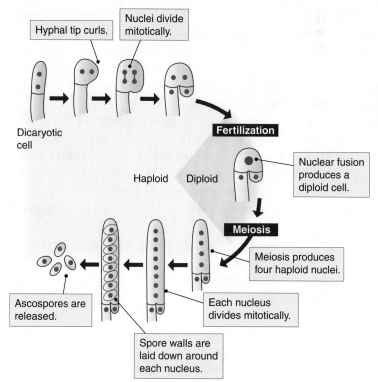

FIGURE 15.41 Overview of ascospore formation.

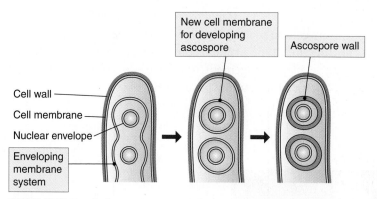

FIGURE 15.42 Detail of ascospore formation.

15.22 Basidiomycetes

The basidiomycetes are certainly the most commonly observed fungi because so many of them make macroscopic fruiting bodies. Although many of the ascomycetes do too, their fruiting bodies are often inconspicuous, and few of the commonly observed mushrooms are ascomycetes; the great majority are basidiomycetes.

Some basidiomycetes grow as yeasts, but most are mycelial, with septate, uninucleate mycelium. Most are saprophytic, but there are some pathogens, most notably of plants. Two groups, the **smuts** and the **rusts,** cause billions of dollars of damage to crop plants annually. Asexual reproduction is similar to that in ascomycetes, with mycelial fragmentation and the production of conidia being common mechanisms.

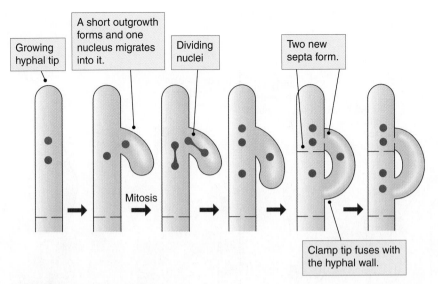

FIGURE 15.43 Formation of clamp connections.

As with ascomycetes, the principal characteristic of the basidiomycetes is the mechanism of formation of haploid spores during sexual reproduction. The process of sexual reproduction is initiated when mycelia of two different mating types come in contact. Cell fusion results in the formation of a dicaryotic cell. Multiplication of this cell results in a dicaryotic mycelium, in which each of the cells has two nuclei, one of each of the parental types. In many basidiomycetes, maintenance of the dicaryotic state involves a specialized mechanism of mitosis and septation, which results in structures termed **clamp connections** (Figure 15.43).

(a)

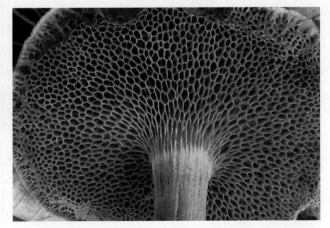

(b)

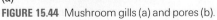

FIGURE 15.44 Mushroom gills (a) and pores (b).

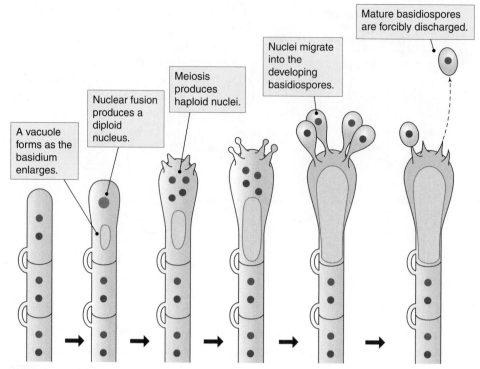

A vacuole forms as the basidium enlarges.

Nuclear fusion produces a diploid nucleus.

Meiosis produces haploid nuclei.

Nuclei migrate into the developing basidiospores.

Mature basidiospores are forcibly discharged.

FIGURE 15.45 Basidiospore formation.

Basidiospores are produced in fruiting bodies, many of which are macroscopically visible and structurally complex. They are, as in the ascomycetes, composed of hyphae matted together, and there may be several different morphologically recognizable tissues. In most cases, a layer of spore-producing tissues lines the surface of hundreds of pores that lead to the outside, or it lines the surfaces of thin plates termed **gills** that hang from the underside of the fruiting body cap (Figure 15.44).

Basidiospores are formed at the tips of hyphae in the spore-forming tissue. Nuclear fusion forms a transitory diploid nucleus, which immediately undergoes meiosis to form haploid nuclei. These migrate into four projections on the surface of the tip cell, now termed a **basidium.** Each projection then develops into a basidiospore, which is released when mature. The release process often involves the forceable ejection of the spore (Figure 15.45).

Summary

In this chapter, we have only touched on the diversity of protists. The complexity of the protistan cell dwarfs that of all other living organisms, and several books the size of this could easily be written on the topic and still not exhaust it. Even so, only a small portion of the protists has been studied in detail at the cellular level, and even fewer at the molecular level. The cell and molecular biology of this group remains one of the most exciting fields in all of biology.

The application of molecular tools to study the phylogeny of the protists is in the process of revolutionizing our understanding of them. As we begin to see the outlines of their evolutionary history, we begin to be able to understand their cell biology better; understanding, for instance, which structural features reflect common descent and which ones represent convergence. It is clear, however, that the evolutionary history of the protists is immensely complex, and major problems remain. It is likely that the phylogeny of this group will see major revisions before a reasonably stable natural classification is possible.

It is clear that among eucaryotic organisms, the vast majority of evolutionary diversity lies within the protists. Many of their lineages rival procaryotic lineages in their antiquity, and there has been much diversification in the billions of years that they have been in existence. For all of the obvious diversity of plants and animals, they are recent newcomers to biodiversity—twigs on the tree of life. Among eucaryotes, as for life in general, it is truly a microbial world.

Study questions

1. What properties characterize the most deeply branching eucaryotic lineages?
2. How are protists related to macroscopic organisms?
3. What types of metabolism characterize eucaryotic microbes?
4. Which groups of eucaryotes contain human pathogens?
5. What different forms of phagocytosis are found in protists?
6. How do the ciliate micronucleus and macronucleus differ?
7. How are motility in flagellated cells, cells with undulating membranes, and dinoflagellates similar and different?
8. Several eucaryotic lineages, like the alveolates and the stramenopiles, include several related sublineages; what characterizes these major lineages?

16 Biogeochemistry and Microbial Ecology

I N MANY WAYS, the earth is a microbial planet. Microbes are the most abundant organisms on earth, constitute the bulk of the earth's biomass, mediate most of the chemical transformations of matter on earth, and maintain the chemical composition of the atmosphere. Plants and animals, for all their visibility, are but twigs on the tree of life and make no essential contribution to the sustainability of life on earth. In this chapter, we survey the microbial contributions to transformation of matter on earth and the habitats largely responsible for them. Microbial habitats are much more diverse than are the habitats of multicellular organisms. They fall into two major categories: free living and symbiotic. We discuss the ecology of symbioses in the next chapter; here we discuss the ecology of free-living microbes and their effects on the global chemistry.

Biogeochemistry

Biogeochemistry is the study of the chemical transformations catalyzed by living organisms and their effect on the earth. The planet we live on is not an inert object inhabited by life; it is a dynamic, interactive system in which many of its properties are the result of the interaction among abiotic forces and living organisms. The chemical composition of the atmosphere, the physical and chemical properties of soil, the salinity of the oceans, and many other features we take as a given are in fact the result of the activities of living organisms. Indeed, many have compared the functioning of the planet as a whole to the functioning of an individual organism, with features that maintain constant temperatures, that detoxify pollutants, that circulate nutrients, and the like. This metaphor is known as the **Gaia hypothesis,** after the ancient Greek goddess Gaia, the earth.

16.1 Chemical transformations on earth are cyclic

All organisms effect chemical changes in their environment. The chemical processes of life generally involve both redox reactions, in which some compounds are oxidized and others reduced, and reactions that reorganize chemical compounds, such as the conversion of simple nutrients into complex polymers. Over the more than three billion years that life has existed here, no essential nutrient has been completely exhausted. This is because for every transformation process, there is a reciprocal process that causes the reverse result.

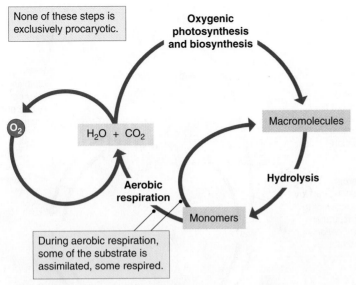

FIGURE 16.1 The aerobic carbon/oxygen cycle.

For instance, carbon and oxygen atoms cycle endlessly through an aerobic carbon cycle: oxygenic photosynthesis causes the conversion of CO_2 and water to macromolecules and O_2; however, hydrolytic enzymes degrade the macromolecules, and aerobic respiration converts the monomers and O_2 back to CO_2 and water. These processes are complementary, and neither can function indefinitely without the other. Together they form a cyclic process that can endure as long as there is energy input from the sun (Figure 16.1).

All transformations of matter on earth are organized into such cycles, and the fact that life has endured for more than 3 billion years without causing a fatal depletion of anything necessary testifies to the intrinsic stability of such arrangements. In addition to the carbon/oxygen cycle, the quantitatively most important element cycles are those of nitrogen and sulfur (Figures 16.2 and 16.3).

16.2 Terrestrial life depends on gaseous or mineral reservoirs

Most of the nutrients used by living organisms are water soluble—nitrogen sources such as amino acids, ammonia, and nitrate; sulfur sources such as hydrogen sulfide, sulfate, and cysteine; and carbon sources such as sugars and amino acids. Because there is a one-way flow of liquid water from the land masses into the oceans, nutrients tend to be removed from soil and transferred to the oceans. Without some mechanism for returning the material to the land, terrestrial life would be impossible, and life on the planet would be confined to oceans and lakes.

For carbon, the mechanism of return is the conversion of organic compounds in the oceans to CO_2, mainly in the sediments of the continental shelf and inshore waters. The CO_2, being gaseous, is available everywhere, including the farthest inland portions of the land masses. Procaryotes are quantitatively the dominant mineralizers of carbon in marine ecosystems.

For nitrogen, return to the land requires a two-step process. First, nitrate, the predominant soluble form of nitrogen in the oceans, is reduced to N_2 by denitrification (an anaerobic respiration) in the sediments of the continental shelf. The N_2, being a gas, is a nitrogenous reservoir available to all of the land masses. Nitrogen fixation

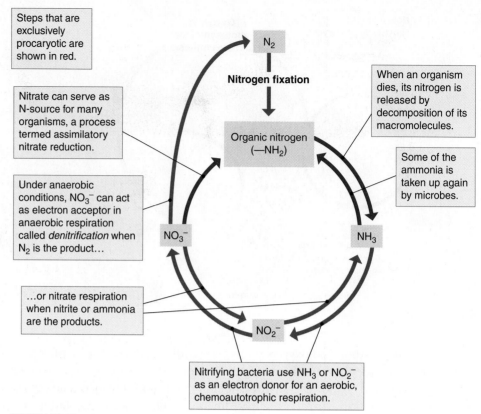

Nitrate can serve as N-source for many organisms, a process termed assimilatory nitrate reduction.

Under anaerobic conditions, NO$_3^-$ can act as electron acceptor in anaerobic respiration called *denitrification* when N$_2$ is the product…

…or nitrate respiration when nitrite or ammonia are the products.

When an organism dies, its nitrogen is released by decomposition of its macromolecules.

Some of the ammonia is taken up again by microbes.

Nitrifying bacteria use NH$_3$ or NO$_2^-$ as an electron donor for an aerobic, chemoautotrophic respiration.

N$_2$

Nitrogen fixation

Organic nitrogen (—NH$_2$)

NO$_3^-$

NH$_3$

NO$_2^-$

FIGURE 16.2 The nitrogen cycle.

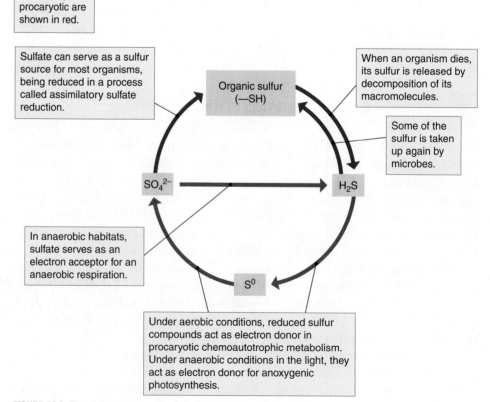

Sulfate can serve as a sulfur source for most organisms, being reduced in a process called assimilatory sulfate reduction.

In anaerobic habitats, sulfate serves as an electron acceptor for an anaerobic respiration.

When an organism dies, its sulfur is released by decomposition of its macromolecules.

Some of the sulfur is taken up again by microbes.

Under aerobic conditions, reduced sulfur compounds act as electron donor in procaryotic chemoautotrophic metabolism. Under anaerobic conditions in the light, they act as electron donor for anoxygenic photosynthesis.

Organic sulfur (—SH)

SO$_4^{2-}$

H$_2$S

S^0

FIGURE 16.3 The sulfur cycle.

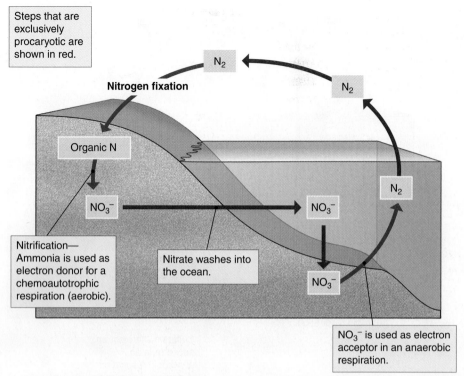

Steps that are exclusively procaryotic are shown in red.

Nitrogen fixation

N₂

Organic N

Nitrification—
Ammonia is used as electron donor for a chemoautotrophic respiration (aerobic).

NO₃⁻

Nitrate washes into the ocean.

NO₃⁻

NO₃⁻

N₂

NO₃⁻ is used as electron acceptor in an anaerobic respiration.

FIGURE 16.4 Return of nitrogen to the land.

in soil returns nitrogen back to terrestrial ecosystems. Both of these processes are done only by procaryotes (Figure 16.4).

Phosphorous exemplifies another type of reservoir: mineral instead of gaseous. Phosphorous is a component of several minerals, and it is released into soil by a solubilization process that is largely the result of acids released by procaryotes and fungi. It ultimately washes into the oceans, where it eventually precipitates as phosphate minerals, to become sedimentary rock. Return to the land takes millions or billions of years of tectonic processes.

For sulfur, the process is more complex; both mineral and gaseous reservoirs contribute to the available sulfur content of soils. Weathering of sulfide-containing minerals and their oxidation by sulfur oxidizing bacteria releases sulfate into soils, with the sulfide minerals being replaced by tectonic processes as for phosphate. In addition, the gas hydrogen sulfide is produced in large quantities by sulfate reducers in the sediments of the continental shelf. Much of the sulfide precipitates as sulfide minerals, but a considerable amount is released into the atmosphere, where it is chemically oxidized to the gas SO_3. This dissolves in raindrops as H_2SO_4 and is returned to the land (Figure 16.5).

16.3 Anaerobic ecosystems are maintained by oxygen consumption

As a result of the great evolutionary success of oxygenic photosynthesis, oxygen is present in earth's atmosphere at a partial pressure of about 0.2 atm. Atmospheric gases penetrate everywhere—into the soil, into the deep oceans, into the tissues of animals, and even deep into the earth's crust, dissolved in water percolating down through networks of fine cracks. Thus, there should in principle be no anaerobic habitats on a planet with such a high atmospheric concentration of O_2.

Despite this, there are abundant anaerobic niches on earth. They fall into two principal categories—the intestinal tracts of animals and marine or freshwater sediments

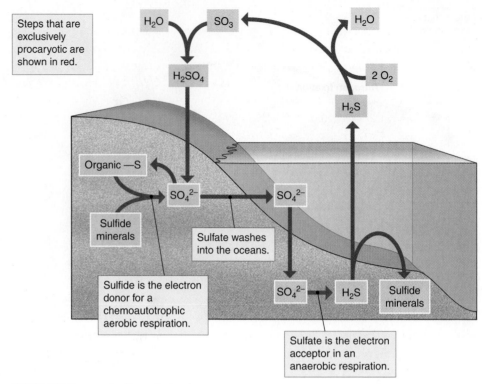

Steps that are exclusively procaryotic are shown in red.

H_2O SO_3 H_2O

H_2SO_4 $2 O_2$

H_2S

Organic —S

SO_4^{2-} SO_4^{2-}

Sulfide minerals

Sulfate washes into the oceans.

Sulfide is the electron donor for a chemoautotrophic aerobic respiration.

SO_4^{2-} → H_2S Sulfide minerals

Sulfate is the electron acceptor in an anaerobic respiration.

FIGURE 16.5 Return of sulfur to the land.

rich in organic material. Both are anaerobic for the same reason—not that they exclude oxygen, but rather that they consume it as fast as it can diffuse in.

In sediments, this process is mediated by a mixture of aerobes and facultative anaerobes concentrated in the narrow interface zone between the aerobic and anaerobic zones. Here aerobic respiration, using the organic nutrients settling from above, and fermentation end products diffusing up from below, consumes oxygen at such a rapid rate that none penetrates through the interface into the anoxic zone below (Figure 16.6).

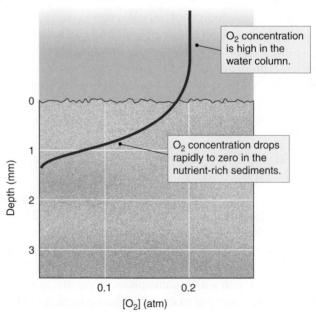

O_2 concentration is high in the water column.

O_2 concentration drops rapidly to zero in the nutrient-rich sediments.

Depth (mm)

$[O_2]$ (atm)

FIGURE 16.6 O_2 profiles in sediment.

In the animal intestinal tract, oxygen continuously diffuses in from the well-vascularized intestinal wall. Peristalsis continually mixes the intestinal contents, and thus, a stable interface zone is hard to maintain. Thus, obligate aerobes cannot really survive there, as they necessarily experience complete anoxia frequently. Even organisms that attach to the intestinal epithelium risk being detached and mixed into the anoxic lumen. Thus, oxygen consumption in the animal intestinal tract is almost entirely due to facultative anaerobes that live mainly by fermentation but that have sufficient concentrations of electron transport components in their membranes that they can make instant use of traces of oxygen when they encounter it. Thus, they keep the oxygen concentration extremely low.

16.4 Some carbon is sequestered in geological formations

Although the animal intestinal system is anaerobic, everything in it is there only temporarily—its contents are regularly emptied to the outside. Thus, the organic matter becomes available as substrate for aerobic respiration and can be mineralized back to CO_2. Sediments, on the other hand, may remain permanently anaerobic. Eventually, sediments become sedimentary rock, sequestering any embedded organic material for geological eons. Under the pressure and temperature of the deep crust, this biomass becomes coal and oil. Eventually it reenters the aerobic carbon cycle when tectonic processes expose it on the surface again.

Even when buried deep in subsurface rock, a mile or more below the surface and long before tectonic process expose sequestered carbon at the surface, biological transformation may begin. Evidence is accumulating that there is a specialized bacterial and archaeal flora in the deep rocks, growing in the places where water percolating down from the surface through fine cracks in the rock meets coal or oil deposits.

16.5 Gaseous products transfer carbon from anaerobic sediments to the aerobic zone

Most of the carbon of anaerobic sediments is returned to the aerobic zone in the form of gases; only a small amount is sequestered in geological formations every year.

CO_2 is a major end product of many fermentations. It comes from decarboxylation reactions, such as in the conversion of pyruvate to acetyl-CoA (which in anaerobes commonly produces CO_2 and hydrogen). As a gas, it readily diffuses out of the anaerobic zone and reenters the aerobic carbon cycle.

Material that is not oxidized to CO_2 in a fermentation is eventually accumulated mainly as acetate. Acetate, in turn, serves as the carbon and energy source for anaerobic respiration: sulfate reduction dominates in marine sediments and methanogenesis dominates in freshwater ones. In sulfate reduction, the ethanol or acetate is oxidized to CO_2, whereas sulfate is reduced to sulfide. The CO_2 can then reenter the aerobic carbon cycle (Figure 16.7).

In freshwater sediments, there is usually not enough sulfate to support very much sulfate reduction, and the most important terminal members of the anaerobic food chain are methanogens. Their end products from acetate are CO_2 and methane. The methane is then oxidized to CO_2 by anaerobic archaea or by aerobic bacteria that live at the interface of the aerobic and anaerobic zones.

Fortunately, as methane is a potent greenhouse gas, these methane-oxidizing bacteria and archaea are efficient and let little methane escape into the atmosphere.

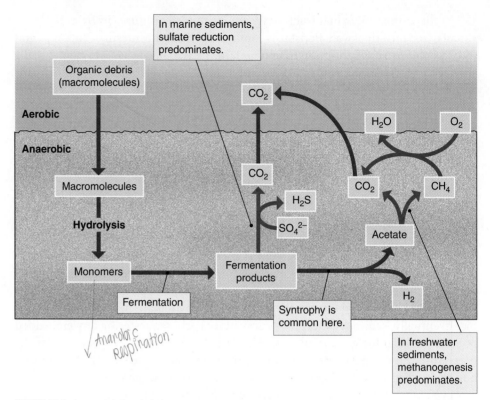

FIGURE 16.7 Anaerobic food chain.

Nevertheless, the methane concentration in the atmosphere is increasing, largely because of the expansion in the numbers of methanogen-containing animals—mainly ruminants such as cattle and wood-eating insects such as termites. Methane produced in the rumen or in the termite hindgut is vented directly into the atmosphere. This methane is a significant contributor to global warming.

16.6 Syntrophy is common in anaerobic ecosystems

A critical step in anaerobic freshwater ecosystems is the conversion of a diverse range of common fermentation end products to acetate. Common fermentation end products include butyrate, propionate, ethanol, succinate, etc. Conversion of these to acetate, the principal substrate for methanogenesis, is done largely by **syntrophic associations.** Syntrophic means "feeding together" and refers to situations in which two organisms can together get energy out of an overall reaction that neither can do individually. This generally involves the production of hydrogen gas by one organism and its consumption by another, and, thus, the process is also termed **interspecies hydrogen transfer.**

The conversion of these fermentation end products into acetate is an oxidative process, and the nicotinamide adenine dinucleotide (NADH) generated has to be reoxidized by hydrogenase, generating H_2 (there is no other available electron acceptor). This reaction is very unfavorable unless the H_2 concentration is kept very low—10^{-5} atm or so. Thus, these reactions, which do occur during the anaerobic degradation of organic material, depend on the presence of methanogens to consume H_2 and to keep its concentration extremely low (Figure 16.8).

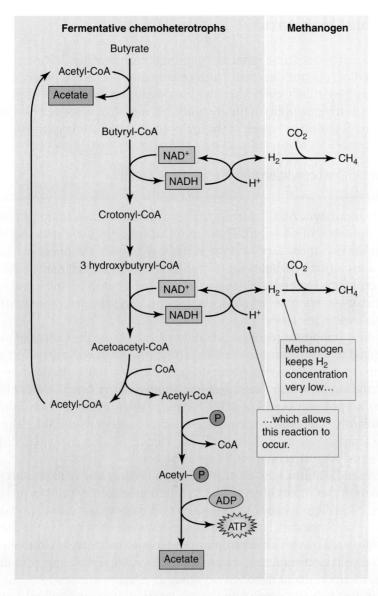

Fermentative chemoheterotrophs   Methanogen

Butyrate

Acetyl-CoA

Acetate

Butyryl-CoA

NAD^+

NADH

H^+

CO_2

H_2 → CH_4

Crotonyl-CoA

3 hydroxybutyryl-CoA

NAD^+

NADH

H^+

CO_2

H_2 → CH_4

Acetoacetyl-CoA

CoA

Acetyl-CoA

Acetyl-CoA

Methanogen keeps H_2 concentration very low…

…which allows this reaction to occur.

P

CoA

Acetyl–P

ADP

ATP

Acetate

$$2 \text{ butyrate} \longrightarrow 4 \text{ acetate} + 4 H_2$$

$$4 H_2 + CO_2 \longrightarrow CH_4 + 2 H_2O$$

$$\overline{2 \text{ butyrate} + CO_2 \longrightarrow 4 \text{ acetate} + CH_4 + 2 H_2O}$$

FIGURE 16.8 Syntrophic oxidation of butyrate to acetate.

Syntrophic associations are seen commonly in anaerobic freshwater ecosystems, where sulfate reducers are not the terminal members of the food chain. In marine sediments, sulfate reducers use many of the fermentation end products directly, and syntrophs are at a competitive disadvantage. Where methanogens are the terminal member of the food chain, however, syntrophy is a necessary step in the turnover of organic material, as the methanogens cannot use organic fermentation end products, other than acetate and formate, as their carbon and energy source. Syntrophy may involve not only a nutritional interdependence, but also a tight physical association, with the two types of cell growing attached to each other, regulating their cell cycles in synchrony. Thus, even in soil and water habitats, symbiosis is encountered.

Syntrophy is common in anaerobic ecosystems

Microbial Habitats

Procaryotic microbes define, as far as we know, the limits of life. Temperatures from −10°C to 120°C, pH from less than 1 to 11, miles under the earth's surface, the depths of the oceans—virtually everywhere we look we find communities of procaryotic microbes. Clearly, the limits of life are much broader than we would have expected on the basis of our observations of the limits on multicellular organisms.

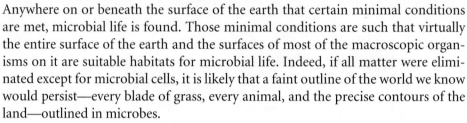

16.7 The minimal conditions for life are quite broad

FIGURE 16.9 Water at 400°C from a geothermal vent.

Anywhere on or beneath the surface of the earth that certain minimal conditions are met, microbial life is found. Those minimal conditions are such that virtually the entire surface of the earth and the surfaces of most of the macroscopic organisms on it are suitable habitats for microbial life. Indeed, if all matter were eliminated except for microbial cells, it is likely that a faint outline of the world we know would persist—every blade of grass, every animal, and the precise contours of the land—outlined in microbes.

The most important requisite is liquid water. Thus, at the surface of the earth, life is limited to temperatures as low as −10°C, where salt concentrations are high enough to depress the freezing point, to approximately 100°C. In the deep oceans, however, the immense hydrostatic pressure (approximately 1 atm for every 10 meters in depth) keeps water liquid at 400°C or more (Figure 16.9). Thus, in the deep oceans, microbes can grow at temperatures that are higher than 100°C. The highest recorded temperature for growth is slightly over 120°C, but many scientists think that the actual limit is as much as 5 degrees higher.

The second requisite is an electron donor and energy source to support primary productivity or the import of organic compounds produced by primary productivity elsewhere. Most illuminated habitats, as long as they are moist, can support oxygenic photosynthesis. Some dark habitats, such as geothermal vents deep in the oceans, can support life based on chemoautotrophic primary productivity, as reduced minerals that can serve as electron donor are carried in the water percolating from the earth, and the sea water contains dissolved oxygen as an electron acceptor.

Other dark habitats depend on the importation of organic material as carbon and energy source for chemoheterotrophs. For instance, the sediments of the continental shelves surrounding the continents are densely populated, but they are mostly too deep to support photosynthesis. Instead, they are fed by a constant fine "rain" of organic debris from the primary productivity in the surface waters. This rain consists of particles of excrement of pelagic animals, discarded exoskeletons of molting arthropods, dead organisms, etc. Thus, the continental shelves are rich habitats due entirely to the importation of organic compounds from the surface, where there is abundant primary productivity.

The third requisite for life is a source of all the various elements that make up biomass. In chemoheterotrophic habitats such as the continental shelf, this is not generally a problem because these habitats are fed by biomass importation, and thus, essential elements are generally present. For habitats that depend immediately on primary productivity, however, the availability of certain minerals may be the factor limiting the growth of phototrophs. The elements most commonly limiting are phosphate and nitrogen. The main reason that the near-shore waters of the oceans, over the continental shelves, are such productive ecosystems is that primary productivity is high due to phosphate and nitrate washed off the land masses and into the oceans.

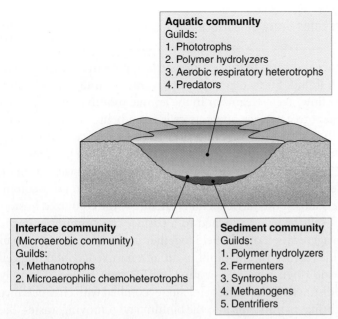

Aquatic community
Guilds:
1. Phototrophs
2. Polymer hydrolyzers
3. Aerobic respiratory heterotrophs
4. Predators

Interface community
(Microaerobic community)
Guilds:
1. Methanotrophs
2. Microaerophilic chemoheterotrophs

Sediment community
Guilds:
1. Polymer hydrolyzers
2. Fermenters
3. Syntrophs
4. Methanogens
5. Dentrifiers

FIGURE 16.10 Community structure of a pond.

16.8 Microbial communities consist of physiological guilds

The microbial community that populates a given habitat generally consists of a number of **guilds**—populations of microbial cells that perform similar metabolic roles in the community. For instance, a freshwater pond might have three principal communities: the aquatic community, the sediment community, and the interface community, each of which is composed of several guilds (Figure 16.10).

Each guild may contain populations of a number of different organisms, each with its own particular niche in the community. For instance, the aerobic phototrophic community in our freshwater pond is likely to contain cyanobacteria, diatoms, chrysophytes, green algae, and others and probably numerous types of each of these.

16.9 Microbial communities create steep chemical gradients

Microbial communities are capable of using nutrients rapidly enough that they create a steep chemical gradient—a vanishingly low concentration near the cells (due to the very high-affinity permeases that procaryotes use for nutrient uptake) and a high concentration a millimeter away. We saw this for oxygen concentration in sediments earlier in this chapter. The reverse is true for waste products (CO_2, H_2S, organic acids, etc.); these compounds can reach very high concentrations in the immediate vicinity of the cells that produce them but drop to low concentrations over the space of a few millimeters. Thus, the growth rate of microbial cells in nature is often limited by the rate of diffusion of nutrients to the cells or of toxic wastes away from them, and the movement of both organic and inorganic compounds in soil and water is driven largely by the concentration gradients that are established by microbes.

16.10 Procaryotes grow on surfaces as biofilms

Much evidence indicates that in nature surface growth is as important as growth suspended in liquid. Surface growth has several advantages. In habitats in which there is water flow, as in streams or in the animal mouth or urinary tract or even in soil, attachment to a surface prevents cells from being washed out of the habitat. Water flow then serves to bring nutrients to the sessile cell. Also, many nutrients adsorb to surfaces, and, thus, their concentration tends to be higher in the vicinity of a surface. Surfaces may be a direct source of nutrients as well, as for decaying bits of organic debris. The surfaces of other organisms are commonly colonized by bacteria that then subsist on nutrients that leak out of the colonized host.

Bacterial cells growing attached to a surface are commonly in the form of a **biofilm**—an aggregation of cells held together in a common matrix of extracellular polysaccharide (Figure 16.11). Biofilms are often mixed, containing cells of multiple species, and often have two or more guilds represented. The polysaccharide matrix has a distinct architecture, being honeycombed with channels through which water flows, bringing nutrients into the biofilm and removing wastes. Biofilm architecture is under active control of the cells, and the continuous remodeling of biofilms involves quorum sensing. It also appears that genetic exchange by both conjugation and transformation increases in frequency in biofilms. Thus, biofilms are active communities that are established, continuously maintained and modified, and even dispersed, under the active control of the constituent cells.

Biofilms make the cells within them more resistant to predation by protozoa (although they do not protect them completely), and they also protect cells against toxic chemicals. Thus, biofilms are resistant to disinfection and even to antibiotics. They can be a serious problem in medicine, particularly when foreign surfaces such as artificial joints, catheters, and the like are placed into the body. These are frequently colonized by bacteria, which are then very difficult to get rid of because of biofilm resistance to antibiotics.

Biofilms are dynamic, as discussed previously. Cells may be motile within a biofilm, and there is continual readjustment of the arrangement of cells and the architecture of the extracellular polysaccharide matrix. Enzymes that break down the matrix are often induced when biofilms get to a certain thickness, and they then result in the release of cells from the biofilm, which presumably can then colonize other surfaces and reinitiate the process.

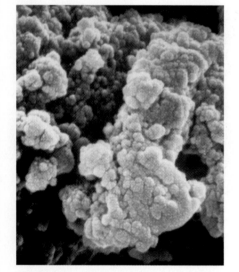

FIGURE 16.11 A biofilm.

16.11 Microbial habitats generally allow periods of rapid growth alternating with starvation, or continuous slow growth

Although there are many different nutritional regimes in microbial habitats, they fall into two general types: Some present nutrients episodically, as when an animal dies and its tissues are decomposed. Many soil organisms experience such momentary surpluses of nutrients, interspersed with long periods of scarcity. A similar variability, but with a more regular rhythm, is experienced by the inhabitants of the human and many other animal intestinal tracts. These animals feed episodically, and their gut microflora gets pulses of nutrients at variable intervals. In contrast, some microbes live in environments that are relatively constant, as in a flowing stream or in the intestinal tracts of grazing animals that feed nearly continuously.

Microbes that alternate periods of growth with long periods in stationary phase clearly are subjected to different selective pressures than those that experience relatively constant conditions. In general terms, they are selected for rapid initiation of growth when nutrients become available (short lag time), rapid growth rates when nutrients are present, and the ability to spend long periods of time in stationary phase. Microbes with more continuous availability of nutrients are selected largely for high-efficiency permeases that can take up nutrients at low concentration and for the ability to sustain very low growth rates without going into stationary phase.

Summary

Procaryotes dominate this planet not only in numbers and biomass, but also in their roles in global transformations of matter. All such transformations must be part of cyclical processes; otherwise, the unidirectional transformation would exhaust precursor materials and accumulate products unable to further sustain life. In these cyclical processes, on which all life is dependent, procaryotes are the dominant players. Not one such process is performed exclusively by eucaryotes, but many are absolutely dependent on procaryotes. Thus, the sustainability of life on earth is dependent on the activities of procaryotes.

Procaryotes also define the limits of physical conditions under which life is possible. No eucaryotic organism, even the versatile protists, can come close to the capabilities of procaryotes to inhabit niches with very high temperature or the extremes of pH. Again, the lesson is clear; earth is fundamentally a planet of procaryotic life.

Study questions

1. Which reactions in the nitrogen cycle are oxidations of the nitrogen, which are reductions, and which involve no valence change?
2. Which reactions in the sulfur cycle are oxidations of the sulfur, which are reductions, and which involve no valence change?
3. What would be the consequences of the extinction of all denitrifying procaryotes?
4. What would be the consequences of the extinction of all nitrogen-fixing procaryotes?
5. Identify the major reservoirs of the principal bioelements carbon, nitrogen, sulfur, and phosphorous.
6. Compare and contrast the anaerobic carbon cycles in marine and freshwater sediments.
7. What are the fundamental requirements for life on earth, and what limits do they set on acceptable microbial habitats?
8. Identify the principal guilds of the marine sediment community.
9. Why are biofilms so common?
10. What are the two basic nutritional patterns seen in microbial communities in nature?

17

Symbiosis

W HEN WE REFER to organisms adapting to their environment, we often think of their adaptations to their physical environment—the temperature extremes in winter and summer, rain and snow, sunlight and shade, the chemistry of the soil, etc. For most organisms, however, the most important elements of their environment are other organisms, and most of their adaptations are to the other organisms present in their habitat. Frequently these adaptations make an organism more or less dependent on the presence of another; when such dependency leads to a stable physical association between the two, we call it **symbiosis**— "living together." We have already seem some examples of this—for instance, the syntrophs, discussed in Section 16.6, live symbiotically with each other, and each would be called a **symbiont.** Symbionts may be approximately the same size, but quite commonly there is a large size difference between the two. In this case, the larger symbiont is termed the **host.** In this chapter, we consider several important symbioses of microbes with other microbes, plants, or animals.

17.1 Symbiosis benefits one partner—the other may benefit, be harmed, or be unaffected

In order for symbiotic associations to be selected through evolutionary time, at least one of the partners must benefit; that is, its survival and reproduction are improved relative to life on its own. Often both partners benefit, and such symbioses are called **mutualistic.** Mutualistic symbioses are often wonderfully complex, as evolution actively selects for adaptations in both partners to make the association more intimate, more stable, and more effective. Many of the associations we describe in this chapter are mutualistic.

In other cases, the second partner may be unaffected, or it may be harmed. Such symbioses are termed **parasitic,** and the symbiont that benefits is termed a **parasite.** Particularly where one partner is harmed, selection favors aggressive adaptations in the parasite and defenses in the host. The result amounts to an evolutionary arms race, in which adaptations in the parasite making it more effective are countered by adaptations in the host, which are in turn countered by new adaptations in the parasite, and so on. The chief category of parasitic symbioses is infectious disease; this is the topic of Chapters 18–20.

The dividing line between mutualistic and parasitic symbioses is not always absolute. A symbiosis may shift from one category to the other, depending on the conditions. For instance, the mammalian intestinal flora is generally considered to be largely beneficial and thus mutualistic (see Sections 18.2 and 18.3). Nevertheless, imbalances in the microbial composition of this flora may cause serious intestinal problems, even death, for the host, and the intestinal flora is a common source of harmful infections of the genital and urinary tracts.

17.2 The benefit of symbiosis to microbial symbionts is usually nutritional

There are a number of types of benefit that symbiosis can confer on symbionts. They may gain protection from adverse environmental influences; for instance, inhabitants of the mammalian intestinal tract experience a constant temperature and are protected from desiccation. They may also provide recognition devices for one of the symbionts, as in the light-emitting organs of some deep-sea fish (discussed later in this chapter).

For microbes, however, symbiosis usually serves a nutritional function—that is, the microbial symbiont normally obtains nutrients from its partner. Of course, that does not preclude other functions from being simultaneously provided. In the case of the intestinal inhabitants of animals, the microbes obtain their nutrients from their host as well as benefiting from the constant physical conditions of the environment. In the light-emitting organs of fish, the microbes receive nutrients as well as protection, and the host uses the microbial bioluminescence as a recognition device or as an aid to capture prey.

Symbiosis may also serve a nutritional function for the host; for instance, no animal has the enzymatic ability to hydrolyze cellulose. Those animals whose diet consists largely of wood or plant leaves, such as termites and grazing and browsing mammals, depend on their microbial symbionts for cellulose digestion.

17.3 In ruminant animals, microbial fermentation is the first step in digesting food

One of the most successful groups of mammals is the ruminants: cattle, sheep, goats, deer, elk, antelope, camels, and the like. These animals graze on grasses or browse on twigs and leaves. They have very complex digestive tracts, characterized by several stomachs, the first of which is essentially a large fermentation chamber—approximately 100 liters in a cow (Figure 17.1). Ruminants feed nearly continuously, and when they are not feeding, they regurgitate small amounts of the rumen contents and re-chew it. This guarantees that the fibrous plant material is finely ground and available for microbial attack.

The rumen is a fermentation chamber, containing vast numbers of microbes (approximately 10^{11} per ml) where hydrolysis of cellulose and other plant polymers occurs, followed by fermentation of the hydrolytic products. The end products are a series of short-chain volatile fatty acids: principally acetate, propionate, and butyrate. Gaseous end products are largely CO_2 and methane. The gas is released by belching, and anything that interferes with the belching reflex can be rapidly lethal, as large amounts of gas are produced and if not released can rupture the rumen, causing a fatal infection of the abdominal cavity. A single cow can produce 60 to 80 liters of gas a day, and this is a significant contributor to global warming, as discussed in Section 16.5.

The fermentation end products are absorbed through the rumen wall and circulated to the animal's tissues by the blood. They serve as energy sources for cellular respiration in these tissues, much as glucose serves as the energy source for other animals.

The rumen is essentially a chemostat, in which there is a continuous input of new plant material mixed with saliva, and there is continuous outflow of rumen

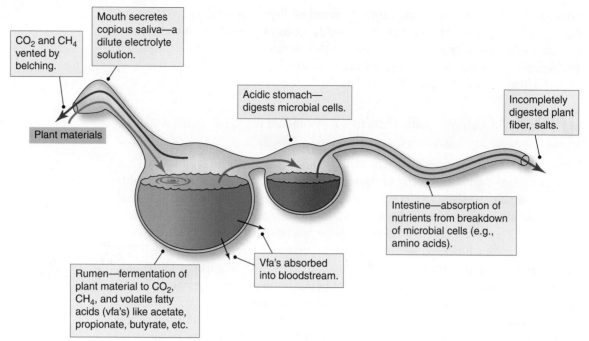

FIGURE 17.1 Simplified diagram of the digestive system of a ruminant.

contents into the second stomach and the lower part of the intestinal tract. There, the microbial cells are digested, and their proteins serve as the nitrogen source for the animal. Thus, ruminants feed their microbial symbionts first, and then they feed on the products of that fermentation: volatile fatty acids and microbial cells. This is in contrast to the digestive systems of most other animals, in which the animal secretes enzymes to break down the food polymers (except cellulose) and absorbs the products, with the microbes getting the leftovers.

The ruminant symbiosis is considered a mutualistic one. Clearly, the host benefits, and although the fate of each individual microbial cell is to be killed and digested, the rumen microbes as a community and each of its constituent species benefits from the continuous source of food and stable physical environment.

17.4 In many symbioses the microbial partner provides fixed nitrogen

The most common limiting nutrient for plant growth is nitrogen, as the most common inorganic nitrogen ion, nitrate, is water soluble and rapidly leaches out of soils. Thus, terrestrial plants depend on nitrogen fixation to replace leached nitrogen. This nitrogen fixation is commonly done by symbiotic associations between plants and nitrogen-fixing bacteria. Although there is some nitrogen fixation by free-living procaryotes, it is quantitatively minor compared to that of symbiotic associations.

There are four major types of nitrogen-fixing symbioses, as shown in Table 17.1.

17.5 The water fern is important in rice cultivation

In the tropics, the symbiosis between the small water fern *Azolla*, and its cyanobacterial symbiont *Anabaena azollae* is of great agricultural importance (Figure 17.2). This little fern, a few centimeters across, floats on the surface of water, with its roots

Table 17.1 Principal Types of Nitrogen-Fixing Symbioses

N₂-Fixing Symbiont	Host	Host Genera or Species	
Cyanobacteria	Bryophytes Ferns Gymnosperms Angiosperms	6 to 8 genera 1 genus 11 genera 1 genus	Particularly important in tropical ecosystems
Cyanobacteria	Fungi	300 to 350 genera	Lichens
Rhizobia	Legumes	100 cultured, and many hundreds of wild species	Agriculturally important
Actinoycetes	Angiosperms	200 species in different families	Particularly important in temperate ecosystems

dangling below. As the leaves develop, they form a pocket on the lower surface, which eventually closes off to form an interior cavity. The cyanobacterial symbionts colonize these pockets and grow to high numbers inside the mature cavities. When growing inside *Azolla* leaves, the *Anabaena* filaments have a large number of heterocysts (specialized N₂-fixing cells; see Chapter 11), and, thus, the rate of nitrogen fixation is considerably higher than necessary to support the microbial growth; the excess is secreted as ammonia and used by the plant.

The *Azolla-Anabaena* symbiosis is cultivated in many parts of Southeast Asia and increasingly in Africa also. The water fern is cultivated in rice paddies and can proliferate to such an extent that it fills the space between rice plants. When the rice is harvested and paddies drained, the Azolla plants decay into the soil and nourish the next rice crop.

17.6 The most important N₂-fixing symbiosis is the rhizobium-legume one

Although the *Azolla-Anabaena* symbiosis is important in some rice-growing parts of the world, a different symbiosis is responsible for nitrogen fixation in dry-land farming and in uncultivated soils. That is the root-nodule symbiosis that involves bacteria of the rhizobium group (members of the proteobacteria—see Section 14.4) with host plants of the legume family. The legumes include about a hundred plants cultivated

(a)

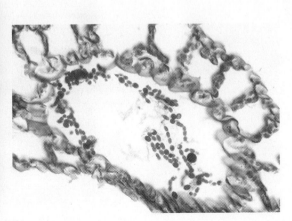
(b)

FIGURE 17.2 Photo of *Azolla* (a) and *Anabaena azollae* seen in a section through an *Azolla* leaf (b).

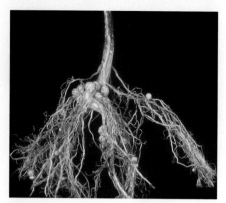

FIGURE 17.3 A legume root nodule.

for the nutritious seeds (such as peas, bean, peanuts), as well as for animal forage (such as clover and alfalfa). In addition, there are many hundreds of species of wild legume. The family has been enormously successful and has spread to virtually all habitats worldwide, in part because of its ability to colonize nitrogen-poor soils.

Nitrogen fixation in legumes occurs in specialized organs termed **root nodules,** which are swellings on the roots that are typically the size of a small pea or smaller (Figure 17.3). They have a reddish tinge when broken open and consist of plant cells that contain hundreds or thousands of bacterial cells each. It is these bacterial cells that are responsible for fixing nitrogen, most of which is exported to the plant. When the plant dies, its tissues are decomposed, and the nitrogen enriches the soil. Even in agricultural systems in which the crop is harvested, the roots remain and enrich the soil. This is the basis of crop rotation, in which a leguminous crop is planted in alternation with nonleguminous crops. Sometimes, if greater enrichment of the soil is desired, the entire plant crop is ploughed into the ground, a practice called **green manuring.**

Because of the wide distribution of wild legumes and because of the ubiquity of legumes as cultivated crops, the legume root nodule is probably the most important source of soil nitrogen on a global basis.

17.7 Establishment of the legume-rhizobium symbiosis involves an exchange of chemical signals

As a legume seedling grows, its roots excrete a mixture of chemical compounds into the soil. The detailed composition of this secretion varies with species, but in most cases includes one or more flavenoids (Figure 17.4). These compounds stimulate the growth of specific strains of rhizobia and cause a chemotactic attraction to the roots. The rhizobia attach to the surface of the roots at the tips of fine root hairs—extensions of the root epidermal cells that increase their surface area. The flavenoids

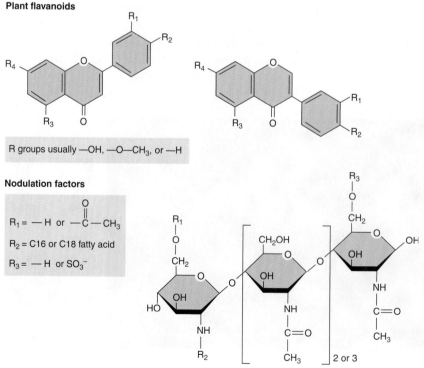

FIGURE 17.4 Flavenoids and nodulation factors.

　　CHAPTER 17 SYMBIOSIS

also induce a series of approximately 30 ***nod* genes** (for nodulation) in the rhizobia, whose products are involved in establishment of the nodule.

Some of the *nod* genes encode enzymes that synthesize a species-specific **nodulation factor,** a short oligosaccharide of modified N-acetylglucosamine residues. This factor induces the plant roots to form an **infection thread**—a cellulose-lined tube from the surface of the root hair into the depths of the root (Figure 17.5). Rhizobial cells migrate down this infection thread to the interior of the root and into cells of the root cortex. There, they multiply inside a large vesicle, and the infected cells multiply as well. The result is a swollen region of the root, consisting of a mass of hundreds or thousands of root cells, each containing thousands of rhizobial cells.

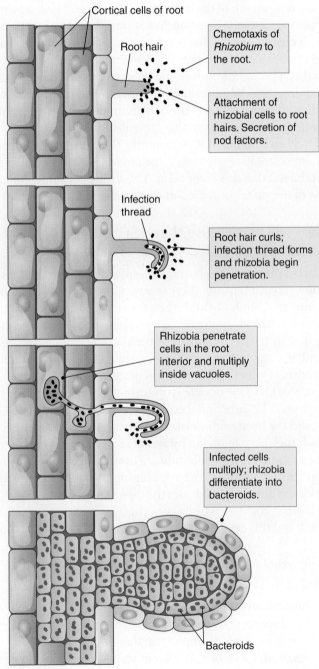

Cortical cells of root

Root hair

Chemotaxis of *Rhizobium* to the root.

Attachment of rhizobial cells to root hairs. Secretion of nod factors.

Infection thread

Root hair curls; infection thread forms and rhizobia begin penetration.

Rhizobia penetrate cells in the root interior and multiply inside vacuoles.

Infected cells multiply; rhizobia differentiate into bacteroids.

Bacteroids

FIGURE 17.5 The nodulation process.

Establishment of the legume-rhizobium symbiosis involves an exchange of chemical signals

The nodule is a dynamic entity—at one end of it, the process of infection of new cells continues, whereas at the other end, the infected cells are dying and lysing. In between, the majority of the nodule mass consists of infected cells in which active nitrogen fixation is occurring. The entire mass is surrounded by an extension of the plant vascular tissue, which transports nutrients to the nodule and fixed nitrogen, in the form of ammonia, from it.

The rhizobia that are active in nitrogen fixation differentiate into swollen, lumpy, distorted rods termed **bacteroids.** Bacteroids are no longer capable of cell division, but remain metabolically active and respire carbon compounds transported to them by the plant vascular tissue (mainly tricarboxylic acid cycle intermediates) and use the ATP and reductant from respiration to fix N_2 into ammonia, which is excreted to the plant.

The ultimate fate of each of the bacteroids is to die and lyse, leaving no descendants. This of course raises the question: What is the benefit of this symbiosis to the rhizobia? This is clearly a mutualistic symbiosis, as both partners have elaborate adaptations to establish and maintain it. The benefit to the plant is obviously the fixed nitrogen that it obtains; however, the benefit to the rhizobia is less clear, as all of the bacteroids in the nodule eventually die, and there is no transmission of infection plant to plant. It is thought that the benefit to the microbial symbiont is not to the cells in the nodule, but rather to their cousins in the **rhizosphere**—the region of soil immediately around the root. Apparently, the plant roots continue to secrete compounds that preferentially stimulate the growth of the particular strain of rhizobium that nodulates that species of plant. Thus, the plant contributes to the competitive success of the rhizobia in the soil ecosystem.

17.8 N_2 fixation requires regulation of oxygen concentration within the nodule

Within the infected plant cell, the bacteroids induce nitrogenase. This induction is a response to low oxygen tension: there is a two-component regulatory system in the cell membrane that binds O_2, and when the concentration drops to a low level, the receptor protein changes conformation and phosphorylates a transcription factor, activating it. This transcription factor then activates transcription of a special sigma factor and, in conjunction with the new sigma factor, activates the transcription of the nitrogenase genes and other proteins necessary to nitrogen fixation.

Thus, the level of O_2 in the nodule is critical—too high, and nitrogenase will not be induced (nor would it function, because of its oxygen sensitivity—see Section 8.24); too low, and the bacteroids cannot fix nitrogen because they become starved for energy, which they have to obtain by aerobic respiration. To maintain a low but steady O_2 concentration, the host cells induce a novel oxygen-binding protein, **leghemoglobin.** This protein binds O_2 tightly, keeping its concentration at about one-tenth that outside the nodule, but releases the O_2 when the concentration drops, thus maintaining a low but constant oxygen concentration. Like vertebrate hemoglobin, the protein is red, and this explains the pink tinge seen in a crushed root nodule.

17.9 The actinomycete *Frankia* forms N_2-fixing symbioses with many host plants

Because of the importance of legumes in agriculture, the legume-rhizobium symbiosis is the most important source of terrestrial fixed nitrogen. In the wild, however, another N_2-fixing symbiosis is probably as important: that of the actinomycete *Frankia* with hundreds of plants, from a number of different plant families, including alder trees in the high latitudes and mountains and ceanothus bushes at lower altitudes and latitudes.

The actinomycete invades the roots of its host plant, forming a nodule that consists of a ball of swollen and branched roots up to several centimeters in diameter (Figure 17.6). Within the nodule, the mycelial bacterium proliferates and forms swollen vesicles at the tips of hyphae, where nitrogen fixation occurs (Figure 17.7). Little is known about the details of the establishment of the symbiosis or of the regulation of nitrogen fixation. It is presumed that the symbiosis is mutualistic, like the legume-rhizobium one, and probably with the same benefits for the partners.

FIGURE 17.6 *Frankia* root nodule.

17.10 Lichens consist of highly coadapted fungi and either algae or cyanobacteria

Lichens are associations of a fungus with a phototrophic organism—either an alga or a cyanobacterium. Each different lichen has a morphologically consistent structure and is given a genus and species name as if it were a single organism; the same name is also used for the fungal symbiont by itself. The two symbionts are highly adapted to life together, but they can usually be cultivated separately from their partner and indeed commonly are found separate in nature. Apparently, the lichen state is a selectively advantageous alternative to separate existence, and each symbiont spends some of its time separate and some in symbiosis.

Although most lichens are two membered, with a single fungal symbiont and a single phototroph, some lichens consist of three or even four symbionts. The most common three-membered lichen consists of a fungus, a green alga that provides most of the CO_2 fixation, and a cyanobacterium that provides fixed N_2.

Lichens are found worldwide, from the arctic to the tropics, from the intertidal to above tree line. They commonly colonize exposed surfaces, such as rocks, tree bark, and the like. There are several different forms: **crustose** lichens adhere tightly to the surface; **foliose** lichens form tough, leafy structures attached at their base to the surface; and **fruticose** lichens form branched, stalk-like structures that commonly hang from the surface (Figure 17.8).

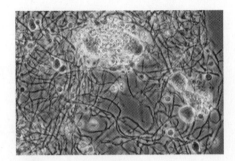

FIGURE 17.7 *Frankia* hyphae.

(a)

(b)

(c)

FIGURE 17.8 (a) A crustose lichen, (b) a foliose lichen, and (c) a fruticose lichen.

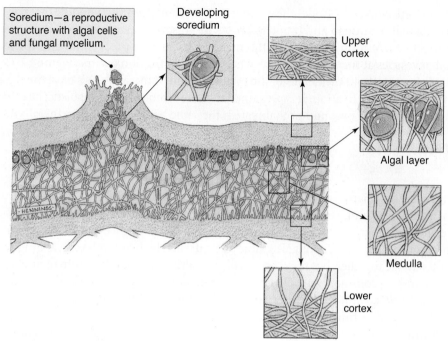

Soredium—a reproductive structure with algal cells and fungal mycelium.

Developing soredium

Upper cortex

Algal layer

Medulla

Lower cortex

FIGURE 17.9 Lichen cross-sections.

The fungal symbiont is usually an ascomycete, although a few basidiomycetes are found in lichens. Because about half of the named fungi are ascomycetes and because more than half of the ascomycetes are components of lichens, about a quarter of the earth's different fungi are components of lichens. Currently, there are about 13,000 to 14,000 species described, each with a different fungal symbiont. The algal or cyanobacterial symbiont is less diverse, however; only about a hundred different species are found in all the lichens.

Within the body of a crustose lichen, there are discrete zones (Figure 17.9). The outer surface is a dense layer of matted fungal hyphae, below which is a layer of algal cells enmeshed in fungal hyphae. Below that is a layer of fungal hyphae, and then the substratum. In foliose and fruticose lichens, the cortex and algal layer are on both sides of a central layer of fungal mycelia. When the phototrophic symbiont is an alga, the fungal cells commonly send haustoria into the algal cell. These are absorptive organs that consist of hyphal penetrations underneath the algal cell wall.

In the lichen symbiosis, the phototrophic symbiont provides fixed carbon to the fungal symbiont. In addition, when cyanobacteria are involved, fixed N_2 may also be provided. In some ecosystems, such as deserts and tundras, lichens containing cyanobacteria are the most important source of fixed nitrogen.

The fungal contribution to the lichen symbiosis appears to be twofold: it provides efficient mineral uptake, and it provides protection against desiccation. Mineral uptake is aided by the secretion of a complex mixture of organic acids, termed lichen acids, that dissolve minerals in the substratum. This is a major cause of rock weathering, and hence of soil formation.

17.11 Mycorrhizae are mutualistic associations of fungi with the roots of angiosperm plants

Most vascular plants form associations with soil fungi called **mycorrhizae** (singular: mycorrhiza). These are associations between a fungus and the roots of the plant, such that the plant provides carbohydrate produced by photosynthesis, and the fungus provides its immense surface area for the uptake of inorganic nutrients.

Mycorrhizae are initiated by the chemotactic growth of the mycorrhizal fungus toward the plant root. When it reaches the root, the fungus penetrates the surface and grows into the interior of the root. Two major groups of mycorrhizae are distinguished by the fine structure of the infected root: In **endomycorrhizae,** the fungus produces haustauria, which penetrate through the plant cell walls and grow into the cell, invaginating the plant cell membrane rather than rupturing it. This produces a large surface area of close contact between the plant and the fungus, where nutrient exchange occurs. Most agricultural crops, ornamental plants, grasses, and cacti fall into this group, and the fungal partner is usually a zygomycete (see Section 15.20).

The other type of mycorrhiza is termed **ectomycorrhizae,** and in this type, the fungal hyphae grow between plant cells and do not penetrate the cell walls. The roots become swollen and covered with a dense mat of fungal hyphae, from which absorptive hyphae radiate out into the soil. Most trees form ectomycorrhizae, usually with basidiomycetes, sometimes with ascomycetes (Figures 17.10 and 17.11).

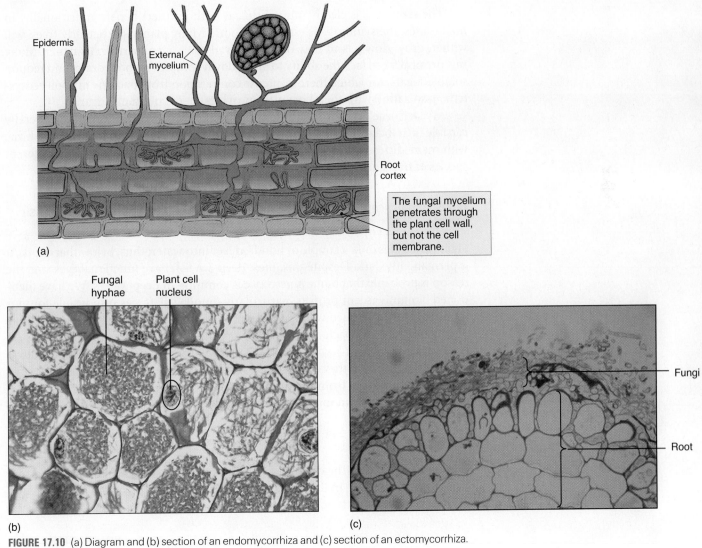

FIGURE 17.10 (a) Diagram and (b) section of an endomycorrhiza and (c) section of an ectomycorrhiza.

Mycorrhizae are mutualistic associations of fungi with the roots of angiosperm plants

FIGURE 17.11 Ectomycorrhizae (two roots on the left) and an uninfected root (right).

The symbiosis is clearly mutualistic, as both partners benefit nutritionally. In many cases, the symbiosis is essential to both. Many plants, particularly trees, will wither, grow slowly, and may die if deprived of their mycorrhizal partner. Many mycorrhizal fungi lack the ability to break down complex carbohydrates and require soluble sugars, for which there is intense competition from bacteria in a soil ecosystem. Association with their host plant provides them abundant sugar.

Mycorrhizae are not highly species specific. Mycorrhizal fungi are generally capable of associating with many species of host, and plants can form mycorrhizae with many different fungal species. A single tree may host as many as 30 different species of fungi on its roots simultaneously.

17.12 Luminescent bacteria form mutualistic associations with deep-sea fish

Light penetrates only a couple of hundred feet into water; thus, below that depth, it is permanently dark. Despite this, most deep-sea fish have functional eyes, and the reason is probably that bioluminescence is common in deep waters. We have mentioned bioluminescent protists, particularly dinoflagellates, as responsible for bioluminescence in surface waters (Section 15.11); in deep waters, the responsible organisms are often bacteria, principally *Photobacterium,* a member of the vibrio group and a relative of *E. coli.* These bacteria are principally inhabitants of the intestinal tract of fish, but they can also be found growing on small particles of organic debris that rain down from the surface waters.

Photobacterium can make an enzyme **luciferase,** which catalyzes this reaction:

$$NADH + H^+ + O_2 + R\text{-}CHO \rightarrow NAD^+ + H_2O + R\text{-}COOH + light$$

The light comes from the decay of an electronically excited FMN prosthetic group that is an intermediate in the reaction. The aldehyde that is involved in the reaction is produced from a fatty acid by this reaction:

$$NADPH + H^+ + ATP + R\text{-}COOH \rightarrow NADP^+ + AMP + PP + R\text{-}CHO$$

Bioluminescence is thus a very expensive activity; it requires two reduced pyridine nucleotides (which could otherwise be used in respiration to produce several molecules of ATP) and two high-energy phosphates (because the PP is rapidly hydrolyzed to two phosphates).

Luciferase synthesis is under the control of a quorum-sensing system. *Photobacterium* produces and excretes a homoserine lactone (a common class of quorum-sensing signal molecules in bacteria), which is required for bioluminescence (Figure 17.12). When cell density is low, so is the lactone concentration, and the cells remain dark. When the cell density is high, so is the lactone concentration, and luciferase is induced. Thus, in the deep sea, a glowing spot of light means a bit of organic debris rich enough to support a dense colony of bacteria. This may help the *Photobacterium* colonize the fish intestinal tract.

One group of fish has exploited the attractiveness of a glowing spot to develop a lure that brings its prey close. These **angler fish** have a specially modified dorsal spine that projects forward and dangles a fleshy mass in front of the fish's mouth (Figure 17.13). The lure is filled with luminescent bacteria, which are fed by nutrients provided by the fish and which emit light because of their high density. Prey are attracted by the lure and then caught by the angler. The cavity in which the bacteria live is open to the outside, and presumably, the organ is colonized by free-living *Photobacterium*.

Other fish, and some squid, have luminous organs that are used in recognition, for courting and schooling behavior. In some cases, the bioluminescence is produced by the tissues of the fish, but in other cases, it is produced by symbiotic luminescent bacteria. For instance, the flashlight fish has light organs beneath its eyes (Figure 17.14). These are filled with *Photobacterium* and glow continuously. In addition, the fish has an eyelid-like structure that it can pull over the organ to turn off the light. There is a great diversity of bioluminescent organs in fish. They may occur in many parts of the body. They may allow the light to be turned off or not. They may open to the outside or to the intestinal tract. In most cases, they involve quite complex anatomical adaptations, suggesting a highly evolved symbiosis.

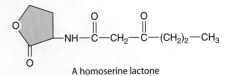

A homoserine lactone

FIGURE 17.12 The homoserine lactone autoinducer of bioluminescence.

A luminescent organ at the top of this stalk lures fish to the angler fish's mouth.

FIGURE 17.13 An angler fish.

Luminescent bacteria form mutualistic associations with deep-sea fish

Open shutter shows the glow of bioluminescent bacteria.

Shutter closed

FIGURE 17.14 Flashlight fish with shutter open and closed.

17.13 Geothermal vent animals host chemoautotrophic symbionts

We have mentioned deep-sea hydrothermal vents before, as the habitats of highly thermophilic bacteria and archaea, defining the upper temperature limits of life. These habitats are interesting for another reason: they are the only habitats on earth that are fueled by chemoautotrophic primary productivity rather than photosynthesis. The hot water percolating through the subterranean rocks dissolve high concentrations of reduced minerals, most importantly H_2S, and thus, the seawater in the vicinity of a vent has a high concentration of these reduced minerals. Their oxidation by aerobic respiration (using the oxygen dissolved in sea water) provides the energy and reductant for chemoautotrophic CO_2 reduction. Despite the novel source of primary productivity for this ecosystem, it is ultimately dependent on oxygenic photosynthesis for the oxygen.

This local primary productivity supports rich communities of microbes and animals. Most of the deep sea floor is sparsely inhabited by animal life consisting of widely spaced scavengers that subsist on the small amount of organic material that drifts down through the immense water column. In the vicinity of hydrothermal vents, however, there are high concentrations of novel invertebrate animals. Dominating these communities are clams, mussels, and tube worms (Figure 17.15). All of these animals have symbiotic associations with bacteria, principally sulfide oxidizers.

The most striking symbiosis is that of the giant (up to 2 meters long) tube worm *Riftia*. This worm completely lacks a digestive system: no mouth or intestinal tract. In place of a gut it has a spongy mass of tissue, termed the **trophosome,** that makes up about 50% of the weight of the animal. Within the trophosome there is a dense mass of chemoautotrophic sulfide-oxidizing bacteria, with more than 10^9 symbionts per gram of tissue. H_2S, O_2, and CO_2 are absorbed by the gills and transported to the trophosome, where the symbionts oxidize the sulfide and reduce oxygen and carbon dioxide. Some of the fixed CO_2 is exported to the animal.

The other large animals of the vent communities also have symbiotic sulfide-oxidizing bacteria, although they are less dependent on them. The clams and mussels have symbiotic bacteria in the gills, but they are filter feeders and can supplement the contribution of their symbionts by filtering free-living bacteria out of the water. There are high numbers of bacteria suspended in the water in the vicinity of hydrothermal vents, many of them also chemoautotrophic.

FIGURE 17.15 The giant tube worm *Riftia*.

Summary

One way that many microbes survive in nature is to live in association with other microbes or with macroscopic organisms. The microbe usually gets nutrients from this association and may be protected from predation or other environmental hazards. Many of these symbioses are mutualistic, and the host also benefits from the association. Hosts may benefit nutritionally also (e.g., from microbial CO_2 fixation, nitrogen fixation, cellulose hydrolysis, or nutrient absorption), or they may gain recognition devices.

Mutualistic symbioses are very common. Most plants enter symbioses with soil fungi, and most animals maintain symbioses with intestinal microorganisms. Thus, almost all macroscopic organisms owe part of their evolutionary success to the symbiotic associations they maintain with microbes. Again, we see the heavy dependence of the biosphere on the microbial world.

Study questions

1. What benefits do microbes and hosts receive in a mutualistic symbiosis?
2. How does the ruminant digestive process differ from that of other animals, such as humans?
3. Coprophagy (the eating of excrement) is common to many animals, particularly young animals that eat excrement of adults of the same species. Why might this behavior have been selected?
4. How is nitrogenase protected from inactivation by oxygen in the Azolla symbiosis?
5. How is nitrogenase protected from inactivation by oxygen in the legume symbiosis?
6. What are the benefits to the host and microbe in the legume symbiosis?
7. Why might a microbe whose principal habitat is the intestine of deep-sea fish find bioluminescence advantageous?
8. Why is bioluminescence under quorum control?
9. Is a sustainable ecosystem based entirely on chemotrophic (rather than phototrophic) primary productivity possible?

18

Host Defenses against Microbial Infection

THE EARTH IS really a microbial planet, with the evolutionary newcomers—macro-organisms—being unessential to the sustainability of the biosphere. Nevertheless, their evolution made a big difference to the microbes because they provide a number of new ecological niches to inhabit. The tissues and surfaces of macro-organisms provide unique habitats, often with abundant nutrients and water, but also with complex mixtures of chemicals toxic to most microbes. Adaptation to these habitats included development of resistance to host defenses, as well as the ability to use the available nutrients with maximum efficiency to outcompete other microbes.

Each individual human, each tree, and each fish are as much an ecosystem as an individual, and the numbers of microbes can outnumber the cells making up the macro-organism. It has been estimated that there are about 10^{13} human cells in an individual human body, which is colonized by about 10^{14} microbial cells. Health and disease can be viewed as ecological balance or disruption of this complex system.

18.1 Host defenses confine microbial growth to the outside surfaces of the body

The principal function of the defenses discussed in this chapter is to prevent microbial colonization of the deep tissues—those that lie beneath the skin and mucous membranes that line the outside of the body. This is a dynamic process, as microbes continually breach the barrier of skin and mucous membranes. Every scratch that bleeds opens the bloodstream to the outside. Every cut can introduce foreign matter contaminated with bacteria into the tissues. Even vigorous chewing can make the teeth move slightly, causing small tears in the tissue below the gum line which allow bacteria to penetrate. Maintenance of sterility in the deep tissues is not so much a matter of preventing access, as it is of limiting access, and dealing with the continual low-level penetration.

In animals, a state of health is characterized by near sterility of the deep tissues, the lungs, the bladder, and the uterus, with microbial colonization restricted to the other body surfaces. The most heavily colonized area is the mucous membranes and luminal space of the intestinal tract, which can support microbial densities of 10^{11} cells per milliliter of intestinal contents. In mammals with intestinal tracts like humans, this is true of the lower small intestine and of the large intestine. The upper small intestine is much more lightly colonized because of secretion of toxic bile, and the stomach is nearly sterile because of its acidity (pH of approximately 2.0). Less heavily, but still abundantly, colonized in humans and other animals are the mouth and vagina. The urinary and respiratory tracts are nearly sterile in their recesses, but contaminated at their openings. The skin is lightly colonized over most of its surface (with about 10^7 bacteria per cm^2); however, in areas that remain moist, such as in the armpits, in the groin, under the foreskin of the penis, and between the toes (when shoes are worn), the levels can be quite a bit higher.

18.2 The normal flora is usually protective to the host

From the first animals and plants in the seas, to the current huge range of biodiversity, macroscopic organisms have been inhabited by microbes. Thus, the evolutionary process has been one of **coadaptation,** in which both the macroscopic and the microscopic symbionts are continuously selected for adaptations that make their association more stable and sustained and less harmful. This has had the effect of reducing the harm that the microflora does to its host because harm is likely to lead to host defensive reactions, or even death of the host, both of which reduce the chances that symbionts can continue living supported by the host. The result has been that every healthy plant and animal carries hundreds of different types of microorganism, each adapted over millions of years of evolution to inhabit a particular site on the host body, without doing its host serious harm, as well as to coexist with the other microbes in the same site. This collection of microbes is called the **normal flora,** or the **resident flora.**

Some species of the normal flora are specific to particular hosts, but most are capable of colonizing a range of related hosts. *E. coli*, for instance, is found in the intestinal tracts of most mammals, but uncommonly in birds or reptiles. This is presumably a reflection of the biochemical unity of large classes of plant and animal because of their descent from a common ancestor.

The normal flora plays a very important protective function, by preventing the body from being colonized by other, less benign, parasites. The eons of selection of the normal flora for minimal harm to the host result in a stable and sustainable symbiosis. In order for an invading pathogen to get a foothold in the ecosystem, it has to first compete with the resident flora and their specific adaptations. Animals delivered aseptically by caesarian section and raised in a sterile environment, so-called **germ-free animals,** are highly susceptible to infections that would be easily resisted by animals raised with their normal flora intact.

18.3 The normal flora occasionally causes disease

That the normal flora is usually protective does not mean that it is universally benign. Some species cause continual low-level harm, not serious enough to reduce the host's fitness sufficiently to be subject to natural selection, but enough to cause discomfort and sometimes more serious long-term consequences. Others can cause serious infections if they get into the wrong place or if conditions remove competitors and allow them to multiply to higher than normal levels.

For example, the normal flora of the mouth contains many fermentative bacteria that live in biofilms on the teeth, or in the gingiva—the crevices between the teeth and gums. In the continuously flowing environment of the animal mouth, being washed into the saliva and swallowed are certain death to a microbe; survival requires surface growth, or growth in the gingival crevices. Daily removal of this biofilm is the purpose of tooth brushing and flossing. If not removed, plaque thickens, and much of the biofilm becomes anaerobic. Anaerobes and facultative anaerobes in the biofilm then produce organic acids as fermentation end products, and these acids erode the enamel of the tooth, causing **dental caries** (cavities) and inflammation and destruction of gum tissue. Untreated, such dental disease is often progressive, invading the jawbone, and eventually causing fatal infections elsewhere in the body. Archaeological records indicate that dental disease has been around as long as there have been humans.

Members of the normal flora may also become serious pathogens if they get to other areas of the body. Most urinary tract infections, for instance, are caused by *E. coli,* a normally benign intestinal inhabitant.

Other members of the normal flora can be serious pathogens even in their normal niche if conditions allow them to proliferate to a greater than normal extent. For instance, yeast vaginitis is commonly caused by the yeast *Candida albicans,* a member of the normal vaginal flora of many women. It is usually is kept to low numbers by the acidity of the vagina, itself a result of the bacterial fermentation of glycogen secreted by cells of the vaginal mucosa. When host physiology is disturbed and less glycogen is secreted, lactic acid production by fermentation decreases, pH rises, and *Candida* multiplies to pathological numbers.

A similar result is sometimes seen with oral antibiotics, which can seriously disrupt the intestinal flora. Death of some members of the normal flora allows the survivors to multiply to fill the vacancies. This can cause life-threatening intestinal infections by a member of the normal flora—for example, *Enterococcus* or *Clostridium.*

Constitutive (Innate) Host Defenses

All multicellular organisms have a variety of defenses against microbial invaders—microbes that breach the skin or mucous membranes to gain access to deep tissues or microbes that have been breathed into the lungs or migrated into bladder, kidney, or uterus. The defenses of mammals are the best known and apparently also the most complex. We focus on these here.

We can divide the defenses into **constitutive defenses** (often called **innate immunity**), which are always activated, and **inducible defenses** (or **adaptive immunity**), which are specifically induced by exposure to an invader. This is a somewhat artificial distinction, as many defenses are constitutive at a low level and are induced to higher levels by invaders, or a complex defense measure may be partly constitutive and partly inducible. Nevertheless, in the interest of simplifying this very complex subject, we divide defenses in this way. In this section, we examine some of the constitutive defenses; in the next two sections, we discuss the immune system and other inducible defenses.

18.4 Many defenses are mechanical

Many of the basic protections against microbial invasion are mechanical structures or activities. Skin is a complex, multilayered tissue with tight adhesions among cells, providing a nearly impregnable barrier if intact. Mucous membranes line the surfaces of all passageways into the body—the alimentary tract, the respiratory tract, the intestinal tract, the genital tract, and the urinary tract. Mucous membranes are not as tough as skin and can be penetrated by some microorganisms, but nevertheless provide a significant barrier to most microbes.

The respiratory tract is lined with cells that not only secrete mucous (a solution of heavily glycosylated proteins), but they are also ciliated. The beating of their cilia sweeps mucous away from the lungs and toward the pharynx, where it is swallowed and digested. Most bacteria suspended in the air get trapped in this mucous lining as

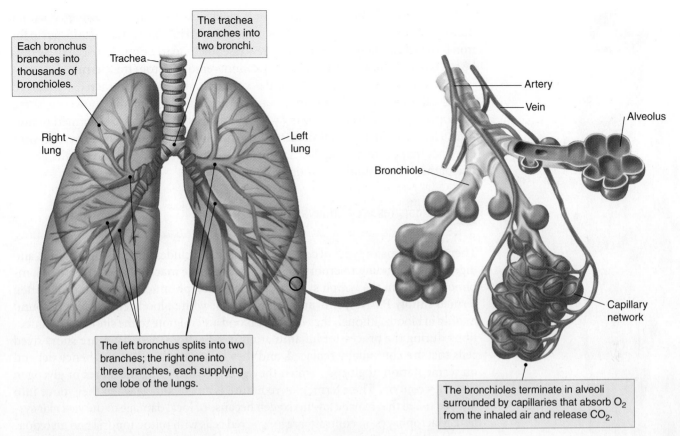

Each bronchus branches into thousands of bronchioles.

The trachea branches into two bronchi.

Trachea

Right lung

Left lung

The left bronchus splits into two branches; the right one into three branches, each supplying one lobe of the lungs.

Artery

Vein

Alveolus

Bronchiole

Capillary network

The bronchioles terminate in alveoli surrounded by capillaries that absorb O_2 from the inhaled air and release CO_2.

FIGURE 18.1 Diagram of the respiratory tree.

they are swept along with a breath, through the ever-smaller tubes of the respiratory tree. In this way, the number of microbes that actually reach the depths of the lungs is quite small. In fact, most particles will not ever reach the lungs; it is only particles in the size range 1 to 5 μm that have much of a chance of avoiding being captured in mucous. This is a very important defense, as most airborne particles are well above 5 μm, and, thus, very few actually reach the very delicate alveoli (Figure 18.1).

Much of the body is swept by flowing liquids. The flow of urine from the kidneys through the bladder and out through the urethra, the flow of saliva that washes everything unattached out of the mouth and into the stomach, and the flow of tears across the cornea of the eye are examples. Similarly, cells of all tissues are bathed by the interstitial fluid, which seeps out of the capillaries of the vascular system, flows slowly through the tissues, enters the lymphatic system (where it is called **lymph**), and then is returned to the vascular system (Figure 18.2). These flows restrict the ability of invading microbes to colonize these places, as the washing action of the moving fluids sweeps contaminants out of the tissues, unless the invaders have some mechanisms to maintain their position in a flowing stream.

18.5 Some nutrients are sequestered in the animal body

The animal body is a concentrated source of nutrients, but they are not all readily available to invading microbial cells. Most are contained within cells of the animal body and are not available unless a pathogen has the ability to lyse host cells to release their nutrients. Even within the blood, lymph, and interstitial fluid, nutrient availability may be a problem. Carbon compounds, such as glucose and amino acids, are abundant in these fluids, but iron is not. Iron is highly insoluble and is transported within the animal body bound to a specific protein, **transferrin.** The affinity of transferrin for iron is so high, around 10^{-12} M, that there is essentially no free iron. Pathogens may thus be starved for iron, simply because they cannot remove it from transferrin.

Of course, most of the iron of the vertebrate body is sequestered within red blood cells, as part of hemoglobin, and is normally unavailable; however, some pathogens have evolved the ability to lyse red blood cells, a process termed **hemolysis.** Hemoglobin degradation then releases the iron. The possession of **hemolysins,** enzymes that cause hemolysis, can thus be an important contributor to the ability of a pathogen to colonize host tissues.

18.6 Phagocytes prey on invading microbes

There are two major types of cells whose function includes phagocytosis, killing, and digestion of invading microorganisms. These are the **macrophages,** a form of lymphocyte (cells in the lymph system), and **polymorphonuclear neutrophils** (often termed simply **PMNs**), a form of leukocyte, or white blood cell. PMNs are found mainly in blood, although they leave the blood and migrate to the site of active infections during the process of inflammation (see Section 18.9). They are short-lived cells that are continually replaced, and they lack mitochondria and hence depend on fermentation of glucose, which they get by hydrolyzing the stores of glycogen that they contain. Their fermentative nature is an advantage when they move into injured tissue that may be low in oxygen because of local damage to the vascular system. Macrophages, in contrast, are long-lived cells with mitochondria and an oxidative metabolism. They are found in lymph, interstitial fluid, the alveoli of the lung, the brain, and virtually everywhere else within the body.

Both types of phagocytic cell move by amoeboid motion and actively seek out invading microbes, which they ingest by phagocytosis. After being ingested, a specialized killing sequence of reactions is initiated—the **oxidative burst,** in which the phagocyte produces a series of toxic oxygen compounds designed to kill the phagocytosed prey (Figure 18.3). Of course, the oxidative burst requires aerobic conditions, and, thus, it is not active in PMNs operating in anaerobic infection sites. Under these conditions, pathogen killing is less effective and depends on an array of digestive enzymes introduced into the phagosome by lysosomal fusion.

This phagocytic capability is particularly important in keeping the very delicate tissue in the alveoli of the lungs free of infection, as there is no resident microflora there to compete with pathogens. Bacteria that make it to the alveoli pose a life-threatening hazard if not disposed of rapidly. Antimicrobial compounds in the secretions of the lung mucosa play a role, but the dominant protective measure is the large number of macrophages that are found there.

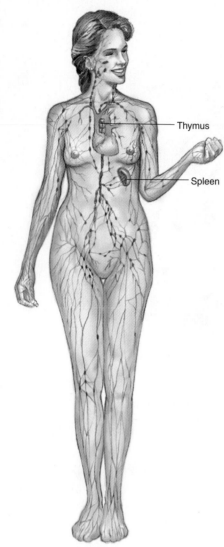

FIGURE 18.2 Diagram of the lymphatic system.

Thymus

Spleen

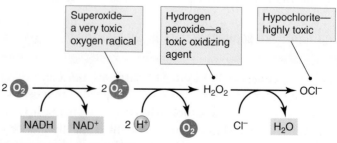

Superoxide— a very toxic oxygen radical

Hydrogen peroxide—a toxic oxidizing agent

Hypochlorite— highly toxic

$$2\ O_2 \longrightarrow 2\ O_2^- \longrightarrow H_2O_2 \longrightarrow OCl^-$$

NADH   NAD$^+$ 2 H$^+$ O$_2$ Cl$^-$ H$_2$O

FIGURE 18.3 The oxidative burst.

In addition to its protective role, phagocytosis of invaders is the first step in several other protective measures, such as inflammation and the immune response, as we see later in this chapter.

18.7 Natural killer cells destroy virus-infected host cells

One type of lymphocyte, termed **natural killer (NK) cells,** seeks out virus-infected host cells. Like macrophages, these cells cruise the body looking, in this case, for virus-infected cells, which are recognized by the viral glycoproteins that appear on their surface as part of the process of budding. When such a cell is encountered, the NK cell binds tightly to it and kills it.

The mechanism by which the target cell is killed is clever, as it takes advantage of a system of programmed cell death in the target cell itself. All cells in multicellular organisms have a system that, when activated, leads to suicide of the cell (a process known as **apoptosis**). Such a pathway is necessary because the process of development of multicellular organisms requires the death of specific cells at specific times and places, and apoptosis is the mechanism by which this happens. NK cells take advantage of this system in their target cells by secreting a protein termed **perforin.** This protein dissolves into the target cell membrane and forms a pore, through which proteins called **granzymes,** also secreted by the NK cell, diffuse. The granzymes activate apoptosis, causing suicide of the target cell. The NK cell then releases its dying prey and resumes its search for other virus-infected cells (Figure 18.4).

Although this process results in death of host cells, these can be replaced by division of adjacent cells as part of the normal process of tissue regeneration. If the virus life cycle is interrupted before progeny virions are produced, however, the infection is aborted. Even if NK cells do not completely short circuit the infection, every infected cell that is killed reduces the severity of the infection somewhat.

18.8 Antimicrobial peptides of many kinds are found in all animals

The mammalian body has a wide range of chemical defenses against invading microbes, some inducible and some constitutive. Among the constitutive defenses are antimicrobial proteins secreted by a variety of different tissues. They fall into two classes: enzymes and antimicrobial peptides. Antimicrobial enzymes include lysozyme, an enzyme that hydrolyzes murein, and phospholipases that hydrolyze the ester linkages in phospholipids.

More important even than these enzymes are a set of **antimicrobial peptides (AMPs)** that are secreted by probably all animals (and plants). They are thus of great antiquity and may have been essential to the transition to multicellular life. AMPs are generally small proteins (under 100 amino acids). They are amphipathic, with separate regions of polar and nonpolar amino acids. They thus have an affinity for membranes, dissolving into them with the polar region on the surface and the nonpolar region in the hydrophobic membrane interior. After dissolving into a membrane, some AMP proteins aggregate to form a pore, thus destroying the membrane (Figure 18.5). In other cases, it appears that the dissolved AMPs simply interfere with lipid packing, thus making the membrane leaky. Host membranes are resistant to the effects of AMPs, probably because their membranes have no overall net charge, unlike microbial membranes, which are usually negatively charged.

AMPs are present on nearly all body surfaces and in all of the tissues. Each plant and animal makes many different ones, probably dozens.

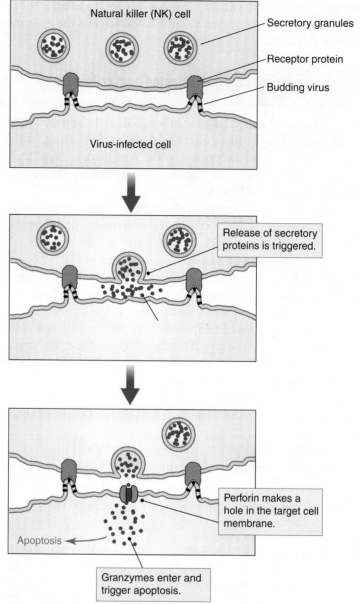

FIGURE 18.4 NK cell action.

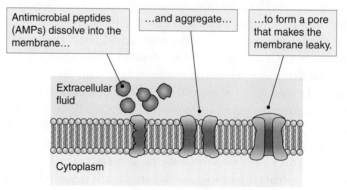

FIGURE 18.5 Creation of a pore by aggregation of defensins.

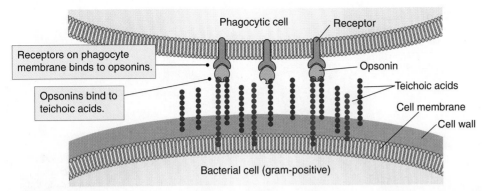

FIGURE 18.6 Opsonization.

Another category of protective proteins is **opsonins,** proteins that enhance phagocytosis of invading microbes. The first step in phagocytosis is the binding of the phagocyte membrane to the microbe to be phagocytosed, and opsonins are proteins that enhance that binding. A number of proteins made by animals bind nonspecifically to structures commonly found on the surfaces of bacteria, like the lipopolysaccharide of gram-negative bacteria or the teichoic acids of gram-positive bacteria. These proteins thus coat the surfaces of invading bacteria. The phagocytic cells in turn have a receptor that specifically binds to other sites on the opsonins, thus attaching the bacteria to the surface of the phagocyte and initiating phagocytosis (Figure 18.6).

Inducible (Adaptive) Host Defenses—Inflammation, Interferon, and RNA Interference

In addition to the constitutive defenses, mammals have a variety of defensive mechanisms that are induced by invading microbes. Some of these defenses are highly specific for the particular strain of microbe involved; others are relatively nonspecific. The best known of these is the **immune system,** which makes mammals resistant to infection by microbes that caused a prior infection; however, there are many other inducible defenses, some of which we describe here.

18.9 Inflammation increases host defenses at the site of infections

The inflammatory process is a mechanism for concentrating antimicrobial substances at the site of an infection. Its results are familiar to all—swelling, redness, and tenderness, sometimes with formation of pus. The redness is due to dilation of the capillaries that brings increased blood flow to the site of infection. The swelling is due to increased leakiness of the capillaries, which allows serum (carrying antimicrobial peptides) to leak into the infected tissue. The pus is formed of large numbers of dead phagocytic cells that invaded the tissue in search of microbial cells.

Initiation of inflammation is via the **complement system,** a group of about a dozen proteins in blood, lymph, and interstitial fluid. Complement proteins are denoted with a capitol "C" followed by a number to indicate the specific protein. The complement system also plays a role in the inducible immune system, and we return to it again there. For now, we focus on only part of it.

The key player in the complement system is C3. C3 can be hydrolyzed by a protease into two fragments, C3a and C3b. This hydrolysis happens continually at a low

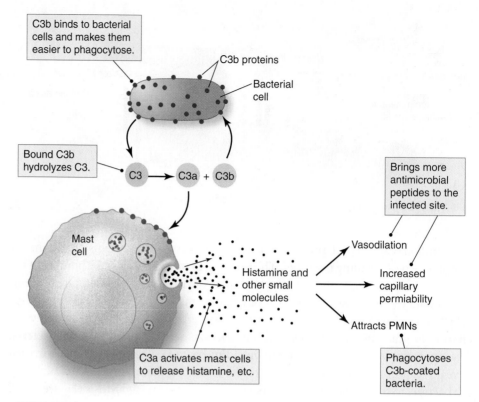

C3b binds to bacterial cells and makes them easier to phagocytose.

C3b proteins

Bacterial cell

Bound C3b hydrolyzes C3.

C3 → C3a + C3b

Brings more antimicrobial peptides to the infected site.

Mast cell

Vasodilation

Increased capillary permiability

Histamine and other small molecules

Attracts PMNs

C3a activates mast cells to release histamine, etc.

Phagocytoses C3b-coated bacteria.

FIGURE 18.7 Role of C3 in initiating inflammation.

rate in the body, and thus, there is a continual low-level presence of C3a and C3b; however, normally they are rapidly degraded. C3b can bind to the surface of bacterial cells or virus particles, binding to LPS or teichoic acids, or the repeated patterns of viral surface proteins. When bound to such a surface, it becomes stabilized and is not broken down. This has two major consequences.

The first is to increase the rate of hydrolysis of C3, thus producing more C3a and C3b. This happens because C3b itself is a protease that can hydrolyze C3, producing more of itself. Thus, when C3b is stabilized by binding to a bacterial cell, the rate C3 hydrolysis goes up very rapidly, insuring that all bacteria in the vicinity become coated with C3b and that the concentration of C3a in the infected tissue rises.

C3a is an activator of **mast cells**—cells in the tissues that release specific compounds that mediate inflammation. When C3a binds to mast cells, they release a number of compounds, including histamine. These compounds act to dilate capillaries and to increase capillary permeability, which brings more antimicrobial peptides to the site. They also attract chemotactic PMNs in the blood to migrate out of the capillaries and into the tissue. Thus, C3a activates the principal features of inflammation (Figure 18.7).

Meanwhile, the C3b that remains attached to the cells of microbial invaders helps in their phagocytosis. Phagocytic cells have a C3b receptor protein in their cell membrane and thus can easily bind to microbes coated with C3b. Because attachment is the first and most difficult step in phagocytosis, the bound C3b greatly increases the susceptibility of microbial cells to phagocytosis; in other words, C3b is another type of opsonin.

18.10 Interferon reduces the ability of viruses to spread

When host cells are infected by a virus, the synthesis of a protein termed **interferon** is often induced. Induction is by double-stranded RNA, which is an intermediate in

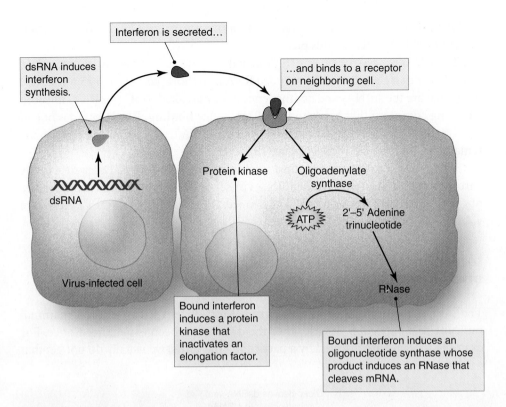

Figure contains the following labels:

dsRNA induces interferon synthesis.

Interferon is secreted...

...and binds to a receptor on neighboring cell.

dsRNA

Virus-infected cell

Protein kinase

Oligoadenylate synthase

ATP

2'–5' Adenine trinucleotide

RNase

Bound interferon induces a protein kinase that inactivates an elongation factor.

Bound interferon induces an oligonucleotide synthase whose product induces an RNase that cleaves mRNA.

FIGURE 18.8 Action of interferon.

the replication of RNA viruses, and is also commonly seen in the folded structure of mRNAs made by DNA viruses. Thus, the presence of double-stranded RNA in significant amounts signals the presence of replicating viruses.

Interferon has little effect on the virally infected cell itself; instead, it is secreted and makes neighboring cells temporarily incapable of protein synthesis and, therefore, less able to support viral replication. Thus, when virions produced by the first cell are released and penetrate neighboring cells, their replication is blocked, and they are ultimately degraded. After a few days in this state, the interferon-affected cells revert to their normal state and resume their own protein synthesis.

The mechanism by which interferon depresses protein synthesis is twofold. When it binds to the surface of a target cell, it induces the synthesis of two enzymes inside the target cell. One is a protein kinase that specifically phosphorylates one of the elongation factors necessary for ribosome movement along mRNAs, thus blocking protein synthesis.

The other enzyme synthesizes an oligonucleotide that consists of three adenosine residues linked 2' to 5'. This adenine trinucleotide activates an RNase that cleaves mRNA, thus further depressing protein synthesis. Cells affected by interferon are thus crippled in their ability to make protein and are thus very poor hosts for viral replication (Figure 18.8).

18.11 RNA interference can destroy viral mRNA

Eucaryotic cells contain another system of defense against RNA viruses—a system called **RNA interference (RNAi),** which is mediated by a set of proteins and short double-stranded RNA molecules called **small interfering RNA (siRNA).** Uninfected cells also use components of the siRNA pathway for normal regulation of gene expression, but we do not discuss this process.

The RNAi process involves two principal enzyme complexes: an endonuclease called **Dicer** that cleaves double-stranded RNA into short pieces (20 to 25 nucleotide pairs) that are the siRNAs and another endonuclease called **RISC** (for **RNA-induced silencing complex**) that binds to the siRNA. After binding, RISC degrades one of the strands of the siRNA and then binds to any mRNA that is complementary to the remaining strand of the siRNA (Figure 18.9).

This system thus protects against viral infection in two ways. Dicer cleaves the chromosomes of double-stranded RNA viruses and the replicating chromosomes of single-stranded RNA viruses. Furthermore, RISC cleaves viral mRNA to reduce viral gene expression.

Inducible Host Defenses—The Immune System and Humoral Immunity

The immune system takes its name from the fact that it can make a mammal **immune** to a particular disease, after the disease has been experienced once. For example, after we are infected by a particular cold virus, we usually do not get that

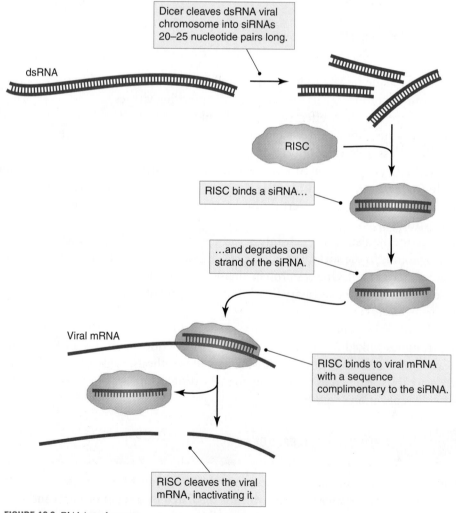

FIGURE 18.9 RNA interference.

CHAPTER 18 HOST DEFENSES AGAINST MICROBIAL INFECTION

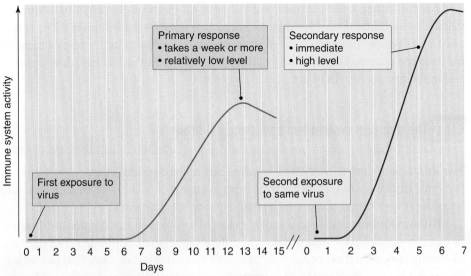

FIGURE 18.10 Immune system activity on first and second exposure.

particular cold again. We may pass the cold on to others in our household, but they do not pass it back to us because we have become immune. (The principal reason we can get so many colds over the course of a lifetime is that there are many different viruses that cause colds.)

Not only is the immune system responsible for preventing recurrent infections, it is also often responsible for curing us of the initial infection. Thus, when we become infected with a cold virus, the viral multiplication in our upper respiratory tract causes the symptoms that we experience. As the immune system kicks in, we begin to get over the cold. Thus, the immune system is a major defense against disease on both first and subsequent exposures. It requires a week or so to become active the first time infection by a specific pathogen occurs, and it is during this period when the immune system is becoming activated that disease symptoms are manifest. This first response is termed the **primary response.** After it has been activated once for a specific pathogen, its reactivation requires only a day or so when the pathogen is encountered again, and this brief period is not long enough for sufficient multiplication of the pathogen to cause disease. This second, more rapid response is called a **secondary response,** and the ability to mount a secondary response is called **immunological memory** (Figure 18.10).

18.12 The immune system recognizes only foreign molecules

The immune system targets foreign materials, termed **antigens,** usually virions or virus-infected host cells, bacterial or protozoal pathogens, or protein toxins produced by pathogens. The foreign materials recognized by the immune system are then destroyed in several different ways, which we describe in subsequent sections. Obviously, it is critical that the immune system be able to distinguish between foreign and native molecules; although it is protective to destroy a protein toxin or invading microbe, to destroy a native host protein or host cell can be damaging. Indeed, there are a number of diseases in which the immune system reacts to one or more native antigens. Such **autoimmune diseases** can be severely debilitating or fatal.

There are no systematic differences between foreign and native materials—they consist of the same types of molecules and often have regions of similar sequence.

Thus, the immune system of each individual must learn not to target native molecules. The inability of the immune system to target native molecules is called **tolerance,** and the development of tolerance is one of the first steps in the development of the immune system. The induction of tolerance occurs during late fetal development and in the first few weeks of life.

18.13 The immune system has two principal branches

The immune system is really two separate but related systems that respond to the presence in the mammalian body of foreign materials. The two branches of the immune system target these invaders by separate mechanisms.

Humoral immunity is mediated by a class of soluble proteins termed **antibodies,** or **immunoglobulins.** There are several different classes of antibodies, but they are all distinguished by their possession of highly specific binding sites that can attach the antibody to a small region of the surface of a bacterial cell or a virus, binding to the specific sequence of sugars of the lipopolysaccharide, or amino acid sequences of surface proteins. Antibodies can also bind to small regions of soluble proteins, such as the toxins that are secreted by some bacteria.

The other branch is termed **cell-mediated immunity,** and it is discussed after humoral immunity.

18.14 Antibodies are proteins that have several identical, highly specific binding sites

There are several different kinds of antibodies, but they all are characterized by having more than one identical binding site at the tips of their "arms." Each arm is composed of two different protein subunits, and both subunits are involved in forming the binding site (Figure 18.11). Just like an enzyme active site, these binding sites are highly specific and can bind with high affinity to only a few closely related targets—for instance, a sequence of three or four amino acids on the surface of a protein or a few sugar residues in a polysaccharide. Because of their high specificity, the immune system needs literally millions of different antibody molecules to be able to be assured that it will be able to target sites on any of the pathogens or toxins that a mammal might be exposed to during its lifetime. How this great diversity of different antibodies is made is briefly discussed later.

The objects to which antibodies bind are called **antigens,** and most antigens have many different sites on their surface to which an antibody can bind. Each of these specific sites is termed an **epitope** (sometimes the older term **antigenic determinant** is used) (Figure 18.12).

18.15 Antibody binding to antigens protects by enhancing phagocytosis and by preventing antigen entry into host cells

The binding of antibodies to an antigen protects the host in several ways: it stimulates the complement system (discussed in the next section), it facilitates phagocytosis of the antigen, and it can prevent viral or toxin antigens from entering host cells.

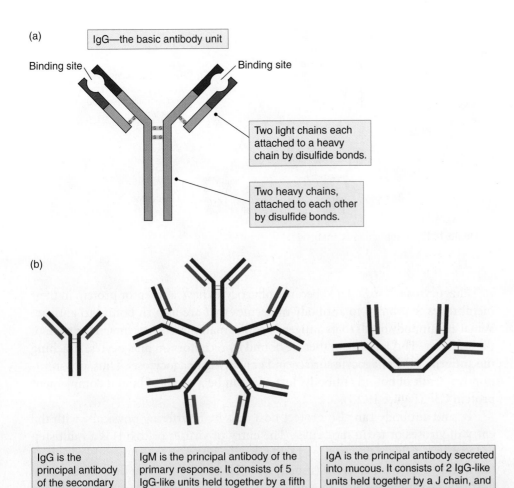

(a)

IgG—the basic antibody unit

Binding site

Binding site

Two light chains each attached to a heavy chain by disulfide bonds.

Two heavy chains, attached to each other by disulfide bonds.

(b)

IgG is the principal antibody of the secondary response.

IgM is the principal antibody of the primary response. It consists of 5 IgG-like units held together by a fifth subunit—the "J chain."

IgA is the principal antibody secreted into mucous. It consists of 2 IgG-like units held together by a J chain, and bound to a secretory subunit.

FIGURE 18.11 Diagram of the structure of the principal antibody classes.

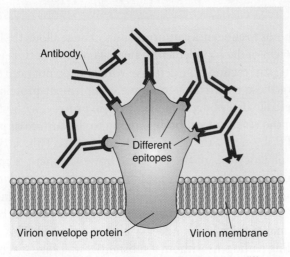

Antibody

Different epitopes

Virion envelope protein

Virion membrane

FIGURE 18.12 Binding of different antibodies to the different epitopes on the surface of a virion.

Antibody binding to antigens protects by enhancing phagocytosis and by preventing

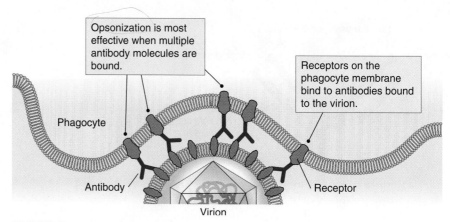

Opsonization is most effective when multiple antibody molecules are bound.

Receptors on the phagocyte membrane bind to antibodies bound to the virion.

Phagocyte

Antibody

Receptor

Virion

FIGURE 18.13 Opsonization by antibody.

Phagocytosis is stimulated because phagocytes have a receptor protein in their membranes that binds to antibody molecules that are in turn bound to antigen. When an antibody binds to its antigen, a conformational shift occurs in the stem of the antibody. This region can then be bound by receptors on phagocytes, attaching the antigen to the phagocyte surface and enhancing phagocytosis. Thus, opsonization is a result of bound antibody, just as it can be a result of bound complement protein C3b (Figure 18.13).

Bound antibody can also protect host cells by interfering physically with the entry of viruses or toxin molecules. The entry of viruses or toxins is a multistep process, in which the first step is a highly specific interaction between the virion or toxin and a particular receptor (usually a membrane protein) on the surface of the host cell. If antibodies bind to epitopes at or near the site that interacts with the receptors, the presence of the antibody can then block the binding to host cells, thus **neutralizing** the virion or toxin (Figure 18.14).

18.16 Antibody binding activates the complement system

When antibody binds to antigen, a conformational change allows the bound antibody to attach to a protein of the complement system. This protein (C1) is then activated and initiates a series of reactions, the details of which we do not need to describe. The ultimate result of these reactions is to cleave complement protein C3, which we already encountered in describing the inflammation process (see Section 18.9).

Thus, one of the results of antibody binding to the surface of pathogens is the accumulation of C3b on the surface of the invading cells. This, as described previously, enhances phagocytosis of the invaders; however, there is another result of bound C3b. It initiates another series of reactions involving several more of the proteins of the complement system, with the final result that a **membrane attack complex** is formed. This is a pore through the membrane formed by complement proteins, which makes the membrane leaky and kills the cell. This attack complex is effective in killing pathogens without a substantial wall (like mycoplasmas, many protozoal pathogens, and even enveloped viruses) and pathogens with a relatively thin wall (like gram-negative bacteria); however, the thick walls of gram-positive bacteria and fungi prevent the membrane attack complex from accessing the cell membrane, and these kinds of pathogens are relatively resistant to the killing effect of complement (although they are still susceptible to the opsonizing effect).

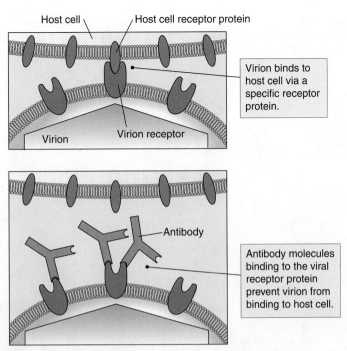

Host cell

Host cell receptor protein

Virion binds to host cell via a specific receptor protein.

Virion

Virion receptor

Antibody

Antibody molecules binding to the viral receptor protein prevent virion from binding to host cell.

FIGURE 18.14 Neutralization of a virion.

18.17 Antibody-producing cells are generated by a process of clonal selection

The mammalian body is capable of producing on the order of a million different antibody molecules, each with a different amino acid sequence in the region of the binding site; however, there are clearly not several million different genes involved. Indeed, the human genome appears to contain only about 25,000 to 30,000 different protein-coding genes. How does the diversity of antibody proteins arise?

It turns out that there are only a handful of genes for antibodies. During embryonic development of the immune system, the developing fetus produces precursors of antibody-producing cells, called **B cells,** a type of **lymphocyte** (cells found in large numbers in the lymphatic system). Part of that differentiation involves recombination events between the different antibody genes. Because there is a nearly infinite variety of different recombination events, differing in the exact site of the recombination, the result is that a great diversity of different cells are produced, each of which has a different antibody gene. Thus, when a mammal is born, it possesses small populations of several million different B cells. Each of these B cells presents on its surface multiple copies of membrane-bound antibody, each with the same epitope specificity.

The process of generation of B-cell diversity contains much randomness in the recombination process, and thus, many antibodies are generated that can bind to native proteins. In order to achieve tolerance of these native epitopes, those B cells that express antibodies to native epitopes are killed during the developmental process. Thus, normally, an animal at birth will only be capable of making antibodies to foreign epitopes.

It is likely that most of the epitopes to which these many antibodies are directed will never be encountered, and the small numbers of potential antibody-producing cells remain arrested in their cell cycle throughout life. However, cells that make antibodies that bind to epitopes that are actually encountered are stimulated to multiply when they encounter their cognate epitope. Thus, the mammalian body specifically amplifies the ability to produce only those antibodies that are directed against epitopes that have actually been encountered (Figure 18.15).

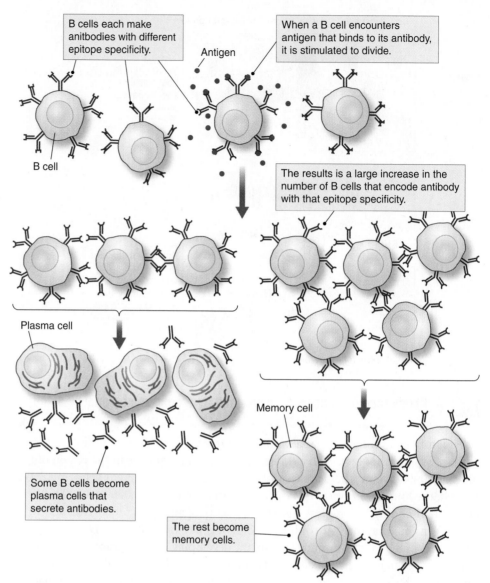

B cells each make antibodies with different epitope specificity.

When a B cell encounters antigen that binds to its antibody, it is stimulated to divide.

Antigen

B cell

The results is a large increase in the number of B cells that encode antibody with that epitope specificity.

Plasma cell

Some B cells become plasma cells that secrete antibodies.

Memory cell

The rest become memory cells.

FIGURE 18.15 Clonal selection.

As a B-cell clone is multiplying during the primary response, some of the daughter cells differentiate into **plasma cells,** cells that secrete antibodies. Part of the process of plasma cell differentiation involves alternate splicing of one of the antibody genes, such that an exon at the C-terminal end is spliced out of the mature mRNA, removing the hydrophobic membrane-binding domain. Thus, in B cells, antibody remains attached to the cell membrane by this hydrophobic region, whereas in plasma cells, the absence of the hydrophobic region allows the antibody to be released from the cell. A plasma cell can secrete several thousand molecules of antibody per second. Most plasma cells last for a few days to a few weeks so that antibody levels in blood and tissue peak near the end of an infection and decline fairly rapidly thereafter.

This multiplication process takes a number of days—typically 5 to 10 days before there are enough antibody-producing cells to control an invading pathogen. This is why the first encounter with a pathogen can lead to disease before the primary response.

Most of the B cells that accumulate as a result of this process of clonal selection do not differentiate into plasma cells, but rather become long-lived **memory cells.** Thus, after exposure to a particular epitope, the body has hundreds of times more

CHAPTER 18 HOST DEFENSES AGAINST MICROBIAL INFECTION

B cells with that particular epitope specificity than before. A second exposure to the same epitope results in a very rapid production of high levels of antibody because there are so many B cells able to differentiate immediately into plasma cells. This is the basis of the secondary response and of immunological memory, discussed at the beginning of this part of the chapter.

18.18 Induction of B-cell multiplication requires bound antigen and helper T cells

The stimulation of B cells to begin replication is mediated by another cell type, **helper T cells,** or **T_H cells.** These are a category of T cell, or T lymphocyte, whose principal role is to stimulate the immune system. Their role in humoral immunity is complex, but their most important job is to stimulate the multiplication of B cells when their cognate epitope is present.

T_H cells recognize B cells via a complex recognition system that includes two principal membrane proteins: **major histocompatibility complex II** (usually referred to as **MHC II**) on the B cell and **T-cell receptors** (**TCRs**) on the T_H cells. The MHC II is a membrane protein found on the surface of B cells, macrophages, and a few other cell types. It has a groove on its external portion in which are bound short peptides derived from proteins that have been ingested by endocytosis.

In addition to MHC II, B cells have bound antibody on their surface. When an antigen, for instance, a protein, bacterial cell, or a virus, binds to the surface antibody, this stimulates the B cell to ingest the antigen. The antigen is then digested, and fragments are then presented on the cell surface bound to MHC II (Figure 18.16).

The other part of the system is the T-cell receptor on the surface of T_H cells, which has the role of recognizing foreign peptides bound in MHC II, very much like antibodies recognize foreign epitopes. Indeed, it is clear that T-cell receptors, anti-

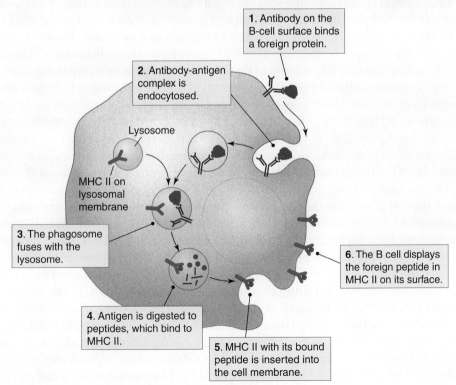

1. Antibody on the B-cell surface binds a foreign protein.

2. Antibody-antigen complex is endocytosed.

Lysosome

MHC II on lysosomal membrane

3. The phagosome fuses with the lysosome.

4. Antigen is digested to peptides, which bind to MHC II.

5. MHC II with its bound peptide is inserted into the cell membrane.

6. The B cell displays the foreign peptide in MHC II on its surface.

FIGURE 18.16 Transport of peptides to the cell surface on MHC II.

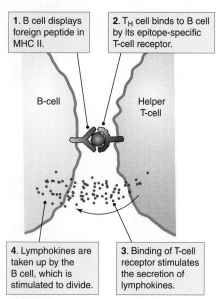

1. B cell displays foreign peptide in MHC II.

2. T$_H$ cell binds to B cell by its epitope-specific T-cell receptor.

B-cell

Helper T-cell

4. Lymphokines are taken up by the B cell, which is stimulated to divide.

3. Binding of T-cell receptor stimulates the secretion of lymphokines.

FIGURE 18.17 Binding of TCR to MHC II.

bodies, MHC proteins, and a number of other immune system proteins evolved from a single primordial protein and thus are homologous. The genes that encode these various proteins make up the **immunoglobulin gene superfamily.**

Each T$_H$ cell has many identical T-cell receptors in its cell membrane, which have on their external surface a binding site that recognizes a particular foreign peptide bound into the groove on MHC II. Thus, a T$_H$ cell can bind to a B cell only if the peptides displayed by MHC II on its surface are foreign and match the epitope specificity of the T$_H$ cell's receptor. When this happens, the T$_H$ cell is stimulated to secrete a number of proteins called **lymphokines,** which are regulators of many functions of the immune system. One of the effects of some of these lymphokines is to stimulate B-cell multiplication (Figure 18.17).

T$_H$ cells are like B cells in that each different type of T$_H$ cell starts out as a small population that expresses a single epitope specificity in its surface T-cell receptor. A process of clonal selection operates to increase the numbers of T$_H$ cells of a given epitope specificity when that epitope is encountered, as described later. Thus, during the primary response, both B cells and T$_H$ cells are multiplying and differentiating, and afterward, there are memory cells of both.

Inducible Defenses—Cell-Mediated Immunity

The cell-mediated arm of the immune system complements humoral immunity and is particularly important in combating viral infections. Cell-mediated immunity is mediated by specific cells termed **cytolytic T lymphocytes,** or **CTLs,** closely related to T$_H$ cells, but dramatically different in their effects.

18.19 CTLs target cells that are making foreign proteins

Like T$_H$ cells, CTLs have epitope-specific T-cell receptors on their surface. CTLs seek out cells in the body that are making foreign proteins and then destroy these cells. This is principally a defense against virus-infected cells, as most cells make only native proteins. However, virus-infected cells make virus proteins in addition to, or instead of, the native proteins. Destruction of these cells helps to contain the virus infection.

CTLs recognize their targets via a complex recognition system that includes a different form of major histocompatibility complex protein on the surface of the target cell, termed **MHC I,** and T-cell receptors on the CTLs. The MHC I is a membrane protein very similar to MHC II, with a groove on its external portion in which are bound short peptides. In this case, however, the peptides are derived from intracellular proteins, rather than from proteins taken in from outside by phagocytosis. Almost all cells in the mammalian body display MHC I on their surface, and they all break down a small portion of their internal proteins continuously. This protein turnover results in a mixture of peptides derived at random from all the different proteins of the cell. These peptides are then displayed on the cell surface bound to MHC I. Thus, the thousands of molecules of MHC I on the surface of a cell display a random sample of peptides derived from all of the protein being made inside. If the cell is normal, all of these will be from native proteins, but if the cell is infected by a virus, some MHC I molecules will display peptides that are foreign and can be recognized by the CTLs (Figure 18.18).

The other part of the system is the T-cell receptor, which functions exactly the same way the T-cell receptors do for T$_H$ cells, by recognizing foreign peptides bound in MHC I. Thus, a CTL can bind to a target cell only if some of the peptides displayed by MHC I on its surface are foreign and recognized by the epitope-specific T-cell receptor. When the CTL binds, it is activated to secrete the proteins perforin and granzymes

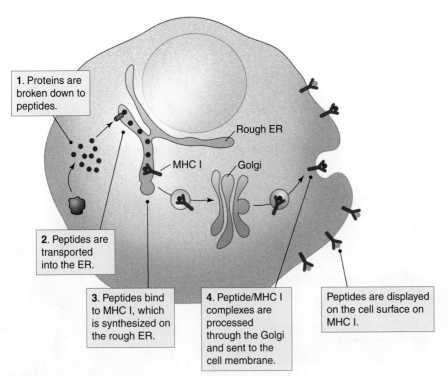

1. Proteins are broken down to peptides.

Rough ER

MHC I

Golgi

2. Peptides are transported into the ER.

3. Peptides bind to MHC I, which is synthesized on the rough ER.

4. Peptide/MHC I complexes are processed through the Golgi and sent to the cell membrane.

Peptides are displayed on the cell surface on MHC I.

FIGURE 18.18 Transport of peptides to the cell surface on MHC I.

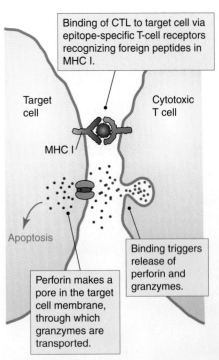

Binding of CTL to target cell via epitope-specific T-cell receptors recognizing foreign peptides in MHC I.

Target cell

Cytotoxic T cell

MHC I

Apoptosis

Perforin makes a pore in the target cell membrane, through which granzymes are transported.

Binding triggers release of perforin and granzymes.

FIGURE 18.19 Binding of TCR to MHC I.

into the gap between the two cells, lysing the target cell by the same mechanism used by NK cells. The CTL then disengages and moves on to seek out other targets (Figure 18.19).

18.20 CTLs and T$_H$ cells are produced by a process of clonal selection similar to the production of B cells

As with B cells, T-cell differentiation (both CTLs and T$_H$ cells) involves gene rearrangements in the developing fetus, leading to small populations of many different types of T cells, each capable of recognizing different epitopes presented by MHC I or II (depending on the type of T cell). Also, as with B cells, clones that have T-cell receptors that recognize native epitopes are destroyed at this time.

With T cells, there is also a multiplication process that is stimulated by encounter with recognized epitopes presented by antigen-presenting cells. In the case of T cells, the antigen-presenting cells are often macrophages. The number of T cells able to bind a particular antigen may only be a few hundred before encountering that antigen, but within a week can increase a million-fold or more. Thus, clonal selection operates to amplify the cell-mediated arm of the immune system, just as with humoral immunity (Figure 18.20).

Also, like B cells, some of the many T cells that result from clonal selection differentiate into cells that actively attack invaders (CTLs), and some differentiate into T$_H$ cells that assist in B-cell multiplication, whereas the rest differentiate into long-lived memory cells. Thus, a second exposure to an antigen normally leads to rapid induction of large numbers of CTLs and T$_H$ cells, in contrast to the slow and lower level induction of the primary response.

Unlike B-cell development, however, the stimulation of T-cell multiplication and differentiation is much less dependent on helper T cells. Nevertheless, most antigens stimulate both humoral and cell-mediated immunity, and, thus, the secre-

CTLs and T$_H$ cells are produced by a process of clonal selection similar to the production

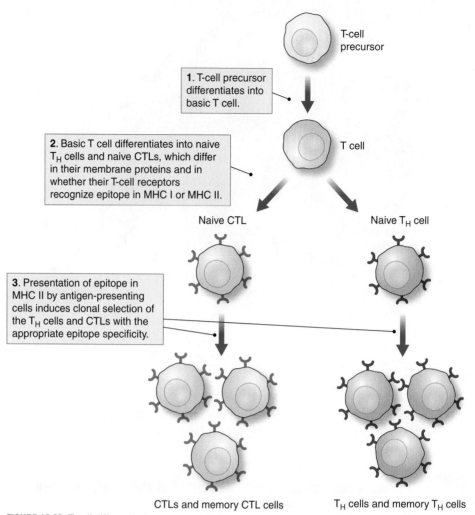

1. T-cell precursor differentiates into basic T cell.

2. Basic T cell differentiates into naive T_H cells and naive CTLs, which differ in their membrane proteins and in whether their T-cell receptors recognize epitope in MHC I or MHC II.

3. Presentation of epitope in MHC II by antigen-presenting cells induces clonal selection of the T_H cells and CTLs with the appropriate epitope specificity.

T-cell precursor

T cell

Naive CTL

Naive T_H cell

CTLs and memory CTL cells

T_H cells and memory T_H cells

FIGURE 18.20 T-cell differentiation.

tion of lymphokines by T_H cells helps to activate T-cell proliferation as well as B-cell proliferation, although the effect on B cells is much greater.

18.21 Much of the immune system activity occurs in lymph nodes

The lymphocytes that mediate the immune system activities—B cells, plasma cells, T cells, T_H cells, CTLs, antigen-presenting cells, etc.—migrate continuously throughout the body; however, they tend to spend quite a bit of their time in **lymph nodes,** which are specialized organs that filter the flowing lymph and that scan it for foreign materials. The lymph node is filled with a meshwork of fibers of connective tissue, and a variety of lymphocytes attach to these fibers (Figure 18.21). Lymph flows into a lymph node on one side and exits on the other; in between, it flows through a dense network of lymphocytes, thus maximizing the exposure of the lymphocytes to any foreign antigens that might be there.

Lymph nodes bring together all of the components of the immune system. They contain a large portion of the body's memory B cells and T cells, and they transport antigens to these cells in the flowing lymph. They are thus a principal site for immune system activation. Indeed, during the course of an active infection, multiplication of lymphocytes in a lymph node can be so active that the node swells significantly and

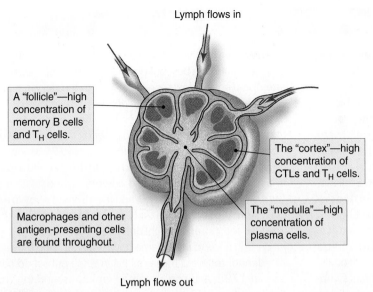

Lymph flows in

A "follicle"—high concentration of memory B cells and T_H cells.

The "cortex"—high concentration of CTLs and T_H cells.

Macrophages and other antigen-presenting cells are found throughout.

The "medulla"—high concentration of plasma cells.

Lymph flows out

FIGURE 18.21 A lymph node.

can become quite tender. For instance, during upper respiratory infections such as colds, the lymph nodes in the neck just under the jaw can be felt as tender lumps, whereas they are normally undetectable. Lymph from the head and neck passes through these lymph nodes, and this is where the principal contact occurs between cold viruses and B cells and T cells. Other principal sites of lymph nodes are in the groin, through which lymph from the legs passes, and in the armpit, through which lymph from the arms passes; these too can become noticeably swollen during the course of an infection. There are also a number of lymph nodes within the body cavity, but they are too deep to ever become detectable, even when badly swollen.

18.22 Vaccination is the deliberate stimulation of the immune system

It has been known for many centuries that those who survived certain diseases became immune to getting them again. The ability to evoke this immunity without suffering the disease, however, is a fairly recent innovation. The first efforts were to prevent serious cases of smallpox by deliberately transmitting the disease, usually by scratching some material from a pustule on someone with active smallpox into the skin of the recipient. This was called **variolation,** and it usually caused a much less serious disease than naturally acquired smallpox—typically, a few percent died, in contrast to about 30% who acquired the disease naturally. The reason for this is not known, nor do we know how variolation was discovered. It apparently originated in Turkey and was brought to Europe in the 1700s. It became widely employed, both to reduce the chance of dying and to reduce the scarring that was characteristic of smallpox. It also became a major tool of military medicine, as a smallpox outbreak could seriously reduce the effectiveness of armies. However, variolation raised the specter of transmission of smallpox from a variolated person to the general population, causing an outbreak. For that reason, it was always controversial.

One of the major advances in the history of medicine occurred in the late 18th century, with the development of **vaccination** (see MicroTopic on page 402). Vaccination was the deliberate infection of people with a virus (cowpox), which led to immunity to a related, but much more serious, virus (variola, or smallpox). Vaccination was rapidly adopted (although not without some controversy).

Smallpox has always been the most feared human disease. It is quite contagious, very painful, causes permanent facial scarring, and has a high mortality rate (30% in Eurasia, where it evolved and has been around for several thousand years, and much higher when it was introduced into North America and the Pacific Islands). So it was a matter of great importance when Edward Jenner announced, in 1798, the discovery of a method to prevent the disease.

Jenner was a doctor in rural Gloucestershire, England, where he had the opportunity to see many cases of the rare disease **cowpox.** Cowpox, a viral disease of wild rodents in England and parts of Europe, is occasionally transferred to domestic livestock and can afflict all members of a dairy herd when milkers go from cow to cow. It reduces milk yields significantly. From cows, cowpox can spread to the people milking them. The initial symptom is the development of an open, ulcerating sore, usually on the hand. This is followed by fever, malaise, pains, headache, swollen lymph nodes, and vomiting. Although the disease can be quite severe, the fatality rate is essentially zero.

Jenner appears to have been impressed with the resemblance of the disease to smallpox, the major differences being the nature and number of pocks and the overall mortality. In the early 1790s he noted that people who got cowpox appeared not to get smallpox, and by 1796 he had compiled of list of nearly a dozen cases in which a previous case of cowpox prevented later smallpox, either natural or from variolation, deliberate smallpox infection.

In May of 1796, Jenner performed the first of two major experimental tests of his theory: he transmitted cowpox deliberately from a sore on the hand of a girl who had milked diseased cows to an 8-year-old boy who had had neither cowpox nor smallpox, by scratching material from the girl's sore into the boy's arm.

The boy got a normal case of cowpox and recovered fully. Jenner then variolated the boy, but no smallpox ensued. Variolation was attempted again a few months later, with the same result.

Jenner's experiments were delayed for a year, because of the lack of cowpox outbreaks, but he did three experimental variolations of farmhands who had had cowpox the previous year, and none showed symptoms. Then, in March of 1798, new cases of cowpox occurred, which allowed him to do a second experiment. In this instance he transmitted cowpox by inoculation directly from a cow to a 5-year-old boy, and then from this boy to a second child, from the second to a third, third to fourth, and fourth to fifth. The first, second, and last children were then challenged by variolation, and none of them came down with smallpox.

Jenner reported these results in a self-published pamphlet later in 1798, and copies spread rapidly around the world. Although it took 180 years, Jenner's method eventually led to the remarkable result that smallpox was driven extinct in the wild, one of the greatest human achievements of all time.

Although Jenner's experiments look ethically questionable to the modern scientist, it is worth remembering the stakes—a safe and effective preventative for the most feared human disease. Furthermore, the disease that Jenner infected the children with was unpleasant for a few days but was nonfatal, caused no known permanent disability, and promised to protect them against a much worse disease to which they would almost certainly be exposed at some time in their lives. The challenge by variolation was an accepted, although controversial, medical practice, that would itself confer benefits if Jenner had been wrong and the children were not protected. The reason that Jenner used mainly children, incidentally, is that the majority of adults had been either infected or variolated with smallpox.

Vaccinators spread through Europe, and later the Americas, inoculating people with material taken from the lesions of previously vaccinated people. Vaccine was also prepared from cows. Somewhere along the line, a different virus, called **vaccinia,** contaminated the vaccine, and, thus, what is now used as a smallpox vaccine is no longer cowpox virus, but the closely related vaccinia virus. No natural host for vaccinia is known, nor is it known where or when it supplanted cowpox. Fortunately, vaccinia causes much milder disease than cowpox and thus is preferable.

Because of vaccination, the proportion of people immune to smallpox gradually grew, and the disease became less serious as an epidemic disease. The process took a long time, but a century and a half after Edward Jenner most of the developed world had essentially eradicated smallpox as a serious public health problem. This led to the audacious proposal to eradicate smallpox from the entire world, an effort that was coordinated by the World Health Organization (WHO). Over a 10-year period, from the late 1960s, vaccinators spread throughout the developing world. In the early stages of the program, the effort was to vaccinate everyone. Later, vaccinators focused on stopping outbreaks by simply identifying cases of smallpox and vaccinating all of those who had come into contact with the patient, a method known as **ring vaccination** (because it established "ring" of immune people around each active smallpox case). Eventually, in 1978, the last natural case of smallpox was identified. In 1980 WHO declared smallpox eradicated. This is one of the greatest human achievements ever—to eradicate completely the disease that had caused the most cumulative suffering and death in all of human history.

Smallpox had several features that allowed its eradication. It infected only humans, and, thus, there was no animal "reservoir" of the disease to transmit infection into the human population. There were no "silent" infections—infections that were unrecognized because they did not cause symptoms but which could still transmit disease to others. The symptoms were distinctive and easily recognizable. There was an effective vaccine available that could be administered without specialized medical training, and the vaccine would cause effective immunity even if administered a few days after exposure (most vaccines have to be given weeks before exposure to be effective).

Buoyed by this success, WHO is attempting to eradicate other diseases—measles and polio are the targets of current campaigns, and there is a reasonable chance of success for these two diseases.

18.23 Vaccines may be live or dead organisms or inactivated toxins

The original vaccine against smallpox was a live, unmodified virus. It worked because it had two features: it was closely enough related to the smallpox virus that immunity to it also conferred immunity to smallpox, and it did not itself cause fatal disease. This particular combination of characteristics is rare among pathogens, so that this approach did not work to stimulate immunity to other pathogens. After it was recognized that immunity could be artificially induced, however, other approaches were developed. There are now three principle ways of stimulating immunity: killed vaccines; live, attenuated vaccines; and inactivated toxins.

Killed vaccines consist of pathogens that have been killed, usually by chemical treatment. This commonly lyses cells. These vaccines thus consist of a mixture of all of the proteins, nucleic acids, and polysaccharides of the pathogen, and antibodies to all of these compounds are stimulated. Of course, in a real infection, most of these antibodies are ineffective because the antigens to which they bind are intracellular. Thus, most modern vaccines are not of this type. The original Salk polio virus, for instance, was of this type.

Live vaccines consist of pathogens whose virulence has been **attenuated**—that is reduced to a level where they cause infection but cause no symptoms, or only minor symptoms. This approach is most common with viruses, and attenuation is commonly done by allowing the virus to replicate for many generations in another organism than humans or in tissue culture with nonhuman cells. During the course of such multiplication, the virus adapts to its new host and in so doing becomes less adapted to multiplying in humans. The Sabine polio vaccine was of this type.

One major advantage of live vaccines is that they can spread to unvaccinated people, thus increasing the number of immune people without having to vaccinate everyone. This has been particularly common with the live polio vaccine and has been of particular importance in the developing world, where vaccine coverage is often very spotty. This ability, however, comes at some risk that the virus will revert to a virulent form as it spreads among humans or that rare humans may be susceptible to serious disease even from the vaccine strain. Since WHO has targeted polio for eradication, wild polio viruses have been completely eliminated from the Western Hemisphere. That, however, does not mean that there are no cases of polio in North America or South America; there are still sporadic cases, but they are caused by the vaccine strain. After polio is eradicated from the world and vaccination is discontinued, the expectation is that the vaccine strain will die out naturally.

Most bacterial pathogens cause harmful symptoms as a result of producing protein toxins (see Chapter 19). This provides another possible target for vaccination, using toxin molecules that have been denatured so that they are no longer capable of causing harmful effects. Such inactivated toxins are called **toxoids,** and they stimulate the production of antibodies that can bind to, and neutralize, the native toxin, thus preventing the harmful effects of an infection, even though they do not prevent the infection itself. The tetanus vaccine, for example, is a toxoid.

Sometimes it is not necessary to inactivate a toxin in order to use it as a vaccine. This is the case for the AB toxins. These are toxins with two subunits—the toxin itself (the A subunit) and a second subunit (the B subunit) that is not itself toxic, but which is required to transport the toxic subunit into target cells. Thus, the purified B subunit can be used in its native conformation as a vaccine. The antibodies that are then produced will prevent the action of the toxin because they will prevent the A and B subunits from associating. The anthrax vaccine is of this type. *Bacillus anthracis* produces two separate A subunit toxins that share a single B subunit. Vaccine against the B subunit thus prevents the action of two different toxins and thus prevents disease.

Summary

The mammalian body has a many-layered, complex system of defense against microbial colonization. Some protections are mechanical—the barriers of skin and mucous membranes, the flow of tears, saliva, and interstitial fluid, the mucociliary blanket, and the like. Others are chemical—the antimicrobial proteins of many kinds that are made by many different cells. Still others have a cellular basis—the professional phagocytes that patrol the body, the cytotoxic T cells and the antibody-secreting plasma cells that are the basis of the immune system.

The discovery of vaccination opened an entirely new opportunity for humans to prevent disease. At one time thought to presage the eradication of all infectious diseases, the reality has been more modest; however, vaccination did succeed in eradicating smallpox, the most devastating human disease ever, and the eradication of polio and measles is in sight. Many other serious diseases are controlled, at least in the developed world, by vaccination.

Study questions

1. List the locations on the human body that support a normal flora and those that do not and are normally bacteria-free.
2. What are the advantages and disadvantages of the normal flora to the host?
3. Summarize the major constitutive, or innate, defenses.
4. Compare and contrast macrophages and PMNs.
5. How do the various antimicrobial proteins function?
6. How does the complement system function to combat infection?
7. How do the interferon and RNAi systems function to combat infection?
8. How is antibody binding protective to a foreign cell or protein?
9. What is the basis of immunological memory?
10. Compare and contrast MHC I and MHC II.

Microbial Pathogenesis

IN THE LAST CHAPTER, we discussed the defenses that macroscopic organisms have to prevent their tissues from serving as nutrients for microbial invaders. Especially in mammals, there is a truly impressive array of defenses, such that it is a wonder that any microbes can breach them and establish an infection. Over the long span of evolutionary time, however, there has been what amounts to an endless seesaw, in which each improvement in host defenses is countered by new adaptations in microbial pathogens. If the full range of defenses had come into being all at once, no microbe could have coped, but primitive multicellular organisms developed primitive defenses, which spurred simple microbial countermeasures. Then, over time, both defensive and offensive adaptations became increasingly sophisticated, to the point at which we now have very complex, co-adapted hosts and parasites. In this chapter, we describe some of the microbial adaptations that allow them to colonize the mammalian body and the processes that lead to damage to the host and the development of disease symptoms. We focus here on bacterial and viral pathogens.

19.1 Many pathogens infect surfaces only

Many pathogens normally cause surface infections and rarely or never penetrate to the underlying deep tissues. Surface growth reduces the exposure of pathogens to the array of host antimicrobial defenses but does not avoid them completely. There are many antimicrobial compounds secreted by skin and mucous membranes, including antibodies in mucous secretions. Phagocytes patrol the membranes of the lungs, and, of course, colonization of surfaces often requires bacterial pathogens to compete successfully with the normal flora.

One major class of such surface infections is the bacterial gastrointestinal infections. These result when pathogenic bacteria are ingested and establish themselves in the intestinal tract. To do so, they typically have to be able to prevent themselves from being eliminated rapidly by the movement of material through the gut. This they do by competing with the normal flora for attachment to the epithelial cells that line the intestinal tract (Figure 19.1). Attachment is usually via pili whose tips bear specialized proteins that bind to receptors on the host cells—most commonly the oligosaccharides on cell membrane glycoproteins. Urinary tract infections also require that the pathogen attach to epithelial cells to avoid being washed out.

Such infections are really external; however, other surface infections may penetrate into the surface layers of the body, but without going all the way through

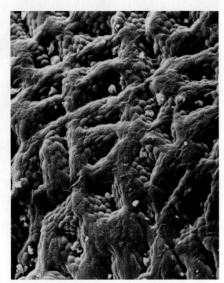

FIGURE 19.1 Microbes attached to the gut epithelium.

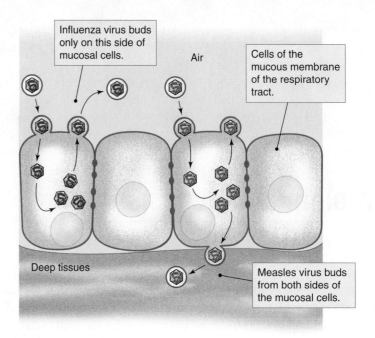

Influenza virus buds only on this side of mucosal cells.

Air

Cells of the mucous membrane of the respiratory tract.

Deep tissues

Measles virus buds from both sides of the mucosal cells.

FIGURE 19.2 Budding of influenza virus and measles virus.

them. For instance, skin infections such as acne or boils are commonly infections of glands within the skin that secrete sweat or the various chemicals that act to lubricate and protect skin.

Because viruses only replicate within host cells, viral infections always involve at least limited penetration of host surfaces; however, many viruses are essentially restricted to surfaces, including many that infect the respiratory tract or the intestinal tract. This is largely a result of the pattern of virus release from host cells. In viruses that cause surface infections, budding of progeny viruses occurs only on the surfaces of the host cell that faces the exterior—the air spaces of the respiratory tract or the lumen of the intestine (Figure 19.2).

19.2 Bacterial adaptations may increase invasiveness or toxigenicity

As a matter of convenience, we can distinguish between properties of bacterial pathogens that increase **invasiveness** (the ability of a pathogen to invade the body, to spread within it, and to establish an infection) and **toxigenicity** (the ability of a pathogen to do harm by producing toxic chemicals). Together, these two determine the **virulence** (capacity to cause disease) of a pathogen. The seriousness of a disease is normally a function of the effectiveness of both invasiveness and toxigenicity, although in some cases one will predominate. For instance, the organism that causes diphtheria (*Corynebacterium diphtheriae*) colonizes the throat, and its growth is restricted to external surfaces in that location. Its invasiveness is thus low; however, it produces a potent toxin—a protein that inhibits protein synthesis on eucaryotic ribosomes. The toxin is secreted and absorbed through the mucous membranes into the bloodstream, which disseminates it around the body. When absorbed by host cells, the toxin acts as an enzyme to modify one of the elongation factors (EF2) necessary for protein synthesis. Because enzymes act catalytically, a single molecule of toxin can completely inactivate all molecules of EF2 in a cell, thereby completely stopping protein synthesis. If enough cells are affected, organ failure can result, especially heart failure. Diphtheria is frequently fatal if untreated; fortunately, vaccination has brought the disease under control in the developed world.

In contrast, the organism that causes gangrene (*Clostridium perfringens*) is highly invasive, but only weakly toxigenic. Gangrene is a wound infection in which tissue in the vicinity of the wound is progressively killed. An aggressive case of gangrene can spread from a single site, for instance a toe, progressively destroying the foot, the lower leg, the thigh, and ultimately, penetrating the trunk and causing death. Before antibiotics, gangrene was a common cause of fatalities following war wounds or industrial accidents; amputation of affected limbs was the only treatment. It was also a common cause of fatality for women who underwent illegal abortions in the United States before Roe v. Wade allowed abortion to be performed in hygienic medical facilities.

Gangrene occurs when clostridial spores are introduced into a wound that is poorly supplied with fresh blood (*Clostridium* is an anaerobe and will not grow in a wound that is well oxygenated). This is more likely in wounds with extensive tissue damage (such as gunshot or shrapnel wounds) than in clean cuts (as from a sharp sword or a scalpel). After the spores germinate and the clostridia begin to grow, they secrete enzymes that diffuse into the surrounding tissue and lyse cells. Most important is an enzyme termed α-toxin, an enzyme that hydrolyzes cell membrane lipids. Host cell lysis provides nutrients for clostridial growth, destroys the vascular system to maintain anaerobiosis, and allows the clostridial cells to invade new tissue. These enzymes are responsible for the invasiveness of the pathogen, but they are only locally effective so that toxigenicity is quite low.

Invasiveness

Invasiveness includes the ability of a pathogen to gain access to the site where it replicates and the ability to defeat host defenses long enough to reproduce sufficiently to produce the symptoms of disease.

19.3 Pathogens are transferred among hosts by several routes

The initial encounter of a host with a pathogen usually is the result of inhaling contaminated air, ingesting contaminated food or water, directly contacting another infected host, or being bitten by an infected organism (such as a rabid dog, a mosquito, or a tick). These four routes define four large categories of disease—**respiratory diseases** are caused by inhaling air contaminated with pathogens; **intestinal diseases** are caused by ingestion of contaminated food or water (the **fecal-oral** route); and **sexually transmitted diseases** (**STDs**) are the major category of disease transmitted by direct contact. In the tropics, a number of skin diseases are also transmitted by direct contact. Finally, the **vector-borne diseases** are transmitted by the bite of an infected **vector** (or carrier), usually an arthropod such as tick or mosquito that became infected by previously biting another infected host (Figure 19.3).

Although respiratory diseases are usually considered to be transferred by respiratory droplets—droplets of saliva or mucous that are released by sneezing, coughing, or even just talking—direct contact is now known to be an equally important route. Most people touch their nose and mouth many times an hour and thus can

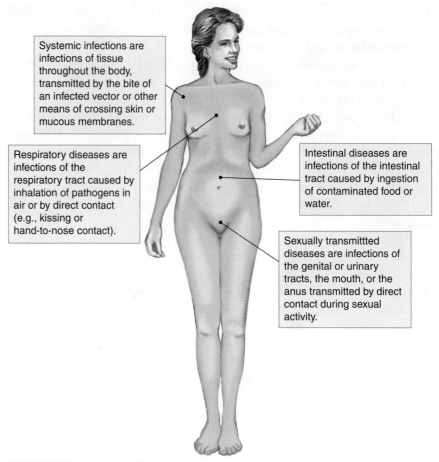

Systemic infections are infections of tissue throughout the body, transmitted by the bite of an infected vector or other means of crossing skin or mucous membranes.

Respiratory diseases are infections of the respiratory tract caused by inhalation of pathogens in air or by direct contact (e.g., kissing or hand-to-nose contact).

Intestinal diseases are infections of the intestinal tract caused by ingestion of contaminated food or water.

Sexually transmittted diseases are infections of the genital or urinary tracts, the mouth, or the anus transmitted by direct contact during sexual activity.

FIGURE 19.3 Disease categories.

transfer respiratory pathogens to their hands. These can be transferred to others by shaking hands or holding hands and then transferred to the new respiratory tract when the new host puts hand to lips or nose.

Most pathogens thus initially contact the host epidermis, either skin or the mucous membranes that line the respiratory, gastrointestinal, and genitourinary tracts. Often this is the principal site of infection, as discussed previously. Access to the deeper tissues is normally prevented by the tight adhesions between skin cells and, to a lesser extent, cells of mucous membranes. Vector-borne pathogens are an exception; they are injected by their vector directly into the deeper tissues.

Despite the barrier that skin and mucous membranes present, many pathogens have developed the ability to penetrate the epidermis and invade deeper tissues. Thus, for these pathogens, the mechanism of transfer does not determine the site of infection. Viruses such as measles or smallpox, for instance, are transmitted by the respiratory route but cause systemic infections. Among the tissues affected are the mucous membranes of the throat and upper respiratory system; this is the source of the respiratory droplets that transmit the infection, even though the most noticeable symptoms are the rash or pustules that result from infection of cells in the vascular system.

Some diseases are not contagious, but rather are the result of disruption of the normal flora. The normal flora includes a number of opportunistic pathogens, whose invasiveness is low when they must compete with all of the other members of the normal flora. If they are introduced to another site with little or no competition, however, they can multiply sufficiently to cause disease. Many urinary tract infections, particularly among women, are caused by *E. coli* strains from the intestinal tract.

Similarly, many abscesses and many wound infections are caused by normal skin or intestinal bacteria. Infections may also follow antibiotic treatment, when the normal flora is suppressed by the antibiotic, leaving the more antibiotic-resistant members to multiply to higher numbers than is normal. Serious intestinal infections by *Enterococcus*, for instance, can follow a course of oral antibiotics.

19.4 Establishing an infection is often an improbable event

Although pathogens often have elaborate adaptations that partially neutralize host defenses, most encounters of pathogen and host result in a failure of the pathogen to establish an infection. The effectiveness of a pathogen in establishing infection is expressed as its **50% infectious dose,** or **ID_{50}**. This is the number of pathogen cells that will lead to infection in 50% of exposed hosts. Some pathogens are highly infectious, but most are not. *Francisella tularensis*, for instance, has an ID_{50} of only a few cells when inhaled (leading to the disease tularemia, or "rabbit fever"). Inhalational anthrax, in contrast, has an ID_{50} of about 10^4. A high number like this does not mean that it takes that many cells to establish an infection; rather, it means that exposure to lower numbers will result in a lower percentage of hosts becoming ill. If 1 million people were exposed to an aerosol of anthrax spores such that the average numbers of spores inhaled were 10,000, 0.5 million would become ill. However, if they inhaled only a few spores each, a handful would still become ill. Thus, the larger the number of inhaled spores, the higher the probability that one of them will evade host defenses and cause an infection. The relationship between exposure and percentage that become infected is generally logarithmic (Figure 19.4).

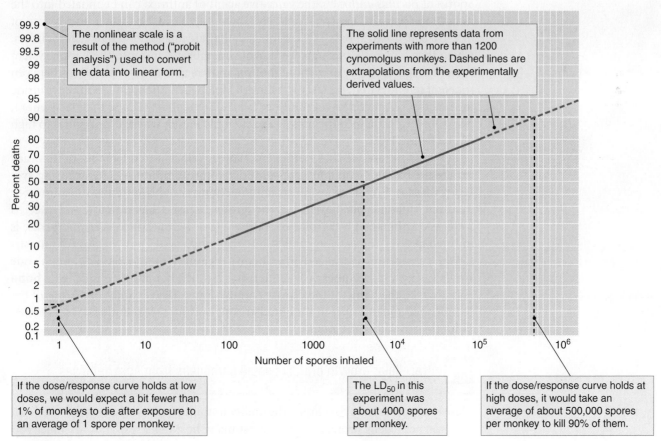

FIGURE 19.4 Dose curve for anthrax spores in monkeys.

19.5 Access to deep tissues requires wounds, or active penetration of mucous membranes

Few microbes can breach intact skin, but when the skin barrier is broken by a wound, pathogens can gain access to the subcutaneous tissue. This, however, is not a predictable event, and no obligate pathogen can depend on a fortuitous wound to allow it to transfer from one host to another. Thus, wound infections are generally caused by **opportunistic pathogens,** microbes with the potential to cause infection, but that can also multiply independently. One exception is rabies, a virus that infects the central nervous system and causes behavioral changes that increase the likelihood that the infected animal will bite another animal. This insures transmission, as by the time these behavioral changes occur, viral replication is advanced and there are high concentrations of virions in the saliva. Rabies virus will infect most mammals and is endemic in most of the world. Humans are generally accidental hosts, but there are about 35,000 human cases per year, all fatal.

Vector-borne pathogens do routinely penetrate through the skin, either injected directly into the subcutaneous tissues or later through the wound left by the arthropod bite.

Mucous membranes, unlike skin, can be crossed by highly invasive pathogens. Generally, the pathogen enters and destroys mucosal cells by the methods described in the next several sections. This ulceration of the mucosa deepens until it allows penetration into the underlying tissue, into the lymph system, or into the blood.

Sometimes a pathogen uses host cells to carry it across the mucous membrane. Spores of *Bacillus anthracis,* the causative agent of anthrax, can be inhaled into the lungs if the particles are small enough (about 1 to 5 μm, equivalent to a clump of one to a half-dozen spores). When inhaled, they are phagocytosed by macrophages, which then penetrate through the mucous membrane and into the lymph system (macrophages move by amoeboid motion and are able to force their way between mucosal cells to enter and leave the lungs). Meanwhile, the spore germinates inside the macrophage, and the vegetative cells produce toxins that lyse the macrophage, releasing the bacilli into the lymph and then the blood, where it multiplies to high levels (Figure 19.5).

Once in the deep tissues, pathogens may cause localized infections, or they may invade the bloodstream and be disseminated throughout the body, causing **systemic disease.** Frequently, a localized infection precedes systemic infection. For instance, bubonic plague is a vector-borne disease transmitted from infected rodents to humans by the bite of a flea. When the pathogen, *Yersinia pestis,* is injected into the tissues, it is carried into the lymph system by the flow of interstitial fluid and causes a localized infection in a regional lymph node. The lymph node swells visibly and becomes very tender; this swollen lymph node is called a **bubo** (from which the name of the disease comes). Later, the yersinias break out of the lymph node and multiply in the bloodstream, causing a frequently fatal septicemia if not treated with antibiotics.

19.6 Intracellular growth protects some pathogens from host defenses

Many pathogens have developed the ability to multiply inside host cells, which for bacterial pathogens protects them against many host defenses. They are most vul-

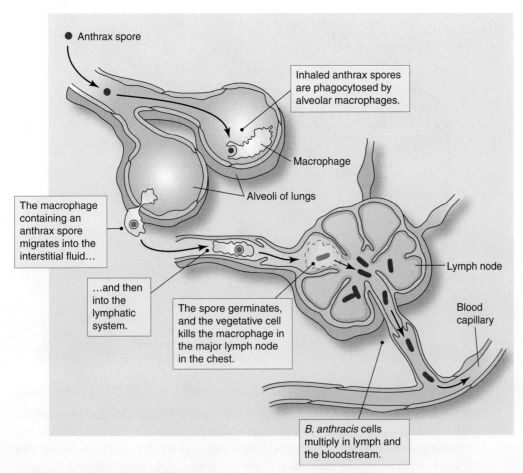

● Anthrax spore

Inhaled anthrax spores are phagocytosed by alveolar macrophages.

Macrophage

Alveoli of lungs

The macrophage containing an anthrax spore migrates into the interstitial fluid…

…and then into the lymphatic system.

The spore germinates, and the vegetative cell kills the macrophage in the major lymph node in the chest.

Lymph node

Blood capillary

B. anthracis cells multiply in lymph and the bloodstream.

FIGURE 19.5 Infection process of *B. anthracis*.

nerable to defenses when they pass from cell to cell, as they are then exposed to antimicrobial compounds, and to attack by the immune system.

To survive inside host cells, however, requires that they be able to resist those host-cell systems that normally kill ingested bacteria. Most pathogens that grow intracellularly enter their host cells by phagocytosis. Thus, they would normally be killed by the oxidative burst and digested by enzymes in lysosomes. Intracellular pathogens deal with this threat in several ways. Some, such as *Rickettsia*, *Shigella*, and *Listeria*, produce enzymes that degrade the phagosome membrane, releasing them into the host cell cytoplasm before lysosomal fusion. They then multiply within the host cytoplasm, protected from both intracellular and extracellular defenses. These pathogens co-opt the host cell cytoskeletal system to facilitate their transfer to adjacent host cells. They induce the formation of a bundle of actin microfilaments that propels them to the cell surface, where they bulge into adjacent cells, ultimately being transferred by membrane fusion. This further protects them from host defenses, as they do not have to traverse the extracellular environment to spread within the tissue (Figure 19.6).

Other intracellular pathogens, such as *Mycobacterium* and *Chlamydia*, secrete proteins that insert into the phagosome membrane and prevent lysosomal fusion. Other secreted proteins make the phagosome leaky so that small molecules leak into it from the cytoplasm, providing nutrients for the pathogens.

Viruses, of course, always multiply inside host cells; however, this is not as protective as it is for cellular pathogens, as mammals have evolved a system for detect-

Intracellular growth protects some pathogens from host defenses

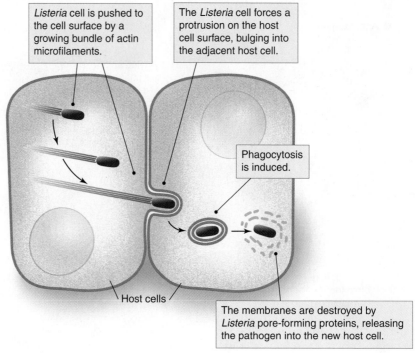

Listeria cell is pushed to the cell surface by a growing bundle of actin microfilaments.

The Listeria cell forces a protrusion on the host cell surface, bulging into the adjacent host cell.

Phagocytosis is induced.

Host cells

The membranes are destroyed by Listeria pore-forming proteins, releasing the pathogen into the new host cell.

FIGURE 19.6 Cell-to-cell transfer of *Listeria*.

ing virus-infected cells by displaying viral peptides on the infected cell surface (the MHC I system—see Chapter 18). This allows the infected cell to be destroyed by the immune system. Bacterial pathogens do not lead to these displays, as their proteins are made within the parasite cells, and not in the host cell cytoplasm.

19.7 Intracellular pathogens may induce phagocytosis

The mammalian body has several different "professional phagocytes"—cells that are specialized for phagocytosis and destruction of foreign cells; macrophages are an example. Some intracellular bacterial pathogens infect these cells. In this case, the host cell's ability to phagocytose is used as the means of entry. However, most mammalian cells do not normally engage in phagocytosis. For pathogens infecting these cells—for instance, cells of the gastrointestinal tract—the pathogen has to induce the host cell to engage in an activity that it is not programmed to do. There are two basic ways that this is done.

In the first, the pathogen binds to the surface of the host cell, using a surface protein that binds with high specificity to a specific surface protein on the host cell. The host cell surface proteins are ones that are normally involved in binding host cells together (the cells of multicellular organisms adhere to each other via these proteins). Binding of the pathogen induces a conformational change in the host protein that causes a local rearrangement of the actin cytoskeleton, pushing a pseudopod around the pathogen and inducing phagocytosis.

The second mechanism involves a specific type of secretion system that actually injects a dozen or more specific proteins into the host cell. The injected proteins then catalyze the local arrangement of the actin cytoskeleton to induce phagocytosis. This secretion system, called a **type 3 secretion system** (**T3SS**), or an **injectisome,** is evolutionarily related to the bacterial flagellar basal body. It consists of a proteinaceous tube that extends from the cell membrane, through the murein and the outer mem-

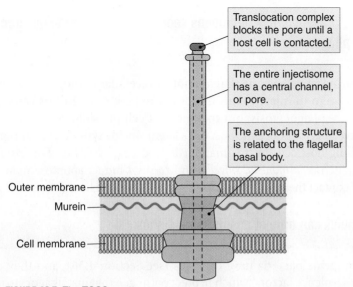

Translocation complex blocks the pore until a host cell is contacted.

The entire injectisome has a central channel, or pore.

The anchoring structure is related to the flagellar basal body.

Outer membrane

Murein

Cell membrane

FIGURE 19.7 The T3SS.

brane. It is supported by rings of protein in the cell and outer membranes (Figure 19.7). The portion of the tube that extends beyond the cell surface can penetrate the cell membrane of a host cell and act as an injection device. The system is quite complicated, consisting of nearly twenty different proteins, and is found in a number of gram-negative bacteria (Figure 19.8).

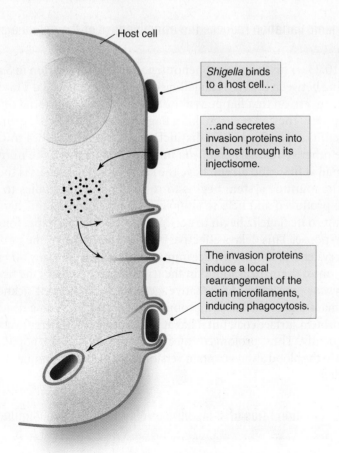

Host cell

Shigella binds to a host cell...

...and secretes invasion proteins into the host through its injectisome.

The invasion proteins induce a local rearrangement of the actin microfilaments, inducing phagocytosis.

FIGURE 19.8 The process of induced phagocytosis.

19.8 Many gram-negative pathogens secrete defensive proteins directly into phagocytes

Many gram-negative bacteria that are not intracellular pathogens also use a type 3 secretion system to inject toxins directly into host cells. In this case, the goal is not to induce phagocytosis, but to alter host cell physiology in other ways. One common result is to induce the host cell's latent suicide system. One important function of killing host cells in this manner is to disarm defenses. For instance, some pathogens, such as *Salmonella, Shigella,* and *Yersinia,* induce apoptosis in macrophages when they contact them, thus neutralizing this important host defense.

19.9 Capsules can protect against phagocytosis

Many pathogenic bacteria have capsules (see Section 4.36), and they are clearly important virulence factors, which protect pathogens against phagocytosis. Mutant strains that lack the ability to make capsules are often avirulent. There are probably several mechanisms by which capsules act to reduce phagocytosis, but one of them is by preventing access of the C3b component of the complement system to the bacterial cell surface. Remember that this protein binds to lipopolysaccharide or teichoic acids on the bacterial cell surface and then, in turn, binds to phagocytes. This specific binding to the phagocyte surface greatly enhances the efficiency of phagocytosis, and in its absence, phagocytosis is very much less efficient.

19.10 Antigenic variation reduces the effectiveness of the immune system

In Section 10.23 we discussed the phenomenon of phase variation in *Salmonella*—the switching between different flagellin subunits of the flagella. This is regulated by a genetic inversion that happens about once every 10^5 bacterial divisions, and when it happens, the cell starts making flagella with different flagellins. The different flagellins are antigenically distinct, and, thus, antibodies to one will not bind to the other. Thus, as *Salmonella* initiates an infection, the number of cells remains small at first, and all will have the same type of flagella. As the pathogens multiply, the immune system begins to respond, and antibodies to the flagella begin to be produced and help to suppress the infection. As this happens, some cells with alternate flagella begin to accumulate, and they escape, for a while, the immune response. This delays effective immune response to the pathogen.

Similar systems of altering surface immune targets exist in many other pathogens. Perhaps the most dramatic is found in the trypanosome protists, the best studied of which is *Trypanosoma brucei,* the causative agent of African sleeping sickness (so called because coma is one of the late symptoms, immediately preceding death). *T. brucei* has a glycoprotein surface coat, but it has at least a thousand different versions, which switch frequently. Thus, prolonged infection is possible, as the population of trypanosomes in the blood always contain versions to which the immune system has not yet responded.

19.11 Rapid mutation rates of some viruses slow an effective immune response

A similar variation is seen in many viruses, particularly the RNA viruses. The polymerases that replicate RNA viruses lack the proofreading function of cellular DNA polymerase and thus make more frequent mistakes. This results in populations of progeny viruses that differ from each other in small ways. Most of the mutations make no difference to the host defenses, but mutations that alter epitopes on envelope glycoproteins can change virions sufficiently that they appear to be new variants to the immune system. Thus, a viral infection may be prolonged by the successive emergence of new antigenic versions of the virus. This is a prominent feature of HIV, for example, which replicates its RNA chromosome in a two-step process: first reverse transcriptase produces a DNA version, and then after integration of this DNA copy into the host chromosome, new viral RNA chromosomes are produced by host RNA polymerase. Neither reverse transcriptase nor RNA polymerase proofread, and, thus, both make relatively frequent mistakes. So, over the course of an HIV infection, which can last years, many different versions of the virus are produced. Indeed, the variation can be so significant that each human HIV infection is distinctive, with its own unique mixture of viral genotypes.

19.12 Some pathogens block the action of antibodies

When antibodies bind to the surface of a pathogen, a conformational change exposes receptors on the stem portion of the antibody molecule that can activate complement proteins and that can bind the pathogen to phagocytes and facilitate phagocytosis. Some pathogens, such as *Staphylococcus aureus,* secrete proteins that bind to the stem of antibodies and physically block the reaction with complement or phagocytes. Other pathogens, such as *Streptococcus pneumoniae,* secrete enzymes that cleave antibodies in the hinge region that links the binding sites to the stem. In either case, the pathogen's surface is covered with bound antibody (or fragments of antibodies), but the binding sites for complement and phagocytes are missing or concealed (Figure 19.9).

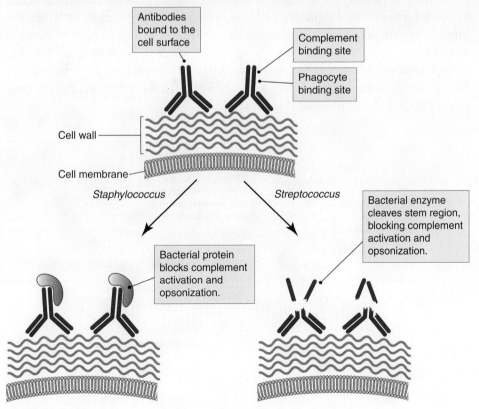

FIGURE 19.9 *Staphylococcus* and *Streptococcus* defenses against antibodies.

19.13 Many pathogens disrupt defenses by blocking cellular communication

A number of pathogens, both viruses and bacteria, disrupt the cellular communication networks on which host defenses depend. A good example is the pox viruses, such as smallpox, monkeypox, vaccinia virus, and others. Smallpox was once the single most devastating infectious disease of humans, killing and maiming more than any other disease. It was eradicated in the late 1970s and exists now in only a few laboratory freezers (see Section 18.22).

One reason that smallpox, and many other pox viruses, can be such serious pathogens is their sophisticated countermeasures against host defenses. These are mainly in the form of binding proteins that the virus-infected cell produces. The proteins are encoded by the viral genome, but they seem evolutionarily to have been derived from host cell receptor proteins. In other words, it appears that the viruses have, over the millions of years of coevolution with their hosts, picked up some of the cellular genes for membrane receptor proteins involved in host defenses. The viral version of the genes lacks the transmembrane region of the protein, and, thus, the resulting protein is soluble rather than membrane bound. It is able to bind the signaling molecule, preventing them from reaching their target receptors (Figure 19.10).

The smallpox virus, for example, has a genome with nearly 200 genes, of which about 20 encode different secreted proteins that interfere with host functions, usually by binding to them as just described. These include proteins that bind interferon, complement proteins, and signal molecules in the inflammation response. They also make an intracellular protein that binds double-stranded RNA, preventing it from inducing interferon.

Other pox viruses (but not smallpox) can reduce their visibility to host defenses by blocking the MHC I signaling pathway, which is the mechanism by which the immune system detects virus-infected cells (see Section 18.19). This interference is done by a protein encoded by the viral genome, which appears to insert into the endoplasmic reticulum and to prevent the MHC I from making it to the cell surface, probably by redirecting it to lysosomes. Thus, from the outside, the cell does not appear to be virus infected (Figure 19.11).

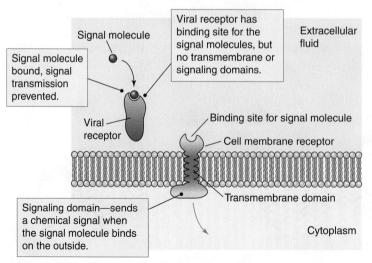

FIGURE 19.10 Soluble and membrane-bound receptor molecules.

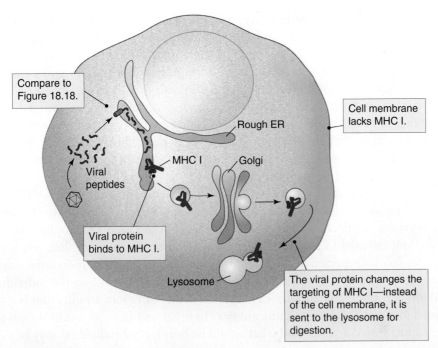

Compare to Figure 18.18.

Rough ER

Cell membrane lacks MHC I.

Viral peptides

MHC I

Golgi

Viral protein binds to MHC I.

Lysosome

The viral protein changes the targeting of MHC I—instead of the cell membrane, it is sent to the lysosome for digestion.

FIGURE 19.11 MHC I diversion to lysosomes.

Most poxviruses also have genes that encode growth hormones that induce host cells to replicate. Thus, the virus-infected cells are induced to multiply, a mechanism of producing more than one virus-infected cell from each initial infection. They also have genes that produce proteins that block apoptosis, whether induced by the infection itself, or induced by CTLs or NK cells (see Chapter 18).

19.14 Antibiotic resistance is an increasingly common defense

Antibiotics are chemical compounds produced by one microbe that are toxic to others. They are produced in very low concentrations in natural habitats such as soil and serve to give the producing organisms a competitive advantage. Of course, this has selected for microbes with genes that confer resistance to antibiotics as well; however, the low concentrations and patchy distribution of antibiotics in nature has meant that antibiotic resistance has not been an important selective advantage outside of soil ecosystems, and antibiotic resistance was rare before about 1950. That changed rapidly, however, with the introduction of antibiotics into medicine. Initially it was thought that antibiotics could eliminate bacterial disease; however, the bacteria responded by rapidly acquiring resistance genes by plasmid transfer. Most such genes are thought to have originated in soil organisms.

In most cases, the antibiotic-resistance genes are on plasmids, often carrying genes conferring resistance to several different antibiotics. Most of these genes encode enzymes that cleave the antibiotics or that modify them chemically into an ineffective form. Pathogens usually obtain resistance genes by plasmid transfer, not by mutation. Some pathogens are now resistant to so many of the commonly used antibiotics that antibiotic therapy is very difficult. Indeed, there is concern in medical circles that antibiotics could be come much less useful than they have been for the last half century, leading to a serious decrease in our ability to treat infections. For instance, a number of gram-positive cocci, such as *Staphylococcus, Streptococcus,* and *Enterococcus,* cause serious human infections. In the 1950s, these infections were routinely cured with penicillin; however, the frequency of penicillin resistance increased rapidly, forcing doctors to switch to methicillin. In the 1980s, however, methicillin resistance became a problem, and the drug of choice for gram-positive cocci became vancomycin. Now, vancomycin resistance is becoming a serious problem (Figure 19.12). Infections with gram-positive cocci that are simultaneously resistant to penicillin, methicillin, and vancomycin are now a serious medical problem and difficult to treat, as other antibiotics are less effective for these organisms.

How to cope with this problem raises serious issues in medical ethics. Clearly, it would be useful to decrease the amount of antibiotics used to reduce the selective pressure on microbes to develop resistance. For instance, the prophylactic use of antibiotics (used to prevent, rather than cure, an infection), or their use when an infection is suspected but not certain, could be limited. The individual physician's responsibility, however, is to give the best possible care to the individual patient, and not using antibiotics prophylactically or before an infection is confirmed may put patients at significant risk. In other words, what would be best for the population as a whole and what would be best for the individual may be very different.

Low levels of antibiotics are also used in food for domestic animals, a practice that keeps the animals healthier and increases their rate of growth. This, however, also leads to widespread exposure of animal pathogens and the animal's normal flora to antibiotics used in human medicine. Because a number of human pathogens are also animal pathogens or commensals, this practice increases the likelihood of antibiotic-resistant human pathogens emerging. Indeed, several recent outbreaks of intestinal disease have been traced to antibiotic-resistant ani-

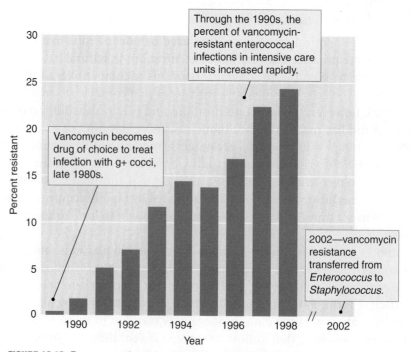

FIGURE 19.12 Frequency of vancomycin-resistant enterococci.

CHAPTER 19 MICROBIAL PATHOGENESIS

mal pathogens that contaminated meat during slaughter. Thus, there is good reason to ban the use of antibiotics as animal food supplements (at least those used in human medicine); however, this is opposed by farmers and the food industry, as discontinuing use would cost them significantly—another set of competing interests.

Toxigenicity

Toxigenicity is a function of how much toxin pathogens make, how potent the toxin is, and how critical the toxin's target is. As described previously, the localized growth of *Corynebacterium diphtheriae* in the throat can produce enough of a highly potent toxin to kill a person. In contrast, *Bacillus anthracis* produces two toxins, but fatal concentrations of toxin are not reached until multiplication of the pathogen has created a massive septicemia (blood infection). Numbers of cells can reach 10^9 cells per milliliter of blood.

Interestingly, few bacteria produce small molecule toxins. Many insect and reptile venoms and plant toxins are small molecules (molecular weight under 1,000). Bacteria, in contrast, rarely produce such toxins—their toxins are macromolecules; however, many bacteria do produce low-molecular weight compounds that are toxic to other bacteria, as an aid in the incessant competition in nature. We call these compounds antibiotics, and they are discussed in Chapter 21.

19.15 Exotoxins are proteins with specific enzymatic activity

Medical microbiologists traditionally distinguish two principal types of toxin: exotoxins and endotoxins. **Exotoxins** are proteins that are secreted sometimes during active pathogen growth, sometimes during sporulation of the pathogen, and sometimes in response to a specific stimulus (such as binding to a host cell). There is a wide range of different exotoxins, with different modes of action. Some have already been mentioned—the α toxin of *Clostridium perfringens* is a hydrolytic enzyme; diphtheria toxin is an enzyme that modifies an elongation factor.

Exotoxins are categorized on the basis of the tissue that they primarily target. Those whose effect is primarily on the nervous system, such as the toxins involved in tetanus and botulism, are called **neurotoxins.** Those whose effect is largely on the intestinal tract, such as the toxins involved in cholera and other bacterial diarrheas, are termed **enterotoxins.** Exotoxins that kill cells of multiple tissues, such as diphtheria toxin, are termed **cytotoxins.**

Several exotoxins have a common mechanism. They split NAD^+, adding the ADP-ribose portion of it to a target molecule and releasing the nicotinamide base (Figure 19.13). The effects of this addition depend on the target molecule. Diphtheria toxin and *Pseudomonas* exotoxin A target Elongation Factor 2 in many different tissues, and the result is inactivation of the target protein, leading to cell death throughout the body. In contrast, cholera toxin ADP ribosylates a protein that regulates adenylate cyclase in intestinal epithelial cells. The alteration permanently and irreversibly activates the regulatory protein, which in turn activates adenylate cyclase, causing cAMP concentration to rise. This triggers the cell to pump water from the bloodstream into the intestinal tract, causing diarrhea (often so severe as to lead to death from dehydration).

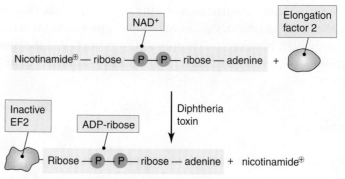

FIGURE 19.13 Reaction catalyzed by diphtheria toxin.

19.16 Many exotoxins have two subunits, one of which is a transporter

There are a great variety of different exotoxins, with a wide variety of mechanisms of action. Many of them, however, share the property of consisting of two different subunits, one of which is the toxin (the A subunit) and the other of which has a role in transporting the toxin into the target cell (the B subunit). These are called **AB toxins.** Most toxins that act inside host cells but are produced by bacteria growing outside are AB toxins. This includes most neurotoxins, enterotoxins, and toxins that inhibit protein synthesis, such as diphtheria toxin and cholera toxin.

The details of how the B subunit transports the A subunit into the cytoplasm are not clear. There appear, however, to be two different mechanisms. Some AB toxins bind to receptor molecules on the host cell membrane, and the B subunit then transfers the A subunit into the cell, remaining outside itself. Others are endocytosed after binding, and the A and B subunits separate after the vesicle pH becomes acidic. The A subunit is then transported into the cytoplasm, and the B subunit is transported back to the cell surface (Figure 19.14).

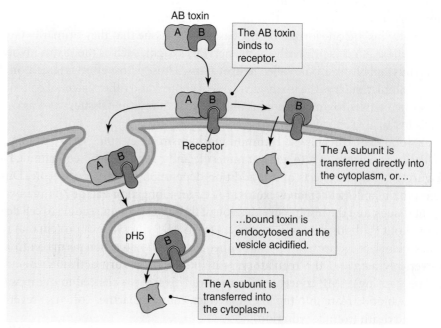

FIGURE 19.14 AB toxins.

19.17 Some diseases are intoxications

Most diseases caused by microbes are the result of infection; that is, the development of disease requires multiplication of the pathogen in host tissues. A few diseases, however, are the result of ingestion of a microbial toxin, and they do not require the pathogen to establish itself in the host. The most serious of these is botulism, caused by the ingestion of botulinum toxin in food. The toxin is formed when cells of the anaerobe *Clostridium botulinum* grow in the food. After ingestion, the toxin (an AB exotoxin) is absorbed into the bloodstream and taken up by nerve cells, where it prevents the release of neurotransmitters at neuromuscular synapses (junctions of nerve cells with muscle cells). Exactly how it does this is unclear, but it appears that the toxin is a protease and that it cleaves proteins required for the release of secretory vesicles. The result is the inability to contract the muscle, causing flaccid paralysis. Death results when the chest muscles are affected sufficiently to prevent breathing. Treatment includes putting the patient on a mechanical ventilator for as long as necessary for the toxin to be broken down and cleared from the body, often a month or more.

Botulinum toxin is among the most potent poisons known; the **50% lethal dose**, or **LD_{50}** (the dose that will kill 50% of exposed organisms), is about 1 ng/kg of body weight for most species when the toxin is injected. The oral dose is much higher, however, as the toxin is sensitive to proteases in the intestinal tract.

Most cases of botulism are now caused by ingestion of home-canned vegetables that are not too acidic or fermented fish dishes popular in Asia. It is a relatively rare disease. Interestingly, although *C. botulinum* is unable to colonize the adult human intestinal tract, it can establish itself in babies. The result is a disease termed *infant botulism*, a true infection.

A fairly common disease that is an intoxication, rather than an infection, is staphylococcal food poisoning. Many organisms can cause food poisoning, most of which are infections; however, *Staphylococcus* forms several potent toxins that cause vomiting, and staphylococcal growth in food can produce sufficient toxin to cause the disease when ingested.

One of the toxins, staphylococcal enterotoxin B, was developed into a weapon by the United States when it had an active biological warfare program (in the 1950s and 1960s). The intent was to incapacitate enemy troops with it. This toxin was attractive as a weapon because it is unusually stable, unlike most proteins that are hard to disseminate in an aerosol cloud without inactivating them. Botulinum toxin was also developed as a weapon, lethal in this instance, by the United States and other countries. Neither of these was ever used (see Chapter 21).

19.18 Some exotoxins disrupt host cell membranes

Many cytolytic toxins kill host cells from the outside, in contrast to AB toxins. There are two principal types: pore-forming proteins and phospholipases. The pore-forming toxins are proteins that can insert themselves into membranes and create a pore that allows cytoplasmic contents to leak out, thus killing the cell. These toxins probably are principally aimed at killing phagocytes, but pathogens have no way to precisely target them, and, thus, they kill whatever host cells are in the vicinity. A number of pathogenic *Streptococcus* species make this type of toxin.

The other type of cytolytic toxin consists of enzymes that damage membrane lipids, usually phospholipases that split the phosphate group off of membrane lipids. This destabilizes the membrane, and cell death ensues. The clostridial α toxin is an example of this. Not only does it break down host tissue allowing the pathogen to spread, but it also is a very effective defense against the PMNs that invade the anaerobic infection site.

19.19 Endotoxin is the lipopolysaccharide of the gram-negative wall

In contrast to exotoxins, **endotoxin** is a structural component of the cell: the lipopolysaccharide of the gram-negative wall (the toxic part of the molecule is the lipid A). Small amounts of LPS are continually released by growing cells, but much more is released if the cells are killed by host defenses or by antibiotics. Endotoxin is released in complexes with outer membrane protein and phospholipid, and these are then taken up by phagocytes. Their subsequent effects are the result of the **cytokines** (signaling proteins) released by the phagocytes. The cytokines stimulate different cells to react in various ways. Endotoxin thus has multiple effects, unlike the highly specific effects characteristic of exotoxins. Principal effects include high temperature (a result of cytokine stimulation of the hypothalamus gland), low blood pressure, low blood sugar, and diarrhea. Endotoxin poisoning can be a serious consequence of systemic infection with gram-negative pathogens, particularly if a large number of them die simultaneously, as can happen when an ongoing systemic infection is treated with antibiotics. Although all gram-negative bacteria have LPS in their outer membrane, the LPS varies greatly in its toxicity.

19.20 Genes for virulence factors are often clustered in "pathogenicity islands" or located on plasmids or prophages

Genes for **virulence factors**—features that enhance virulence—appear to have been transferred frequently among pathogens. It thus appears that an evolutionary innovation that improves a pathogen's ability to resist host defenses, spread more effectively, or in other ways to be a more successful pathogen, tend to spread among pathogens. The arrangement of genes for virulence factors suggests that they have been recently acquired in many pathogens. Genes for exotoxins are frequently located on plasmids or on prophages integrated into the chromosome. When virulence factor genes are chromosomally located and not part of a prophage, they are often found clustered with other virulence factor genes. These regions are called **pathogenicity islands.** Frequently, pathogenicity islands have a ratio of G + C to A + T base pairs that is significantly different from the rest of the genome, suggesting that they have been transferred from another organism. Also, the DNA sequence of genes encoding virulence factors often show a high degree of similarity to similar genes in distantly related organisms, again suggesting that they have been acquired by transfer, not by descent. This is similar to the widespread lateral transfer of genes for antibiotic resistance, discussed previously.

19.21 Host damage in viral infections is mainly caused by host defenses

Bacterial pathogens commonly damage their hosts primarily through the toxins they produce. Viruses, on the other hand, do not ordinarily produce toxins. The damage they do is largely a result of two things: the tissue damage they do by killing host cells and the aggressive defenses they provoke. The latter is the most important. Although host defenses are designed to protect the host, they often do substantial damage to host cells as well as to pathogens. At the site of viral replication, inflammatory and immune modulators are present at high concentration and provoke responses that damage healthy cells as well as virus-infected ones. They can also diffuse from the sites of infection to produce more generalized, systemic effects (such as fever).

Summary

In the last chapter we described the impressive range of defenses the mammalian body has to prevent bacterial or viral infections. Yet microbes have developed equally impressive countermeasures that preserve their ability to colonize the body. The rapid spread of antibiotic resistance provides a model for how many of these abilities might have evolved. It is clear that in the case of antibiotic resistance, the resistance genes were present in small numbers of soil bacteria, and when high concentrations of antibiotics began to be used in medicine, this established strong selection for antibiotic-resistant pathogens. Transposons and plasmid transfer were responsible for transferring resistance genes from soil bacteria to pathogens. Such transfers must have occurred repeatedly in the past, but until the selective conditions favored antibiotic resistance, the plasmids did not persist. Other genes involved in pathogenesis also appear to have been transferred to pathogens from other organisms. Thus, bacterial genetic exchange enables acquisition of novel abilities on a time scale that is very rapid in evolutionary terms.

Study questions

1. What are the components of bacterial virulence?
2. How do intracellular pathogens gain access to the interior of their host cells, and how do they protect themselves from being killed in the phagolysosome?
3. What is the type-3 secretion system, and how do pathogens use it?
4. How is antigenic variation caused, and why is it valuable to pathogens?
5. Besides antigenic variation, how do pathogens defeat the immune system?
6. Compare and contrast endotoxin and exotoxins.
7. What are AB toxins, and how do they work?
8. What are the major categories of exotoxins?

20

Epidemiology and Human Disease

W̲E̲ ̲B̲R̲I̲E̲F̲L̲Y̲ ̲D̲I̲S̲C̲U̲S̲S̲E̲D̲ in the last chapter how pathogens are transferred from one host to another. Here we approach this question in more detail and extend the discussion to how pathogens spread through populations. This is, in an evolutionary point of view, the most important aspect of pathogen biology. Every time a host dies or clears an infection by becoming immune, all of the pathogens die. This represents extinction of this particular clone of pathogen, unless before that event some of them have successfully colonized a new host. Thus, the pathogens that are alive today are ones that have an unbroken history of successful transfer before host death or immunity. That does not mean that every infected host has spread its infection to another, for all of history; rather, it means that among all infected hosts at a given time, at least one successfully transferred the pathogen to a new host. The study of the processes whereby pathogens are transferred and of the frequency of disease and its spread through a population is termed **epidemiology.**

20.1 An epidemic is the occurrence of disease at higher than expected frequency

Many diseases are present in populations at fairly constant levels, although there may be frequently seasonal variations. For instance influenza (flu) commonly infects millions of Americans every year, being most common in the winter. Typically 30,000 to 40,000 die every year. Periodically, however, a new variant of the flu virus appears as a result of genetic recombination among different strains of flu virus, and this new variant spreads, infecting far more people than normal. This is called an **epidemic.** When an epidemic spreads to multiple continents, as new variants of flu routinely do, it is called a **pandemic** (Figure 20.1).

Epidemics do not have to involve large numbers of cases, as in flu epidemics. When disease prevalence is normally very low, or even zero, a few cases can constitute an epidemic; however, epidemiologists commonly use the term **outbreak** for localized epidemics.

20.2 Disease at expected frequency is endemic

When disease is found in a population at frequencies consistent with historical experience, it is said to be **endemic.** *Endemic* means "within a population," and, thus, the term is also used to refer to the geographical distribution of disease. For instance, malaria is endemic in the tropics worldwide, but is rare in the temperate zones.

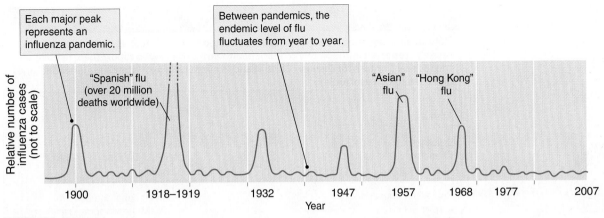

FIGURE 20.1 Flu prevalence.

Outbreak of a disease outside its endemic range constitutes an epidemic, as does a significant increase of disease frequency within the endemic range.

20.3 Distinguishing between endemic and epidemic disease requires background information derived from surveillance

Recognizing an epidemic requires prior knowledge of the normal frequency of disease. That, however, requires years of collection of disease statistics. Fortunately, all countries conduct some degree of disease surveillance, and international organizations conduct global surveillance of some diseases (the World Health Organization for human diseases, the Organization International des Epizooites for animal diseases, and regional Plant Protection Organizations for plant diseases). In the developed world, disease surveillance is quite thorough, but in the developing world it is much less so.

Disease surveillance information can come from many sources. For some diseases, physicians are required by law to report every case that they see to state or national authorities. These are called **reportable diseases,** and they are usually ones that are relatively uncommon, but potentially serious public health threats (Table 20.1). For diseases that are more common, too much work would be required to report every case, and information is collected from hospital admissions and deaths records. Flu, for instance, infects millions of Americans every year, and it would be an immense burden of paperwork to have to record every case; however, statistics are collected from hospitals on the deaths due to flu, and these provide a baseline to which current trends can be compared. Alternatively, statistics can be collected from diagnostic laboratories on the identification of specified pathogens.

When such data are collected over a number of years, certain patterns emerge. Of particular interest are long-term trends and seasonal trends. Tuberculosis, for instance, declined in incidence over most of the 20th century, and, thus, even the modest increases seen recently are regarded with alarm (Figure 20.2). Flu, like many other diseases, shows a marked seasonal pattern, with a peak in the winter and a trough in summer (largely because of increased transmission when people are crowded indoors in winter). A sharp increase in flu deaths in January would normally not cause much alarm, but even a slight increase during the summer would get a lot of attention.

Table 20.1   Reportable Diseases

Human diseases that are reportable in the United States. All cases of these diseases must be reported immediately to state health authorities, who report them in turn to the federal Centers for Disease Control and Prevention (many states require additional diseases to be reported).

AIDS	Lyme disease
Anthrax	Malaria
Botulism	Measles
Brucellosis	Mumps
Chancroid	Pertussis (whooping cough)
Chicken pox	Plague
Cholera	Polio
Coccidioidomycosis	Rabies (animal and human)
Cryptosporidiosis	Rocky Mountain spotted fever
Diphtheria	Rubella
Enterohemorrhagic *E. coli* infection	Salmonellosis
Giardiasis	Shigellosis
Gonorrhea	*Streptococcus* infection, invasive
Haemophilus influenzae pneumonia	Syphilis
Hansen's disease (leprosy)	Tetanus
Hantavirus pulmonary syndrome	Toxic shock syndrome
Hepatitis (viral, types A, B, and C)	Tuberculosis
Legionellosis	Typhoid fever
Listeriosis	Yellow fever

Human diseases that are reportable to the World Health Organization—all countries must immediately report cases of the following diseases to the World Health Organization.

Smallpox	Flu (new subtype only)
Polio (wild-type only)	SARS

Animal diseases that are reportable to the Office International des Epizooites— all countries must immediately report cases of the following diseases to the Office International des Epizooites.

Diseases of mammals and birds

African horse sickness	Newcastle disease
African swine fever	Peste des petits ruminants
Bluetongue	Rift Valley fever
Bovine pleuropneumonia	Rinderpest
Foot and mouth disease	Sheep pox and goat pox
Highly contagious avian flu	Swine vesicular disease
Hog cholera (classical swine fever)	Vesicular stomatitis
Lumpy skin disease	

Fish diseases

Epizootic hematopoietic necrosis	Spring viremia of carp
Infectious hematopoietic necrosis	Viral hemorrhagic septicemia
Oncorhynchus masou virus disease	

Mollusk diseases

Bonamiosis	Mikrocystosis
Haplosporidiosis	Perkinsosis
Marteiliosis	

An additional 67 animal diseases must be reported on an annual basis, including 12 diseases of birds and 5 of bees.

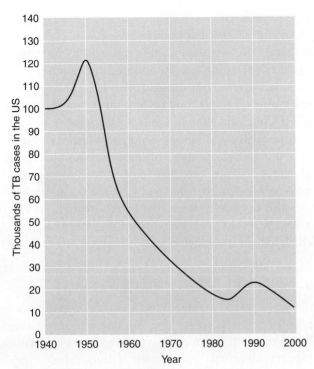

FIGURE 20.2 Tuberculosis incidence.

In addition to long-term and seasonable trends, some diseases show a periodicity measured in several years. Flu, again, is an example; every several years, genetic drift generates a version of the virus sufficiently different from previous ones that it can infect more people. Thus, flu shows unpredictable but unsurprising spikes every 2 or 3 years.

20.4 Epidemic detection requires alert physicians

Detection of an epidemic requires physicians to notice something unusual—either an unusually high number of cases of a known disease or an unusual constellation of symptoms that does not exactly match a known disease. Undoubtedly, many small epidemics go undetected, either because they are not sufficiently above the background to be noticed or because although they are a new disease they resemble a known disease sufficiently to be routinely misdiagnosed.

A good example of the key role of the individual physician is the detection and identification of the new disease **Hantavirus pulmonary syndrome** (**HPS**). This is a severe viral pneumonia caused by a member of the Hantavirus group. Other viruses in this group cause diseases known as hemorrhagic fevers, a very different kind of disease. In 1993, a physician at an Indian Health Service clinic in New Mexico treated a patient who died of **acute respiratory distress syndrome** (**ARDS**), a catch-all term for patients with severe lower respiratory problems regardless of the causative agent. There are very large numbers of ARDS cases in the United States every year, and many of them are never diagnosed more specifically. This case, however, struck the doctor as unusual, as the patient was a healthy man in his 30s, and he died rapidly. He thus took the unusual step of contacting other physicians in the region to see whether they had seen similar cases and found that there were four other cases in healthy young Native Americans. Without that alertness and the intuition that comes with experience in a community, this outbreak would not have been identified, or its identification would have been significantly delayed.

It turns out that the HPS virus is endemic throughout the Western Hemisphere in a variety of rodents, including the white-footed deer mouse. In 1993, there were heavy rains, breaking a several-year drought. This led to a large crop of piñon nuts (seed of the piñon pine), a staple food of deer mice. The deer mouse population increased about 10-fold, and human contact with mouse droppings and urine increased accordingly. Because Native Americans also harvest piñon nuts and because deer mice often invade homes, the opportunities for contact were numerous. In the end, about 50 cases were identified, spread throughout the West, but concentrated in New Mexico, Arizona, and Colorado. The fatality rate was 50%. Since then, sporadic cases have been identified throughout North and South America, and serological studies have confirmed that there had been sporadic cases before 1993, which had escaped diagnosis.

20.5 Epidemic investigation follows standard procedures

After an outbreak has been detected, a decision has to be made whether to investigate further. There are insufficient resources to investigate every outbreak, and, thus, only a small proportion is actually investigated. If the outbreak is important in some way, however, the decision will be made to send a team to investigate. In the United States, outbreak investigation is largely a state responsibility, but for large or unusual or particularly serious outbreaks, the Centers for Disease Control and Prevention (CDC) in Atlanta, Georgia, will send epidemiologists and infectious disease specialists to assist state epidemiologists. When they arrive, they are briefed by the local or state health personnel, and they review hospital records and any other relevant material. They then formulate a **case definition,** a description of what symptoms are necessary to classify a particular case as part of the outbreak. A case definition in an outbreak of salmonellosis (a food poisoning caused by *Salmonella*) might, for instance, specify fever above 38°C, chills, and vomiting or diarrhea.

After the case definition is decided, hospital records are searched for additional cases that match the definition but that were not previously recognized as part of the outbreak. Local physicians might be contacted, inquiring whether they had seen any patients that met the case definition. As the number of cases expands, the case definition is sometimes refined. Typically, the case definition will be rather broad at the beginning of an investigation so as to not miss any cases. As experience accumulates, however, it is often clear that some cases were really a different disease. The case definition is then made more specific, and cases not meeting the new definition are dropped from the list.

As data are accumulated, a graph of new cases versus time is made to aid determination of how the outbreak started and how it is spreading. For instance, an outbreak of salmonellosis caused by contaminated food at a picnic will show a rapid rise in cases to a peak very early, followed by a fairly rapid decline. In contrast, an outbreak of the same disease caused by a chronically infected food handler at a restaurant will sporadically infect customers (the ID_{50} for *Salmonella* is above 10^6 cells, and, thus, many people who ingest pathogens do not get sick). Cases are also analyzed as to age, sex, occupation, place of residence, place of work, etc. This will often allow a tentative hypothesis about the origin of the outbreak to be formulated (Figure 20.3).

Many outbreak investigations end at this point. If important issues are still unclear, however—for instance, if the evidence suggests that a food handler is continuing to infect people in the community—then further investigations will be

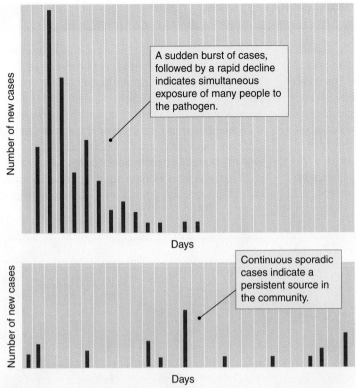

A sudden burst of cases, followed by a rapid decline indicates simultaneous exposure of many people to the pathogen.

Continuous sporadic cases indicate a persistent source in the community.

FIGURE 20.3 Pointsource and community-acquired *Salmonella* outbreaks.

made. Usually this involves interviewing cases to determine their movements over the few days in which they must have come in contact with the pathogen: where they ate, what they ate and drank, whether they had contact with any other cases, etc. This will often allow the identification of common sources of infection—certain foods eaten at a picnic by those who became ill, certain restaurants eaten at, sexual partners in common, travel to a particular foreign country, etc. This information will often lead to identification of the source of the outbreak or will at least narrow it down considerably (see MicroTopic on page 430).

20.6 The rate of transmission determines the course of epidemics

The local course of an epidemic of contagious disease is largely a function of the **rate of transmission**—the average number of secondary cases that are caused by exposure to an infected individual. If the average rate of transmission is over 1.0, the outbreak will be expanding. When the rate drops below 1.0, the peak is past and the outbreak is dying out. The rate of transmission is the outcome of a complex mixture of factors, some of which are pathogen properties (e.g., the number of live pathogens in secretions or hemorrhages), some are host properties (such as overall health or the strength of constitutive and inducible defenses), some are population properties (urban versus rural, population density, percent of population that is immune), and some are seasonal (whether people spend a lot of time crowded indoors or have contact with wildlife).

Diseases characteristically have an **incubation period,** the period between becoming infected and the onset of symptoms. This can be as little as 6 hours for some intestinal diseases and upper respiratory diseases, or it can be years for diseases

In the 19th century, cholera was one of the most feared diseases. It caused vomiting and violent diarrhea, so severe that it could cause death from extreme dehydration in a few hours. The death rate was very high, often over 50%. In the 19th century, six successive pandemics swept out of India (cholera's original home, where it is endemic), reaching Europe in five of these instances, and the Americas in three of them. Each pandemic lasted two or three years before cholera disappeared from the West, until the next pandemic. It broke out in epidemics first in port cities, then inland, affecting all neighborhoods but concentrating in crowded slums. Sometimes family members and caregivers would become sick, indicating contagion, but oftentimes they would not. New cases could pop up at a distance from others. How it was transmitted was a complete mystery, and, therefore, doctors were at a loss to combat its spread. The prevailing theory was that the stench of rotting organic material (garbage, sewage, stagnant water, etc.) caused the disease, but this theory fit poorly with the observed distribution of cases.

The man who worked out the principal mode of transmission of cholera was John Snow, a London physician (an anes-

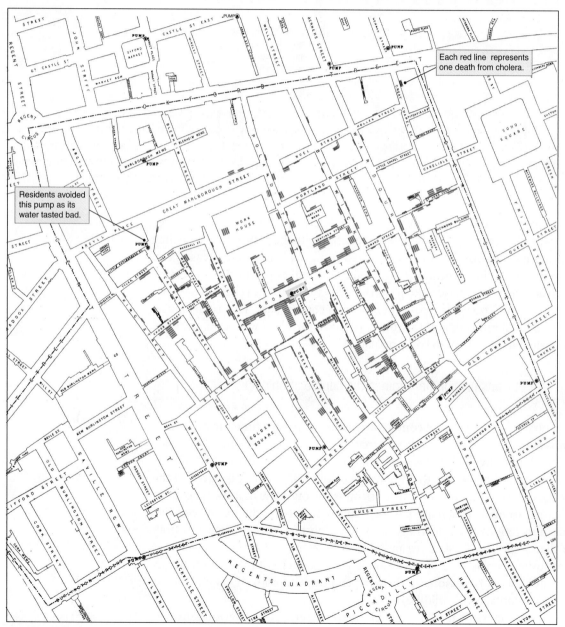

FIGURE 20B.1 Snow's map.

Each red line represents one death from cholera.

Residents avoided this pump as its water tasted bad.

thesiologist) who became interested in the question during the 1848–1849 epidemic. He thought that because the disease appears to affect directly the digestive tract, it would be reasonable to presume that the "poison" enters through the mouth rather than being breathed in through the nose. That thought led him to imagine the fecal-oral route of transmission, in which disease is the result of consumption of water contaminated with "the evacuations of a cholera patient," or transmission of fecal material on the hands to the mouth, which can happen to caregivers. He recognized that the causative agent multiplied in cholera patients, and he observed, "For the morbid matter of cholera having the property of reproducing its own kind, must necessarily have some sort of structure, most likely that of a cell."

Observation of several outbreaks of cholera in 1849 convinced Snow that water-borne transmission was a plausible explanation for the spread of cholera. However, his best-known study was not made until the next pandemic, in 1854. There was a sudden outbreak of cholera in the Soho district of London that killed about 600 people in a few days. In this area of the city, residents obtained their water from a series of wells, from which they carried buckets to their houses. Snow immediately recognized this as a point-source outbreak and plotted the place of residence of cholera patients on a map (Figure 20B.1). It was clear that the outbreak centered on a particular pump, on Broad Street. He then interviewed family members of a sample of about 90 victims to determine where they got their water. Nearly all of them got their water from the Broad Street pump, even when other pumps were closer (Broad Street had a reputation for better-tasting water). In the few households that got their water elsewhere, the cholera victims walked to their work or school, passing the Broad Street pump on the way and probably taking a drink. From the city engineers, he knew that a sewer line ran past the well only about two meters away. The line may have been leaking, although Snow did not excavate to see. There were also overflowing cesspools in the immediate vicinity of the well, so there was ample opportunity for fecal contamination of the Broad Street well. On the basis of this evidence, Snow prevailed on the authorities to remove the handle from the Broad Street pump to prevent access to the water. By then, however, the epidemic was essentially over, and it is not clear that removing the pump handle had any effect (Figure 20B.2).

Snow's account of this outbreak and his other evidence for water-borne transmission were not accepted for quite a few years; eventually, however, they prevailed. Since the end of the 19th century, many cities made sustained efforts to engineer their sewage and water systems to prevent cross-contamination. This engineering infrastructure underlies much of the relative freedom from disease of the developed world.

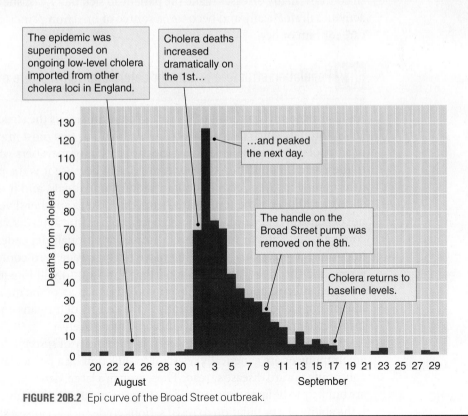

The epidemic was superimposed on ongoing low-level cholera imported from other cholera loci in England.

Cholera deaths increased dramatically on the 1st…

…and peaked the next day.

The handle on the Broad Street pump was removed on the 8th.

Cholera returns to baseline levels.

FIGURE 20B.2 Epi curve of the Broad Street outbreak.

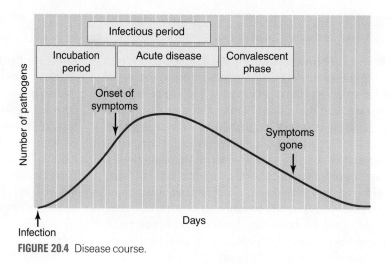

FIGURE 20.4 Disease course.

such as leprosy, tuberculosis, or AIDS. During the incubation period, the pathogen is multiplying and invading its target tissue but remains at low enough levels that symptoms are not yet apparent. Often the pathogen levels become high enough for transmission before they are high enough to cause symptoms. Thus, an individual who is not yet sick can be actively transmitting the disease to others because the **infectious period**—the period during which those infected can transmit the disease to others—begins during the incubation period (Figure 20.4).

Related to this is the extent to which a pathogen debilitates its host. A person suffering from a cold usually does not restrict his or her movements or social contacts and thus can continue to infect others as long as he or she remains infectious. In contrast, many diseases make the patient so sick that he or she curtails his or her activities dramatically and becomes a source of infection principally for those who care for him or her.

20.7 Population structure is a major determinant of the rate of transmission

Another major determinant of the rate of transmission is the structure of the population through which the pathogen is spreading. The two most important properties of a population are its density and the proportion of members who are susceptible to the disease. Population density is important because it is the principal factor in determining the frequency of contact among individuals, and it is this contact that allows a pathogen to be transmitted from one host to another. Even diseases that do not depend on close contact between hosts, such as intestinal diseases transmitted by contaminated water or vector-borne diseases, are strongly dependent on density. The higher the population density, the more likely sewage is to contaminate drinking water, and the higher the likelihood that a mosquito will bite another host after biting an infected one. One of the serious consequences of the massive growth of the human population, and consequently of the density of humans on the planet, is that major epidemics, even pandemics, become more likely.

What is really important, however, is the density of *susceptible* hosts. Hosts that are immune or are genetically resistant to infection by a particular pathogen do not contribute toward disease spread. Thus, even in a large, dense population, epidemics are unlikely to be large or sustained if a large proportion is immune. **Herd immunity** is the ability of a population to resist serious epidemics because a significant proportion of its population is immune. Achieving herd immunity is an important goal of vaccination programs. It is not necessary for every individual in a population to be immunized to prevent epidemics. After the proportion of immunized individuals reaches a certain threshold level, there may be sporadic cases of the disease among unvaccinated individuals; however, the rate of transmission remains below 1.0, and the nascent epidemic fizzles. Of course, the exact proportion will vary for different diseases, but is often above 80%. One of the persistent worries of public health officials is public resistance to vaccination, particularly among subpopulations (immigrants, religious groups, and alternative health groups). Even if the number of these is not sufficient to make the overall population fall below the threshold for herd immunity, it can make a subgroup fall below that threshold and invite a serious outbreak within that particular community.

20.8 High-speed transportation allows global dissemination of diseases

The spread of epidemics is strongly dependent on the travel patterns and speed that characterize a population. When there is limited travel and when the speed of travel is slow, epidemics remain localized or spread very slowly. For instance, in the 14th century it took the Black Death, a pandemic of plague, about 7 years to spread from the Black Sea throughout Europe (Figure 20.5). The rate of spread across Europe averaged about 5 km per day—a reasonable rate for families fleeing the outbreak on foot or in oxcarts.

When travel is more common and travel speed faster, however, larger and more widely dispersed epidemics can occur. For human populations, both the number of people who travel more than a few miles from their homes and the rate of their travel have generally increased over history, especially since the beginning of the industrial revolution. Now, as we enter the 21st century, millions of people travel long distances daily, and there are very few places on earth that cannot be reached in less than the incubation time of most diseases. Thus, diseases can be transmitted over large distances very rapidly, and disease outbreaks anywhere are a matter of concern to the entire world.

A good example is the recent outbreak of **severe acute respiratory syndrome (SARS)**. This was a new disease, causing a severe breathing difficulty and high fever. It has a fatality rate of about 10% on average, but is much greater in older patients (50% in those over 60 years old). SARS apparently originated when the causative virus was transferred from animals to humans. The SARS virus infects several species of wild animal that are sold as pets or as sources of folk medicines in Chinese markets, and it is thought that these animals were the source of the human virus. The SARS virus, however, is significantly different from viruses isolated from these animals, and thus, it appears to have evolved rapidly after being transferred.

The outbreak began in November 2002 in southern China and spread widely, infecting over 5000 people. The Chinese health authorities were slow to recognize the disease as novel and did not notify the World Health Organization promptly. The international community thus did not wake up to the global health threat until the disease had been transmitted by travelers to Vietnam, Singapore, Taiwan, and Hong Kong. Hong Kong was hit particularly hard, with nearly 2000 cases (Figure 20.6).

Fortunately, it turns out that SARS is transmissible for only a short period at the onset of symptoms. Thus, when the disease had been recognized, prompt isolation of cases and the use of masks and eye protection by health-care workers brought a

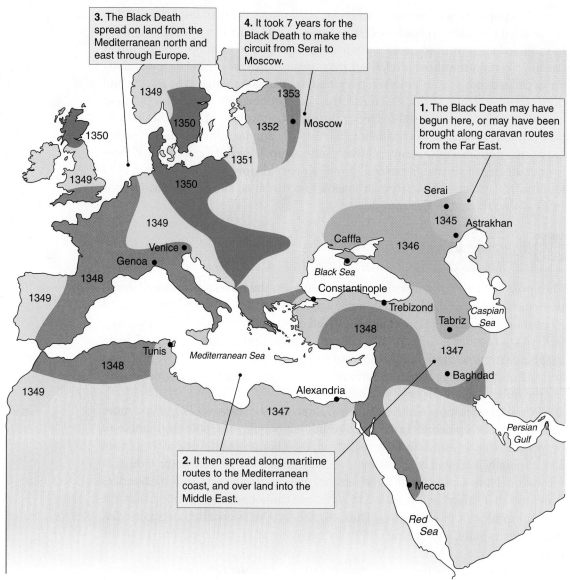

3. The Black Death spread on land from the Mediterranean north and east through Europe.

4. It took 7 years for the Black Death to make the circuit from Serai to Moscow.

1. The Black Death may have begun here, or may have been brought along caravan routes from the Far East.

2. It then spread along maritime routes to the Mediterranean coast, and over land into the Middle East.

1349
1350
1350
1349
1349
1350
1351
1353
1352
Moscow
Serai
1345 Astrakhan
1346
Cafffa
Black Sea
Venice
Genoa
Constantinople
Trebizond
1348
Caspian Sea
1349
Tabriz
1348
1347
Baghdad
Tunis
Mediterranean Sea
1348
1349
Alexandria
1347
Mecca
Persian Gulf
Red Sea

FIGURE 20.5 Spread of the Black Death.

rapid end to local outbreaks. In most countries outside of Asia, secondary transmission was limited by these prompt infection control measures. Indeed, in the United States, nearly all cases (probably less than 50 and maybe less than 10; the exact number is uncertain as SARS is difficult to diagnose with certainty) were primary cases in international travelers; there was virtually no secondary transmission. In contrast Canada had an outbreak with significant secondary transmission, and this was limited to hospitals (as it was in Singapore and Vietnam). By the end of June 2003, the epidemic was over. A total of over 8000 cases were identified, with nearly 800 deaths.

Although it was relatively limited, its travel from its origin in China to Europe and North America in a few weeks demonstrates the vulnerability of the modern world to pandemics. If SARS had been anywhere near as communicable as, for instance, influenza, the outbreak would certainly have developed as a global pandemic, infecting hundreds of millions of people and killing tens of millions. The world was fortunate in this instance.

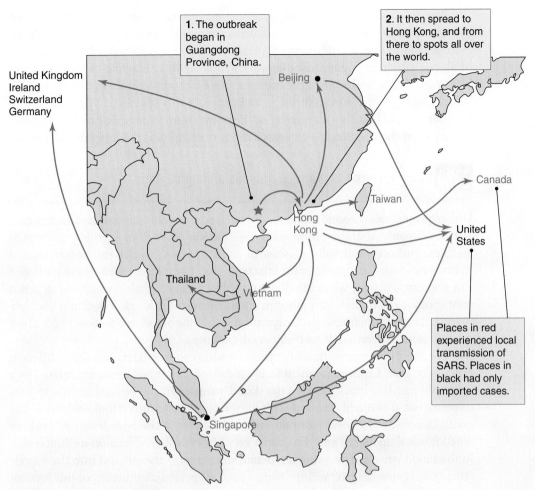

1. The outbreak began in Guangdong Province, China.

2. It then spread to Hong Kong, and from there to spots all over the world.

United Kingdom
Ireland
Switzerland
Germany

Beijing

Canada

Taiwan

Hong Kong

United States

Thailand

Vietnam

Places in red experienced local transmission of SARS. Places in black had only imported cases.

Singapore

FIGURE 20.6 Spread of SARS.

Some animal diseases are communicable to humans

Some human diseases are actually diseases of various animals that can be transmitted to humans; these diseases are called **zoonoses** (singular is **zoonosis**). A good example is **plague,** a disease of rodents that can be transmitted to humans but in its most common form (**bubonic plague**) is poorly transmissible from one human to another. The pathogen (the bacterium *Yersinia pestis*) is transmitted from one host to another by fleas. For zoonoses, we usually term the principal animal population that maintains the disease the **reservoir,** or **reservoir host.** In the case of plague, the ancestral reservoir is thought to be gerbils of central Asia. In gerbils, plague is a mild disease that can become chronic. Thus, the animals can maintain the pathogen for long periods and transmit it to other gerbils without it causing much mortality. From gerbils, the disease can be transmitted to other rodents, such as rats, ground squirrels, marmots, and prairie dogs, for which plague is as deadly as it is for humans. Plague is now endemic in rodents in many parts of the world. Most of these secondary reservoirs are colonial burrowing rodents such as prairie dogs, ground squirrels, and marmots (ground hogs). Human outbreaks of bubonic plague usually follow outbreaks among urban rats, and, thus, an early warning of an impending plague outbreak is numerous dead rats. When the rats die in large numbers, their fleas seek alternate hosts and can thus transmit the disease to humans.

Plague is normally transmitted by a flea bite, but it can sometimes be transferred by the respiratory route. When the infection penetrates into the lungs, causing hemorrhages in the alveoli, pathogens are present in sputum in high numbers. This form of plague is termed **pneumonic plague,** and it is highly contagious among humans. It is nearly 100% fatal. One of the most serious epidemics of human history was the "Black Death," which spread throughout Europe, North Africa, the Middle East, and possibly Asia, killing about a quarter of the population. Contemporary descriptions make it clear that the Black Death was a mixture of bubonic and pneumonic plague.

20.10 Many zoonoses are not contagious among humans

Unlike plague, many zoonoses can be transferred from animals to humans, but cannot be transmitted among humans. HPS (Section 20.4) is an example; anthrax is another. Anthrax is normally a disease of grazing animals, which can be transmitted to humans when they handle or eat infected tissue. Like plague, the anthrax pathogen can also cause disease when inhaled, but unlike plague, inhalational anthrax is not contagious. Even though the pathogen enters through the lungs, it does not multiply there, but rather penetrates into the lymph and then the blood (see Section 19.5). Thus, sputum does not contain large numbers of pathogens.

The three different portals of entry for anthrax spores determine three different types of disease. **Cutaneous anthrax** is a localized skin infection characterized by a large, but painless, black ulcer on the skin. Untreated, it can be fatal in 10% to 20% of cases, but it is readily treated with antibiotics. **Gastrointestinal anthrax** is the result of consuming meat from an animal that was in the later stages of anthrax when it was slaughtered and thus had a very high number of *Bacillus anthracis* cells in the blood stream. The pathogen penetrates through the gut and into the blood-stream and causes a systemic infection. There are periodic outbreaks of this form of anthrax throughout the developing world and in Russia. The mortality rate is probably 50% to 60%. Finally, **inhalational anthrax** results from the inhalation of anthrax spores. This is a systemic infection with very high mortality—probably over 80% if untreated and 50% or so even with aggressive antibiotic treatment. Fortunately, this is a rare disease, as most inhaled spores are destroyed by the body's defenses and the ID_{50} is thus quite high (estimated at 10^4 spores for humans).

20.11 Species-specific strains of influenza virus cause avian and human flu

Influenza is a respiratory infection that infects birds and mammals, including humans. It is not a true zoonosis, however, for two reasons. First, it is endemic in humans, rather than periodically penetrating human populations from an animal reservoir (we saw the annual pattern of flu in humans at the beginning of this chapter). Second, the viral strains that infect humans are distinct from those that infect birds. Avian strains of flu can occasionally cause human disease, but rarely transmit among humans. Thus, there are two distinct populations of flu viruses circulating among avian and human populations.

Unfortunately (for humans), these two populations occasionally mix. Flu viruses have what is termed a **segmented genome,** by which is meant that they have several chromosomes. They are single-stranded RNA viruses with eight separate molecules of RNA, each carrying one or two genes. If two different flu viruses infect the same cell, the progeny viruses package a random selection of the two parental types of chromosome, a process known as **reassortment.** This is distinct from the normal process of mutation during replication, which produces genetic variants that differ from the parental type at one or a few single bases. In the case of reassortment, one or more entire segments is replaced by the homologous segments from a different parental type. The major pandemics of flu that sweep the world a few times a century are thought to arise when reassortment between avian and human strains generates a novel human strain (Figure 20.7).

A major source of reassorted flu viruses appears to be pigs. Pigs can support the replication of both human and avian viruses and thus constitute a possible location for reassortment. Alternatively, humans (and maybe birds) may be rarely infected simultaneously with avian and human strains and thus produce reassorted viruses. China has been the source of several pandemic flu strains, and it is thought that this is because it is common in rural China to encounter intensive animal husbandry in which geese, ducks, and pigs are found in high numbers. Endemic human flu strains can thus mix in pigs with avian flu strains introduced into the geese and ducks by migrating waterfowl.

We tend to think of flu as a nuisance, but it is in fact a very serious disease. Although the fatality rate is low (typically about 0.1%), the disease is so contagious that it infects a very large number of people. Typically, 30,000 to 40,000 Americans die every year from the flu. The very old and the very young are the most susceptible.

Perhaps the most lethal human disease epidemic ever was the 1918 flu epidemic, caused by an avian virus that had accumulated a series of point mutations that allowed it to replicate in humans. There appear to have been two waves. The first, beginning in the winter, and apparently originating in the central United States, was relatively mild, with a low lethality typical of flu. An additional mutation occurred in the summer, however, probably in France, that made the virus considerably more lethal and conferred the unusual property of high lethality for young adults. Aided by the troop crowding and the major population movements associated with the First World War,

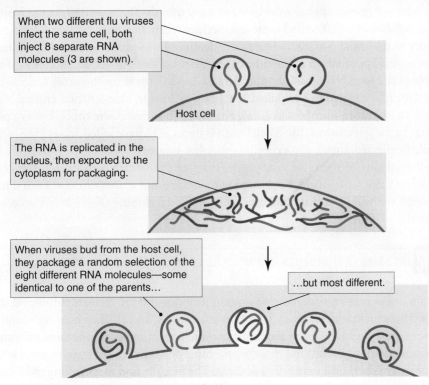

When two different flu viruses infect the same cell, both inject 8 separate RNA molecules (3 are shown).

Host cell

The RNA is replicated in the nucleus, then exported to the cytoplasm for packaging.

When viruses bud from the host cell, they package a random selection of the eight different RNA molecules—some identical to one of the parents...

...but most different.

FIGURE 20.7 The reassortment process in flu viruses.

Species-specific strains of influenza virus cause avian and human flu

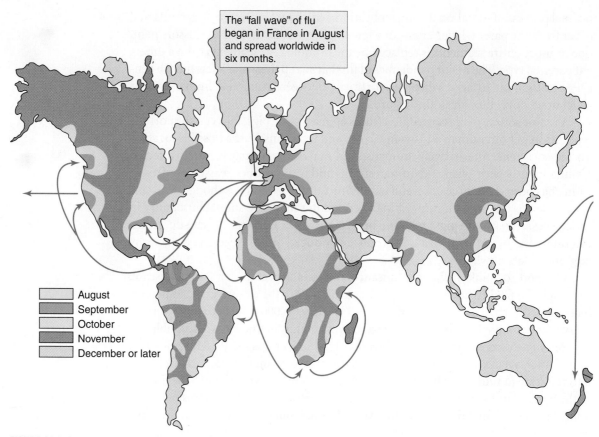

The "fall wave" of flu began in France in August and spread worldwide in six months.

August
September
October
November
December or later

FIGURE 20.8 Spread of the 1918 flu.

both waves of disease spread worldwide in a few months (Figure 20.8). In the United States, 600,000 to 700,000 died—more Americans than died in all the wars of the 20th century (a bit under 500,000). The global death toll was probably about 50 million, out of a world population at the time of less than 2 billion, or about 2% to 3% of the people in the world. In absolute numbers of deaths, this is on the same scale as the Black Death (estimated to have killed 25 million in Europe, North Africa, and the Near East, with unknown numbers in Asia). Flu deaths, however, were much less as a percentage of the population (the Black Death is thought to have killed about 25% of the people in affected areas). Flu experts expect that sooner or later a new pandemic flu with the lethality comparable to or greater than the 1918 strain will occur. Let us hope that when it does, the world is prepared with vaccine and antiviral drugs. Even if it is, the death toll will be substantial, especially in the developing world where access to medical care is limited.

20.12 Diseases that provoke immunity become childhood diseases

Certain diseases are most common among children and are termed **childhood diseases.** These include diseases such as measles, mumps, and chicken pox. Although in the developed world most childhood diseases are now disappearing because of immunizations, in prior generations, most people came down with all of these diseases while they were children, and many died of them. The predilection for infecting children is not due to any particular susceptibility of children or to any preference on the part of

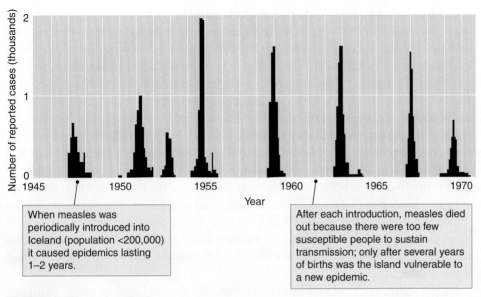

When measles was periodically introduced into Iceland (population <200,000) it caused epidemics lasting 1–2 years.

After each introduction, measles died out because there were too few susceptible people to sustain transmission; only after several years of births was the island vulnerable to a new epidemic.

FIGURE 20.9 Measles in Iceland.

the pathogen. Rather, it is a function of the immunity that these diseases confer to survivors. The diseases were highly contagious, and the pathogens did not vary in the way that flu does. Thus, these are disease that people got only once and then became immune to reinfection for life. The diseases were endemic at such a high frequency that they would sweep through communities every few years. Consequently, the only susceptible people would be children born since the last time the disease swept through and the few who had fortuitously escaped infection previously. Thus, the only significant pool of susceptible people was children, and the diseases came to be thought of as childhood diseases.

For such a disease to become endemic in a population, therefore, the population needs to be large enough for the birth rate to supply sufficient numbers of susceptible children every year for an unbroken chain of transmission to occur. For measles, for instance, that has been estimated at nearly a million people. Thus, in populations of less than a million, measles will be an epidemic disease that sweeps through the population then dies out until the next time it is introduced from outside (Figure 20.9). In general, this will be true of any disease that leads to lasting immunity in survivors. That includes many diseases, even the common cold. Although our daily experience is that we can get many colds in a lifetime—even well over a hundred—we also know that we do not generally get a cold a second time after we have just had it. In fact, there are probably nearly 200 different viruses that can infect the human upper respiratory tract and cause the symptoms that we call the "common cold." Some of them are capable of reinfecting people multiple times; however, most lead to immunity, and these colds we only get once. One of the few benefits of aging is that many people get fewer colds as they pass into middle age, and at least one of the reasons is that they are immune to many of the endemic cold viruses.

20.13 Many human diseases are evolutionarily recent phenomena

Because of immunity, small, isolated populations are typically free of childhood diseases and colds, except when they come into contact with the outside world, as with measles in Iceland. This has had major effects on human history. For instance, disease was a major source of mortality for indigenous peoples in the Western Hemisphere and the Pacific Ocean islands when they came into contact with Europeans, who communicated Eurasian disease that were completely absent from the Americas and Pacific Islands. It has been estimated that about 90% of Native Americans died of introduced disease within a century or two of the first contact. Some scholars believe that the Europeans would have been unable to colonize the New World without the depopulation that their diseases caused. The Native Americans were very numerous, and many tribes were skilled and aggressive warriors. The Europeans had several thousand miles of ocean to cross in small ships and, thus, could not transport more than a few hundred soldiers or settlers at a time. For instance, when the Pilgrims landed at Plymouth rock, they encountered a native population demoralized by having lost 90% to 98% of their people to a disease outbreak a couple of years earlier, probably transmitted by a shipwrecked sailor. Contemporary accounts describe the woods as littered with the decomposing bodies of Indians who had died of disease.

Of course, for most of human evolution, people lived in small bands of hunters and gatherers, typically numbering fewer than 200 and often fewer than 100. These bands probably only had only sporadic contact with other bands, and when they did make contact, the interactions were often hostile (we infer these patterns from study of hunting-gathering tribes that survived into the 20th century). Such a population structure would not have supported the disease that we now consider typical of humans—measles, mumps, smallpox, influenza, colds, etc. Thus, for most of human evolution, disease patterns were very different than they have been for the last 5,000 to 8,000 years, since humans began building cities and human populations became large enough to support disease that lead to immunity.

The diseases that probably afflicted our forebears were thus different than those we now have. Zoonoses caught from prey animals would have been common. Otherwise, chronic infections that can drag on for years, and infections that can become latent and reactivate years later would have been the principal types of infections of primitive humans. These would have included some intestinal diseases such as *Salmonella* infections, which can persist for years, some skin diseases that provoke very little immunity, perhaps tuberculosis (which can last for many years before killing its host), and herpes viruses, which can become latent. A good example of the last is chicken pox. It causes an acute infection that can prostrate a child for a week or two and that is characterized by pustules on the skin. When it clears, however, the child will not get chicken pox again. In about a third of those who recover from chicken pox, however, there are copies of the virus genome latent in certain nerve cells. These viruses can reactivate later in life and cause a skin disease called shingles. Shingles looks nothing like chicken pox, as it typically shows large patches of very painful skin rash. It is infectious, however, and when children contact an adult with shingles, they can come down with chicken pox. Thus, the chicken pox virus can be transmitted from generation to generation, even in a small community.

20.14 Many of the diseases of urbanized humans came from domestic animals

If primitive humans had patterns of infectious disease characterized by zoonoses, chronic disease, and diseases with long latent periods, where did the modern array of diseases come from? It is probable that most of the major human diseases that provoke lasting immunity came from domestic animals, especially ones derived from wild herd animals with large populations. When human populations started to become large and urbanized, their herds of domestic animals became large as well. The large animal populations, like their ancestral wild cattle herds, allowed disease to be maintained stably in their animal hosts. Thus, transmission of rare animal pathogen variants that could infect humans became more likely. Many of these diseases may have started out as zoonoses that infected humans periodically and became transmissible among humans when the pathogen mutated. Most of the major human pathogens have close relatives that infect domestic animals—cowpox and smallpox, for example, or measles and rinderpest.

This hypothesis for the origin of the major human diseases also helps to explain why the New World appears not to have had its own set of lethal diseases that would have had the same effect on Europeans as the European diseases had on Native Americans. Not a single disease native to the New World is known, with the possible exception of syphilis. Syphilis is a sexually transmitted disease that is thought to have first appeared in Europe in the early 1500s, and many have speculated that it was brought back to Europe by early explorers in the Americas. If so, it would be the only disease known to have originated in the Americas, in contrast to the many that Europeans brought with them. Why was this? If the major source of new disease for newly urbanized human societies was their domestic animals, then the paucity of uniquely American diseases makes sense, as neither of the major American civilizations—the Inca and the Aztec—had any domestic animals in large numbers.

20.15 New diseases are continuously emerging

We tend to think of diseases as fixed phenomena, with a finite number of different diseases and with each disease characterized by a particular specific set of symptoms and a specific set of pathogenic mechanisms; however, this is not really true. Disease is the result of the interaction of pathogen and host and is subject to natural selection on both short and long time scales. Thus, infectious disease represents a continually changing set of ecological relationships. Tuberculosis, for instance, declined markedly in severity and incidence over the course of the 19th and 20th centuries, as we saw earlier in this chapter. Syphilis, a sexually transmitted disease caused by the spirochete *Treponema pallidum,* was a rapidly lethal infection when it was first introduced into European populations in the 16th century. But by the 19th century, it was causing infections that could last decades, with a very slow progression through several defined stages. It could still be fatal, but the time between infection and death had lengthened from weeks to decades.

This trend toward reduced virulence is a fairly common one, and it makes sense. If the host lasts longer, there are more opportunities for disease transmission. Thus, variants of a pathogen that allow the host to survive longer are expected to increase in frequency as the pathogen evolves, and diseases that are initially fatal in a short time may become increasingly less lethal, or the time to death may extend. Of course, disease is a very complex phenomenon, and this simple characterization does not always apply.

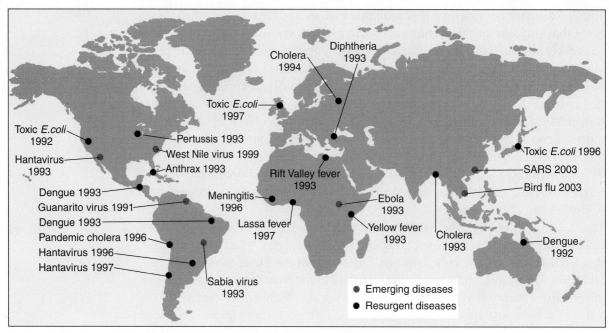

FIGURE 20.10 Recent disease emergences.

Not only do known diseases change, but new diseases are continuously appearing. New diseases are termed **emerging diseases,** as they are ones that generally start with a few small outbreaks and then become increasingly frequent. Disease emergence is a complex phenomenon, but one of the major underlying reasons for it is the increasingly rapid environmental disruption that humans are causing. This brings humans into contact with animals with which they previously had little contact and thus encourages the transfer of pathogens from wild animal populations to humans. It is likely that this process will accelerate over the next couple of decades. A representative sample of new disease emergences over the last couple of decades is shown in Figure 20.10.

20.16 Infectious disease is a major cause of death worldwide

Despite the advances of medicine, with vaccines and antibiotics, infectious disease remains a major cause of death in the world. In 2002, the last year for which final figures were available at the time this chapter was written, the world population was 6.2 billion, and there were 57 million deaths during the year. Diseases caused by microbial infection ranked second as a cause of death, causing nearly 11 million deaths, or nearly 20% of the total. The most lethal are influenza and bacterial pneumonia, followed by AIDS, diarrheal disease, tuberculosis, and malaria, all of which cause more than a million deaths annually (Tables 20.2 and 20.3).

All five of these diseases would be in the top 10 causes of death worldwide if they were listed separately, rather than combined with all other infectious diseases. In the developed world, in contrast, only one or two individual diseases would be in the top 10 causes of death, and infectious diseases combined, although still in the top 10, are typically fourth or fifth rather than second (Table 20.4).

Table 20.2 Causes of Death Worldwide in 2002

Cause	Millions of Deaths	Percentage
Heart disease	16.7	29.3
Microbial disease	10.9	19.1
Cancer	7.2	12.8
Injuries[1]	5.2	9.1
Respiratory disease[2]	3.7	6.5
Perinatal[3]	2.5	4.3
Digestive disease	2.0	3.5
Neuropsychiatric disease	1.1	1.9
Diabetes	1.0	1.7
Genitourinary disease[4]	0.8	1.5
All other causes	8.9	10.4

[1]Includes accident (3.5 million), suicide (0.87 million), violence (0.56 million), and war (0.17 million).
[2]Mainly smoking related.
[3]Maternal death in childbirth and stillbirths.
[4]Mainly kidney failure.

Table 20.3 Principal Infectious Diseases as Causes of Death Worldwide in 2002

Disease	Millions of Deaths
Influenza and pneumonia	3.9
AIDS	2.8
Diarrheal disease	1.8
Tuberculosis	1.6
Malaria	1.2
Measles	0.61
Pertussis (whooping cough)	0.29
Meningitis	0.17
Syphilis	0.16
Hepatitis B and hepatitis C	0.15

Table 20.4 Causes of Death in the United States in 2002

Cause	Hundreds of Thousands of Deaths
Heart disease	2,443
Cancer	697
Stroke	163
Respiratory disease	125
Accidents	107
Diabetes	73
Influenza and pneumonia	67
Alzheimer's disease	59
Kidney failure	41
Septicemia	34

Infectious disease is a major cause of death worldwide

Summary

Infectious disease is essentially an ecological phenomenon. The ability of an infectious agent to maintain itself over evolutionary time requires mechanisms of transmission that ensure that at least some minimal percentage of infected hosts pass the infection on to another host. All of the adaptations that underlie pathogenesis will make no long-term difference if the genes that encode them do not get replicated and passed down the generations. Thus, the transmission of pathogens from host to host is one of the critical requirements for evolutionary success. Understanding disease dynamics at the population level is how we investigate these features of the disease process.

Study questions

1. What is an epidemic disease? An endemic disease?
2. How is disease surveillance done?
3. What are the steps of an epidemic investigation?
4. Flu normally shows peaks of disease incidence on an annual, an every 2-to-3 year, and a several-decade basis. Explain.
5. What are "childhood diseases," and why are they confined to urbanized populations?
6. Compare and contrast the diseases of humans living in hunting-gathering communities with those of urbanized humans.

Human Exploitation of Microbes

21

I̲N̲ T̲H̲E̲ P̲A̲S̲T̲ several chapters we have examined the damage that microbes can do to humans by causing infectious disease. Additional harm is done by microbial food spoilage and by damage to pipes, wooden structures, and other human constructions. Not all microbial effects on humans are negative, however; many of their activities benefit humans. We have already mentioned the protective effect of the normal flora, and it appears that the normal flora also benefits humans nutritionally by synthesizing one or more vitamins. Beyond this, humans have learned to use microbes to produce foods or other valuable products, to kill insect and rodent pests, and to dispose of our sewage. These applications are the subject of this chapter.

Food Microbiology

Human and animal foods consist of natural products: the flesh of other animals, animal secretions or excretions (like milk and honey), and plant material, including roots, shoots, leaves, fruits, and seeds. Most of these, with the exception of seeds, are moist enough that they spoil rapidly. Thus, humans have, until the recent invention of refrigeration, faced a problem of food storage. For much of human evolution, there was little option for storage of excess food. When certain trees set fruit, for instance, humans could gorge themselves, but after the fruit started decaying, there was not much they could do. Perhaps they learned to dry some fruit in the sun, but it is unlikely that they could have stockpiled very much.

Some food fermentations began as means of extending the life of perishable foods; others transform foods into something with entirely different qualities—wine and beer, for instance, are very different from the juices from which they are derived. Over the eons, humans have learned to exploit many different fermentations to make a wide variety of foods. We discuss the most important of them here.

21.1 The lactic acid fermentation preserves food

The earliest deliberate use of microorganisms was almost certainly for food preservation. Although microbial growth in food materials normally spoils it—giving it offensive smells and tastes, and sometimes making it harmful—some specific fermentations alter food without spoiling it. Once altered in this way, the foods last longer before spoilage.

The principal fermentation used in this way is the lactic acid fermentation. The lactic acid bacteria are a group of aerotolerant anaerobes that ferment sugars to lactic acid as the sole or principal fermentation end product. The accumulation of lactic acid inhibits the growth of other organisms, protecting the food from spoilage, at least for a short period of time.

The earliest food fermented in this way was probably milk, and the fermentation was probably initially simply permitted, rather than actively encouraged. Milk that is not pasteurized will often undergo spontaneous lactic acid fermentation, as there are many lactic acid bacteria that grow on mammalian skin, including on the udders of cattle, goats, and sheep. Thus, when raw milk spontaneously ferments, it is often a lactic acid fermentation that predominates. The products of a lactic acid fermentation of milk include buttermilk, yogurt, kefir, and others (which differ mainly because different lactic acid bacteria are involved).

A similar preservation of vegetable matter can be accomplished by packing the material in brine. Soluble sugars leech out of the material and are then fermented by lactic acid bacteria that grow naturally on the surfaces of plants. As with milk fermentations, the low pH that results from the accumulation of lactic acid slows the growth of other microorganisms. Foods such as sauerkraut, pickles, green olives, and kim-chee are prepared this way, as is silage—plant material stored as animal food. Silage is prepared by chopping green and moist plant material, such as corn stalks, and storing them in a deep pile, normally in a cylindrical silo. A lactic acid fermentation then acidifies the material, which remains nutritious for cattle or other animals for the winter.

21.2 Cheese production usually involves secondary microbial transformations

Cheese is a milk product that contains precipitated milk protein and milk fat. It is prepared initially by a lactic acid fermentation of milk, at the end of which the pH is low enough that the milk curdles—the protein and fat (the **curd**) separate from the liquid, or **whey.** The curd is then separated from the whey, and residual liquid is removed by putting it under pressure—by wrapping in cheese cloth and putting a weight on top—or by putting it in a strainer for a few days, allowing gravity to dry it. After a few days, the product is a raw cheese, such as cottage cheese or ricotta.

The curdling process is sometimes aided by adding **rennin** (or **rennet**)—a protease that cleaves a specific bond in milk protein, which aids curdling, as well as contributing to the subsequent ripening.

Most cheeses are **ripened** by secondary microbial transformations. One process involved in all ripened cheeses is the gradual hydrolysis of milk protein, producing amino acids, some of which may be further fermented to organic acids and ammonia. The hydrolysis of proteins is generally the result of proteases released from cells of dying lactic acid bacteria, as well as by proteases produced by other microbes. Hard cheeses, such as Cheddar or Swiss cheese, undergo a limited amount of proteolysis—only 25% to 35% of the protein is broken down. In soft cheeses, like Brie and Limburger, all of the protein has been broken down.

Other kinds of transformations involve the active multiplication of different groups of microbes. In Swiss cheese, for instance, a secondary fermentation takes place in which lactic acid is fermented to a mixture of propionic and acetic acids and CO_2. The accumulating CO_2 forms the holes in the cheese, and the taste is significantly affected by the accumulation of propionic acid.

Several soft cheeses are ripened by a surface growth of white mold. Cheeses such as Brie and Camembert have a soft coating of the mold *Penicillium camembertii*, which produces enzymes that hydrolyze the milk fats and partially oxidize the fatty acids, activities that have major effects on flavor and texture.

The blue cheeses also are mold ripened, but in this case, the mold spreads through the cheese, following cracks and lines of weakness in the curd. The major ripening mold for blue cheeses is *Penicillium roquefortii,* a blue-colored relative of *P. camembertii,* which produces many of the same types of flavor agents.

21.3 The ethanol fermentation of fruit juice produces wine

The ethanol fermentation is performed by yeast under anaerobic conditions. Many different yeasts grow on the surface of ripening fruit, and, thus, fruit juice often has a high population of yeast. If left undisturbed it will undergo fermentation, producing ethanol and CO_2 from sugars in the juice. This was discovered independently by all major prehistoric civilizations, and by the time records began being kept in written form, wine production was centuries old.

A variety of fruits, vegetables, and plant saps can be used to make alcoholic beverages, but by far the most important is the grape. In making wine, grapes are harvested when ripe and then crushed to form a thick slurry called a **must.** The sugar concentration in the must is typically 10% to 20% and will yield an alcohol concentration of about 12% at the end. White wines are made from either red or white grapes by filtering the must to remove the grape skins. Because the color of a grape is in the skin only (not in the flesh), the resulting wine is white. Red wines are made by conducting the fermentation in the presence of the skins; the alcohol leeches the pigment from the skin and colors the wine.

There are many different yeasts on the surface of grapes, members of at least half a dozen different genera. The fermentation of wines using the natural yeasts (often called **wild yeasts**) is thus a very complex process, with different yeasts dominating during different parts of the fermentation. *Saccharomyces cerevisiae* comes to predominate late in the fermentation, when the high alcohol concentrations kill most of the others. To gain more control of this highly complex process, modern wine industries often kill the wild yeasts by exposing the must to the toxic gas SO_2 (sulfur dioxide) and then add cultures of *S. cerevisiae.*

Wines are not only high in alcohol, but they are quite acidic as well because of an abundance of organic acids, particularly malate and tartrate. The pH is typically around 3. The low pH, combined with the high alcohol content, makes them very resistant to further spoilage; however, the intense acidity can adversely affect taste, particularly in some red wines. In these wines, a secondary fermentation by members of the lactic acid bacteria, *Lactobacillus oenos* and a few others, reduces acidity by converting malic acid to lactic acid and CO_2 (converting a dicarboxylic acid to a monocarboxylic acid (Figure 21.1)). This is known as the **malo-lactic fermentation** and is one of the most difficult aspects of wine making to control. It is desirable in some wines, mainly high-acidity reds, and undesirable in others, such as low-acid reds and most white wines. It is often encouraged where desired by adding cultures of *L. oenos,* but preventing it when it is not desired is difficult.

Although grapes are overwhelmingly the most important fruit used for wine production, many other fruits, and sometimes even vegetables, can be used. Wine made from apples is called **cider.** Other wines are named after the fermentable substrate: blackberry wine, Loganberry wine, rhubarb wine, etc. In the tropics, a wine-like drink is made from palm tree sap and is called **palm wine.** In Central America, the fermentation of a mash prepared from the agave plant (a succulent) produces **pulque.** Another wine-like drink is **mead,** made by fermenting diluted honey.

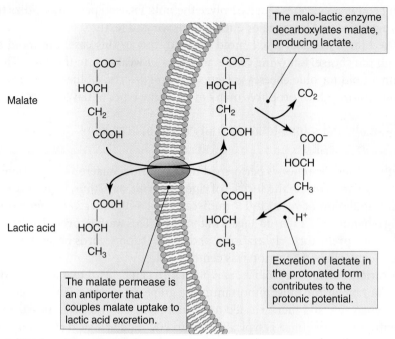

The malo-lactic enzyme decarboxylates malate, producing lactate.

Malate

Lactic acid

The malate permease is an antiporter that couples malate uptake to lactic acid excretion.

Excretion of lactate in the protonated form contributes to the protonic potential.

FIGURE 21.1 The malo-lactic fermentation.

The only other type of wine that approaches grape wine in importance is rice wine, made throughout Asia. It is best known by its Japanese name, **sake.** Sake is made from steamed rice, on which mold growth (*Aspergillus oryzae*) hydrolyzes starch to produce soluble sugar. *S. cerevisiae* grows simultaneously, producing the ethanol.

21.4 Beer is the other major beverage made by an alcohol fermentation

Beer is a beverage made by fermenting grains, usually barley or a mixture of barley and corn or rice. Unlike grapes, grains have little soluble sugar; their carbohydrate consists of the polysaccharide starch. Yeasts typically do not hydrolyze polysaccharides, and thus, as with the production of sake from rice, some means of hydrolyzing the starch to release sugar has to be part of beer brewing. The traditional way is to moisten the grain, which induces germination (grains are seeds of grasses). When the grain germinates, it induces hydrolases that break down starch and protein. This step is termed **malting** because the partially germinated grain is called **malt.** After malting, the malt is dried in a kiln, which halts the germination process and stops the hydrolysis process while the starch is still largely intact. The drying, however, is done at a low enough temperature that the hydrolytic enzymes are not inactivated.

The next step in brewing is the preparation of the sugar-rich liquid that is the substrate for fermentation, called **wort.** To prepare the wort, the malt is ground to break up the individual grains. If other grains, such as corn or rice, are to be included, they are added at this stage. Unmalted barley is often added as well. After milling, the grain is mixed with a small amount of hot water to form a thick **mash,** which is held for several hours for the hydrolytic enzymes to hydrolyze the starch.

After most of the starch is hydrolyzed, the soluble sugars and other nutrients (such as amino acids) are extracted from the mash by passing water through it. This **sweet wort** is then boiled. The boiling serves multiple purposes. It kills microbes that might compete with the yeast and impart undesirable tastes; it denatures soluble proteins, including the hydrolytic enzymes, and it drives off volatile compounds that might spoil the taste.

Dried **hops** are also added to the sweet wort during boiling. Hops are the female flowers of the hop vine and are a major flavor component of beer. The principal flavor constituents are compounds termed **α-acids,** and these are the main source of the bitter taste of beer. They also inhibit some bacteria that might otherwise grow in the beer and spoil it (Figure 21.2).

The hopped sweet wort is then fermented. The yeast used was traditionally yeast that had been saved from a previous batch of beer. Thus, brewing yeasts were selected over centuries for desirable characteristics (rapid fermentation, good taste, easy separation from the beer when fermentation was over, etc.). Now brewers prepare an inoculum from stored pure cultures of their own special yeasts. Typically, four or five batches of beer will be brewed with the yeast added from the previous batch; then the yeast is discarded, and a fresh inoculum prepared from the stored pure culture. This prevents the accumulation of mutations and contaminants and helps the repeatability of the process.

The fermentation normally is conducted in large vats (100,000 to 200,000 liters) with conical bottoms. The bottoms allow settled yeast to be drawn off easily at the end of the fermentation, without losing too much beer (the same kinds of tanks are used in making wine, for the same reason). Stirring is not necessary, as the rapid production of CO_2 produces bubbles that rise to the top. The upward flow of CO_2 creates currents that keep the mixture well stirred (Figure 21.3).

When the fermentation is complete, the yeast is removed, and the beer is aged in the cold in any of several different ways, depending on the kind of beer desired and the local brewing traditions. It is at this stage that the carbonation is developed in traditional brewing systems. Residual yeast cells ferment the small amount of remaining sugar to carbonate the beer. This may be aided by adding some extra sugar and is usually done in sealed tanks to trap the carbonation. The aging also encourages proteins and other large molecules to aggregate, making them easier to remove by filtration at the time of bottling.

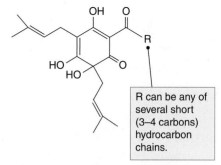

R can be any of several short (3–4 carbons) hydrocarbon chains.

FIGURE 21.2 Structure of α-acids.

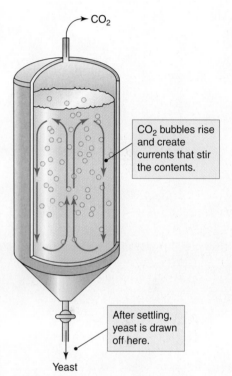

CO₂ bubbles rise and create currents that stir the contents.

After settling, yeast is drawn off here.

Yeast

FIGURE 21.3 A typical industrial fermenter.

21.5 Distillation of wines and beers produce liquor

All of the different kinds of wines and beers can be distilled to concentrate the alcohol. During the distillation process, a large number of additional compounds with significant vapor pressure at the distillation temperature of ethanol will also be concentrated, giving each of the different distilled beverages a distinctive taste. Whiskey, for instance, is made by distilling an unhopped beer, brandy by distilling wine, apple brandy (calvados) by distilling cider, tequila by distilling pulque, and rum by distilling a mead-like drink made by fermenting diluted molasses.

21.6 A partial oxidation of wine produces vinegar

Wine or cider is also the starting material for the production of vinegar (from the French "vin aigre," which means "sour wine"). It is produced when bacteria in the group known as **acetic acid bacteria** grow in the wine, using aerobic respiration to oxidize the ethanol to acetic acid (Figure 21.4). Some members of this group lack a TCA cycle and therefore cannot oxidize the acetate further, and, thus, it accumulates. Others are capable of oxidizing acetate, but do so more slowly than they produce it, and thus, they too produce acetate, although transiently.

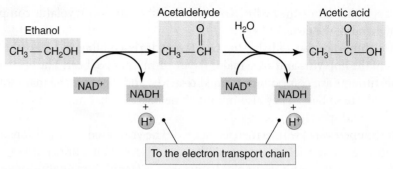

FIGURE 21.4 Acetic acid formation by acetic acid bacteria.

As long as wine is in an unopened bottle, the ethanol is stable because there is not enough air in the bottle to allow aerobic respiration; however, after opening, the wine will spoil slowly. Traditional vinegar production in the home simply involved allowing an opened bottle of wine to oxidize for several weeks. Traditional industrial processes were not much more complicated: half-full barrels of wine were inoculated with fresh vinegar and incubated. Acetic acid bacteria added with the fresh vinegar then oxidized the ethanol in the aerobic upper layers of wine.

These traditional methods were slow because much of the wine was anaerobic, and oxidation only occurred in the aerobic zone in contact with air. Modern industrial methods speed the oxidation by active aeration of the wine and can produce a finished product in a couple of days.

21.7 The ethanol fermentation is also used to raise bread

The manufacture of beer and wine takes advantage of the production of ethanol by fermenting yeast, and in some cases, the CO_2 is also desired (beer and sparkling wines). In the case of bread, the CO_2 is the desired product. Ancient breads were **unleavened** (i.e., not raised) and were very dense. At some point, leavening was discovered, perhaps by adding yeast (left over after beer or wine production) to the flour and water that are the basic ingredients of all bread. Until recently, most bakeries got their yeast from breweries; in the last 200 years, a separate industry dedicated to producing yeast for baking has replaced the traditional brewer's yeast as a source for bakers.

The CO_2 that is produced by yeast during fermentation is trapped by the matrix of bread proteins, or **gluten,** and causes the entire mass to expand and become less dense. It is this textural change that is the primary goal of leavening, but there are subtle changes in taste as well. In the late 19th century the German chemist Liebig invented baking powder as a substitute for yeast. Baking powder is a mixture of sodium bicarbonate with an acid that releases CO_2 when moistened. It was intended to simplify baking by replacing the complex natural leavening process by a simple, easily controllable chemical one. This, however, did not happen, because breads raised with baking powder have less attractive texture and taste than naturally raised breads. The detailed reasons are not clear. Part of it, though, has to do with the slower pace of CO_2 evolution in the yeast fermentation, which allows time for the maturation of the gluten network.

The soluble sugars that yeast requires for fermentation are traditionally produced from the flour by the action of amylase, a starch-splitting enzyme that is present in small amounts in flour; however, most flour produced in industrialized nations is now highly refined and bleached, and the amylases have been inactivated. For that reason, small amounts of sugar are usually added to the bread dough.

Of course, the fermentation also produces ethanol. Most of that, however, is driven off during baking, and bread contains so little ethanol as to be undetectable either by taste or by effect.

21.8 Soy fermentations are common in the Far East

A number of different Asian foods are made from fermented soy products. Examples known in the West include soy sauce and tempeh, but there are many more. Most of them are mixed fermentations, in which the principal agent is a mold (usually *Aspergillus*), with bacteria as secondary agents (usually lactic acid bacteria).

Soy sauce is made by incubating mashed, partially cooked soybeans with a starter culture containing the mold and bacteria, usually at 30°C for about 3 days, after which a salt brine is added. The mixture is then incubated at cooler temperatures for months. A number of other fermented sauces similar to soy sauce are made throughout Asia, differing in the details of preparation and in the species of microbes involved.

Tempeh is an Indonesian fermented soy product that is generally sliced and fried. It is made by inoculating boiled (but not fully cooked) soybeans with the mold *Rhizopus* and incubating for about a day—sufficient time to allow the mold mycelium to spread throughout, but not so long that the mold sporulates.

21.9 Microbes are used to produce many food additives

Modern food production often involves the addition of compounds to alter the flavor, stability, texture, or color of foods. Many of these compounds are produced by microorganisms in industrial fermentations. The term *fermentation* is used in industrial microbiology to refer to any process of growing microbial cells in order for them to produce a product. The fermentations that produce wine and beer are fermentations in this sense, as well as being fermentations in the biological sense— anaerobic growth using substrate-level phosphorylation to produce ATP. Many industrial fermentations are aerobic, however, and the cells grow by respiration. These are not fermentations in the biological sense.

Many of the food additives produced by microorganisms are normal intermediates or products of metabolism—amino acids, vitamins, organic acids, etc. In many cases, these fermentations use mutants that overproduce the compound, for instance by having lost the ability to repress enzyme synthesis and to feedback-inhibit pathway enzymes.

Food additives produced by fermentation include acids to acidify foods, amino acids, antioxidants, emulsifiers, flavoring compounds, sweeteners, vitamins, and many other types of compounds. A partial list is given in Table 21.1.

21.10 Microbial cells themselves can be food

In rare cases, the microbial biomass is used directly as human or animal food. Most often the microbial cells are used to increase the protein content of a diet—in this case, it is often termed **single-cell protein.** In other cases, the microbial cells are a delicacy or are used as a supplement (yeast cells, e.g., are often used as a supplement for vitamins).

Table 21.1	Food Additives Produced by Industrial Fermentations		
Amino acids Glutamic acid Lysine Tryptophan **Vitamins** Vitamin C Vitamin B$_2$ Vitamin B$_{12}$	**Antioxidants** Ascorbic acid **Sweeteners** Aspartame Thaumatin **Flavors and enhancers** Vanillin Monosodium glutamate	**Thickeners** Glycan Pullulan Xanthan **Colors** β-carotene Astaxanthin Zeaxanthin Lycopene	**Acidifiers** Citric acid Lactic acid Malic acid

The direct consumption of microbes is rare, with the exception of the fungi. Many fungi produce edible fruiting bodies (**mushrooms**), and these are used as food by many human cultures and by many animals. Outside of the fungi, however, microbial cells rarely accumulate in masses that are attractive as food. One exception is the cyanobacterium *Spirulina,* whose spiral filaments can form mats on the surface of alkaline ponds. These mats have traditionally been harvested and dried and then used as food by the 16th-century Aztec in Mexico and by contemporary inhabitants of Chad in central Africa. In both places, alkaline lakes with high mineral content and bright sunlight allows *Spirulina* to grow to a thick gelatinous surface scum that the natives harvest and dry. The dried microbial mat can then be crumbled in other foods or eaten as a snack.

Most single-cell protein is produced as an animal feed supplement, but some is grown for human food. The industrial production of single-cell protein depends on a cheap source of carbon and energy to grow the microbial cells on, as the product owes its commercial competitiveness to low cost. Therefore, the single-cell protein industry is largely linked to other industries that produce high-carbon waste that can be used as a microbial food. Paper mill waste, waste molasses from sugar refineries, citrus peels and pulp from the juice industry, spent beer wort, whey from the cheese industry, feed lot waste, and various hydrocarbons from the petroleum industry are all used as substrates for producing single-cell protein. Because these are waste materials, they make economical substrates on which to cultivate microbes.

Useful Products from Microbes

In addition to their uses to convert food materials into more digestible, palatable, or nutritious foods and to produce food additives, microbes may be used to directly produce products that are useful to humans. There is a wide range of such products—solvents produced as fermentation end products, enzymes that are used in detergents or other applications, and perhaps most importantly, antibiotics. With the advent of genetic engineering, microbes are increasingly being used to produce therapeutic proteins, like hormones and immune mediators. It appears that the industrial use of microbes will be significantly increased in the coming decades.

21.11 Antibiotics are microbial products

Antibiotic is a term for a chemical compound secreted by a microorganism that is toxic to other microorganisms. The first antibiotic discovered was penicillin, from the fungus *Penicillium.* In 1928, the British microbiologist Alexander Fleming was working with bacteria and noticed that a fungal colony that had contaminated one of his plates had prevented bacterial colonies from growing nearby (Figure 21.5). This suggested that the mold was secreting something that killed the bacterial cells. Fleming showed that the substance was highly effective against gram-positive bacteria and nontoxic to laboratory animals. Some years later the Second World War had broken out, and there was a resurgence of interest in penicillin. The war brought an urgent need for agents to control wound infections and sexually transmitted diseases (which removed enormous numbers of men from active duty), and this provoked an effort to develop an industrial process to produce large amounts of penicillin. By 1942, penicillin production began, and by 1943, there was a sufficient production capacity to meet military needs. Penicillin then began to be used in civilian medicine as well. The effect was dramatic. In World War I, the healing rate for bone infections and compound fractures (common in war injuries) was 25%; in World War II, it was 95%, largely because of penicillin.

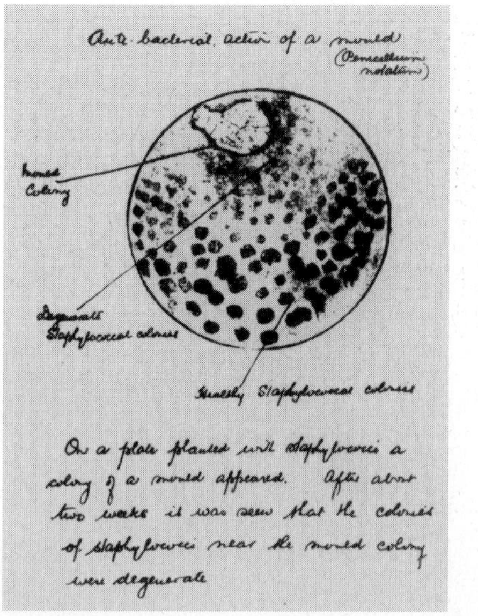

During and after the war, there was an intense search for other antibiotics. The second antibiotic discovered was streptomycin, secreted by the actinomycete *Streptomyces.* This was discovered by Selman Waksman, who patented it by arguing that in nature microbes do not produce antibiotics. He claimed that they only do so in laboratory culture, and, thus, antibiotics are not natural products (which cannot be patented). We now know that he was wrong—antibiotic-producing microbes do indeed secrete antibiotics in nature, and their function is almost certainly to improve the secreting organism's fitness in the ruthless competition among microbes in the wild. After the precedent had been established, however, the practice of patenting antibiotics continued and has very significantly increased the profits that can be made from new antibiotics and thus has encouraged companies to seek them aggressively. The result has been a very large number of antibiotics approved for clinical use (Table 21.2). Antibiotics have transformed the practice of medicine, as they allow effective control of many types of bacterial infections that were previously fatal or led to lasting debility. However, antibiotic resistance is becoming a problem, and some now fear that the age of antibiotics in medicine is coming to an end. Others are

FIGURE 21.5 Fleming's penicillin plate.

Table 21.2	Major Categories of Antibiotics	
Antibiotics	**Producing Organism**	**Target**
B-lactams Penicillin Cephalosporin	Molds	Murein synthesis
Aminoglycosides Streptomycin	Streptomycetes	Protein synthesis (translation)
Tetracyclines	Streptomycetes	Protein synthesis (translation)
Macrolides Erythromycin	Streptomycetes	Protein synthesis (translation)
Rifamycins	Streptomycetes	Protein synthesis (transcription)
Cyclic polypeptides Polymyxin B Bacitracin	Bacilli	Cell membrane Murein synthesis
Polyenes Nystatin	Streptomycetes	Cell membrane

confident that new antibiotics will continue to be discovered, allowing us to keep ahead of emerging antibiotic resistance.

In a way, though, Waksman was right. In nature, the amount of antibiotics secreted is very small. The compounds are complicated organic molecules, and their synthesis is expensive. The only competition that matters to a microbe consists of other microbes in the immediate vicinity—often only a few micrometers away. Thus, there is no need for large amounts to be secreted. For economical industrial production, however, much higher levels need to be attained, and thus, antibiotic production is done by mutant strains of microbes that produce orders of magnitude more antibiotic than their wild-type parents.

21.12 Antibiotics are secondary metabolites

Several features characterize most antibiotics. First, most are made by filamentous organisms: the procaryotic actinomycetes and the eucaryotic fungi. The remainder is mostly produced by *Bacillus* and myxobacteria, both of which go through a complex process of differentiation when populations are starved (see Chapter 11). Second, antibiotics are only secreted as the culture enters stationary phase and as the process of forming spores commences. In the actinomycetes, the reason is thought to be that much of the mycelium lyses, and the released nutrients are what the remaining mycelium uses as energy source for spore formation (Figure 21.6). The antibiotics seem to function to reduce competition for these small amounts of nutrients. Other reasons may apply in other groups.

Because antibiotics are generally produced in stationary phase, they are called **secondary metabolites** to distinguish them from primary metabolites, or compounds produced during the primary growing phase.

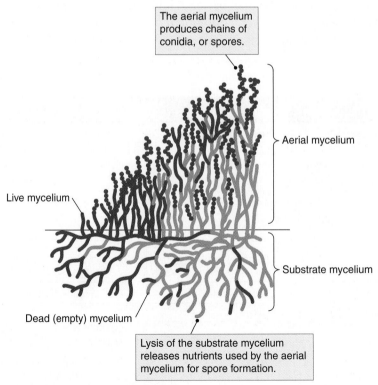

The aerial mycelium produces chains of conidia, or spores.

Aerial mycelium

Live mycelium

Substrate mycelium

Dead (empty) mycelium

Lysis of the substrate mycelium releases nutrients used by the aerial mycelium for spore formation.

FIGURE 21.6 Autolyzing *Streptomyces.*

21.13 Antibiotics are often modified chemically after synthesis

Antibiotics are produced by industrial fermentations because most have chemical structures that are expensive to synthesize chemically. Once the basic compound is available, however, it is often modified chemically to produce entire families of antibiotics (Figure 21.7). For instance, the portion of the penicillin molecule that is toxic to bacteria is the four-membered **β-lactam ring,** but other portions of the molecule may increase or decrease its activity, affect its stability, or affect its ability to get through the gram-negative outer membrane.

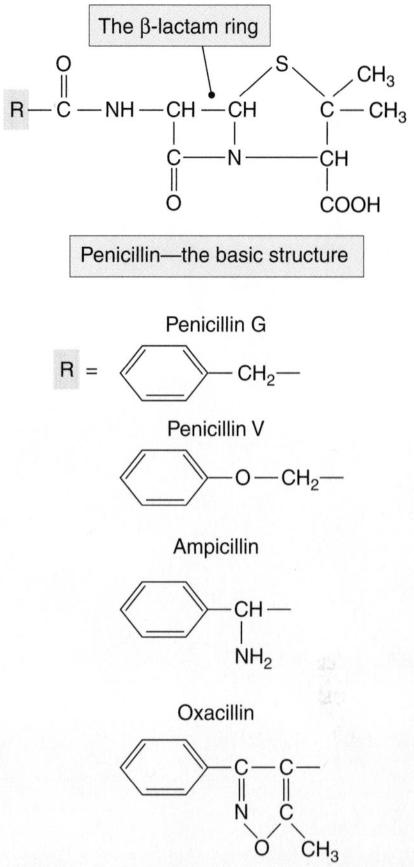

Several natural penicillins are known, varying in the side chain attached to the β-lactam ring. All are produced by *Penicillium,* and which ones are produced at any particular time depends largely on the chemical composition of the growth medium. Thus, by varying the medium, different penicillins can be produced. The most effective of the natural penicillins are the first one produced industrially, penicillin G, and penicillin V.

Many more penicillins have been produced by chemical alteration, however, replacing the side chain that is produced biologically with a synthetically manufactured one. Each semisynthetic penicillin has slightly different biological effects, and, thus, there are therapeutic reasons for producing a number of them. For instance, penicillin G is acid labile and thus is not effective orally (it decomposes in the acid of the stomach). Penicillin V is acid stable and can be used orally. Penicillins G and V are not particularly active against gram-negative bacteria, whereas the semisynthetic ampicillin is (and is acid stable as well). Others, such as oxacillin (another semisynthetic penicillin), were developed to resist attack by **penicillinase,** an enzyme made by some bacteria that cleaves the β-lactam ring and inactivates penicillin.

FIGURE 21.7 Penicillins.

21.14 Antibiotics often bind to uniquely bacterial targets

The selective toxicity of many antibiotics—lethal to bacteria, relatively nontoxic to humans—is due to the fact that their target is a uniquely bacterial one. Two targets are particularly important, as they are the target of a large proportion of all antibiotics: cell wall synthesis and protein synthesis.

Because mammalian cells do not have cell walls, compounds that interfere with cell wall synthesis are generally not toxic to mammals but can be lethal to bacteria. The principal antibiotic target in the complex process of wall synthesis in bacteria is the cross-linking step, in which newly inserted peptidoglycan strands are covalently linked to the existing strands (see Section 4.7). Several classes of antibiotics (the penicillins and the cephalosporins) bind to the enzyme, **transpeptidase,** that catalyzes this step and thus inhibit it. The result is that the bacteria can still make murein, but the wall that is made in the presence of antibiotic is not cross-linked. It becomes progressively weaker as the cell grows, until it cannot resist the turgor pressure and the cell bursts (Figure 21.8).

Of course, all organisms synthesize protein; however, there are significant differences in the machinery of protein synthesis between bacteria and eukaryotes, and many antibiotics target bacterial ribosomes exclusively. For instance, streptomycin binds to a protein on the small subunit of bacterial ribosomes and increases the rate of errors during translation to a lethal level. Tetracycline binds to the large subunit and blocks the binding of charged tRNA in the A-site. Chloramphenicol binds to

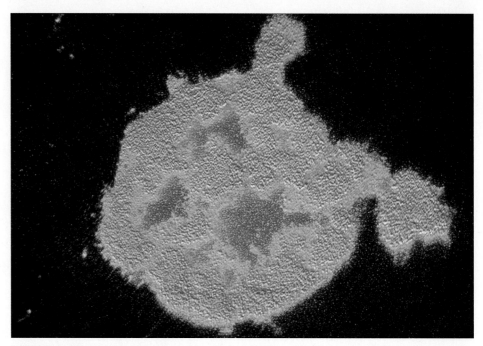

FIGURE 21.8 Bacterium grown in the presence of penicillin.

the large subunit and prevents the movement of the ribosome along the mRNA. None of these affects eucaryotic ribosomes.

Biocontrol

The natural population size of all types of organism is regulated in part by disease. Disease tends to be a density-dependent phenomenon—when the host organisms are plentiful, transmission of pathogens is more effective, and disease becomes a greater problem for the host. This basic ecological fact is the basis for the use of microbes as agents for **biocontrol**—the use of pathogens or predators to control the population size of pests. Unlike the use of chemical pesticides, biocontrol does not aim to eliminate pests or the damage they do, just to limit them to low levels. Probably the most common targets of biocontrol technologies are insects.

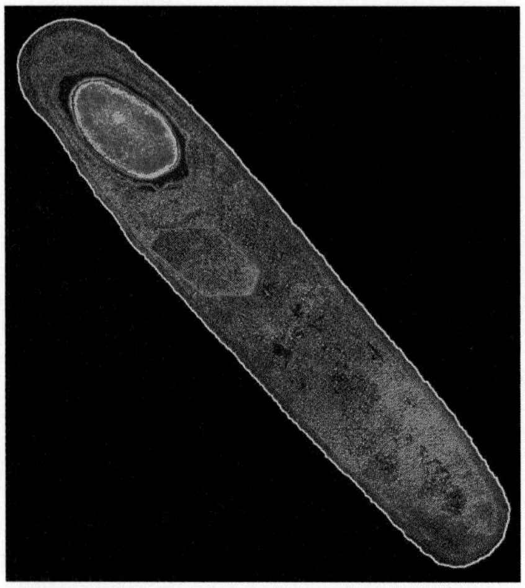

21.15 The most widely used biocontrol agent is *Bacillus thuringiensis*

This organism produces a crystal of protein that forms inside the mother cell alongside the developing spore. Thus, when sporulation is complete, a mixture of spores and these **parasporal crystals** is produced (Figure 21.9). If a caterpillar (larval stage of insects in the family Lepidoptera) ingests this mixture, it is lethal. The protein dissolves in the acidic gut and is cleaved to an active form, producing a toxin that makes the gut permeable. The insect becomes paralyzed by the resulting electrolyte imbalance, while the spores in the gut germinate and invade the hemolymph (the insect equivalent of blood). The insect is basically turned into a sack of nutrients, and the bacillus cells multiply until the nutrients are all gone. They then sporulate,

FIGURE 21.9 Sporulating *Bacillus thuringiensis* cell.

FIGURE 21.10 Rabbits at a water hole in Australia.

forming another crop of spores and crystals that contaminates the local environment. This natural amplification is one of the advantages of *B. thuringiensis* as a biocontrol agent.

There is a worldwide industry of cultivating *B. thuringiensis* for use as an insecticide. It is generally nontoxic to organisms other than insects. Recently, the gene encoding the protein has been cloned, and a number of plants that express this protein in their leaves have been engineered. These genetically engineered plants include several food crops (e.g., corn), plus plants grown for fiber (cotton).

Other biocontrol agents are viruses, which can be used in the same way against insects or other pests and which have the same advantage of biological amplification when they are used.

Like all control agents, from DDT to antibiotics, development of resistance is a problem. It was once thought that the mode of action of toxins and pathogens was too complex for resistance to emerge easily, but that turns out to be incorrect. Resistance to *B. thuringiensis* is now emerging among target insects. It is also possible for biocontrol programs to fail because of changes selected in the pathogen, rather than in the host. One of the most spectacular examples of biocontrol failure was due to this. Rabbits were introduced into Australia in the 19th century and rapidly became a huge ecological problem, as there were no natural predators (Figure 21.10). In an attempt to control them by using disease, the myxoma virus was introduced in the 1950s. This virus is a close relative of the human smallpox virus and causes a highly lethal rabbit disease called **myxomatosis.** When it was introduced, it initially brought the rabbit population down in a spectacular fashion, but soon the rabbits began to rebound. The reason turned out to be that the myxoma virus had lost most of its virulence, probably because the reduced-virulence strains competed better in nature because they had significantly higher likelihood of being transferred before the host died (Figure 21.11). Particularly when the rabbit density had been reduced by myxomatosis, the wild-type virus had difficulty in being transmitted in a timely fashion. In the end, myxomatosis failed as a biocontrol agent, and the rabbits are still a problem. Another attempt is now being made with another virus, rabbit calicivirus; it will be several years before the results will be clear.

The most widely used biocontrol agent is *Bacillus thuringiensis*

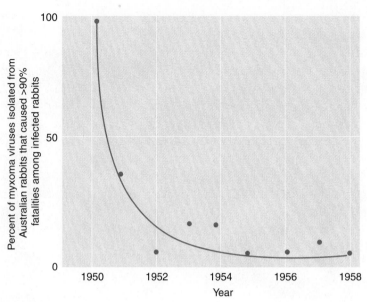

FIGURE 21.11 Virulence of myxoma virus used for biocontrol.

Wastewater Treatment

Humans produce significant amounts of biological waste, which can present a health hazard—principally transmission of intestinal diseases. It can also present an ecological hazard—pollution of water systems, leading to abundant microbial growth resulting in anoxia and killing of fish and other aquatic life. This was not a problem when humans lived in small nomadic bands. They moved frequently, leaving their waste behind. When humans began to settle down in the Neolithic, however, their waste accumulated near their dwellings, and problems began to be serious. When human cultures began to urbanize, with major cities home to tens or hundreds of thousands, the problems became acute, and polluted water, stench, high levels of intestinal diseases, flies, and rodents came to characterize cities. This changed around 1900 in the developed world, as the link between sewage and intestinal disease became clear. One of the principal differences between the developed world, with its modest burden of infectious disease, and the developing world, with its crippling burden of disease, is the presence in the former of major engineering infrastructure that scrupulously separates sewage from drinking water and that oxidizes much of the organic matter in sewage before it is released into the environment.

21.16 Wastewater treatment has two principal goals

Wastewater treatment has two principal goals: the killing of pathogens and the removal of organic material. The former is accomplished in two different ways. First, many pathogens die naturally outside the body, as they are adapted to life in the complex ecosystem of the mammalian body. Outside, they are competitively less fit and have difficulty competing with other organisms adapted to life in dilute aquatic systems. They thus die naturally during the processing of the wastewater. In addition, sewage treatment plants operated by cities and towns normally sterilize the effluent from the plant by **chlorination** (adding hypochlorite, OCl^-) or by irradiating it with ultraviolet light to produce ozone. These powerful oxidants are intended to kill any pathogens remaining after the process.

The second goal, oxidation of organic compounds, is a microbial process. After the particulate matter in wastewater is removed, what remains is a dilute solution of organic matter. In order to oxidize this to CO_2, it is held in contact with a large number of microbial cells (mostly bacteria) for a period of a few days, during which the bacteria scavenge most of the organic matter, oxidizing about half of it and converting the rest to new microbial cells. Wastewater treatment is thus an intrinsically microbial process. Of course, without treatment this oxidation would happen naturally in the lake, river, or soil into which the effluent is dumped; however, the treatment process insures that most oxidation occurs in the plant, protecting the environment from the negative consequences of oxidizing it in the field. Nevertheless, treatment never completely removes organic material from wastewater, and the release of large volumes of treated wastewater can still cause significant environmental degradation.

Because nearly every town and city in the world has some form of wastewater treatment facility, these constitute one of the major and most visible applications of microbiology.

21.17 Primary treatment removes particulate matter from sewage

The most common wastewater treatment consists simply of removing particulate material and then disposing of the liquid without further treatment. Removal of solids from wastewater is done by holding the wastewater in a tank that is large relative to the rate of water flow. Thus, although there is continuous flow into and out of the tank, the liquid spends considerable amount of time there. This allows solids that are denser than water to settle to the bottom and those that are lighter than water to rise to the surface. A scraper moves slowly and continuously over the bottom, pushing the sediment into an effluent pipe. Floating material is skimmed off the top in a similar fashion. Both the sediment and the floating material are pumped to an anaerobic digester, discussed later. This form of treatment is termed **primary treatment,** as it is the first step in what can be a two or three step process. For many municipalities in both developed and developing world, however, this is the only form of treatment. There is considerable ecological damage from the large volumes of wastewater dumped by thousands of cities worldwide, but the costs of upgrading all these plants are huge and have thus far prevented more widespread use of further treatment.

Rural households also normally use only primary treatment as well. When houses are dispersed, a central treatment facility is not economical, and each house has to treat its own sewage. This in done in a **septic tank,** which functions exactly the same as the settling tanks in primary treatment in a wastewater treatment plant. The principal difference is that the sediment is not removed continuously but has to be pumped out as the tank fills up (typically every few years) (Figure 21.12).

The effluent from a septic tank flows out through perforated pipes into the soil, where it is oxidized by soil bacteria as it percolates downward. The area over which the wastewater is distributed into the soil is called a **leach field.** Septic systems can be an effective way to treat sewage, but they are critically dependent on the nature of the soil in the leach field. If the soil is unsuitable, the sewage can rise to the surface and pool, or it can flow downward too rapidly and contaminate the groundwater, which can then contaminate drinking water wells.

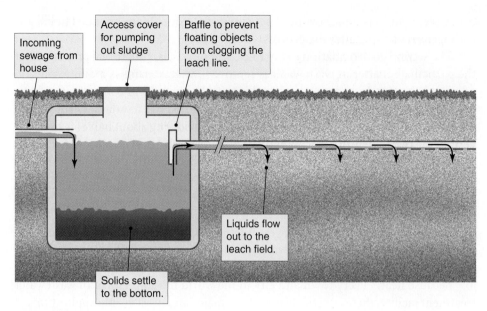

Incoming sewage from house

Access cover for pumping out sludge

Baffle to prevent floating objects from clogging the leach line.

Liquids flow out to the leach field.

Solids settle to the bottom.

FIGURE 21.12 Septic tank and leach field.

21.18 Solids are digested anaerobically

The solids that have been separated from liquid wastewater during primary treatment are pumped into a large tank, termed an **anaerobic digester,** where they are held for weeks. The material pumped into the digester consists predominantly of human solid waste—bacterial cells and undigested fiber. During the weeks in the anaerobic digester, the fiber is degraded; the microbial cells die, and their remains are used by other microbes as food. The result is a succession of different microbes, each using the residue of the former ones for food. Ultimately, much of the material is converted to CO_2 and CH_4, with the remainder consisting of microbial cells of the final successional stages. This material is typically disposed in landfill but is sometimes used as fertilizer (it is rich in nitrogen). If there is any industrial waste in the wastewater, however, the output of the anaerobic digester can contain toxic industrial compounds that resist degradation by microbes, or heavy metals.

The methane that is a major product of anaerobic digestion can, at least in larger plants, be compressed and used as an energy source for plant operations. In smaller plants, however, this is not cost-effective, and the methane is burned as it leaves the digester.

21.19 Secondary treatment oxidizes dissolved organic material

The oxidation of dissolved organic compounds in wastewater is done by exposing the liquid to large numbers of bacteria, which use the organic compounds as source of carbon and energy for growth. The result is the conversion of most organic material in wastewater to either CO_2 or to microbial cells. The cells are then allowed to settle and are disposed of in the anaerobic digester. The process is termed **secondary treatment.**

There are three main systems of secondary treatment. The simplest, called **lagooning,** is to allow the effluent from primary treatment to flow into one end of a large, shallow lagoon and out of the other end. As it flows through the lagoon, microbes oxidize the dissolved organic material. A large enough surface area to keep the shallow layer of liquid oxygenated is critical. Wastewater lagoons combine the oxidation and settling steps, as the microbial cells settle to the bottom of the lagoon, while the treated water flows out of the end. Thus, lagoons have to be drained periodically and the sludge on the bottom scraped out. This system is an option only for small or moderately sized municipalities with sufficient available land.

In many places, land is too valuable for lagooning to be an option. These communities have two options for secondary treatment in a confined space. The oldest type is termed a **trickling filter,** and it consists of a large tank filled with rocks. The wastewater is sprayed on the top and trickles down through the bed of rocks, ultimately flowing out through vents on the bottom. As it flows over the rocks, dissolved organic material is oxidized by bacteria growing in biofilms on the rocks. As the biofilms get thicker, pieces of them break off and flow out in the effluent. These **flocs** are then separated from the water by allowing them to settle in a settling tank identical in principal to the settling tanks of primary treatment. The sludge is pumped to the anaerobic digester (Figure 21.13).

The alternative is to use an **activated sludge digester,** a tank in which the wastewater is actively aerated and stirred (Figure 21.14). This is quite effective and is now the preferred approach. As with a trickling filter, the effluent from an activated sludge digester goes to a settling tank to remove the microbial cells for anaerobic digestion. In this process, however, some of the settled cells are pumped back into the activated sludge digester. This increases the microbial biomass in the digester, thus speeding up the oxidation of dissolved organic compounds. More importantly, though, it provides a strong selection for cells that settle quickly, which makes the settling step more efficient. The result is that cells in an anaerobic digester generally are not individual cells but consist of dozens to hundreds of cells held together in a common capsular matrix. These aggregates are called flocs,

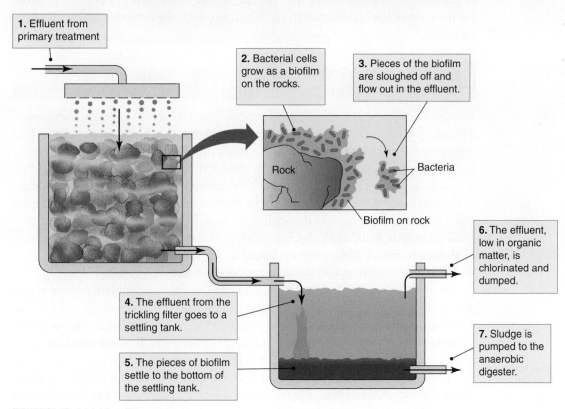

1. Effluent from primary treatment

2. Bacterial cells grow as a biofilm on the rocks.

3. Pieces of the biofilm are sloughed off and flow out in the effluent.

Rock

Bacteria

Biofilm on rock

4. The effluent from the trickling filter goes to a settling tank.

5. The pieces of biofilm settle to the bottom of the settling tank.

6. The effluent, low in organic matter, is chlorinated and dumped.

7. Sludge is pumped to the anaerobic digester.

FIGURE 21.13 A trickling filter.

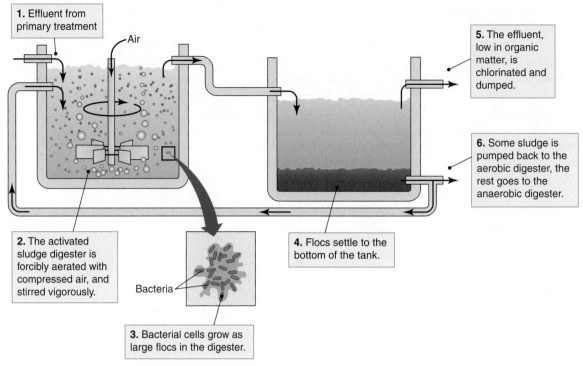

1. Effluent from primary treatment

Air

2. The activated sludge digester is forcibly aerated with compressed air, and stirred vigorously.

Bacteria

3. Bacterial cells grow as large flocs in the digester.

4. Flocs settle to the bottom of the tank.

5. The effluent, low in organic matter, is chlorinated and dumped.

6. Some sludge is pumped back to the aerobic digester, the rest goes to the anaerobic digester.

FIGURE 21.14 The activated sludge process.

and they are critical to the success of the process. If the cells in the digester were single and separate, they would not settle effectively in the settling tank, and the effluent from the plant would still contain significant amounts of organic material (in the form of the cells themselves), defeating one of the two principal goals of the process.

21.20 Wastewater treatment is subject to periodic failures

Wastewater treatment is an empirical process that is subject to periodic disruptions. The operators of treatment plants have no control over the volume in wastewater that enters the plant or over the organic compounds dissolved in it. Variations in flow rate can be significant, as can variations in chemical composition, particularly when industrial wastes are part of the input. These variations can, in turn, lead to variation in the efficiency of the oxidation process. Probably the most serious problems arise when there are disruptions in the settling process. Two types of variation can be particularly serious. **Bulking sludge** is a failure of flocs to settle quickly enough because of abundant growth of filamentous bacteria. The filaments protrude from the flocs and retard their settling, and the flocs are not separated from the liquid.

Even if the flocs settle adequately, they can rise again, a problem termed **rising sludge.** Rising sludge can be caused if the wastewater contains a significant amount of nitrate. When this happens, denitrifying bacteria can grow in the anaerobic sludge at the bottom of the settling tank, producing nitrogen gas. Bubbles of nitrogen get trapped in the settled sludge, causing it to float.

Wastewater treatment plant operators have little control over bulking or rising sludge. When either happens, they have no choice but to continue operating the plant and fiddling with the operating parameters until, hopefully, the problem goes away.

Biological Weapons

Like all other technologies that humans have developed, microbiology has been used for hostile as well as for benign purposes. In most of the wars of history, more people were lost to disease than to enemy action—both soldiers and civilians. It is perhaps natural that some have attempted to harness disease as a weapon. Efforts to do so throughout history were primitive and probably mostly ineffective. After about 1900, however, the revolution in microbiology that led to the understanding of the mechanisms of contagion provided the information necessary to the intelligent use of microbes as weapons. Since then, biological weapons have been feasible at least in principle. Fortunately, their use has been infrequent, and they are now banned by international law.

21.21 Biological weapons were used in both world wars

World War I, from 1914 to 1918, is known principally for its brutal trench warfare and the widespread use of chemical weapons. Less well known were German attempts to use disease as a weapon as well. Germany used two animal diseases—glanders and anthrax—to attempt to sicken animals that were being shipped to Britain, France, or Russia. To do this, they had secret agents working in a series of neutral countries—Romania, Spain, the United States, and Argentina—injecting horses and mules in the holding pens prior to shipment. As far as can be determined, the program had little effect.

Most countries entered World War II with military microbiology programs that were designed to develop biological weapons for use on the battlefield, but none of these programs developed weapons that could be fielded during the course of the war. England, however, produced a stockpile of 5 million cattle cakes—little blocks of pressed linseed meal, each containing 5×10^8 anthrax spores. The cakes were designed to be spread over the German countryside in case Germany used chemical or biological weapons first. Because Germany did not, the stockpile was not used, and it was destroyed after the war.

Of all the belligerents in the war, only Japan used chemical and biological weapons in China against Chinese troops and civilians. The principal biological weapon was plague, distributed by dropping infected fleas from airplanes or by releasing infected rats into cities. It has been estimated that between 100,000 and 200,000 Chinese died of the resulting plague outbreaks, but the number is impossible to confirm. Probably an additional 10,000 Chinese died in Japanese medical experiments, which used Chinese prisoners as human "guinea pigs."

In addition to plague, a number of intestinal diseases, such as typhoid fever and cholera, were used to contaminate wells and food left behind by retreating Japanese troops.

At the end of the war, the Japanese war criminals who had used prisoners in medical experiments and who had used disease as weapons against the Chinese people were given immunity from war crimes prosecution in exchange for cooperation with the U.S. occupying authorities. Probably one principal motivation was that the United States had an active program to develop biological weapons, but it had not used them nor had it done human experiments with lethal diseases. Thus, the Japanese had information that the United States could not get any other way. A second reason was that any prosecution would have forced the United States to share the information it was getting from the Japanese with the Soviet Union, which was a co-prosecutor in the Tokyo War Crimes Tribunal. As the cold war was well under-way by the time the Tribunal began, the United States had good reasons to want to keep this information out of Soviet hands.

21.22 Biological weapons are banned by international law

After World War II, several countries continued their programs of biological weapons development. Principal among them were the United States, the United Kingdom, Canada, France, and the USSR. These programs developed weapons systems, including sprayers, bombs, rocket warheads, and artillery shells that could carry powdered or liquid suspensions of pathogens. Lethal diseases such as anthrax and incapacitating diseases such as Q fever were developed as weapons. Most of these countries convinced themselves that the weapons could be effective, especially against unprotected civilian populations, and some (the United States and USSR) accumulated large stockpiles.

This was despite a long-standing prohibition of using biological or chemical weapons in war—the 1925 Geneva Protocol banned the use of these weapons; however, it was still legal to possess the weapons, as retaliation in kind was widely recognized as a right under international law. In 1969, the United States, by President Richard Nixon's executive order, unilaterally abandoned its offensive biological weapons program and destroyed its stockpiles. The reasons were complicated but revolved around the fact that the United States had nuclear weapons, and there were no clear situations in which it might wish to use biological weapons.

This renunciation cleared the way to a new treaty, the Biological Weapons Convention, that banned development, production, and stockpiling of pathogens and toxins in amounts in excess of what is necessary for peaceful medical and other uses. The treaty was signed in 1972 and took effect in 1975.

Despite this new treaty, the Soviet Union continued and even expanded its biological weapons program. This expansion continued throughout the 1970s and 1980s and was only ended after the Soviet Union had collapsed. Boris Yeltsin, president of the Russian Federation, finally ended the program by decree in 1992.

Several other countries are known to have violated the Biological Weapons Program—most prominently Iraq, but South Africa also. South Africa voluntarily gave up its program after its White apartheid government was replaced by a Black majority government, and Iraq disarmed immediately after the 1991 Gulf War. In addition to these three countries, about a half dozen others are suspected of developing biological weapons, but the evidence is weak and could easily be misleading.

Summary

Microbes have been exploited for human purposes since prehistoric times. By a process of empirical discovery, humans learned to leaven bread, make alcoholic beverages, and preserve food using various microbial fermentations. By the time that scientists began to understand the basis for these activities, in the very late 1800s, a complex and mature industry produced a wide variety of foods by microbial processes. Modern technologies have not changed these industries in any fundamental way, but they have allowed much more rational control and greatly increased the reliability and repeatability of the processes.

After the microbial world became better known, a variety of new industries developed to produce novel microbial products, most notably food additives of many different kinds, and antibiotics. The antibiotic industry has transformed medicine by allowing highly effective treatment and prevention of bacterial infections. These new uses of microbes are expanding rapidly as genetic engineering allows a wide range of new products to be produced by bacteria.

Understanding the microbial basis of infectious disease also laid the foundations for effective control of transmission of intestinal diseases by modern sewage treatment. The infrastructure that scrupulously separates drinking water from sewage now characterizes all of the developed world and underlies much of the higher standard of public health found in these countries.

In these ways, as in so many other ways that we have explored in previous chapters, all life on earth is intertwined. Human well-being is truly dependent on microbes.

Study questions

1. What two fermentations are the most important in the food industry? What foods do they produce, and what organisms are involved?
2. What chemical transformations are involved in cheese ripening, and what is the microbial role in this process?
3. What role does the malo-lactic fermentation play in wine production, and what benefit do the microbes get from it?
4. How do the meanings of the term *fermentation* differ when used by an academic microbiologist and an industrial microbiologist.
5. What are antibiotics, and why do microbes make them?
6. Biocontrol can be quite successful in controlling pest populations, but it can also fail. What are the reasons for lack of success in some biocontrol efforts?
7. What are the goals of wastewater treatment, and what are the principles of the treatment process to achieve these goals?
8. In an activated sludge system, some of the cells that settle out are transferred back into the activated sludge digester. Why is this done, and why isn't it done in a trickling filter system?
9. What countries are known to have made biological weapons in the past century? What countries have used them?
10. What treaties ban biological weapons, and precisely what activities do they prohibit?

Index

Note: page numbers followed by an *f* indicate presence of a figure; a *t* indicates a table.

metabolism, 141, 141*t*
nitrogen fixation, 188
obligate, 43
Anaerobic chamber, 43, 44*f*
Anaerobic digester, in wastewater treatment, 460
Anaerobic ecosystems
oxygen consumption and, 357–359, 358*f*
syntrophy in, 360–361, 361*f*
Anaerobic food chain, 359–360, 360*f*
Anaerobic jar, 43, 44*f*
Anaerobic respiration, 141
Anaerobic respirations, electron acceptors in, 157, 157*t*
Angler fish, 377, 377*f*
Animals
diseases transmitted to humans (*See* Zoonosis (zoonoses))
germ-free, 381
intestinal tract, colonization of, 380–381
intestinal tracts of, oxygen consumption in, 359
lymphatic system, 383, 384*f*
normal flora, 381
nutrient sequestration, 383–384
respiratory tract, 382–383, 383*f*
ruminant, microbial fermentation in, 367–368, 368*f*
Annamoxosome, 62, 316, 316*f*
Annular diaphragm, 29–30, 29*f*
Anoxygenic photosynthesis, 141
Antheridium, 345
Anthrax
contagion, 19–20
cutaneous, 436
gastrointestinal, 436
50% infectious dose in monkeys, 409, 409*f*
inhalational, 436
Anthrax vaccine, 404
Antibiotic resistance, 212, 417–419, 418*f*
Antibiotics
bacterial targets for, 455–456, 456*f*
b-lactam ring, 455, 455*f*
categories, 453, 453*t*
chemical modification after synthesis, 455, 455*f*
definition of, 452
discovery of, 452–454, 453*f*
as secondary metabolites, 454, 454*f*
Antibodies
antigen binding, 392, 393*f*
activation of complement system and, 394
host protection from, 392, 394, 394*f*, 395*f*
sites for, 392, 393*f*
blockage, by pathogens, 415, 415*f*
definition of, 392
neutralization of virion, 394, 395*f*
opsonization by, 394, 394*f*
Antibody-producing cells, generation by clonal selection, 395–397, 396*f*

Antigenic determinant (epitope), 392, 393*f*
Antigens
binding to antibodies, 392, 393*f*
activation of complement system and, 394
host protection from, 392, 394, 394*f*, 395*f*
sites for, 392, 393*f*
definition of, 391
variations, reduction of immune system effectiveness and, 414
Antimicrobial peptides (AMPs), 385, 386*f*, 387, 387*f*
Antisense RNA, prevention of mRNA translation, 229, 229*f*
Antiterminator proteins, 225–226, 227*f*
Antitoxins, 213
Apicomplexans, 336, 336*f*
Apoptosis (programmed cell death), 385, 386*f*
Aquifex, 305
Arabinose operon, 223–224, 223*f*
Archaea, 5*f*, 6, 317–321. *See also* Procaryotes
cell envelopes, 52, 52*f*
cell membrane
lipids in, 62–63, 63*f*
structure of, 62–63, 63*f*
halophilic, 182–183, 183*f*, 207*f*
hyperthermophilic, diglycerol-dibiphytane-tetraether in, 63, 63*f*
kingdoms
Crenarchaoeta, 318*f*, 320–321, 320*f*
Euryarchaeota, 318–320, 318*f*, 319*f*
Nanoarchaeotea, 318*f*, 321, 321*f*
phylogenetic tree, 317–318, 318*f*
pseudomurein, 55–56, 55*f*
reverse TCA cycle, 184–185, 185*f*
ARDS (acute respiratory distress syndrome), 427
Arginine dihydrolase pathway, 149, 149*f*
Artificial taxonomy, of viruses, 136
Ascomycetes, 348–349, 349*f*, 350*f*
Ascospore formation, 349, 350*f*
Ascus, 348
Asexual multiplication (reproduction), 266
eucaryotic, 288
in protists, 116
water mold, 344, 345*f*
Assembly, in viral multiplication process, 122*f*, 132–133, 134*f*
Atmosphere, evolution, oxygen in, 182
ATP
energy metabolism and, 146
yields from microbial respiration, differences in, 157–158, 158*f*
ATP synthase (ATPase), 146
Attenuated vaccines, 403–404
Attenuation, operon control, 224–225, 224*f*–226*f*
Attractants, 242, 242*f*
Autoclave, 40
Autoimmune diseases, 391–392

Bread making, ethanol fermentation for, 450–451
Brightfield microscopy, 27–28, 27f
BSC (biological safety cabinet), 45–46, 46f
BSE (bovine spongiform encephalopathy), 137
BSL (biosafety levels), 44–45, 45t
Bubo, 410
Bubonic plague, 435
Budding
   in endomembrane system, 85, 86f
   in procaryote cell division, 191, 191f
   viral, 134
   yeast, 289
Bulking sludge, 462
Bunsen burner flame, to prevent contamination, 41–42
Burkholderia cepacia, chromosomal organization, 209
Butyrate, assimilation, in photoheterotrophic growth, 174, 174f

C

C3, 387–388, 388f
Calmodulin, 109
Calvin-Benson cycle (ribulose-biphosphate cycle), 161–162,
     183–184, 184f
Cambrian era, 3
Candida albicans, 382
CAP (catabolic activator protein), 232, 233f
Capping, mRNA, 89–90, 90f
Capsids, viral
   assembly, 132–133, 134f
   binal, 120, 120f
   cylindrical or tubular, 120, 120f
   cylindrical tails, 120
   definition of, 118
   envelopes, obtained from budding, 120–121, 121f
   irregular-shaped, 120, 121f
   isosahedral, 120, 120f
   symmetrical forms, 120, 120f
Capsules, procaryotic, 74–75, 74f, 414
Carbon
   aerobic carbon/oxygen cycle, 355, 355f
   in anaerobic sediment, transfer to aerobic zone, 359–360,
     360f
   return to land, 355
   sequestration in geological formations, 359
Carbon dioxide fixation
   autotrophic
     definition of, 177
     NADPH generation for, 179–180, 179f
   pathways
     Calvin-Benson cycle, 183–184, 184f
     hydroxypropionic acid cycle, 186–187, 187f
     Ljungdahl-Wood, 185–186, 186f
     reverse TCA cycle, 184–185, 185f
   reactants
     NADPH, 172–173
     reduced ferredoxin, 172–173

Carotenoid pigments
   protective effects, 206, 207f
   structure, 206, 206f
Carrier proteins, in electron transport system, 152–153, 152f
Case definition, 428
Cassette model, of gene regulation, 289–290, 290f
Catabolic repression, 231, 232, 232f, 233f
Catalase, 207
Caulobacter
   dimorphic cell division, 244–248, 245f–248f
   flagellar genes, transcription of, 247–248, 247f, 248f
   flagellar synthesis, 246–247, 246f
Caulobacter crescentus, as model system, 22, 23f
Cell density, quorum sensing, 250
Cell division, 231
Cell envelopes
   Archaeal, 52, 52f
   definition of, 50
   prevention of osmotic lysis, 50
   types of, 50–52, 50f
Cell-mediated immunity
   cytolytic T lymphocytes
     production by clonal selection, 399–400, 400f
     targeting of cells making foreign protein, 398–399, 399f
   definition of, 392
   lymph nodes and, 400–401, 401f
   vaccination and, 401–404
Cell membrane, procaryotic
   fusion with secretory vesicles, 94
   hopanoids, 60, 60f
   lipid and cell wall synthesis, 61
   organelles, intracellular membrane-bound, 62
   protein content of, 61, 61f
   protein synthesis, 61
   structure, 62–63, 63f
   surface area, invagination and, 61–62, 62f
   as two-dimensional fluid, 60, 60f
Cells. *See specific cell types*
Cellular slime molds, 331, 332f
Cellulolytic myxobacteria, 260
Cell wall
   definition of, 50
   eucaryotic, 110–111
   in gram-positive bacteria, 56, 56f
   synthesis, in endoplasmic reticulum, 92
Central metabolism
   pathways, 142, 143f
   unity of biochemistry and, 142
Centric diatoms, 342, 342f
Centrosomes, 104–105, 105f
Chaperone proteins, 84, 234
CheA, 242, 242f
CheB, 243–244, 243f
Cheese production, 446–447
Chemical concentration gradients

Coadaptation, 381
Coccus/cocci, 49, 49*f*
Coenzyme M, 165, 165*f*
Coevolution, 298
Co-inheritance, 271–272, 271*f*
Colithophorids, cell wall of, 111, 112*f*
Colonies, 20, 42
Common cold, 439
Compatible solutes, 205–206
Complementary area, 33
Complement system
   activation by antibody binding, 394
   initiation of inflammation, 387–388, 388*f*
Complex III. *See* Cytochrome b/c complex
Complex media, 42
Composite transposons, 211–212, 212*f*
Compound microscopes, image formation, 25, 26*f*
Concatemer, 128, 129*f*
Condenser, 14
Condensins, 88
Conditional mutations, 270
Confocal microscope, 31
Conidia, 349, 349*f*
Conjugal pili (sex pili), 275
Conjugate area, 33
Conjugation
   definition of, 269, 275
   DNA transfer, 276, 277*f*
   donor-recipient recognition and binding, 275–276, 275*f*
   F insertion sequences, recombination with chromosome and, 278–279, 278*f*
   F′ plasmid creation, 280–281, 282*f*
   Hfr mating, 279, 279*f*
   repression of transfer genes, 276
Conoid complex, 336, 336*f*
Consensus sequence, 214–215
Contractile vacuole complex, 113–115, 113*f*–115*f*
Contractile vacuole pore, 114, 114*f*
Core, 58
Corepressors, 218–219, 219*f*
Cortex, of spore, 250, 252, 252*f*
Corynebacterium diphtheriae, 406–407, 419
Cotransduction frequencies, 284, 284*f*
Co-translational protein targeting, 83–84, 84*f*
Cowpox, 402
C period, 195–196, 195*f*–197*f*
Crenarchaotes, 318*f*, 320–321, 320*f*
Crossing, in genetic mapping, 271–272, 271*f*
Cross-link, 35
Crossovers, in genetic recombination, 268, 268*f*
CRP (cyclic AMP receptor protein), 232, 233*f*
Crustose lichens, 373–374, 373*f*, 374*f*
Cryptobiotic spore, 249
Cryptomonads, 333–334, 333*f*

CTLs. *See* Cytolytic T lymphocytes
CtrA, 246, 246*f*
Cultures, 39
Curd, 446
Cutaneous anthrax, 436
Cyanobacteria
   Calvin-Benson cycle (ribulose-biphosphate cycle), 183–184, 184*f*
   heterocysts, 49–50, 50*f*, 312, 312*f*
   microfossils, 3, 3*f*
   nitrogen-fixing, 49–50, 50*f*, 188
   as obligate photoautotrophs, 173
   phycobilisomes, 169–170, 171*f*
   thylakoid membranes, 61, 61*f*, 168
   type I/II photosynthesis, 175
      cyclic photophosphorylation, photosystem I for, 180, 180*f*
      noncyclic photophosphorylation, photosystem I and II for, 181, 181*f*, 182
Cyanophycin, 256
Cyclic-3′,5′AMP, 232
Cyclic photophosphorylation
   definition of, 171
   electron donor independence, 172
   steps in, 172*f*
   in type I/II photosynthesis, photosystem I for, 180, 180*f*
   in type II photosynthesis, photosystem II for, 177, 178*f*, 179
   in type I photosynthesis, photosystem I for, 176, 177*f*
Cylindrical capsid (tubular), 120, 120*f*
Cytochrome b/c complex (complex III)
   definition of, 152
   procaryotes lacking in, 153, 154*f*
   in type I photosynthesis, 176, 177*f*
Cytochrome b/f complex, 181, 181*f*
Cytochrome oxidase, 153
Cytochromes, 152
Cytokines, 422
Cytokinesis, bacterial, 70, 70*f*
Cytolytic T lymphocytes (CTLs)
   production by clonal selection, 399–400, 400*f*
   targeting of cells making foreign protein, 398–399, 399*f*
Cytophages, 314
Cytoplasm, procaryotic
   acidocalcisomes, 72, 72*f*
   exchange of small molecules, with mitochondria and chloroplasts, 100
   gas vacuoles, 73, 73*f*
   gas vesicles, 73, 73*f*
   magnetosomes, 73–74, 73*f*
   mRNA translation, 90
   ribosomes in, 70, 70*f*
   storage granules, 71, 71*f*
Cytoplasmic membrane, procaryotic. *See* Cell membrane
Cytoplasmic streaming, 106
Cytoproct, 97, 98*f*

EF2 (elongation factors), 406–407
Effector molecule, 218
"Egg fungi" (oomycetes), 343–346, 345f
Electron acceptors, in anaerobic respirations, 157, 157t
Electron donors, for autotrophic growth, 172–173, 173f
Electron gun, 32
Electron microscope, 15–16, 16f
    image formation, 32–33, 34f
    lens, electromagnetic, 31–33
    resolution, 33–34
    scanning, 38–39, 38f
    shadow casting technique, 35–36, 36f
    specimen preparation, freeze etching, 36–38, 37f
    specimens
        fixation of, 35
        negative staining, 35, 36f
        thin sectioning, 35, 35f
Electron transport chains
    branching of, 154, 155f
    oxidase-negative, 153, 154f
Electron transport system, 154, 155f
    electron entry, 159–160, 162f
    in methanogenesis, 165–166, 165f
    protein complexes in, 152–153, 152f, 153f
    reverse, for chemoautotroph reductant generation, 161–162, 163f
Elementary bodies, 316
Elongation factors (EF2), 406–407
Emerging disease, 422f, 441–422
Endemic disease, 424–425, 426t, 427, 427f
Endocytosis, 95
Endoflagella, 312–313, 313f
Endogenotes, 266
Endomembrane system, eucaryotic, 82, 99
    chromosomes, 87–88, 88f
    definition of, 85
    endoplasmic reticulum, 87, 87f
    exchange
        by budding, 85, 86f
        by vesicle fusion, 86f, 87
    nuclear envelope, 87, 87f
        chromatin in, 87–88, 88f
        pore complexes in, 88–89, 89f
    nucleoplasm, 87, 87f
    vesicles, tracks for sliding, microfilaments as, 106
Endomembrane vesicles, microtubules and, 105
Endomycorrhizae, 375–376, 375f
Endonucleases, 227
Endoplasm, 106
Endoplasmic reticulum
    lipid synthesis, 92
    lumen, 92
    membrane synthesis, 91–92
    protein synthesis, 92
    rough, 92

synthesis
    of cell walls, 92
    of digestive enzymes, 92
    of secretory proteins, 92
Endosome
    fusion with lysosomal vesicles, 94–95, 95f
    recycling, of membrane proteins, 96
Endospore
    activation, 256
    definition of, 248
    development, sigma factors and, 254, 254f, 255f
    durability, 248–249
    formation
        initiation, 253–254, 253f
        morphological stages, 250, 251f, 252–253, 252f
        mother cell, 248
        mother cell DNA rearrangements, 254–255, 255f
        as starvation response, 249–250, 249f
    germination, 255–256, 256f
    longevity, 248–249
    maturation, 253, 253f
    outgrowth, 256, 256f
Endospore-formers, gram-positive bacteria, 310–311, 311f
Endosymbiotic theory, 100, 101f, 102
Endotoxins, 422–423
"Energy currency," 146
Energy metabolism, 145–146
Enhancers, DNA looping and, 221, 221f
Enrichment culture, 21–22
Enteric bacteria, 307, 308f
Enterococci, vancomycin-resistant, 418, 418f
Enterococcus faecalis, transposon-mediated gene acquisition, 212
Enterotoxins, 419, 420
Enzymes. See also specific enzymes
    exonucleases, 91
    hydrolytic, 81
    tailing, 90
Epidemic disease
    definition of, 424
    detection, notification of physicians and, 427–428
    vs. endemic disease, 425, 426t, 427, 427f
    global dissemination, 433–434, 434f, 435f
    incubation period, 429, 432, 432f
    influenza, 436–438, 437f, 438f
    investigation, standard procedures for, 428–429, 429f
    rate of transmission
        course and, 429–432, 430f–432f
        population structure and, 432–433
Epidemiology, 424–444
    definition of, 424
    disease transmission, global, 433–434, 434f, 435f
    epidemic disease, detection, notification of physicians and, 427–428
    frequency of disease, 424–425

investigation procedures, 428–429, 429f
occurrence of disease, 424, 425f
surveillance, 425, 426t, 427, 427f
Epitope (antigenic determinant), 392, 393f
Epulopiscium, 49, 49f
Escherichia coli
amino acid pools, 9
arg transcriptional units, 222–223, 223f
baby machine, 193, 193f, 194f
cell division, 70
chemotaxis system, 242–244, 242f, 243f
dry weight, 7, 7f
electron transport systems in, 157–158, 158f
fluorescence micrographs, 30f
growth rate, 193, 195, 235–236, 235t
lac operon, 222, 222f
lipoprotein, 59
macromolecules, 7–8, 7f, 9
micrographs, 16f, 30f
as model system, 22, 23f
molecular weight, 9
porins, 59
proteins, 61
protein synthesis, 235–236, 235t
replication, 5
respiration, ATP yields from, 157–158, 158f
ribosomes, 8–9, 8f
sigma factor, 213–214, 214f, 234
size, 4
successive cell cycles, overlapping of, 195–196, 195f–197f
surface area, 59
Ethanol-acetate fermentation, 148f
Ethanol fermentation
for bread making, 450–451
for wine production, 447–448, 448f
Eucaryotes (Eucarya), 5f, 6, 322–353. *See also specific eucaryotes*
cells, 83f
characteristics, 4
cytoskeletal system, 82
DNA viruses of, replication in nucleus, 127
endomembrane system (*See* Endomembrane system, eucaryotic)
RNA viruses of, replication in cytoplasm, 127
signal sequences for protein targeting, 83
ultrastructure, 3–4, 4f
viruses, polyproteins of, 132, 133f
cell wall, 110
inorganic salts in, 111, 112f
polysaccharides in, 111, 111f
chromosomes, replication origins, 194
classification, 81
complexity of, 9
contractile vacuole complex, 113–115, 113f–115f
coupled transcription and capping, 90

cytoskeletal system, 103–106, 104f, 105f
microfilaments, 103–104
microtubules, 103–104
cytoskeleton proteins
actin, 105–106, 105f
tubulin, 70, 104, 104f
endocytosis, 95
endosymbiotic theory evolutionary origin, 100, 101f, 102, 102f
genetic recombination, 288–290, 289f, 290f
growth, 195
mating type switching, 289–290, 289f, 290f
motility, 106–110, 107f–110f
number of, 5
pellicle, 111–112, 112f
phagocytosis and, 81, 82
phylogenetic tree, 322, 323f
phylogeny, 81, 82f
vs. procaryotes, 81
protist (*See* Protists)
ribosome composition, 294, 294t
sublineages, 322
unicellular, morphologically indistinguishable from mating types, 288–289
Euglenids, 327–328, 327f
Euryarchaeotes
halophilic, 318f, 319–320, 319f
methanogens, 318–319, 318f, 319f
Evolutionary distance calculation, for phylogenetic tree reconstruction, 297
Exocellular hydrolases, 144
Exocytosis, 94, 96–97
Exogenotes, 266
3′ Exonuclease, 227
5′ Exonuclease, 227
Exonucleases, 91
Exotoxins, 419–422, 420f
botulinum toxin, 421
definition of, 419
disruption of host cell membranes, 421–422
mechanism of action, 419, 420f
subunits, 420, 420f
types of, 419
Extreme halophiles, 205, 205f
Extrusomes, 97, 99f
Eyespot, 328

F

F′, 280–281, 282f
F−, 279
F+, 279
Facultative anaerobes, 43, 142
Facultative chemoautotrophs, 142
Facultative organisms, 142
FAD, 154–155
Fecal-oral transmission, of disease, 407

Fusion, vesicle
  lysosomal, 94–96, 95f
  secretory, 94
  steps in, 86f, 87

G

Gaia hypothesis, 354
Galactosyl-glycerol, 222, 222f
Gangrene, 407
Gas pack, for anaerobic jar, 43, 44f
Gastrointestinal anthrax, 436
Gas vacuoles, 73, 73f
Gas vesicles, 73, 73f
Gene conversion, for eucaryotic mate switching, 289–290, 290f
Generalized transduction, 283, 283t
  cotransduction frequencies, 284, 284f
  in gene mapping, 283–284
  from mistakes during packaging, 283, 284f
  vs. specialized transduction, 283t
  three point crosses, 284, 285f
Generation time, 195–196, 195f–197f
Genes, for virulence factors, 422
Genetic mapping, 269–272, 271f
  generalized transduction and, 283–284
Genetics, microbial, 266–291
  conjugation (See Conjugation)
  exchange
    between distant related organisms, 298–299
    plasmid transfer, 268–269
    in procaryotes, 266–269, 267f, 268f
  genome and (See Genome)
  mapping, 269–272, 271f
  multiplication, asexual, 266
  recombination, 266, 288–290, 289f, 290f
  transduction, 269, 281, 283–285, 283t, 284f
  transformation, 272–275, 273f, 274f
Genome
  mapping, 286–288
  procaryotic, 209–238
    chromosomal organization, 209
    coding, 210, 210f
    DNA, 209–210
    global control by modulons (See Global control systems, procaryotic)
    intergenic distances, 210, 210f
    mapping of, 286–288
    nonreplicative transposition, 211
    plasmids in, 212–213
    replicative transposition, 211
  procaryotic, chromosomal organization, 209
  segmented, 436
  viral, 118–120
Germ cell wall, of spore, 251f, 252
Germ-free animals, 381

Germ theory of disease, 19–20, 19f
Giardia, 325, 325f
Gills, 351f, 352
Glassware, sterilization, 42
Gliding motility, of myxobacteria, 78–79, 79f, 260–261, 261f
Global control systems, procaryotic, 230–237
  catabolic repression, 231, 232, 232f, 233f
  glucose permease indirect control of cAMP concentration, 233, 234f
  growth rate control, 231, 235–237, 235f, 235t, 236f
  heat shock response, 231, 233–234, 234f
  modulon, 231
  stringent response, 231, 237, 237f
  for transcriptional units, 217
  types of, 231–232
Glucose permease, indirect control of cAMP concentration, 233, 234f
Glutamate, 206
Gluten, 450–451
Glycerol-biphytane-diethers, 62–63
Glycerol phosphate, 56, 56f
Glycogen granules, 71, 71f
Glycolipids, in mycoplasma cell membrane, 52, 52f
Glycolysis, 142
Glycolytic pathway, 142, 143f
Glycosylation, patterns of, 93, 94f
"Golden age" of microbiology, 15, 15f
Golgi
  cis face, 92
  sorting function, 92–93
  trans face, 92–94
G_1 period, 195, 196
Gradient sensing, sensory adaptation and, 243–244, 243f
Gram-negative bacteria
  cell envelope, 51, 51f
  endotoxins, 422–423
  flagellar basal body, 78, 78f
  murein in, 55
  outer membrane, 57
    impermeability of, 58–59
    lipoprotein of, 59
    structure of, 57–58, 57f
  pathogenic, T3SS system, 414
  periplasm, 59
  transformation, 274–275, 274f
Gram-positive bacteria
  actinomycetes, 310, 311
  cell envelope, 51, 51f
  cell walls, 56, 56f
  endospore-formers, 310–311, 311f
  murein in, 55
  mycoplasmas, 310, 311, 311f
  transformation, 272, 273f, 274
Gram stain, 27–28, 27f

Granzymes, 385, 386f
Green manuring, 370
Green nonsulfur bacteria
  hydroxypropionic acid cycle, 186–187, 187f
  morphology, 306, 307f
  type II photosynthesis, 175, 177
Green sulfur bacteria
  chlorosomes, 170, 171f
  morphology, 313, 313f
  as obligate photoautotrophs, 173
  reverse TCA cycle, 184–185, 185f
  type I photosynthesis, 175, 176, 176f
Growth, microbial, 190–208
  aerobic, protections against toxic effects of oxygen, 206–207, 206f, 207f
  catalysts, 190
  confinement, by host defenses, 380–381
  environmental conditions, 203–208
    osmolarity, 205–206, 205f, 206f
    pH, 204–205, 204f
    temperature, 203–204, 203f, 204f
  exponential, 190, 191f
  population
    balanced, 199–200, 200f
    exponential, 198–199, 198f
    maintenance energy for stationary phase, 200
    spectrophotometric measurement of, 197–198, 198f
    stationary phase, 199, 200f
  synthesis, 190
  temperature range, 203–204, 204f
Growth factors, 42, 270
Growth rate
  control, 231
  pH and, 204–205, 204f
  temperature and, 203, 203f
Growth rate constant, 198
GTP, 146
Guilds, 363, 363f

H

Habitats, microbial, 362–365
  alternating growth periods in, 364–365
  biofilms, 364, 364f
  chemical gradients in, 363
  chemoheterotrophic, 362
  dark, 362, 362f
  guilds, 363, 363f
  minerals requirements and, 362, 362f
  organic material and, 362
  procaryotic, 6
hagA operon, 229–230, 230f
hagAR operon, 229–230, 230f
hagB operon, 229–230, 230f
hagR operon, 229–230, 230f
Halococcus, cell wall, 52, 52f

Halophiles, 205, 205f
Halophilic archaea, 319–320, 319f
Hantavirus pulmonary syndrome (HPS), 427–428
Haustoria, 344
Headful mechanism, for packaging of bacteriophage DNA, 133, 134f
Heat-sensitive mutations, 270
Heat-sensitive solutions, sterilization by filtration, 40–41
Heat shock modulon, 234
Heat shock response, 231, 233–234
Helicobacteria, type I photosynthesis, 175, 176, 176f
Helper T cells (T_H cells), 397
Hemes, 152, 153f
Hemolysins, 384
Hemolysis, 384
HEPA filters, 46, 46f
Herd immunity, 432–433
Hershey-Chase experiment, 125, 125f
Heterocysts
  definition of, 256
  development, DNA rearrangements in, 258–259, 259f
  morphological differentiation, 256–257, 256f, 257f
  nitrogen fixation, 256–257, 257f
  oxygen-evolving complex loss, 256
  spacing, 258, 259f
Heterodisulfide reductase, 165–166
Heteroduplex region, in gram-positive transformation, 272, 273f
Heterokont, 338, 340f
Heteropolymers, 7
Heterotrophs, 140
Hfr strains
  insertion sequences in F′ plasmid creation, 280–281, 282f
  to map genes on bacterial chromosome, 279–280, 280f
  mating, 279, 279f
Hierarchical classification, 292
Histone-like proteins, 65
Histones, 65
History of microbiology, 11–47
  discovery of first microbes, 11–12, 12f
  germ theory of disease, 19–20, 19f
  microscope improvements and, 12–16, 13f–16f
  model systems, 22–23, 23f
  pure culture technique and, 20–21
  spontaneous generation controversy, 16–19, 17f, 18f
HIV, rapid mutation rates, 415
Holdfast, 22, 245
Homolactic fermentation, 148f
Hook, 77
Hopanoids, 60, 60f
Hops, 449
Host, in symbiosis, 366
Host defenses
  cell membrane disruption, by exotoxins, 421–422
  confinement of microbial growth, 380–381

Osmotic shock, 231
Outer membrane, 50, 50f
Outgrowth, endospore, 256, 256f
Oxidase complex (complex IV), 152
Oxidases, alternate, for different electron receptors, 155–157, 156f
Oxidase test, 153
Oxidation
    NADH, 150–151
    of organic compounds, reduction to NAD+, 151–152
Oxidative burst, 384, 384f
α-Oxoglutarate dehydrogenase, in reverse TCA cycle, 185, 185f
Oxygen
    concentration regulation for nitrogen fixation, 372
    consumption, anaerobic ecosystems and, 357–359, 358f
    in evolution of earth's atmosphere, 182
    microbial metabolism and, 141
    sediment profiles, 358, 358f
    toxic effects, protections for aerobic growth, 206–207, 206f, 207f
Oxygenic photosynthesis, 141
Oxyglutarate dehydrogenase, in aerobic chemoheterotrophs, 143–144

P

Palm wine, 447
Pandemic disease, 424, 435f
Parabasalids, 325–327, 326f
Paracoccus
    denitrification in, 156–157, 156f
    electron transport system, 155f
    respiration, ATP yields from, 157
Paralytic shellfish poisoning, 335
Parasite, 366
Parasitic symbiosis, 366–367
Parasporal crystals, biocontrol with, 456–457, 456f–458f
Parsimony methods, for phylogenetic tree reconstruction, 297
Partial diploid production, from specialized transduction, 285, 287f
Pasteur, Louis, 17, 17f, 18f
Pathogenesis, microbial, 405–423
Pathogenicity islands, 422
Pathogens. *See also specific pathogens*
    causing surface infections, 405–406, 406f
    containment facilities, special, 44–47, 45t, 46f
    deep tissue access, 410, 411f
    definition of, 44
    establishment of infection, probability of, 409, 409f
    gram-negative, T3SS system, 414
    intracellular, induction of phagocytosis, 412–413, 413f
    invasiveness, increased, adaptations for, 406–407
    opportunistic, 410
    protection from host defenses
        antibiotic resistance, 417–419, 418f
        antibody blockade, 415, 415f
        blockage of cellular communication, 416–417, 416f, 417f
        capsules, 414
        intracellular growth, 410–412, 412f
    toxigenicity, 419–423
        endotoxins, 422–423
        exotoxins, 419–422, 420f
        increased, adaptations for, 406–407
    transmission routes, 407–409, 408f
    vector-borne, 407, 408, 410
    virulence of, 406
PatS, heterocyst spacing and, 258, 259f
Pauling, Linus, 293
Pellicle
    euglenid, 327–328, 328f
    protist, 111–112, 112f
Penetration, viral
    definition of, 122, 122f
    Hershey-Chase experiment, 125, 125f
    by membrane fusion, in enveloped viruses, 124, 126f
    by nucleic acid injection, in bacteriophage, 123–124, 123f, 124f
    in plants
        by insect injection, 124, 126
        by wounds, 124, 126
Penicillin
    bacterial growth and, 455, 456f
    chemical modification after synthesis, 455, 455f
    discovery of, 452–453, 453f
Penicillinase, 455, 455f
Penicillium camembertii, 446–447
Penicillium roquefortii, 447
Pennate diatoms, 342, 342f
Peptidoglycans, 50, 53
Perforins, 385, 386f
Periplasm, 59
Permeases, 61
Permissive temperature, heat-sensitive mutations and, 270
Peroxidase, 207
Pertrichous flagellation, 77
Petri plates, 42
Phages. *See Bacteriophages*
Phagocytes, 384–385, 384f
Phagocytosis
    definition of, 394
    eucaryotes and, 81, 82
    induction by intracellular pathogens, 412–413, 413f
Phagolysosome
    definition of, 96
    exocytosis, 96–97, 98f
Phase-contrast microscope, 28–29
    condenser, annular diaphragm of, 29, 29f
    image formation, 32–33, 32f, 33f
    light path, 29–30, 29f
    phase plate, 30
    phase ring, 30
Phase-contrast microscopy, 15

Phase plate, 30, 32, 33*f*
Phase ring, 30
Phase variation, 229–230, 230*f*
  in *Salmonella*, 229, 230*f*, 414
PHB (poly-β-hydroxybutyrate), 71, 71*f*
Pheophytins, 168, 169*f*
Phosphate
  "high-energy," 147, 147*f*
  "low-energy," 147, 147*f*
  starvation, 231
Phosphodiesterase, 232
Phosphoenolpyruvate, 146
Phospholipids, glycerol-biphytane-diethers, 62–63, 63*f*
Phospholipid transfer proteins, 100
Phosphorescence, in dinoflagellates, 335
Phosphoribulokinase, in Calvin-Benson cycle, 184, 184*f*
Phosphorus, return to land, 357
Phosphorylation, substrate-level, in fermentation, 147, 147*f*
Phosphotransferase system, sugar uptake, 145, 145*f*
Photoautotrophic growth. *See* Photosynthesis
Photoautotrophs, 140
Photobacterium, 376–377, 377*f*
Photoheterotrophic growth, 173–175
  assimilation
    of butyrate, 174, 174*f*
    of lactate, 173, 173*f*
    of pyruvate, 174, 174*f*
  photosynthesis (*See* Photosynthesis)
Photoheterotrophs, 140
Photons, 31
Photophosphorylation
  alternate mechanism, 182–183, 183*f*
  cyclic, 171, 172, 172*f*
  noncyclic, autotrophic growth and, 172–173, 173*f*
Photosynthesis, 168–183
  anoxygenic, 182
  in bacteria, 175–176, 176*f*
  chlorophyll cycle, 170–171, 171*f*
  in chloroplasts, 176
  mediation, by membrane-embedded reaction centers, 168
  oxygenic
    advantage of, 182
    extinction and, 206
  in procaryotes, 168, 169*f*
  terminology, 168
  type I
    cyclic photophosphorylation using photosystem 1, 176, 177*f*
    distribution in bacterial tree, 175–176, 176*f*
    noncyclic photophosphorylation using photosystem 1, 177, 178*f*
  type II, 177
    cyclic photophosphorylation using photosystem II, 177, 178*f*, 179
    distribution in bacterial tree, 175–176, 176*f*

reverse electron transport required for noncyclic photophosphorylation, 179–180, 179*f*
  type I/II
    characteristics of, 180
    distribution in bacterial tree, 175–176, 176*f*
    noncyclic photophosphorylation uses both photosystem I and II, 181, 181*f*
Photosystem I
  in cyclic photophosphorylation
    for type I/II photosynthesis, 180, 180*f*
    for type I photosynthesis, 176, 177*f*
  in noncyclic photophosphorylation for type I photosynthesis, 177, 178*f*
Photosystem II
  in cyclic photophosphorylation for type II photosynthesis, 177, 178*f*, 179
  loss, in heterocyst formation, 256
  for type I/II photosynthesis, electron donor for, 182
Phototrophs
  anaerobic, 173
  cyclic photophosphorylation in, 171, 172*f*
  definition of, 140
Phycobilins, 169, 170*f*, 312
Phycobilisomes, 169–170, 171*f*, 312
Phylogenetic tree
  of Archaea, 317–318, 318*f*
  of Bacteria, 305, 306*f*
  DNA/DNA hybridization and, 301, 302*f*
  of eucaryotes, 322, 323*f*
  hyperthermophilic kingdoms, 305–306, 306*f*
  reconstruction, 296–298
  universal, 299, 300*f*
Phylogeny
  confusion, by genetic exchange between distant related organisms, 298–299
  definition of, 294
  reconstruction, rRNA for, 294, 294*t*
  species concept, 302–303
Physicians, notification of epidemic disease detection, 427–428
Pigment antennas, 168–170, 170*f*, 176
Pilin, 75
Pilus (pili)
  definition of, 75–76, 76*f*
  gliding motility and, 78–79, 79*f*
Pit-plug, in red algae, 337–338, 338*f*
Plague, 435
Planctomyces, cell envelope, 52, 52*f*
Planctomyces group, 315–316, 315*f*, 316*f*
Plants, transmission, of viroids, 137
Plant viruses, transmission within same plant, 126–127
Plasma cells, 396
Plasmalemma, prokaryotic. *See* Cell membrane
Plasma membrane, prokaryotic. *See* Cell membrane
Plasmids
  antibiotic resistance and, 212
  antibiotic-resistance genes and, 212, 417

Respiration
  in chemoheterotrophs, 144
  steps in, 150–151, 151*f*
Respiratory diseases, transmission, 407–408
Respiratory tract
  anatomy, 383, 383*f*
  mucous membranes, 382–383
Restrictive temperature, heat-sensitive mutations and, 270
Reticulate bodies, 316
Retinal, 182
Retroviruses, replication, 130, 131*f*, 132
Reverse electron transport, for noncyclic photophosphorylation, in type II photosynthesis, 179–180, 179*f*
Reverse TCA cycle, 184–185, 185*f*
Reverse transcriptase, for retrovirus replication, 130, 131*f*, 132
Rhizobium-legume symbiosis, 369–372, 370*f*, 371*f*
Rhizoids, 346
Rhizosphere, 372
Rho, blockage, by antiterminator proteins, 225–226, 227*f*
Rhoptries, 336
Ribitol phosphate, 56, 56*f*
Ribosomal proteins, 8–9, 8*f*
Ribosomes
  assembly, 89, 89*f*
  composition, procaryotic *vs.* eucaryotic, 294, 294*t*
  export, 89, 89*f*
  functions, 61, 70, 92
  morphology, 8–9, 8*f*, 70*f*
  supply, regulation of procaryotic growth rate, 235
Ribulose-biphosphate cycle (Calvin-Benson cycle), 183–184, 184*f*
Ribulose-bisphosphate carboxylase/oxygenase (RuBisCO), 184, 184*f*
Rice cultivation, water fern in, 368–369, 369*f*
Rich media, 42
Rickettsia, 310, 310*f*, 411
Riftia, 378, 378*f*
Ring vaccination, 402–403
Ripened cheese, 446
RISC (RNA-induced silencing complex), 390
Rising sludge, 462
RNA
  in virion, 118–119
  viroid, 136–137, 137*f*
RNA-binding proteins, Rho, 215–216, 216*f*
RNAi (RNA interference), 389–390, 390*f*
RNA-induced silencing complex (RISC), 390
RNA interference (RNAi), 389–390, 390*f*
RNA polymerase
  binding
    prevention of, 220, 220*f*
    transcription factors and, 220, 220*f*
  promoter recognition, 214
  subunits, 214

RNA replicase, 129, 130*f*
RNA viruses
  in eucaryotic cells, replication in cytoplasm, 127
  negative strand, replication steps for, 130, 131*f*
  positive strand, replication steps for, 129, 130*f*
  rapid mutation rates, 414–415
  single-stranded
    negative strand, definition of, 129
    positive strand, definition of, 129
    retroviruses (*See* Retroviruses)
  spread, reduction by interferon, 388–389, 389*f*
Rock
  dating, 3
  layers of, 3
Rods, 49, 49*f*
Rolling circle replication, 127–129, 128*f*, 129*f*
Rood nodules, 369–370, 370*f*
Root-nodule symbiosis, 369–370, 370*f*
Rough endoplasmic reticulum, 92
rRNA
  phylogenies, major lines of decent or domains, 299, 300*f*
  in phylogeny reconstruction, 294, 294*t*
  secondary structure, sequence alignment and, 294–296, 295*f*, 296*f*
rrn operons, 236–237, 236*f*
RuBisCO (ribulose-bisphosphate carboxylase/oxygenase), 184, 184*f*
Rumen, fermentation in, 367–368, 368*f*
Run, tactic responses and, 240
Rusts, 350

S

Saccharomyces cerevisiae, 23, 23*f*, 447
Salinity, microbial growth and, 205–206, 205*f*
Salmonella
  flagellin protein, 229–230, 230*f*
  infections, 440
  outbreak, investigation of, 428–429, 429*f*
  phase variation, 229, 230*f*, 414
SARS (severe acute respiratory syndrome), 433–434, 435*f*
SASPs (small acid-soluble proteins), 252, 252*f*
Saxitoxin, 335, 335*f*
Scanning electron microscopy, 38–39, 38*f*
Scoring, in genetic mapping, 271–272, 271*f*
Secondary F' strains, 281
Secondary metabolites, antibiotics as, 454, 454*f*
Secondary treatment, of wastewater, 460–462, 461*f*, 462*f*
Secretory proteins, synthesis, in endoplasmic reticulum, 92
Secretory vesicles, 94
Segmented genome, 119, 436
Selection
  clonal (*See* Clonal selection)
  in genetic mapping, 271–272, 271*f*
  process of, 293

Sensing
    of attractants or repellants, 241–242, 242f
    gradient, sensory adaptation and, 243–244, 243f
Sensor, in two-component regulatory system, 241–242, 242f
Sensory adaptation, gradient sensing and, 243–244, 243f
Septic tank, 459, 460f
Septum, in procaryotic cell division, 190–191, 191f
Severe acute respiratory syndrome (SARS), 433–434, 435f
Sex, in ciliates, 334–335
Sexually transmitted disease (STD), 407
Sexual reproduction
    in diatoms, 343, 344f
    in protists, 116
Shadow casting technique, for electron microscopy,
        35–36, 36f
Shigella, 411
Shine-Dalgarno sequences, 213
Shuttle streaming, 332
Sigma factors, in sporulation, 254, 254f, 255f
Sigma subunits, in bacteria, 214, 214f
Signal sequences, for protein targeting, in eucaryotic cells, 83
Signal transduction, by surface receptor, 95, 96f
Single-cell protein, 451–452
Singlet oxygen, 206
siRNA (small interfering RNA), 389–390
Skin infections, 405–406
S-layer, 74, 74f
Slime layers, 75
Slime molds, 331–333, 332f, 333f
Sludge
    bulking, 462
    rising, 462
Slug, 331–332
Small acid-soluble proteins (SASPs), 252, 252f
Small interfering RNA (siRNA), 389–390
Smallpox vaccine, development, 401, 402, 403
Smallpox virus, 416–417, 416f, 417f
S motility, of myxobacteria, 261, 261f
Smuts, 350
Snow, John, 430–431
Solid media
    development of, 21
    in isolating pure cultures, 42, 42f
SOS response, 231
Soy fermentation, 451
Soy sauce, 451
Specialized transduction, 283, 283t
    vs. generalized transduction, 283t
    partial diploid production from, 285, 287f
Species concept, 302–303
Spectrophotometry, of microbial population growth, 197–198,
        198f
Spirilum/spirilla, 49, 49f
Spirochetes, 312–314, 313f
Spirulina, 452

Spliceosome, 91
SpoIVCA, 255
Spongiome, 114–115, 114f, 115f
    in contractile vacuole complex, 114–115, 114f, 115f
Spontaneous generation controversy, 16–19, 17f, 18f
Spo0A, initiation of sporulation and, 253–254, 253f
Spo0F, initiation of sporulation and, 253–254, 253f
Sporangiole, 263
Spore coat, 250, 252
Spores, cyanobacterium, 49–50, 50f
Sporulation
    initiation, 253–254, 253f
    morphological stages, 250, 251f, 252–253, 252f
    mother cell, DNA rearrangements in, 254–255, 255f
    in *Myxococcus,* 259–264, 259f–261f, 263f, 264f
    sigma factors in, 254, 254f, 255f
    as starvation response, 249–250, 249f
Staining, 15
Stalked cell, 244–246, 245f
Staphylococcus aureus, antibody blockage, 415, 415f
Starvation response
    as aggregation trigger, 262, 263f
    for sporulation, 249–250, 249f
Stationary phase
    definition of, 231
    of microbial population growth
        description of, 199, 200f
        maintenance energy for, 200
STD (sexually transmitted disease), 407
Sterility
    definition of, 40
    exponential microbial death and, 200, 201f
Sterilization
    of glassware by dry heat, 42
    of heat-sensitive solutions, by filtration, 40–41
    methods
        autoclaving, 40
        filtration, 40–41
Stickland fermentation, 148f
Stomatocysts, 340, 342f
Storage granules, in procaryotic cytoplasm, 71, 71f
Stramenopiles, 338, 340, 340f–342f
Streaking, onto solid media, 42, 42f
Streptococcus, disruption of host cell membranes,
        421–422
Streptococcus pneumoniae, antibody blockage, 415, 415f
Streptomyces
    autolyzing, 453, 453f
    life cycle, 191, 192f
Stringent response, 231, 237, 237f
Substrate mycelium, 311
Succinate, 154, 155f
    dehydrogenase, 154, 155f
Sucrose, transport, into heterocyst, 256–257, 257f
Sugar uptake, by phosphotransferase system, 145, 145f

Sulfate-reducing bacteria
anaerobic chemoautotrophic respiration, 164–165, 164*f*
morphology, 309, 309*f*
Sulfur cycle, 355, 356*f*
Sulfur globules, 71, 71*f*
Sulfur-reducing bacteria, anaerobic chemoautotrophic
respiration, 164–165, 164*f*
Sulfur, return to land, 357, 358*f*
Supercoiled chromosomes, procaryotic, 65, 65*f*
Supercoiled DNA, 209–210
Superoxide, 206
dismutase, 207
Surface exclusion, 275
Surface receptor, signal transduction, 95, 96*f*
Surface receptors, recycling
endocytosis and, 95
in endosome, 96, 97*f*
Surveillance, disease, 425, 426*t*, 427, 427*f*
Swarmer cell, 245, 245*f*
Sweet wort, 448
Symbionts, 366, 378, 378*f*
Symbiosis, 366–379
Azolla-Anabaena, 368–369, 369*f*
definition of, 366
fungal, 373–376, 373*f*–376*f*
host, 366
lichens, 373–374, 373*f*, 374*f*
mutualistic, 366–367
luminescent bacteria and, 376–377, 377*f*, 378*f*
mycorrhizae, 374–376, 375*f*, 376*f*
nitrogen-fixing, 368, 369*t*
actinomycete, 372–373, 373*f*
oxygen concentration regulation for, 372
rhizobium-legume, 369–372, 370*f*, 371*f*
nutritional benefits from, 367
parasitic, 366–367
ruminant, 367–368, 368*f*
symbionts, 366
chemoautotrophic, 378, 378*f*
parasite, 366
Syntrophic associations, 360–361, 361*f*
Syntrophy, in anaerobic ecosystems, 360–361, 361*f*
Systematics, microbial. *See* Classification of microbes
Systemic disease, 410

T

Tactic responses
biased random walk and, 240, 240*f*
definition of, 239
directed random walk and, 240, 240*f*
negative, 239–240
positive, 239–240
run and, 240
tumbling, 240–241, 241*f*
two-component regulatory system, 241–242, 242*f*

Tailing enzyme, 90
Tailing, mRNA, 90–91, 91*f*
Tails, capsid, 120
Taxonomists, 293
Taxonomy, viral, 136
TCA cycle. *See* Tricarboxylic cycle
T-cell receptors (TCRs), 397–399, 398*f*, 399*f*
T cells. *See* T lymphocytes
TCRs (T-cell receptors), 397–399, 398*f*, 399*f*
Teichoic acids, in gram-positive bacterial walls, 56, 56*f*
Tempeh, 451
Temperate phages, 135
Temperature
maximum, 203, 203*f*
minimum, 203, 203*f*
optimum, 203, 203*f*
permissive, 270
range, for procaryotic growth, 203–204, 204*f*
restrictive, 270
water requirements for microbial life and, 362, 362*f*
Terminally differentiated cell type, 254–255, 255*f*
Termination, regulation, by antitermination, 225–226, 227*f*
Terminus, 194
Test (shell), of testate amoebas, 329, 330*f*
Testate amoebas, 329, 330*f*
Tests, 112–112, 113*f*
Thallus, 346
T helper lymphocytes (T$_H$ cells), production by clonal
selection, 399–400, 400*f*
Thermactinomyces, spores, 249
Thermodesulfobacterium, 306, 306*f*
Thermomicrobium, 307, 307*f*
Thermophiles, 204
Thermotoga, 306, 306*f*
Thin sectioning, of electron microscope specimens, 35, 35*f*
Thiospirillum, 240
Three point crosses, 284, 285*f*
Thylakoid membranes
cyanobacteria, 61, 61*f*, 62
definition of, 99
photosystem II electron donor on, 182
phycobilisomes, 169–170, 171*f*
protein targeting to, 84–85, 86*f*
Time-of-entry, interrupted Hfr mating and, 279–280, 280*f*
Tissue culture, 41
T lymphocytes
cytolytic
production by clonal selection, 399–400, 400*f*
targeting of cells making foreign protein, 398–399, 399*f*
differentiation, 399–400, 400*f*
helper, production by clonal selection, 399–400, 400*f*
Toga, 306, 306*f*
Toxigenicity, 419–423
definition of, 419
endotoxins, 422–423
exotoxins, 419–422, 420*f*

Virions
  attachment to host cell, 122, 122f, 123
  chromosomes in, 118–119
  enveloped, release of, 134
  inert, biologically, 120
  neutralization by antibody, 394, 395f
  structure of, 118, 119f
  unenveloped, release of, 134, 134f
Viroids, 136–137, 137f
Virtual image, 25, 26f
Virulence
  enhancement by virulence factors, 422
  of pathogens, 406
Virulence factors, 422
Viruses
  capsids (*See* Capsids, viral)
  DNA
    replication in nucleus, 127
    rolling circle replication of, 127–129, 128f, 129f
  envelopes, obtained from budding, 120–121, 121f
  eucaryotic, polyproteins of, 132, 133f
  genome of, 118–120
  infections
    host damage in, 422–423
    intracellular growth, 411–412
    penetration of host surfaces, 406, 406f
    spread, reduction by interferon, 388–389, 389f
  life cycle of, 118, 119f
    acellular, 118, 119f
    cellular, 118, 119f
  multiplication process, 121–122, 122f
    assembly, 122f, 132–133, 134f
    attachment, 122, 122f, 123
    latency, 122f, 135–136
    macromolecular synthesis, 122f, 127–130, 128f–131f, 132
    penetration (*See* Penetration, viral)
    release, 122f, 134, 134f
  penetration of (*See* Penetration, viral)
  protein synthesis, 132
  rapid mutation rates, 414–415
  replication of, 118
  RNA (*See* RNA viruses)
  size of, 118, 119f
  structure of, 118, 119f
  taxonomy, 136
Volutin granules, 72, 72f

W

Waksman, Selman, 453–454
Wall teichoic acids, 56, 56f
Wastewater treatment, 458–462, 460f–462f
  anaerobic digester, 460
  goals, 458–459
  periodic failures, 462
  primary, 459, 460f
  secondary, 460–462, 461f, 462f
Water fern, in rice cultivation, 368–369, 369f
Water molds, 343–346, 345f
Water requirements, for microbial life, 362, 362f
Whey, 446
Wild yeasts, 447
Wine
  distillation, 449
  partial oxidation for vinegar production, 449–450, 450f
  production, ethanol fermentation for, 447–448, 448f
Winogradsky, Sergius, discovery of chemoautotrophic growth, 160, 160f, 161f
Woese, Carl, 301
Working distance, 26
World Health Organization (WHO), 402, 403
World wars, biological weapons in, 463–464
Wort, 448

Y

Yeasts
  budding, 289
  daughter cell, 289
  growth/reproduction, 348–349, 349f, 350f
  mating type switching, 289, 289f
  mother cell, 289
  wild, 447
Yersinia pestis, 410, 435

Z

Zoonosis (zoonoses)
  contagious, 435–436
  noncontagious, 436
  origin of major human disease and, 440–441
Zoospores, 344
Zukerkandle, Emile, 293
Zygomycetes, 347–348, 348f
Zygospore formation, 347–348, 348f

Photo Acknowledgments

All Chapter Opener Images: © Photos.com

Table of Contents
page x © National Library of Medicine. **page xi** Courtesy of James Gathany/CDC. **page xiii** Courtesy of Mark Wheelis, University of California, Davis. **page xiv** Courtesy of Dr. Edwin P. Ewing, Jr./CDC. **page xv** Courtesy of RG Johnsson/NPS Photo. **page xvi** Courtesy of the National Human Genome Research Institute. **page xvii** © Dynamic Graphics Group/IT Stock Free/Alamy Images. **page xix** Courtesy of Harold Evans. **page xxii** Courtesy of Harlan Kredit/NPS Photo.

Chapter 1
1.3 © EM Unit, VLA/Photo Researchers, Inc. **1.B1** Reproduced from The Iron-Formation on Belcher Islands, Hudson Bay, by E.S. Moore, *The Journal of Geology*. Vol. 26 412–438 (1918). **1.B2** © Dr. J. William Schopf, UCLA.

Chapter 2
2.1 © National Library of Medicine. **2.2** © Science VU/Visuals Unlimited. **2.4a–b** © Brian J. Ford. **2.6** By permission of Oxford University Press. Figure 24 (p. 191), 'The History of Bacteriology' by Bulloch, William (1938). **2.7a–b** © Wim van Egmond/Visuals Unlimited. **2.7c** Reprinted with permission from *Journal of Molecular Biology* 112(1). Normal-to-curly flagellar transitions and their role in bacterial tumbling. Stabilization of an alternative quaternary structure by mechanical force, pp. 1–30, © 1977, with permission from Elsevier. Photo courtesy of May Macnab, Yale University. **2.8a** Courtesy of Mark Wheelis, Ph.D., University of California, Davis. **2.8b** © Biophoto Associates/Photo Researchers, Inc. **2.9** © National Library of Medicine. **2.10** © National Library of Medicine. **2.11** © National Library of Medicine. **2.13** © National Library of Medicine. **2.14** © Dr. Michael Gabridge/Visuals Unlimited. **2.15** Courtesy of Dr. Lesley Robertson, on behalf of the Delft School of Microbiology Archives at Delft University of Technology, The Netherlands (http//www.beijerinck.bt.tudelft.nl). **2.16a** Courtesy of Mark Wheelis, University of California, Davis. **2.16b** © John Durham/Photo Researchers, Inc. **2.16c** Courtesy of Ellen Quardokus, Indiana University. **2.16d** © ISM/Phototake.

Chapter 3
3.6 © Dr. Jack Bostrack/Visuals Unlimited. **3.8** Reprinted from *Proc Natl Acad Sci USA* 1977;74(1). Bacterial flagella rotating in bundles a study in helical geometry, pp. 221–225. Photo courtesy of May Macnab, Yale University. **3.10** © William Margolin, Ph.D., University of Texas Medical School at Houston. **3.13** © Dr. Dennis Kunkel/Visuals Unlimited. **3.15** © Dr. Hans Gelderblom/Visuals Unlimited. **3.17** © Thomas Deerinck/Visuals Unlimited. **3.19** © Dr. Tony Brain/Photo Researchers, Inc. **3.20** Reprinted with permission from the American Society for Microbiology (N Gotoh, H Wakebe, E Yoshihara, T Nakae and T Nishino. *J. Bacteriol.*, 1989 February. 171(2): 983–990.) Photo courtesy of Doctor Naomasa Gotoh, Kyoto Pharmaceutical University. **3.23** Courtesy of Pall Corporation. **3.24** Courtesy of James Gathany/CDC. **3.25** Courtesy of Mark Wheelis, University of California, Davis. **3.26** Courtesy of Dr. James Stone, Biogeochemistry Core Facility, South Dakota School of Mines & Technology. **3.27** © Jack Bostrack/Visuals Unlimited.

Chapter 4
4.2a © Mariona Hernandez Marine, University of Barcelona. **4.2b** © Esther R. Angert, Ph.D./Phototake. **4.4** © J.C. Meeks, University of California. **4.10** © Eye of Science/Photo Researchers, Inc. **4.11** © Dr. R. Rosenbusch, Iowa State University, College of Veterinary Medicine. **4.21** © Dr. Kari Lounatmaa/Photo Researchers, Inc. **4.22** Reprinted with permission from the American Society for Microbiology. (Watson, S.W. and Murray, R.G. E. *J. Bacteriol.*, 1965 June. 89(6) 1594–1609.). **4.26** © Dr. George Chapman/Visuals Unlimited. **4.30** Courtesy of Rut Carballido-López, University of Oxford. **4.31** Reprinted, with permission, from the *Annual Review of Biophysics and Biomolecular Structure*, Volume 33 ©2004 by Annual Reviews, www.annualreviews.org. Figure provided by Jan Löwe, Ph.D., The Medical Research Council, Laboratory of Molecular Biology. **4.32** © X. Wang and

D. Sherratt. **4.33** Reprinted from *Trends in Cell Biology*, Vol. 7, Harold P. Erickson, FtsZ, a tubulin homologue in prokaryote cell division. pp. 362–367, Copyright (1997), with permission from Elsevier and *Proc. Natl. Acad. Sci.* 93(23) Xiaolan Ma, David W. Ehrhardt, and William Margolin, Colocalization of cell division proteins FtsZ and FtsA to cytoskeletal structures in living *Escherichia coli* cells by using green fluorescent protein. pp. 12998–13003. Copyright (1996) National Academy of Science, U.S.A. Photo courtesy of William Margolin, Ph.D., University of Texas Medical School at Houston. **4.34** © Dr. Elena Kiseleva/Photo Researchers, Inc. **4.35** © Brent Selinger, University of Lethbridge. **4.37** Courtesy of Roger Burks (University of California at Riverside), Mark Schneegurt (Wichita State University) and Cyanosite (www.cyanosite.bio.purdue.edu). **4.40** © Michael Abbey/Visuals Unlimited. **4.41** © Ann Auman, Pacific Lutheran University. **4.42** Reprinted with permission from the American Society for Microbiology. (Walsby, A.E. *Bacteriological Reviews*, 1972, 36:1–32.) Photo courtesy of Mark Walsby, University of Bristol. **4.43** © Dennis Bazylinski, University of Nevada, Las Vegas. **4.45** © Uwe B. Sleytr, Center for Nanobiotechnology, University of Natural Resources and Applied Life Sciences Vienna. Image provided by Dietmar Pum, Ph.D. **4.46** © Dr. Jack M. Bostrack/Visuals Unlimited. **4.48** © Phototake/Alamy Images. **4.50** © Scientifica/Visuals Unlimited. **4.54** © Science VU/Dr. Julius Adler/Visuals Unlimited.

Chapter 5

5.19a © Wim van Egmond/Visuals Unlimited. **5.19b** © Biophoto Associates/Photo Researchers, Inc. **5.23** Reprinted with permission from Blackwell Publishing. Tae Oh Cho, Suzanne Fredericq, Sung Min Boo (2003). *Ceramium inkuii* sp. Nov. (Ceramiaceae, Rhodophyta), from Korea, a new species based on morphological and molecular evidence. *Journal of Phycology*, 39(1): 236–247. **5.25** © Lynne Cassimeris, Ph.D. Lehigh University. **5.30** © RMF/Scientifica/Visuals Unlimited. **5.35** © Dr. Charles W. Stratton/Visuals Unlimited. **5.36** © Joe Scott/Visuals Unlimited. **5.37a** © Phototake/Alamy Images. **5.37b** © Steve Gschmeissner/Photo Researchers, Inc. **5.38** © Dr. Dennis Kunkel/Visuals Unlimited. **5.39** © Dr. Richard Kessel & Dr. Gene Shih/Visuals Unlimited. **5.40** © Dr. Dennis Kunkel/Visuals Unlimited. **5.41a–b** © Michael Abbey/Visuals Unlimited.

Chapter 6

6.4 bottom Courtesy of Dr. T. Moravec, Danforth Center, St. Louis. **6.4 top** Courtesy of the CDC. **6.5** © J. Carson/Custom Medical Stock. **6.6a** © Scott Camazine/Photo Researchers, Inc. **6.6b** © F. A. Murphy/Visuals Unlimited. **6.9** © Eye of Science/Photo Researchers, Inc. **6.19** Reprinted with permission from the American Society for Microbiology. (Bradley, D.E.; *Bacteriological Reviews*, 1967, 31:230–314.) Photo courtesy of David E. Bradley, Ph.D.

Chapter 7

7.B1 Courtesy of Roger Burks (University of California at Riverside), Mark Schneegurt (Wichita State University), and Cyanosite (www.cyanosite.bio.purdue.edu).

Chapter 8

8.1 © Science VU/DOE/Visuals Unlimited. **8.6** © Science VU/Dr. Elizabeth Gentt/Visuals Unlimited. **8.7** © Dr. T.J. Beveridge/Visuals Unlimited.

Chapter 9

9.2 © Dr. David Phillips/Visuals Unlimited. **9.3** Reprinted from *Trends in Cell Biology*, Vol. 7, Harold P. Erickson, FtsZ, a tubulin homologue in prokaryote cell division. pp. 362–367, Copyright (1997), with permission from Elsevier and *Proc. Natl. Acad. Sci.* 93(23) Xiaolan Ma, David W. Ehrhardt, and William Margolin, Colocalization of cell division proteins FtsZ and FtsA to cytoskeletal structures in living *Escherichia coli* cells by using green fluorescent protein. pp. 12998–13003. Copyright (1996) National Academy of Science, U.S.A. Photo courtesy of William Margolin, Ph.D., University of Texas Medical School at Houston. **9.4** © S. Langille and R. Weiner. **9.22** © Christine Case/Visuals Unlimited.

Chapter 10

10.14 Reprinted from *Proc Natl Acad Sci USA*. 1990 July 87(14):5504–5508. Su W, Porter S, Kustu S, Echols H. DNA-looping and enhancer activity association between DNA-bound NtrC activator and RNA polymerase at the bacterial glnA promoter. pp. 5504–5508 and Weiss, D.S. et al (1992). Prokaryotic Transcriptional Enhancers. In S.L. McKnight and K.R. Yamamoto (Eds.), *Transcriptional Regulation*. Cold Spring Harbor, New York Cold Spring Harbor Laboratory Press. Copyright (1992) Cold Spring Harbor Laboratory Press. Photo courtesy of Sydney Kustu, Ph.D., University of California, Davis. **10.27** © C. L. Woldringh, Ph.D., University of Amsterdam.

Chapter 11

11.7 © John Smit and Steve Smith, University of British Columbia. **11.11** Courtesy of Mark Wheelis, University of California, Davis. **11.19** Reprinted with permission from Horsburgh, MJ, Thackray, PD, and Moir, A, Transcriptional responses during outgrowth of *Bacillus subtilis* endospores. *Microbiology*. 147, 2933–2941. © 2001 SGM. Photo provided by Anne Moir, University of Sheffield. **11.20** Courtesy of Mark Wheelis, University of California, Davis. **11.21** © Dr. T. J. Beveridge/Visuals Unlimited. **11.25** © Michiel Vos, Max-Planck-Institute for Developmental Biology. **11.26** Reprinted with permission from the American Society for Microbiology. (Burchard, R.P. *J. Bacteriol*, 1982 October. 152(1): 495–501). **11.27** Reprinted from *Proc Natl Acad Sci USA* 1979; 76(11). Social gliding is correlated with the presence of pili in *Myxococcus xanthus*, 5952–5956. Photo provided by Dale Kaiser, Ph.D, Stanford University. **11.28** Courtesy of Kei Inouye, Kyoto University. **11.29** Courtesy of Mark Wheelis, University of California, Davis. **11.30** Courtesy of Dr. Sumiko Inouye, Robert Wood Johnson Medical School.

Chapter 12
12.7 © Dr. Dennis Kunkel/Visuals Unlimited.

Chapter 14
14.3 © Prof. Dr. K.O. Stetter and Dr. Reinhard Rachel, University of Regensburg, Germany. **14.4** © Science VU/DOE-Herter/Visuals Unlimited. **14.6** Courtesy of Dr. Michael Daly, Department of Pathology at Uniformed Services, University of the Health Sciences. **14.7a** © PHOTOTAKE Inc./Alamy Images. **14.7b** © Dr. Hans Ackermann/Visuals Unlimited. **14.8** Photograph by Gary Gaard. Courtesy of Dr. A. Kelman (Department of Plant Pathology, University of Wisconsin-Madison). **14.9** © Science VU/S. W. Watson/Visuals Unlimited. **14.10** Courtesy of Roger Burks (University of California at Riverside), Mark Schneegurt (Wichita State University), and Cyanosite (www-cyanosite.bio.purdue.edu). **14.11** © Scientifica/Visuals Unlimited. **14.12** © M. I. Walker/Photo Researchers, Inc. **14.13** Courtesy of Dr. Ed Ewing, Jr./CDC. **14.14** Courtesy of Mark Wheelis, University of California, Davis. **14.15** © Michael Gabridge/Visuals Unlimited. **14.16** © Michael Abbey/Visuals Unlimited. **14.18** Courtesy of Dr. Edwin P. Ewing, Jr./CDC. **14.19** © Dr. T.J. Beveridge/Visuals Unlimited. **14.20** Courtesy of Dr. V. R. Dowell, Jr./CDC. **14.21a** Courtesy of James Staley, University of Washington. **14.21b** Courtesy of John A. Fuerst, University of Queensland. Lindsay MR, Webb RI, Strous M, Jetten MS, Butler MK, et al. 2001. Cell compartmentalization in planctomycetes novel types of structural organisation for the bacterial cell. *Arch. Microbiol.* 175413–175429. With permission from Springer Science and Business Media. **14.23** Reprinted with permission from the American Society for Microbiology. (Gutter, B., Asher, Y. and Becker, Y.J.; *J. Bacteriol.*, 1973 August. 115(2): 691–703.). **14.25a–c**Zeikus JG, Bowen VG. Comparative ultrastructure of methanogenic bacteria. Feb 1975. *Canadian Journal of Microbiology* (21): 121–129. **14.26a** © Dr. Dennis Kunkel/Visuals Unlimited. **14.26b** © Eye of Science/Photo Researchers, Inc. **14.27a** Courtesy of OAR/National Undersea Research Program (NURP)/NOAA. **14.27b** Courtesy of R.G. Johnsson/NPS. **14.27c** Courtesy of Yellowstone National Park/NPS. **14.28** © Dr. Harald Huber, Ulf Kueper, and Dr. Reinhard Rachel, all at the University of Regensburg, Lehrstuhl fuer Mikrobiologie, Regensburg, Germany.

Chapter 15
15.2 © Aaron Bell/Visuals Unlimited. **15.3a** Courtesy of Dr. Sumiko Inouye, Department of Biochemistry, Robert Wood Johnson Medical School. **15.3b** Courtesy of Dr. Elizabeth Canning, Imperial College London. **15.5** Courtesy of Robert L. Owen, *Gastroenterology* 76: 759–769 (1979). **15.7** © L. Amaral Zettler, L. Olendzenski, and D. J. Patterson. MBL (micro*scope). **15.8b** © Michael Abbey/Visuals Unlimited. **15.10** © Science VU/Fred Marsik/Visuals Unlimited. **15.11**

© Arthur Siegelman/Visuals Unlimited. **15.12a–c** © Wim van Egmond/Visuals Unlimited. **15.13a** © Phototake Inc./Alamy Images. **15.13b** Courtesy of D. B. Scott/NOAA Paleoclimatology. **15.13c** © Alfred Pasieka/Peter Arnold, Inc. **15.14a** © Wim van Egmond/Visuals Unlimited. **15.14b** © Dr. Richard Kessel & Dr. Gene Shih/Visuals Unlimited. **15.16** © Carolina Biological Supply Company/PhotoTake. **15.19** © Manfred Kage/Peter Arnold, Inc. **15.20** Courtesy of the National Human Genome Research Institute. **15.21** © Dr. David M. Phillips/Visuals Unlimited. **15.27a–b** © Wim van Egmond/Visuals Unlimited. **15.31a** © Dr. Peter Siver/Visuals Unlimited. **15.31b** © Dr. Dennis Kunkel/Visuals Unlimited. **15.32** Courtesy of Jason K. Oyadomari, Michigan Technological University. **15.33a–b** © Eduardo A. Morales, Ph.D., The Academy of Natural Sciences of Philadelphia. **15.39** © Dr. Dennis Kunkel/Visuals Unlimited. **15.40** By permission of Oxford University Press; Figure 2.3a, "*Neorospora*: Contributions of a Model Organism" by Davis, Rowland (2000). **15.44a** Courtesy of J. Schmidt/NPS. **15.44b** © Jupiterimages/Polka Dot/Alamy Images.

Chapter 16
16.9 Courtesy of OAR/National Undersea Research Program (NURP)/NOAA. **16.11** Bacterial Biofilm in a Chronic Respiratory Tract Infection. © Hiroyuki Kobayashi. Licensed for use, ASM MicrobeLibrary.org.

Chapter 17
17.2b © Michael Abbey/Visuals Unlimited. **17.3** Courtesy of Harold Evans. **17.6** © David Benson, University of Connecticut. **17.7** © Dr. R. Howard Berg/Visuals Unlimited. **17.8a** Courtesy of R.G. Johnsson/NPS. **17.8b** Courtesy of Harlan Kredit/NPS. **17.8c** Courtesy of the U.S. Fish and Wildlife Service, Alaska Image Library. **17.10b** Courtesy of James D. Mauseth, University of Texas, Austin. **17.10c** Photo courtesy of R.L. Peterson. **17.11** © Science VU/Visuals Unlimited. **17.13** © E. Widder/HBOI/Visuals Unlimited. **17.14a** © Alex Kerstitch/Visuals Unlimited. **17.14b** © Tom McHugh/Photo Researchers, Inc. **17.15** © Woods Hole Oceanographic Institution.

Chapter 19
19.1 © Dr. Richard Kessel/Visuals Unlimited.

Chapter 20
20.B1 Courtesy of Frerichs, R. R. John Snow website: http://www.ph.ucla.edu/epi/snow.html, 2006.

Chapter 21
21.5 © St. Mary's Hospital Medical School/Photo Researchers, Inc. **21.8** © CNRI/Photo Researchers, Inc. **21.9** © George Chapman/Visuals Unlimited. **21.10** Courtesy of the National Archives of Australia (A1200, L44186).